Lecture Notes in Networks and Systems   **654**

The series "Lecture Notes in Networks and Systems" publishes the latest developments in Networks and Systems—quickly, informally and with high quality. Original research reported in proceedings and post-proceedings represents the core of LNNS.

Volumes published in LNNS embrace all aspects and subfields of, as well as new challenges in, Networks and Systems.

The series contains proceedings and edited volumes in systems and networks, spanning the areas of Cyber-Physical Systems, Autonomous Systems, Sensor Networks, Control Systems, Energy Systems, Automotive Systems, Biological Systems, Vehicular Networking and Connected Vehicles, Aerospace Systems, Automation, Manufacturing, Smart Grids, Nonlinear Systems, Power Systems, Robotics, Social Systems, Economic Systems and other. Of particular value to both the contributors and the readership are the short publication timeframe and the world-wide distribution and exposure which enable both a wide and rapid dissemination of research output.

The series covers the theory, applications, and perspectives on the state of the art and future developments relevant to systems and networks, decision making, control, complex processes and related areas, as embedded in the fields of interdisciplinary and applied sciences, engineering, computer science, physics, economics, social, and life sciences, as well as the paradigms and methodologies behind them.

Indexed by SCOPUS, INSPEC, WTI Frankfurt eG, zbMATH, SCImago.

All books published in the series are submitted for consideration in Web of Science.

For proposals from Asia please contact Aninda Bose (aninda.bose@springer.com).

Leonard Barolli

Editor

Advanced Information Networking and Applications

Proceedings of the 37th International Conference on Advanced Information Networking and Applications (AINA-2023), Volume 2

 Springer

Editor
Leonard Barolli
Department of Information and Communication
Engineering, Faculty of Information Engineering
Fukuoka Institute of Technology
Fukuoka, Japan

ISSN 2367-3370   ISSN 2367-3389 (electronic)
Lecture Notes in Networks and Systems
ISBN 978-3-031-28450-2   ISBN 978-3-031-28451-9 (eBook)
https://doi.org/10.1007/978-3-031-28451-9

This Springer imprint is published by the registered company Springer Nature Switzerland AG
The registered company address is: Gewerbestrasse 11, 6330 Cham, Switzerland

Welcome Message from AINA-2023 Organizers

Welcome to the 37th International Conference on Advanced Information Networking and Applications (AINA-2023). On behalf of AINA-2023 Organizing Committee, we would like to express to all participants our cordial welcome and high respect.

AINA is an International Forum, where scientists and researchers from academia and industry working in various scientific and technical areas of networking and distributed computing systems can demonstrate new ideas and solutions in distributed computing systems. AINA was born in Asia, but it is now an international conference with high quality thanks to the great help and cooperation of many international-friendly volunteers. AINA is a very open society and is always welcoming international volunteers from any country and any area in the world.

AINA International Conference is a forum for sharing ideas and research work in the emerging areas of information networking and their applications. The area of advanced networking has grown very rapidly, and the applications have experienced an explosive growth especially in the areas of pervasive and mobile applications, wireless sensor and ad hoc networks, vehicular networks, multimedia computing, social networking, semantic collaborative systems, as well as IoT, Big Data and Cloud Computing. This advanced networking revolution is transforming the way people live, work, and interact with each other and is impacting the way business, education, entertainment, and health care are operating. The papers included in the proceedings cover theory, design, and application of computer networks, distributed computing, and information systems.

Each year AINA receives a lot of paper submissions from all around the world. It has maintained high-quality accepted papers and is aspiring to be one of the main international conferences on the information networking in the world.

We are very proud and honored to have two distinguished keynote talks by Dr. Leonardo Mostarda, Camerino University, Italy, and Prof. Flávio de Oliveira Silva, Federal University of Uberlândia, Brazil, who will present their recent work and will give new insights and ideas to the conference participants.

An international conference of this size requires the support and help of many people. A lot of people have helped and worked hard to produce a successful AINA-2023 technical program and conference proceedings. First, we would like to thank all the authors for submitting their papers, the session chairs, and the distinguished keynote speakers. We are indebted to Program Track Co-Chairs, Program Committee Members, and Reviewers, who carried out the most difficult work of carefully evaluating the submitted papers.

We would like to thank AINA-2023 General Co-Chairs, PC Co-Chairs, and Workshops Co-chairs for their great efforts to make AINA-2023 a very successful event. We have special thanks to Finance Chair and Web Administrator Co-Chairs.

We do hope that you will enjoy the conference proceedings and readings.

Organization

AINA-2023 Organizing Committee

Honorary Chair

Makoto Takizawa   Hosei University, Japan

General Co-chairs

Mario A.R. Dantas   Federal University of Juiz de Fora, Brazil
Tomoya Enokido   Rissho University, Japan
Isaac Woungang   Toronto Metropolitan University, Canada

Program Committee Co-chairs

Victor Ströele   Federal University of Juiz de Fora, Brazil
Flora Amato   University of Naples "Federico II", Italy
Marek Ogiela   AGH University of Science and Technology, Poland

International Journals Special Issues Co-chairs

Fatos Xhafa   Technical University of Catalonia, Spain
David Taniar   Monash University, Australia
Farookh Hussain   University of Technology Sydney, Australia

Award Co-chairs

Arjan Durresi   Indiana University Purdue University in Indianapolis (IUPUI), USA

Fang-Yie Leu   Tunghai University, Taiwan
Kin Fun Li   University of Victoria, Canada

Publicity Co-chairs

Markus Aleksy   ABB Corporate Research Center, Germany
Omar Hussain   University of New South Wales, Australia

Lidia Ogiela	AGH University of Science and Technology, Poland
Hsing-Chung Chen	Asia University, Taiwan

International Liaison Co-chairs

Nadeem Javaid	COMSATS University Islamabad, Pakistan
Wenny Rahayu	La Trobe University, Australia
Beniamino Di Martino	University of Campania "Luigi Vanvitelli", Italy

Local Arrangement Co-chairs

Regina Vilela	Federal University of Juiz de Fora, Brazil
José Maria N. David	Federal University of Juiz de Fora, Brazil

Finance Chair

Makoto Ikeda	Fukuoka Institute of Technology, Japan

Web Co-chairs

Phudit Ampririt	Fukuoka Institute of Technology, Japan
Kevin Bylykbashi	Fukuoka Institute of Technology, Japan
Ermioni Qafzezi	Fukuoka Institute of Technology, Japan

Steering Committee Chair

Leonard Barolli	Fukuoka Institute of Technology, Japan

Tracks and Program Committee Members

1. Network Protocols and Applications

Track Co-chairs

Makoto Ikeda	Fukuoka Institute of Technology, Japan
Sanjay Kumar Dhurandher	Netaji Subhas University of Technology, New Delhi, India
Bhed Bahadur Bista	Iwate Prefectural University, Japan

TPC Members

Admir Barolli	Aleksander Moisiu University of Durres, Albania
Elis Kulla	Fukuoka Institute of Technology, Japan
Keita Matsuo	Fukuoka Institute of Technology, Japan
Shinji Sakamoto	Kanazawa Institute of Technology, Japan
Akio Koyama	Yamagata University, Japan
Evjola Spaho	Polytechnic University of Tirana, Albania
Jiahong Wang	Iwate Prefectural University, Japan
Shigetomo Kimura	University of Tsukuba, Japan
Chotipat Pornavalai	King Mongkut's Institute of Technology Ladkrabang, Thailand
Danda B. Rawat	Howard University, USA
Amita Malik	Deenbandhu Chhotu Ram University of Science and Technology, India
R. K. Pateriya	Maulana Azad National Institute of Technology, India
Vinesh Kumar	University of Delhi, India
Petros Nicopolitidis	Aristotle University of Thessaloniki, Greece
Satya Jyoti Borah	North Eastern Regional Institute of Science and Technology, India

2. Next-Generation Wireless Networks

Track Co-chairs

Christos J Bouras	University of Patras, Greece
Tales Heimfarth	Universidade Federal de Lavras, Brazil
Leonardo Mostarda	University of Camerino, Italy

TPC Members

Fadi Al-Turjman	Near East University, Cyprus
Alfredo Navarra	University of Perugia, Italy
Purav Shah	Middlesex University London, UK
Enver Ever	Middle East Technical University, Northern Cyprus Campus, Cyprus
Rosario Culmone	University of Camerino, Italy
Antonio Alfredo F. Loureiro	Federal University of Minas Gerais, Brazil
Holger Karl	University of Paderborn, Germany
Daniel Ludovico Guidoni	Federal University of São João Del-Rei, Brazil
João Paulo Carvalho Lustosa da Costa	Hamm-Lippstadt University of Applied Sciences, Germany
Jorge Sá Silva	University of Coimbra, Portugal

Apostolos Gkamas	University Ecclesiastical Academy of Vella, Greece
Zoubir Mammeri	University Paul Sabatier, France
Eirini Eleni Tsiropoulou	University of New Mexico, USA
Raouf Hamzaoui	De Montfort University, UK
Miroslav Voznak	University of Ostrava, Czech Republic
Kevin Bylykbashi	Fukuoka Institute of Technology, Japan

3. Multimedia Systems and Applications

Track Co-chairs

Markus Aleksy	ABB Corporate Research Center, Germany
Francesco Orciuoli	University of Salerno, Italy
Tomoyuki Ishida	Fukuoka Institute of Technology, Japan

TPC Members

Tetsuro Ogi	Keio University, Japan
Yasuo Ebara	Osaka Electro-Communication University, Japan
Hideo Miyachi	Tokyo City University, Japan
Kaoru Sugita	Fukuoka Institute of Technology, Japan
Akio Doi	Iwate Prefectural University, Japan
Hadil Abukwaik	ABB Corporate Research Center, Germany
Monique Duengen	Robert Bosch GmbH, Germany
Thomas Preuss	Brandenburg University of Applied Sciences, Germany
Peter M. Rost	NOKIA Bell Labs, Germany
Lukasz Wisniewski	inIT, Germany
Angelo Gaeta	University of Salerno, Italy
Graziano Fuccio	University of Salerno, Italy
Giuseppe Fenza	University of Salerno, Italy
Maria Cristina	University of Salerno, Italy
Alberto Volpe	University of Salerno, Italy

4. Pervasive and Ubiquitous Computing

Track Co-chairs

Chih-Lin Hu	National Central University, Taiwan
Vamsi Paruchuri	University of Central Arkansas, USA
Winston Seah	Victoria University of Wellington, New Zealand

TPC Members

Hong Va Leong	Hong Kong Polytechnic University, Hong Kong
Ling-Jyh Chen	Academia Sinica, Taiwan
Jiun-Yu Tu	Southern Taiwan University of Science and Technology, Taiwan
Jiun-Long Huang	National Chiao Tung University, Taiwan
Thitinan Tantidham	Mahidol University, Thailand
Tanapat Anusas-amornkul	King Mongkut's University of Technology North Bangkok, Thailand
Xin-Mao Huang	Aletheia University, Taiwan
Hui Lin	Tamkang University, Taiwan
Eugen Dedu	Universite de Franche-Comte, France
Peng Huang	Sichuan Agricultural University, China
Wuyungerile Li	Inner Mongolia University, China
Adrian Pekar	Budapest University of Technology and Economics, Hungary
Jyoti Sahni	Victoria University of Technology, New Zealand
Normalia Samian	Universiti Putra Malaysia, Malaysia
Sriram Chellappan	University of South Florida, USA
Yu Sun	University of Central Arkansas, USA
Qiang Duan	Penn State University, USA
Han-Chieh Wei	Dallas Baptist University, USA

5. Web-Based and E-Learning Systems

Track Co-chairs

Santi Caballe	Open University of Catalonia, Spain
Kin Fun Li	University of Victoria, Canada
Nobuo Funabiki	Okayama University, Japan

TPC Members

Jordi Conesa	Open University of Catalonia, Spain
Joan Casas	Open University of Catalonia, Spain
David Gañán	Open University of Catalonia, Spain
Nicola Capuano	University of Basilicata, Italy
Antonio Sarasa	Complutense University of Madrid, Spain
Chih-Peng Fan	National Chung Hsing University, Taiwan
Nobuya Ishihara	Okayama University, Japan
Sho Yamamoto	Kindai University, Japan
Khin Khin Zaw	Yangon Technical University, Myanmar
Kaoru Fujioka	Fukuoka Women's University, Japan

Kosuke Takano	Kanagawa Institute of Technology, Japan
Shengrui Wang	University of Sherbrooke, Canada
Darshika Perera	University of Colorado at Colorado Spring, USA
Carson Leung	University of Manitoba, Canada

6. Distributed and Parallel Computing

Track Co-chairs

Naohiro Hayashibara	Kyoto Sangyo University, Japan
Minoru Uehara	Toyo University, Japan
Tomoya Enokido	Rissho University, Japan

TPC Members

Eric Pardede	La Trobe University, Australia
Lidia Ogiela	AGH University of Science and Technology, Poland
Evjola Spaho	Polytechnic University of Tirana, Albania
Akio Koyama	Yamagata University, Japan
Omar Hussain	University of New South Wales, Australia
Hideharu Amano	Keio University, Japan
Ryuji Shioya	Toyo University, Japan
Ji Zhang	The University of Southern Queensland, Australia
Lucian Prodan	Universitatea Politehnica Timisoara, Romania
Ragib Hasan	The University of Alabama at Birmingham, USA
Young-Hoon Park	Sookmyung Women's University, South Korea
Dilawaer Duolikun	Cognizant Technology Solutions, Hungary
Shigenari Nakamura	Tokyo Metropolitan Industrial Technology Research Institute, Japan

7. Data Mining, Big Data Analytics, and Social Networks

Track Co-chairs

Omid Ameri Sianaki	Victoria University, Australia
Alex Thomo	University of Victoria, Canada
Flora Amato	University of Naples "Frederico II", Italy

TPC Members

Eric Pardede	La Trobe University, Australia
Alireza Amrollahi	Macquarie University, Australia
Javad Rezazadeh	University of Technology Sydney, Australia

Farshid Hajati	Victoria University, Australia
Mehregan Mahdavi	Sydney International School of Technology and Commerce, Australia
Ji Zhang	University of Southern Queensland, Australia
Salimur Choudhury	Lakehead University, Canada
Xiaofeng Ding	Huazhong University of Science and Technology, China
Ronaldo dos Santos Mello	Universidade Federal de Santa Catarina, Brazil
Irena Holubova	Charles University, Czech Republic
Lucian Prodan	Universitatea Politehnica Timisoara, Romania
Alex Tomy	La Trobe University, Australia
Dhomas Hatta Fudholi	Universitas Islam Indonesia, Indonesia
Saqib Ali	Sultan Qaboos University, Oman
Ahmad Alqarni	Al Baha University, Saudi Arabia
Alessandra Amato	University of Naples "Frederico II", Italy
Luigi Coppolino	Parthenope University, Italy
Giovanni Cozzolino	University of Naples "Frederico II", Italy
Giovanni Mazzeo	Parthenope University, Italy
Francesco Mercaldo	Italian National Research Council, Italy
Francesco Moscato	University of Salerno, Italy
Vincenzo Moscato	University of Naples "Frederico II", Italy
Francesco Piccialli	University of Naples "Frederico II", Italy

8. Internet of Things and Cyber-Physical Systems

Track Co-chairs

Euripides G. M. Petrakis	Technical University of Crete (TUC), Greece
Tomoki Yoshihisa	Osaka University, Japan
Mario Dantas	Federal University of Juiz de Fora (UFJF), Brazil

TPC Members

Akihiro Fujimoto	Wakayama University, Japan
Akimitsu Kanzaki	Shimane University, Japan
Kawakami Tomoya	University of Fukui, Japan
Lei Shu	University of Lincoln, UK
Naoyuki Morimoto	Mie University, Japan
Yusuke Gotoh	Okayama University, Japan
Vasilis Samolada	Technical University of Crete (TUC), Greece
Konstantinos Tsakos	Technical University of Crete (TUC), Greece
Aimilios Tzavaras	Technical University of Crete (TUC), Greece

Spanakis Manolis	Foundation for Research and Technology Hellas (FORTH), Greece
Katerina Doka	National Technical University of Athens (NTUA), Greece
Giorgos Vasiliadis	Foundation for Research and Technology Hellas (FORTH), Greece
Stefan Covaci	Technicak University of Berlin (TUB), Germany
Stelios Sotiriadis	University of London, UK
Stefano Chessa	University of Pisa, Italy
Jean-Francois Méhaut	Université Grenoble Alpes, France
Michael Bauer	University of Western Ontario, Canada

9. Intelligent Computing and Machine Learning

Track Co-chairs

Takahiro Uchiya	Nagoya Institute of Technology, Japan
Omar Hussain	UNSW, Australia
Nadeem Javaid	COMSATS University Islamabad, Pakistan

TPC Members

Morteza Saberi	University of Technology Sydney, Australia
Abderrahmane Leshob	University of Quebec in Montreal, Canada
Adil Hammadi	Curtin University, Australia
Naeem Janjua	Edith Cowan University, Australia
Sazia Parvin	Melbourne Polytechnic, Australia
Kazuto Sasai	Ibaraki University, Japan
Shigeru Fujita	Chiba Institute of Technology, Japan
Yuki Kaeri	Mejiro University, Japan
Zahoor Ali Khan	HCT, UAE
Muhammad Imran	King Saud University, Saudi Arabia
Ashfaq Ahmad	The University of Newcastle, Australia
Syed Hassan Ahmad	JMA Wireless, USA
Safdar Hussain Bouk	Daegu Gyeongbuk Institute of Science and Technology, South Korea
Jolanta Mizera-Pietraszko	Military University of Land Forces, Poland
Shahzad Ashraf	NFC Institute of Engineering and Technology, Pakistan

10. Cloud and Services Computing

Track Co-chairs

Asm Kayes	La Trobe University, Australia
Salvatore Venticinque	University of Campania "Luigi Vamvitelli", Italy
Baojiang Cui	Beijing University of Posts and Telecommunications, China

TPC Members

Shahriar Badsha	University of Nevada, USA
Abdur Rahman Bin Shahid	Concord University, USA
Iqbal H. Sarker	Chittagong University of Engineering and Technology, Bangladesh
Jabed Morshed Chowdhury	La Trobe University, Australia
Alex Ng	La Trobe University, Australia
Indika Kumara	Jheronimus Academy of Data Science, The Netherlands
Tarique Anwar	Macquarie University and CSIRO's Data61, Australia
Giancarlo Fortino	University of Calabria, Italy
Massimiliano Rak	University of Campania "Luigi Vanvitelli", Italy
Jason J. Jung	Chung-Ang University, South Korea
Dimosthenis Kyriazis	University of Piraeus, Greece
Geir Horn	University of Oslo, Norway
Gang Wang	Nankai University, China
Shaozhang Niu	Beijing University of Posts and Telecommunications, China
Jianxin Wang	Beijing Forestry University, China
Jie Cheng	Shandong University, China
Shaoyin Cheng	University of Science and Technology of China, China

11. Security, Privacy, and Trust Computing

Track Co-chairs

Hiroaki Kikuchi	Meiji University, Japan
Xu An Wang	Engineering University of PAP, P.R. China
Lidia Ogiela	AGH University of Science and Technology, Poland

TPC Members

Takamichi Saito	Meiji University, Japan
Kouichi Sakurai	Kyushu University, Japan
Kazumasa Omote	University of Tsukuba, Japan
Shou-Hsuan Stephen Huang	University of Houston, USA
Masakatsu Nishigaki	Shizuoka University, Japan
Mingwu Zhang	Hubei University of Technology, China
Caiquan Xiong	Hubei University of Technology, China
Wei Ren	China University of Geosciences, China
Peng Li	Nanjing University of Posts and Telecommunications, China
Guangquan Xu	Tianjing University, China
Urszula Ogiela	AGH University of Science and Technology, Poland
Hoon Ko	Chosun University, Republic of Korea
Goreti Marreiros	Institute of Engineering of Polytechnic of Porto, Portugal
Chang Choi	Gachon University, Republic of Korea
Libor Měsíček	J. E. Purkyně University, Czech Republic

12. Software-Defined Networking and Network Virtualization

Track Co-chairs

Flavio de Oliveira Silva	Federal University of Uberlândia, Brazil
Ashutosh Bhatia	Birla Institute of Technology and Science, Pilani, India
Alaa Allakany	Kyushu University, Japan

TPC Members

Rui Luís Andrade Aguiar	Universidade de Aveiro (UA), Portugal
Ivan Vidal	Universidad Carlos III de Madrid, Spain
Eduardo Coelho Cerqueira	Federal University of Pará (UFPA), Brazil
Christos Tranoris	University of Patras (UoP), Greece
Juliano Araújo Wickboldt	Federal University of Rio Grande do Sul (UFRGS), Brazil
Yaokai Feng	Kyushu University, Japan
Chengming Li	Chinese Academy of Science (CAS), China
Othman Othman	An-Najah National University (ANNU), Palestine
Nor-masri Bin-sahri	University Technology of MARA, Malaysia
Sanouphab Phomkeona	National University of Laos, Laos
Haribabu K.	BITS Pilani, India

Shekhavat, Virendra   BITS Pilani, India
Makoto Ikeda   Fukuoka Institute of Technology, Japan
Farookh Hussain   University of Technology Sydney, Australia
Keita Matsuo   Fukuoka Institute of Technology, Japan

AINA-2023 Reviewers

Admir Barolli
Ahmed Bahlali
Aimilios Tzavaras
Akihiro Fujihara
Akimitsu Kanzaki
Alaa Allakany
Alba Amato
Alberto Volpe
Alex Ng
Alex Thomo
Alfredo Navarra
Anne Kayem
Antonio Esposito
Arcangelo Castiglione
Arjan Durresi
Ashutosh Bhatia
Asm Kayes
Bala Killi
Baojiang Cui
Beniamino Di Martino
Bhed Bista
Bruno Zarpelão
Carson Leung
Chang Choi
Changyu Dong
Chih-Peng Fan
Christos Bouras
Christos Tranoris
Chung-Ming Huang
Darshika Perera
David Taniar
Dilawaer Duolikun
Donald Elmazi
Elis Kulla
Eric Pardede
Euripides Petrakis

Evjola Spaho
Fabian Kurtz
Farookh Hussain
Fatos Xhafa
Feilong Tang
Feroz Zahid
Flavio Corradini
Flavio Silva
Flora Amato
Francesco Orciuoli
Gang Wang
Goreti Marreiros
Hadil Abukwaik
Hiroaki Kikuchi
Hiroshi Maeda
Hiroyoshi Miwa
Hiroyuki Fujioka
Hsing-Chung Chen
Hyunhee Park
Indika Kumara
Isaac Woungang
Jabed Chowdhury
Jana Nowaková
Ji Zhang
Jiahong Wang
Jianfei Zhang
Jolanta Mizera-Pietraszko
Jörg Domaschka
Jorge Sá Silva
Juliano Wickboldt
Julio Costella Vicenzi
Jun Iio
K. Haribabu
Kazunori Uchida
Keita Matsuo
Kensuke Baba

Kin Fun Li
Kiplimo Yego
Kiyotaka Fujisaki
Konstantinos Tsakos
Kouichi Sakurai
Lei Shu
Leonard Barolli
Leonardo Mostarda
Libor Mesicek
Lidia Ogiela
Lucian Prodan
Luciana Oliveira
Makoto Ikeda
Makoto Takizawa
Marek Ogiela
Marenglen Biba
Mario Dantas
Markus Aleksy
Masakatsu Nishigaki
Masaki Kohana
Masaru Kamada
Mingwu Zhang
Minoru Uehara
Miroslav Voznak
Mohammad Faiz Iqbal Faiz
Nadeem Javaid
Naohiro Hayashibara
Neder Karmous
Nobuo Funabiki
Omar Hussain
Omid Ameri Sianaki
Paresh Saxena
Pavel Kromer
Petros Nicopolitidis
Philip Moore Fatos Xhafa
Purav Shah

Rajdeep Niyogi
Rodrigo Miani
Ronald Petrlic
Ronaldo Mello
Rui Aguiar
Ryuji Shioya
Salimur Choudhury
Salvatore Venticinque
Sanjay Dhurandher
Santi Caballé
Satya Borah
Shahriar Badsha
Shengrui Wang

Shigenari Nakamura
Shigetomo Kimura
Somnath Mazumdar
Sriram Chellappan
Stelios Sotiriadis
Takahiro Uchiya
Takamichi Saito
Takayuki Kushida
Tetsuya Oda
Tetsuya Shigeyasu
Thomas Dreibholz
Tomoki Yoshihisa
Tomoya Enokido

Tomoyuki Ishida
Vamsi Paruchuri
Wang Xu An
Wei Lu
Wenny Rahayu
Winston Seah
Yoshihiro Okada
Yoshitaka Shibata
Yusuke Gotoh
Zahoor Khan
Zia Ullah

AINA-2023 Keynote Talks

Blockchain and IoT Integration: Challenges and Future Directions

Leonardo Mostarda

Camerino University, Camerino, Italy

Abstract. Massive overhead costs, concerns about centralized data control, and single point of vulnerabilities are significantly reduced by moving IoT from a centralized data server architecture to a trustless, distributed peer-to-peer network. Blockchain is one of the most promising and effective technologies for enabling a trusted, secure, and distributed IoT ecosystem. Blockchain technology can allow the implementation of decentralized applications that not only perform payments but also allow the execution of smart contracts. This talk will investigate the state of the art and open challenges that are related to IoT and blockchain integration. We review current approaches and future directions.

Toward Sustainable, Intelligent, Secure, Fully Programmable, and Multisensory (SENSUOUS) Networks

Flávio de Oliveira Silva

Federal University of Uberlândia, Uberlândia, Brazil

Abstract. In this talk, we will discuss and present the evolution of current networks toward sustainable, intelligent, secure, fully programmable, and multisensory (SENSUOUS) networks. The evolution of networks happens through these critical attributes that will drive the next-generation networks. Here networks consider data networks capable of transmitting audio and video in computer or telecommunication systems. While there is an established process for the evolution of telecommunication networks, regarding computer networks, this area is still open and has several challenges and opportunities. So far, networks can transmit audio and video data, which sensitize only part of our senses. Still, new senses must be considered in the evolution of networks, expanding the multisensory experience. SENSUOUS networks will shape and contribute to scaling our society's sustainable, smart, and secure digital transformation.

Contents

Integration and Evaluation of Blockchain Consensus Algorithms for IoT Environments

Anderson Melo de Morais[1]($^{(\boxtimes)}$), Fernando Antonio Aires Lins[2],
and Nelson Souto Rosa[1]

[1] Centro de Informática, Universidade Federal de Pernambuco, Recife, Brazil
{amm6,nsr}@cin.ufpe.br
[2] Universidade Federal Rural de Pernambuco, Recife, Brazil
fernandoaires@ufrpe.br

Abstract. IoT devices are increasingly used, leading to the need to ensure the privacy and security of user data. Blockchain can assure the security and immutability of data generated by IoT devices, sensors, and actuators. In Blockchain, consensus algorithms provide the security of creating and recording new blocks. However, choosing which algorithms can be used in the context of IoT is a challenge as they have advantages (e.g., security) and limitations (e.g., long response time). Thus, integrating consensus algorithms into a single solution can be a viable alternative, as it allows taking advantage of the strengths of more than one algorithm and mitigating the weaknesses. Therefore, this paper proposes integrating some consensus algorithms that can be used in IoT. These integrations are evaluated regarding CPU and memory consumption, network traffic, and time to create new blocks. Finally, the feasibility of using integrations of consensus algorithms for IoT is discussed.

1 Introduction

The Internet of Things (IoT) has gained increasing importance in society by transforming everyday things into intelligent and autonomous devices [13]. Many devices handle users' data, and concerns arise related to security and privacy, as well as the way data are processed and stored. In turn, one technology that can provide security to IoT environments is Blockchain, which uses cryptography and decentralized data logging to achieve data reliability and security [6].

Blockchain in IoT presents challenges due to resource limitations of IoT devices (e.g., memory and processing power), which makes it challenging to implement the cryptographic mechanisms used by traditional Blockchains. A fundamental aspect of Blockchain is the consensus algorithm, which validates new blocks before they are inserted into the chain. Many approaches have been proposed to adapt or extend Blockchain elements to make them more suitable for IoT [2].

L. Barolli (Ed.): AINA 2023, LNNS 654, pp. 1–13, 2023.
https://doi.org/10.1007/978-3-031-28451-9_1

The adoption of Blockchain-based solutions has tradeoffs like increasing security accompanied by scalability losses or being highly scalable but energy-intensive. To overcome some of these issues, new approaches [2,17] suggest the integration of multiple consensus algorithms within the same IoT solution. These integrations can benefit IoT as they allow to leverage of the strengths of each algorithm.

In this context, this paper proposes integrating some consensus algorithms that can be used in IoT. Evaluations of these integrations are carried out in terms of CPU and memory consumption, network traffic, and time to create new blocks and discuss the feasibility of integrations for the IoT context.

In this paper, some integrations already proposed in the literature were implemented. Then, existing and new integrations were compared to assess their performance. Algorithms considered safe, such as PoW and PoS [7], and algorithms known for their performance, such as PoA and PBFT [3], were chosen.

The rest of this paper is organized as follows. Section 2 introduces the basic concepts needed to understand the proposed solution. Section 3 presents the proposed integrations of consensus algorithms and their evaluations. Section 4 describes and analyses existing solutions. Finally, Sect. 5 presents the final considerations of the work.

2 Background

This section introduces basic concepts of Blockchain and consensus algorithms, essential to comprehend what is being proposed.

2.1 Blockchain

Blockchain consists of a system of transaction records distributed and shared through nodes in a peer-to-peer (P2P) network [11]. Blockchain presents itself as a safe environment for recording transactions. As once a new block is added, it cannot be deleted or changed without this being noticed by the other nodes.

Some nodes, called miners, have an essential role: validating new blocks through a consensus algorithm to confirm transactions and produce blocks for the chain. In the context of IoT, Blockchain can be used to authenticate, authorize and audit data generated by devices [13]. Because it is decentralized, there is no need to trust a regulatory institution, which reduces the risk of malicious security attacks [6].

2.2 Consensus Mechanisms

Consensus mechanisms are procedures that allow all Blockchain peers to reach a joint agreement on new blocks to be added. This procedure ensures trust and reliability among unknown peers [5].

The first Blockchain, proposed by Nakamoto for Bitcoin [11], adopted the Proof of Work (PoW) algorithm, which has high energy consumption and computational resources such as memory and processing. Other mechanisms have been developed, such as Proof of Stake (PoS) and Proof of Authority (PoA), among others [16]. In this work, the following consensus algorithms were considered, taking into account their security and performance:

- **Proof of Work (PoW).** Any node in the system can verify information that will be added to new blocks. Through a competition mechanism, some nodes can solve a computational puzzle and obtain the reward through the process known as mining. The bigger the computational puzzle to be solved, the higher the Blockchain difficulty [2].
- **Proof of Stake (PoS).** It seeks to ensure security and reduce consumed resources and improve transaction performance. In PoS, each node has a certain amount of digital currencies used between peers in the network. Any network node with a sufficient amount of coins can use them to validate transactions [7].
- **Proof-of-Authority (PoA).** It operates in rounds in which a node is elected mining leader and is responsible for proposing new blocks. Unlike other algorithms, PoA requires fewer message exchanges, providing good performance [8].
- **Practical Byzantine Fault Tolerance (PBFT).** PBFT provides high throughput, and low latency in node creation [1]. A node creates a request containing the data that will be inserted in a new node and sends it to others. After three verification rounds, the remaining nodes will verify the data and return a response.

Consensus mechanisms that take a long to process and create each block can become unfeasible for use in IoT environments. Their data needs to be processed and recorded quickly to avoid possible tampering or loss of important information.

3 Integration of Consensus Algorithms and Evaluations

This section describes the proposed integrations of consensus algorithms and presents the integrations' performance evaluations.

3.1 Integrations

The criterion used to choose the consensus algorithms to be integrated looked for those that presented good tolerance to attacks and good performance at the time of creating new blocks. The purpose of the integrations was to combine these two characteristics and thus obtain efficient performance results, providing security to IoT data.

- **PoA and PoW:** This integration aims to increase security (PoW) and reduce transaction confirmation time (POA). The PoA algorithm chooses the mining node with the highest authority or "reputation"; then, the chosen node runs the PoW to create a block and adds it to the chain. This way, the PoW runs at a lower difficulty level and still has good security as a 'trusted' node is running it.
- **PoA and PoS:** The objective of this integration is to make the creation of new blocks more secure and agile. For this, the mining node is chosen among those with the highest reputation (PoA) and offers the most significant amount of coins (Stake). This integration does not overload the most reputable node, as others can be chosen depending on their amount of stakes.
- **PoS and PoW:** This integration combines the security of PoW with the excellent performance of PoS. Then, this solution chooses the mining node among those that offer the highest coins (Stakes). Next, the selected node runs PoW to create a new block and add it to the chain.
- **PoA and PBFT:** This integration aims to increase the performance and security of PBFT by combining it with PoA. The initial node to be chosen should have the highest authority or "reputation". Then the chosen node runs PBFT to create a new block. PBFT creates a new node faster than the others, so its combination with PoA allows choosing a node considered 'trusted' to start the validation.
- **PoS and PBFT:** This integration aims to make the PBFT home node more secure. Thus, integration with PoS is used, which chooses the initial node among those offering the highest amount of coins (Stake). Then the chosen node runs PBFT to create a new block. This integration increases the security of the PBFT, as it reduces the possibility of choosing a malicious node to start the validation.

Some of the integrations have already been proposed in the literature, such as the use of the PoA with the PoW [3], and also the PoS integrated with the PoW [4]. However, in this paper, the integrations were made using a different approach. For the evaluation, the simulation of Blockchain in Docker was used, which allows for evaluating its behaviour by varying different parameters and metrics [14].

3.2   Evaluation

The methodology used for performance evaluation was inspired on Raj Jain's approach [9] and adopted the following steps:

1. **Definition of objectives and system:** To evaluate the consensus algorithms, a Blockchain simulation was performed using Docker, where each container represents a Blockchain node. Consensus algorithms have been implemented in Node.js. The objective was to observe the time to create new blocks for each integration, as well as the use of RAM and CPU.

2. **Metrics:** The metrics chosen for this evaluation included block creation time, CPU consumption, RAM consumption, and network traffic.
3. **Elaboration of the list of parameters:** consists of the parameters that affect the performance. In this study, load demand parameters were observed, such as CPU, memory, and network consumption.
4. **Selection of factors for performance:** Factors are the parameters that, when varied, will influence more intensity of system performance. In this evaluation, the number of Blockchain nodes varied to 5, 10, and 20.
5. **Selection of factors for performance:** Factors are the parameters that, when varied, will influence more intensity of system performance. In this evaluation, the number of Blockchain nodes varied to 5, 10, and 20.
6. **Selection of the evaluation technique:** The simulation technique was used, where a Blockchain was simulated in Docker. The tests were performed on an Intel Core i5-10210U 1.60GHz processor with 8GB of RAM.
7. **Load selection:** For the tests, it was defined that a new block should be created every 5 min. That is, this is the time when consensus mechanisms should run.
8. **Experiment planning:** for the experiments, a new block was created using each of the performed integrations. To calculate the time to create new blocks, the procedure was repeated 30 times, and the average time was calculated.
9. **Data analysis and interpretation:** Finally, the data obtained in the evaluations were organized and evaluated, as described in the following sections.

The HC-SR04 sensors, which monitor distances, and the DHT11 temperature and humidity sensor were used for the evaluation environment. The data is collected by a Raspberry Pi and stored in a Firebase database, where the Blockchain periodically collects it to register new blocks.

3.2.1 PoA and PoW

The time to create new blocks was compared by varying the number of Blockchain nodes (5, 10, and 20) and the block mining difficulty (3, 4, and 5). The higher the mining difficulty, the safer, but the longer the process. Figure 1 shows the comparison of block creation time in this integration.

According to the graph in Fig. 1, as the PoW mining difficulty increases, the time to create and insert new blocks increases considerably. Metrics of CPU and RAM consumption and network traffic were also evaluated. Figure 2 shows the graphs generated when creating a new block using PoA and PoW integration. The information in the graph corresponds to the central node running PoW at that moment. It is possible to notice that the CPU consumption was above 80%.

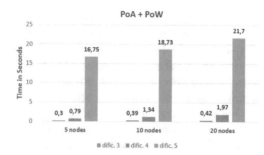

Fig. 1. Comparison of time to create new blocks using PoA and PoW.

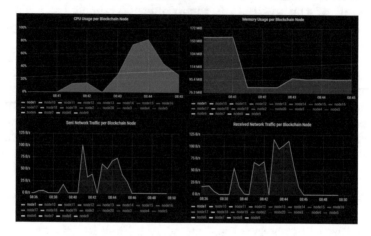

Fig. 2. Metrics analyzed when creating a new block using PoA and PoW.

3.2.2 PoS and PoW

Metrics of CPU and RAM consumption and network traffic were also evaluated. Figure 2 shows the creation of a new block using PoA and PoW integration. According to Figure Fig. 1, as the PoW mining difficulty increases, the time to create and insert new blocks increases considerably. The information in the figure corresponds to the central node running PoW. It is possible to notice that the CPU consumption was above 80%.

Similar to the previous integration, as the mining difficulty of PoW increases, the time to create and insert new blocks on the Blockchain has considerably grown. Figure 4 shows the results of creating a new block using the PoS and PoW integration, considering the central node that was in use at that moment. When evaluating the metrics of CPU, memory, and network traffic consumption, it can be seen that there was a lower CPU usage but a similar consumption of memory and network (Fig. 3).

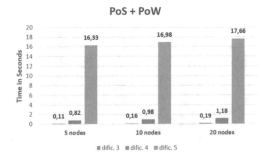

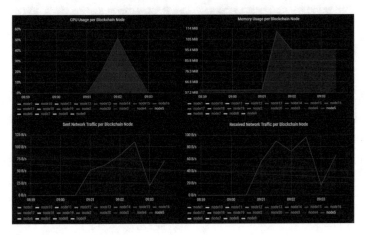

Fig. 3. Comparison of time to create new blocks using PoS and PoW.

Fig. 4. Metrics analyzed when creating a new block using PoS and PoW.

3.2.3 PoA and PoS

Another integration combined PoA and PoS. For the evaluation, the time to create new blocks was compared by varying the number of nodes. Figure 5 shows the result of this comparison, where the times obtained are considerably shorter than in the previous integrations.

The CPU and memory consumption metrics delivered better results for this integration, and the network traffic presented an increase. Figure 6 shows the results for creating a new block using the PoA and PoS integration, considering the primary node in use.

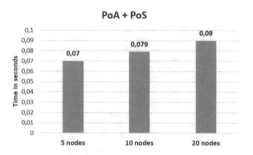

Fig. 5. Comparison of time to create new blocks using PoA and PoS.

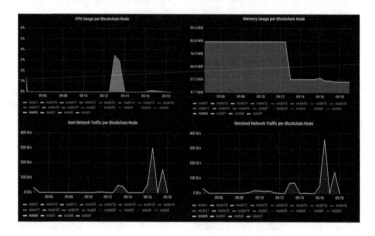

Fig. 6. Metrics analyzed when creating a new block using PoA and PoS.

3.2.4   PoA and PBFT

Integrations using PBFT were also tested. The integration of PBFT and PoA created new blocks in a faster time than in previous integrations. Figure 7 shows the comparison of the creation time of a new block. For this test, the number of Blockchain nodes varied in values of 5, 10, and 20 nodes.

For this integration, the CPU and memory metrics showed lower results, 0.4%, and 62 MB, respectively, compared to integrations that used algorithms such as PoW. However, the network traffic increased to 500 bytes/s. This result happened because, in PBFT, the block needs to be validated by most nodes before being inserted into the Blockchain. Figure 8 shows the results obtained when creating a new block using the PoA and PBFT integration.

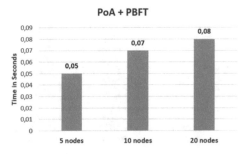

Fig. 7. Comparison of time to create new blocks using PoA and PBFT.

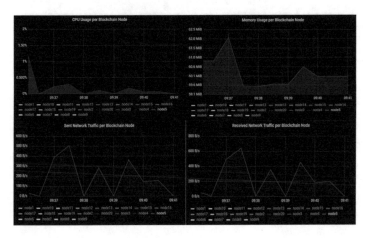

Fig. 8. Metrics analyzed when creating a new block using PoA and PBFT.

3.2.5 PoS and PBFT

PBFT was also integrated with PoS, and the result was similar to the previous integration. However, the block creation time was faster than in integrations not using PBFT. Figure 9 shows the result of the block creation time of this integration.

Also, in this integration, the CPU and RAM consumption metrics showed lower results than those not using PBFT (0.5% and 58.8 MB, respectively). However, the network traffic presented higher values, reaching 500 bytes/s. Figure 10 shows the results obtained when creating a new block using PoS and PBFT.

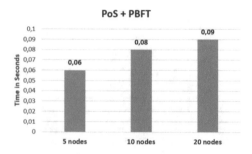

Fig. 9. Comparison of time to create new blocks using PoS and PBFT.

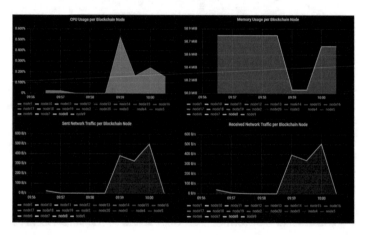

Fig. 10. Metrics analyzed when creating a new block using PoS and PBFT.

Finally, the data was gathered in Table 1, which shows the creation times of new blocks and the values of the metrics analyzed in each integration.

Table 1. Comparison of evaluations performed on integrations

Integrations	Block creation time[a]	Security level	CPU consumption	Memory consumption	Network traffic
PoA + PoW	21,7 s	High	80%	155 MB	100 B/s
PoS + PoW	17,6 s	High	50%	110 MB	90 B/s
PoA + PoS	0,09 s	Medium	3,5%	85,8 MB	300 B/s
PoA + PBFT	0,08 s	Medium	0,4%	62 MB	500 B/s
PoS + PBFT	0,09 s	Medium	0,5%	58,8 MB	500 B/s

[a]Considering Blockchain with 20 nodes and difficulty 5 in PoW

In IoT, most devices have resource limitations, such as processing power and memory. Thus, consensus algorithms that demand high computational requirements and block creation time, such as those based on PoW, may not be feasible.

As shown in Table 1, the integrations that used PBFT showed a better time to create new blocks and performance, which may indicate that they are viable for IoT.

4   Related Work

Alrubei [3] proposed the integration of PoW and PoA. The objective was to take advantage of the attack tolerance presented by the PoW and reduce the transaction confirmation time, integrating it with the PoA, which has a good execution speed.

Bai et al. [4] integrate PoW and PoS to create a solution that presents high tolerance to attacks using PoW and good scalability using PoS. The authors propose a two-tier consensus-based Blockchain architecture optimized for IoT. IoT Devices form the Base-Layer. The Top-Layer consists of the Blockchain, formed by nodes that run a non-Byzantine fault tolerance algorithm to consensus.

Lunardi et al. [10] propose a Blockchain to support different consensus algorithms in a modular design. According to the authors, classical consensus algorithms are not suitable for IoT, and the decision of which consensus algorithm would be most suitable is left open in the literature. In this paper, the authors integrate PBF and a witness-based consensus (WBC) to reduce latency and throughput.

Rasolroveicy and Fokaefs [12] designed a self-adaptive mechanism that dynamically chooses the consensus algorithm to be used in the IoT Blockchain. Raft [5], Proof of elapsed time (PoET) [5], and PBFT algorithms are integrated to manage network cost, performance, and security, as well as reduce power consumption.

This paper made advances to the state-of-the-art by realising new integrations. Integrations were carried out between the PoA and the PBFT and between the PoS and the PBFT. The performance evaluation of these integrations showed that they could be indicated for use in IoT due to their low time to create new blocks and low CPU and memory consumption.

5   Conclusions and Future Work

This work implemented some integrations between consensus algorithms and evaluated their performance and time to create new blocks. The performance was evaluated regarding CPU and memory consumption and network traffic. Unpublished integrations in the literature were proposed, such as PoA with PBFT and PoS with PBFT. Finally, the viability of these integrations for IoT was discussed.

Blockchain has the potential to solve IoT security challenges. However, integrating the two areas still presents many challenges as Blockchain consensus algorithms need to meet IoT requirements. Therefore, it is possible to integrate consensus algorithms, taking advantage of the positive characteristics of each

A. M. de Morais et al.

one. The integrations proposed in this work are indicated as viable for IoT due to their speed in creating new blocks, low CPU cost, and memory consumption.

In future work, it is possible to perform new integrations between consensus algorithms and evaluate their performance and feasibility in IoT. It is also desirable to assess the proposed integrations in specific IoT domains.

References

1. Ali, M., Vecchio, M., Pincheira, M., Dolui, K., Antonelli, F., Rehmani, M.: Applications of blockchains in the internet of things: a comprehensive survey. IEEE Commun. Surv. Tutor. **21**, 1676–1717 (2018)
2. Al Sadawi, A., Hassan, M., Ndiaye, M.: A survey on the integration of blockchain with IoT to enhance performance and eliminate challenges. IEEE Access **9**, 54478–54497 (2021)
3. Alrubei, S., Ball, E., Rigelsford, J.: Securing IoT-blockchain applications through honesty-based distributed proof of authority consensus algorithm. In: 2021 International Conference on Cyber Situational Awareness, Data Analytics and Assessment, pp. 1–7 (2021)
4. Bai, H., Xia, G., Fu, S.: A two-layer-consensus based blockchain architecture for IoT. In: 2019 IEEE 9th International Conference on Electronics Information and Emergency Communication (ICEIEC), pp. 1–6 (2019)
5. Bodkhe, U., Mehta, D., Tanwar, S., Bhattacharya, P., Singh, P., Hong, W.: A survey on decentralized consensus mechanisms for cyber physical systems. IEEE Access **8**, 54371–54401 (2020)
6. Chicarino, V., Jesus, E., Albuquerque, C., Antônio, A.: Uso de blockchain para privacidade e segurança em internet das coisas. Sociedade Brasileira De Computação (2017)
7. Dai, H., Zheng, Z., Zhang, Y.: Blockchain for internet of things: a survey. IEEE Internet Things J. **6**, 8076–8094 (2019)
8. De Angelis, S., Aniello, L., Baldoni, R., Lombardi, F., Margheri, A., Sassone, V.: PBFT vs proof-of-authority: applying the CAP theorem to permissioned blockchain. University of Southampton Institutional Repository (2018)
9. Jain, R.: The Art of Computer Systems Performance Analysis. Wiley, Hoboken (1991)
10. Lunardi, R., Michelin, R., Neu, C., Nunes, H., Zorzo, A., Kanhere, S.: Impact of consensus on appendable-block Blockchain for IoT. In: 16th EAI International Conference on Mobile and Ubiquitous Systems: Computing, Networking and Services, pp. 228–237 (2019)
11. Nakamoto, S., Bitcoin, A.: A peer-to-peer electronic cash system. Bitcoin (2008). https://bitcoin.org/bitcoin.pdf
12. Rasolroveicy, M., Fokaefs, M.: Dynamic reconfiguration of consensus protocol for IoT data registry on Blockchain. In: Proceedings of the 30th Annual International Conference on Computer Science and Software Engineering, pp. 227–236 (2020)
13. Reyna, A.M., et al.: On blockchain and its integration with IoT. Challenges opportunities. Futur. Gener. Comput. Syst. **88**, 173G190 (2018)
14. Yadav, R., Sousa, E., Callou, G.: Performance comparison between virtual machines and docker containers. IEEE Lat. Am. Trans. **16**, 2282–2288 (2018)
15. Wang, B., Li, Z., Li, H.: Hybrid consensus algorithm based on modified proof-of-probability and DPoS. Future Internet **12**, 122 (2020)

16. Wu, M., Wang, K., Cai, X., Guo, S., Guo, M., Rong, C.: A comprehensive survey of blockchain: from theory to IoT applications and beyond. IEEE Internet Things J. **6**, 8114–8154 (2019)
17. Xu, R., Chen, Y., Blasch, E., Chen, G.: Microchain: a hybrid consensus mechanism for lightweight distributed ledger for IoT. arXiv Preprint arXiv:1909.10948 (2019)
18. Zhipeng, F.: Research on blockchain hybrid consensus algorithm based on internet of things. Int. J. Internet Things Big Data (2019)

Nexus: Proxy Service for the Web of Things

Isidoros Paterakis and Euripides G. M. Petrakis[✉]

School of Electrical and Computer Engineering,
Technical University of Crete (TUC), Chania, Greece
ipaterakis1@tuc.gr, petrakis@intelligence.tuc.gr

Abstract. The W3C Web of Things (WoT) architecture is a model for handling Things (i.e. devices) as Web pages. The WoT proxy is a central building block of the architecture. Its primary role is to translate the devices' communication protocols to HTTP(S) so that they can connect to the application server (in the cloud). Nexus promotes this idea even further and proposes that the WoT proxy be converted into a Multi-Context Service Oriented Architecture (SOA) for all device-specific operations that would otherwise be performed on the application server. The advantage is the decoupling of the Proxy and application server roles. WoT proxy becomes application-independent. Different applications can use Nexus as a single point of access to their network devices. Nexus can manage users, devices, and the environment of various applications according to their roles and access rights. Experimental results show that Nexus scales well and responds in real time under heavy workloads.

1 Introduction

The Web of Things (WoT) initiative [7] aims to unify the world of interconnected devices over the Internet. A Thing may refer to any device: a temperature or pressure sensor, a window or door actuator, a smart coffee machine, or a smart car. In response to this requirement, the Web of Things (WoT) architecture model of W3C [11] defines a framework for integrating devices into the Web. Devices may implement any protocol from a wide range of IoT or application-specific protocols (e.g. Bluetooth, MQTT, ZigBee, LoRa, etc.) to communicate and transfer data to the server of an application in the cloud. That communication should be protocol-independent. A way around this is to translate IoT protocols to a common Web protocol (e.g. HTTP). This is the role of a WoT Proxy, that runs on the cloud (or on a gateway).

Nexus is not merely a protocol adaptation service. It is a Service Oriented Architecture (SOA) that implements Things-specific functionality. Any application can connect to Nexus to access the network of IoT devices (through an API) to issue telemetry and remote control functions. Nexus decouples application-specific features on the server from WoT functionality while being secure by design.

L. Barolli (Ed.): AINA 2023, LNNS 654, pp. 14–25, 2023.
https://doi.org/10.1007/978-3-031-28451-9_2

An essential part of Nexus is the directory that holds the virtual image of each device in JSON. In a recent paper [22] we proposed to describe Things similarly to Web services using OpenAPI [13]. OpenAPI Thing Descriptions (TD) provides information about service endpoints, message formats, and conditions for calling the service. New devices can join an application's IoT network at any time, and their OpenAPI descriptions are generated using the OpenAPI Thing Description (TD) generator. Searching for devices (by name, functionality properties) or WoT services is supported by OAQL [2] a SQL language for OpenAPI.

The advantages of Nexus do not end with the above. Nexus is a Multi-Context application that allows different applications to share the same WoT Proxy to access their IoT network. Its role is also to support the operation of different user categories, such as infrastructure owners to install new devices, application developers who subscribe to devices to build applications, and customers (clients) who subscribe to use applications. Nexus administrators approve the registration of new users and monitor and maintain control of all system operations [10].

Access to services and devices is granted only to authorized users (or other services) based on context, user roles, and access policies. Nexus implements the multi-context functionality of the different applications connected to it, including security and context management services that allow devices to publish information, and users (or other services) to subscribe to devices and receive notifications about the availability of this information. The Nexus concept is new and aspires to be an attractive solution for future Web of Things architectures.

Related work and the Web of Thing background are discussed in Sect. 2. Nexus design and functionality are presented in Sect. 3. The evaluation based on synthetic (but realistic data) is discussed in Sect. 4 followed by conclusions and issues for research in Sect. 5.

2   Background and Related Work

The Web of Things (WoT) Architecture of W3C [11] sets the requirements for interacting with Things on the Web. Thing Description (TD) [9] is a central building block of WoT Architecture. It does not bind to a specific application domain, communication protocol, or implementation. The model describes Things exposing metadata showing how a client can interact with Things such as properties, actions, and events. The WoT Proxy is the entry point for discovering services and resources related to a Thing. TDs are hosted in a directory providing a Web interface for registering and searching for Things. Thingweb node-wot[1] is an implementation of TD's API using a JavaScript API similar to the Web browser APIs. It provides an API Interface that allows scripts to interact with Things using Web protocols such as HTTP, HTTPS, CoAP, MQTT, and WebSockets. The WoT Model Service [21, 22] is a more complete implementation.

The idea of using proxies to interconnect Things with the IoT and the cloud (or server) is not new. Integrating Things into the Web of Things is the focus

---

[1] https://github.com/eclipse/thingweb.node-wot.

of the work by Zyrianoff et al. [23]. They propose a method for converting OpenAPI to W3C Architecture Thing Descriptions (TDs). Similar to our previous
work [21,22], the authors highlight that there is no exact match between the
RESTful architecture and the interaction affordances of W3C WoT (e.g. properties, actions) [11]. They attempt to match HTTP methods with WoT interaction
affordances based on their similarities (e.g. HTTP GET with readable properties or readable and writable properties, HTTP POST with actions, etc.). They
observe significant mismatches between generic REST interfaces and the W3C
WoT interface. They also proposed the C3PO tool to translate formal OpenAPI
service descriptions to W3C WoT Thing Descriptions [9] and to instantiate the
API of the service to a WoT proxy.

Complementary to our work, the WoTDL2API tool [15] generates RESTful
APIs of devices from the OpenAPI descriptions using the OpenAPI Code Generator[2]. OpenAPI descriptions of devices are generated from instances of the
WoTDL ontology [14] that models devices as actuators and sensors. The OpenAPI is generated by querying the ontology to extract the HTTP method, query
parameters, URL, and body. This is an iterative and interactive process. HTTP
requests to devices are forwarded to the central WoTDL hub (i.e. IoT Proxy).

In the following, all solutions are compared based on their capacity to offer
functionality similar to Nexus. All comply with the W3C Architecture Recommendation of W3C and use OpenAPI to achieve uniform representation for both
Web services and devices. Nexus is the only solution that supports the decoupling of device and application operations, user identification, and authorization
while being the only Multi-Context Proxy solution to support the interconnection of diverse applications with their IoT network using a single Proxy service.
Table 1 summarizes the results of this comparison.

Table 1. Comparison of IoT mashup solutions.

WoT proxy	Zyrianoff et al. [23]	WoTDL2API [15]	Nexus
Things representation	OpenAPI	OpenAPI	OpenAPI
Things and services directory	Yes	Yes	Yes
Web of things functionality	Yes	Yes	Yes
IoT - app. decoupling	No	No	Yes
Thing description generation	No	No	Yes
User identification - authorization	No	No	Yes
Muliti-context	No	No	Yes

3   Design and Architecture

We followed a state-of-the-art design approach that identified functional and non-
functional system requirements and specifically, (a) functional components and

---

[2] https://openapi-generator.tech.

their interaction, (b) information that is managed and how it is acquired, transmitted, stored and analyzed, (c) different types of users and how they interact with the system, (d) requirements for assuring data, network and user security and privacy. Detail on system design (including a full set of use case, activity, and deployment UML diagrams) can be found in the thesis of the author [19].

Nexus is the single point of access for applications to their IoT network of devices. This allows diverse applications to share the same Proxy. The devices connect to a Gateway whose role is to translate any IoT-specific protocol to HTTP(S) REST and transfer IoT data to the Proxy. This is the only part of Nexus that is affected by the property of a device (e.g. a sensor) to use a specific IoT protocol (e.g. Bluetooth, Zigbee). The rest of the system is protocol agnostic (i.e. data are transferred in JSON). Nexus is deployed on a server (in the cloud) so that it can be scalable and protected. Figure 1 illustrates Nexus topology with connected applications and IoT network.

Each device has a unique identifier (i.e. URI, not necessarily a URL) which is included in its OpenAPI description and stored in the directory in the proxy. To use a device, its identifier has to be discovered. The discovery process is part of the W3C Architecture model of W3C and foresees several Things discovery methods including Direct, Decentralized Identifier (DID), URI, and DNS-based methods [1]. In Nexus, Things discovery as well as protocol translation (from device-specific protocol to HTTP) resorts to IDAS backend IoT management service [5] on the proxy that implements a URI Things discovery method. In the Web of Things realm, Things discovery most likely will be based on DNS methods (i.e. all Things will receive a unique URL and register to the DNS). Hesselman et. al [8] identifies the challenges of interconnecting the Internet of Things with the Domain Name Service (DNS) of the Internet. Integrating DNS with the Internet of Things will facilitate Thing discovery but also expose the network to new security and performance risks if hundreds of thousands of Things communicate with the DNS without human awareness.

All users register to their respective applications and are granted access to Nexus using an OpenID connect [16] service based on *roles*. This allows Nexus to verify the identity of users based on the authentication performed by an authorization server that runs with the application. It allows Nexus to obtain basic information about the users' profiles, and the users to connect to Nexus without exposing their credentials. As a result, users have to register only once (with an application). Each user group is assigned a role encoding authorization to access services or devices. The following user groups and functional requirements associated with each group are identified:

System Administrators: They configure, maintain and monitor Nexus. Except for their competence in providing cloud services, they are responsible for performing Create, Read, Update, Delete (CRUD) operations on (a) users (e.g. they can register new users to the system and define their access rights) and, (b) devices (e.g. they can register new devices to the system). They are responsible for monitoring system operations and users' activities).

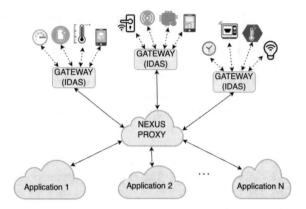

Fig. 1. Nexus ecosystem.

Infrastructure Owners: They are granted permission (by the cloud administrator) to register, configure, monitor, or remove devices. The Web Thing Model API [22] of Nexus provides functionality for connecting and controlling devices. The devices are described similarly to RESTful services. Their descriptions are created using the OpenAPI Thing Generator service [22] and are stored in a directory (a MongoDB).

Application Owners: They subscribe to devices to create applications. Nexus provides OAQL [2] query service to select devices of interest from the directory using device properties such as device type, location, purpose, etc. The applications are RESTful services already and are described using OpenAPI. Their descriptions are also stored in the same directory of devices. An OpenAPI property is used to differentiate between devices and applications.

Customers: Once subscribed to an application they are granted access to the application over the Web. OAQL [2] is used to select applications available for subscriptions from the directory based on criteria such as location, functionality, etc. Customers are granted only access rights to applications.

The class diagram of Fig. 2 illustrates the users' use cases based on roles. It describes users, their properties, and access rights in terms of operations allowed on Nexus entities (i.e. devices and Web services). It is represented as a class (*IS_A*) hierarchy with the most general user class at the top and more specific user classes (i.e. administrations, infrastructure owners, application developers) lower in the hierarchy. Other types of relationships between classes (object properties) are also defined (e.g. *has, manages, owns* relationships). A class is also described by a set of attributes (data properties) together with a set of operations that can be executed on entities of this class.

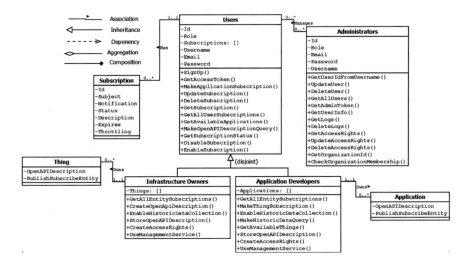

Fig. 2. Nexus users use cases class diagram.

3.1   Nexus Architecture

Figure 3 illustrates Nexus architecture. It is designed as a Service Oriented Architecture (SOA) [3] comprising autonomous RESTful micro-services communicating with each other over HTTP(S). They are organized in groups of services (shown in different colors in Fig. 3). In the following, groups of services implementing the same functionality are discussed together.

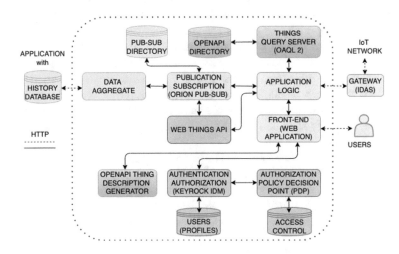

Fig. 3. Nexus architecture.

3.2   Front-End

The *Front-End* services is a Web application that exposes an API and a graphical interface to the Web to accept requests from users. The Front-End is typically used by Nexus administrators to monitor and control the Proxy. Infrastructure owners and application developers access Nexus via OpenID connect [16]. All requests are forwarded to the *Application Logic* service. Device telemetry and control requests are issued directly to the Application Logic service. Its purpose is to orchestrate, control and execute services running on Nexus. When a request is received (from a user or service), it is dispatched to the appropriate service.

A user logs in to an application using a name and a password. A new user has to register to an application to obtain a name and a password. The user is assigned a role (by the administrator). The *Authorization* service of the application will authenticate the user and return an *access token* that encodes the user's identity and role. The token remains active during a session and it's attached to the header of every request. A session is initialized at login and remains active during a time interval which is also specified in advance. A new token is issued every time a new session is initiated (e.g. at user login). The Authorization service is implemented using Keyrock IDM [6]. A separate instance of Keyrock IDM is installed in Nexus and in each application.

3.3   Security

All Nexus services are protected from unauthorized access by OAuth2 [17]. An authorization process takes place using OpenID connect [16]. It enables clients (i.e. Nexus in our case) to verify the identity of the user based on the authentication performed by the Authorization service of an application and to obtain basic user profile information. The Authorization service of the application will issue an *id_token* (along with the *access token* above). The id_token is an assertion by the Authorization service of the application to Nexus that the user did indeed authenticate in the application so that, she/he is entitled to request access to Nexus and the IoT network. Nexus uses the access token to receive (from the application) the user's profile information and requested access rights. This information is stored in a database that is searched each time a user requests access to a service to verify that the user has the appropriate access rights (if not, the request is rejected).

Keyrock Identity Management (IDM) [6] is the core of Nexus's security. Alongside *Policy Enforcement Points*[3] (PEP) and authorization *Policy Decision Point*[4] (PDP) service, guarantee that only authenticated and authorized users can access Nexus services and the IoT network. It receives a request for access from a user that includes an OAuth2 access token in the header and sends the user's attributes to a PDP service which takes the decision to permit or deny access based on the user's role. PDP returns its decision to PEP which, in turn,

---

[3] https://github.com/FIWARE/tutorials.PEP-Proxy.
[4] https://github.com/authzforce/restful-pdp.

may permit (or deny) access to the service. As a special case, a PEP is an intermediate in the communication between two services (rather than between a user and a service). The requesting service must include a *master key* in the header of the request, in order to be authorized by PEP to access the service. The master key is a secret that was defined by the administrator during PEP's creation (each PEP has a different master key). The authorization process is explained in [10].

3.4   IoT Management

The *Web Things API* service receives all telemetry requests from devices and telecommand requests from users (or services) to devices. It provides a RESTful API service designed to support the operations foreseen in the W3C Architecture model. It supports all operations for retrieving Thing Descriptions, their properties, and property values. It implements functions that send a command to a Thing (i.e. an actuator) to execute or retrieve actions and action executions, as well as functions that create, retrieve, and delete subscriptions on Web Thing resources. The Web Thing Model service [22] implements the Web Things API. The Thingweb node-wot[5] is the reference implementation of the W3C WoT Working Group in JavaScript and can be used as well.

3.5   Directory

Thing Descriptions are used to expose Thing metadata to the Web so that other Things or clients (i.e. services or users) can interact and reuse them in applications. IoT devices are described similarly to Web services using OpenAPI [22]. OpenAPI [13] is an industry standard endorsed by Linux Foundation and supported by prominent software vendors like Google, Microsoft, and many others. It is a mature framework for human and machine-readable descriptions of Web services. Given an OpenAPI description, a client (i.e. a human or a service) can easily understand the functionality of the service (a device in Nexus) and how to interact with it using minimum implementation logic.

The OpenAPI Generator TD [22] is the mechanism that generates the OpenAPI description of a Thing from user input. It is a RESTful service in Python Flask and is available on Github[6]. The input comprises (a) the standard OpenAPI Thing Description template that applies to all Things, (b) a payload in JSON with the user settings (e.g. security settings), and (c) the Thing characteristics that will be instantiated to the template. The user (e.g. the device manufacturer) specifies the information that characterizes the device and the functionality it supports (e.g. the properties it provides, the actions it performs, etc.). The output of this mechanism is the OpenAPI description of the Thing (in YAML or JSON). It applies to any device as long as its functionality can be exposed using REST. The directory stores the OpenAPIs of all Things in a

---

[5]   https://www.thingweb.io.

[6]   https://github.com/Emiltzav/wot_openapi_generator.

MongoDB. This data can be retrieved, updated, or deleted or it can be searched with OAQL. OAQL is syntactically similar to SQL, but independent of OpenAPI syntax. It maintains the expressive power of OpenAPI but hides the complexity of the OpenAPI representation in the query expressions.

Information about Things (i.e. descriptions, properties, measurements, actions, action executions, and, subscriptions to Things) is stored in a MongoDB. The same directory stores also OpenAPI descriptions of applications [18] which are also described using OpenAPI. Its purpose is to facilitate infrastructure owners and application developers in their search for similar devices. OAQL [2] is an SQL query language for OpenAPI documents. It is syntactically similar to SQL, it maintains the expressive power of OpenAPI but hides the complexity of the OpenAPI representation in the query expressions. OAQL's syntax is independent of OpenAPI syntax. To differentiate the search for devices from the search for applications, the user-defined attributes *x-type*, *x- id* and *x-devicesUsed* in the OpenAPI description take values for devices and applications respectively.

3.6   Publication and Subscription

The IoT devices connect to a *Gateway* service that translates all telemetry and telecommand requests to JSON and HTTP(S) REST (or the other way around). IDAS backend IoT management [5] implements this service. It is the only service that is affected by the property of Things (e.g. sensors) to use a specific protocol. Following IDAS, data are communicated to the *Publication and Subscription* service in NGSI-LD [20], a data exchange format based on JSON-LD. It is the standard of the European Telecommunications Standards Institute (ETSI) for handling context information. It describes information being exchanged and entities involved (e.g. sensors that publish measurements and users or services that subscribe to this information). An entity can be the description of a concept or object, a subscription to another entity, or a property of an entity. An NGSI-LD document is a valid JSON-LD document.

Only devices registered to this service can publish data to Nexus. ORION-LD [4] is a reference implementation of this service. It receives measurements from devices registered to IDAS service and makes this information available to other services and users based on subscriptions. IoT devices register to ORION-LD as NGSI-LD entities and users or other services can subscribe to these entities to get notified of value changes or when new values become available. A device connected to IDAS becomes a publisher that creates NGSI-LD entities. A subscriber (i.e. a user or another service) may issue a request to search for an entity and subscribe to it or to attributes of that entity. The subscriber will be automatically notified of possible updates of the entities of interest (push policy). Orion-LD awaits confirmation that messages sent by the publisher have been received successfully. On error, the messages have to be re-routed. NGSI-LD entities (i.e. messages, subscriptions) are stored in the *Pub-Sub Directory* (a MongoDB). The service holds the most recent values from all registered sensors (i.e. current values are stored in a non-SQL database).

History (past) measurements are forwarded to the Data storage service and from there to the *History Database* that is part of an application that connects to Nexus. This service collects data flows (history values) from ORION-LD. The time series created from the history of data is stored in MongoDB as either raw (unprocessed) values or aggregated (processed) values. More specifically, maximum, minimum, and average values over predefined time intervals (e.g. every hour, day, week, etc.) are stored. By default, no historic data is collected. Infrastructure Owners must enable the collection of historic data for their Things using either Cygnus or STH-Comet[7] (left as future work).

4   Evaluation

Nexus is deployed in a Virtual Machine (VM) in the Google Cloud Platform running Ubuntu 18.04 with 6 CPU cores (2.8G Hz, 8192 MB RAM) and 25 GB HDD. The following experiment aims to study the system's performance under high concurrency, allowing up to 150 simultaneous (simulating 150 users issuing requests simultaneously). The IoT network has 7,000 simulated (but realistic) devices transmitting measurements to ORION-LD and the history database. The size of the history database is approximately 1.5 GB. Real devices were not available to us for the experiments. Zyrianoff et al. [23] reported that real devices connected to the Web Proxy could cause a latency increase of up to 0.5 ms in requests accessing the IoT network which is negligible compared to other network delays and application service response times.

Table 2. Average response time (ms) of service calls for varying concurrency.

Service request	Concurrency = 50	Concurrency = 75	Concurrency = 100	Concurrency = 125	Concurrency = 150
Create subscription to device	18	21	19	28	81
Create user entity	96	99	83	129	479
Update user entity	82	86	87	115	355
Get User entity	8	15	9	18	14
Create OpenAPI description	11	11	12	13	14
Get OpenAPI description	17	17	18	25	38
Get device data (from database)	25	30	24	41	78

To simulate the effects of a high workload Locust[8] was installed in a separate server. It accepts as input, the number of service requests, and a number of concurrent requests and computes the average response time per request. Table 2 summarizes the results of this experiment. In all cases, response times improve with the simultaneous execution of requests. As expected, subscription and insertion requests (i.e. create, update requests) are always much slower than simple read requests that simply read property values from a service (e.g. a database). The best performance is obtained for concurrency 100.

---

[7]   https://github.com/FIWARE/tutorials.Short-Term-History.

[8]   https://locust.io.

5   Conclusions and Future Work

Nexus is not merely a Web Proxy service for performing IoT telemetry and telecontrol operations on The IoT network. It is a full-fledged Service Oriented Architecture (SOA) that complies with the Web of Things (WoT) functionality of W3C. It decouples application-specific features on the server from IoT functionality. This enables sharing Nexus by more than one application to connect to the IoT network. It is the Single Sign On (SSO) service for users of each application while being secure by design. Nexus introduces the idea that devices should be described similarly to Web services using OpenAPI. Replacing HTTP with HTTPS for connecting with applications is an important security update. Also introducing Apache Kafka or RabbitMQ as a publication and subscription service would improve the performance of Nexus overall [12].

Acknowledgment. We are grateful to Google for the Google Cloud Platform Education Grants program. The work has received funding from the European Union's Horizon 2020 - Research and Innovation Framework Programme H2020-SU-SEC-2019, under Grant Agreement No 883272 - BorderUAS.

References

1. Cimmino, A., McCool, M., Tavakolizadeh, F., Toumura, K.: Web of Things (WoT) Discovery (2022). https://www.w3.org/TR/2022/WD-wot-discovery-20220810/. W3C Working Draft
2. Apostolakis, I.: Simple querying service for OpenAPI descriptions with Semantic Web extensions. Diploma thesis, School of Electrical and Computer Engineering, Technical University of Crete (TUC), Chania, Crete, Greece (2022). https://dias.library.tuc.gr/view/92123. (submitted to IEEE Trans. on Knowledge and Data Engineering)
3. Erl, T.: SOA Principles of Service Design. Prentice Hall, Upper Saddle River, NJ, USA (2007)
4. FIWARE: ORION-LD Developer Guide (2020). https://hub.docker.com/r/fiware/orion-ld
5. Fiware: IDAS: Interface To Internet Of Things (2022). https://fiwaretourguide.readthedocs.io/en/latest/iot-agents/introduction/
6. González, A.A., Navalón, E.G., Moreno, F.A.F., Huertas, A.P.: Identity Manager GE - Keyrock (2015–2016). https://keyrock.docs.apiary.io/#introduction/preface. Fiware
7. Guinard, D., Trifa, V.: Building the Web of Things. Manning Publications Co., Greenwich, CT, USA (2016). https://webofthings.org/book/
8. Hesselman, C., et al.: The DNS in IoT: opportunities, risks, and challenges. IEEE Internet Comput. **24**(4), 23–32 (2020). https://ieeexplore.ieee.org/document/9133283
9. Kaebisch, S., Kamiya, T., McCool, M., Charpenay, V., Kovatsch, M.: Web of Things (WoT) Thing Description (2020). https://www.w3.org/TR/wot-thing-description/. W3C Recommendation
10. Koundourakis, X., Petrakis, E.G.: iXen: secure service oriented architecture and context information management in cloud. Int. J. Ubiquit. Syst. Pervasive Netw. (JUSPN) **14**(2), 1–10 (2021). https://iasks.org/articles/juspn-v14-i2-pp-01-10.pdf

11. Kovatsch, M., Matsukura, R., Lagally, M., Kawaguchi, T., Toumura, K., Kajimoto, K.: Web of Things (WoT) Architecture (2020). https://www.w3.org/TR/wot-architecture/. W3C Recommendation
12. Lazidis, A., Tsakos, K., Petrakis, E.G.M.: Publish-subscribe approaches for the IoT and the cloud: functional and performance evaluation of open-source systems. Internet Things **19**, 100538 (2022). https://doi.org/10.1016/j.iot.2022.100538
13. Miller, D., Whitlocak, J., Gartiner, M., Ralphson, M., Ratovsky, R., Sarid, U.: OpenAPI Specification v3.1.0 (2021). https://spec.openapis.org/oas/latest.html. OpenAPI Initiative, The Linux Foundation
14. Noura, M., Gaedke, M.: WoTDL: web of things description language for automatic composition. In: IEEE/WIC/ACM International Conference on Web Intelligence (WI 2019), pp. 413–417 (2019). https://dl.acm.org/citation.cfm?id=3352558
15. Noura, M., Heil, S., Gaedke, M.: Webifying heterogenous internet of things devices. In: Bakaev, M., Frasincar, F., Ko, I.-Y. (eds.) ICWE 2019. LNCS, vol. 11496, pp. 509–513. Springer, Cham (2019). https://doi.org/10.1007/978-3-030-19274-7_36
16. Octa: OpenID Connect Protocol (2022). https://auth0.com/docs/authenticate/protocols/openid-connect-protocol
17. Octa: What is OAuth 2.0? (2022). https://auth0.com/intro-to-iam/what-is-oauth-2/
18. Papadopoulos, V.: Flow-Based programming support with OpenAPI in the Web of Things. Diploma thesis, Technical University of Crete (TUC), Chania, Crete, Greece (2022). https://dias.library.tuc.gr/view/93710
19. Paterakis, I.: Web proxy service for the Web ofThings. Diploma thesis, School of Electrical and Computer Engineering, Technical University of Crete (TUC), Chania, Crete, Greece (2022). https://dias.library.tuc.gr/view/160
20. Privat, G.: Guidelines for Modelling with NGSI-LD (2021). https://www.researchgate.net/publication/349928709_Guidelines_for_Modelling_with_NGSI-LD_ETSI_White_Paper. ETSI White Paper
21. Tzavaras, A., Mainas, N., Bouraimis, F., Petrakis, E.: OpenAPI thing descriptions for the web of things. In: IEEE International Conference on Tools with Artificial Intelligence (ICTAI 2021), pp. 1384–1391 (2021). https://ieeexplore.ieee.org/document/9643304
22. Tzavaras, A., Mainas, N., Petrakis, E.G.: OpenAPI framework for the web of things. Internet Things **21**, 100675 (2023). https://www.sciencedirect.com/science/article/pii/S2542660522001561
23. Zyrianoff, I., Gigli, L., Montori, F., Aguzzi, C., Kaebisch, S., Di Felice, M.: Seamless Integration of RESTful web services with the web of things. In: IEEE International Conference on Pervasive Computing and Communications, (PerCom Workshops), pp. 427–432 (2022). https://ieeexplore.ieee.org/document/9767531

ABIDI: A Reference Architecture for Reliable Industrial Internet of Things

Gianluca Rizzo[1,2]([⊠]), Alberto Franzin[3], Miia Lillstrang[1,2,3,4,5,6,7],
Guillermo del Campo[5], Moisés Silva-Muñoz[3], Lluc Bono[3],
Mina Aghaei Dinani[1,2], Xiaoli Liu[6], Joonas Tuutijärvi[4], Satu Tamminen[4],
Edgar Saavedra[5], Asuncion Santamaria[5], Xiang Su[4,7], and Juha Röning[4]

[1] HES SO, Valais, Sierre, Switzerland
{gianluca.rizzo,mina.dinani}@hevs.ch
[2] Universitá di Foggia, Foggia, Italy
[3] Université Libre de Bruxelles (ULB), Brussels, Belgium
{alberto.franzin,moises.silva-munoz,lluc.bono}@ulb.be
[4] University of Oulu, Oulu, Finland
{joonas.tuutijarvi,satu.tamminen,xiang.su,juha.roning}@oulu.fi
[5] Universidad Politecnica de Madrid (UPM), Madrid, Spain
{guillermo.campo,edgar.saavedra,asuncion.santamaria}@upm.es
[6] University of Helsinki, Helsinki, Finland
xiaoli.liu@helsinki.fi
[7] Norwegian University of Science and Technology, Gjøvik, Norway
xiang.su.su@ntnu.no

Abstract. The rationale behind the ever increasing combined adoption of Artificial Intelligence and Internet of Things (IoT) technologies in the industry lies in its potential for improving resource efficiency of the manufacturing process, reducing capital and operational expenditures while minimizing its carbon footprint. Nonetheless, the synergetic application of these technologies is hampered by several challenges related to the complexity, heterogeneity and dynamicity of industrial scenarios. Among these, a key issue is how to reliably deliver target levels of data quality and veracity, while effectively supporting a heterogeneous set of applications and services, ensuring scalability and adaptability in dynamic settings. In this paper we perform a first step towards addressing this issue. We outline ABIDI, an innovative and comprehensive Industrial IoT reference architecture, enabling context-aware and veracious data analytics, as well as automated knowledge discovery and reasoning. ABIDI is based on the dynamic selection of the most efficient IoT, networking and cloud/edge technologies for different scenarios, and on an edge layer that efficiently supports distributed learning, inference and decision making, enabling the development of real-time analysis, monitoring and prediction applications. We exemplify our approach on a smart building use case, outlining the key design and implementation steps which our architecture implies.

© The Author(s), under exclusive license to Springer Nature Switzerland AG 2023
L. Barolli (Ed.): AINA 2023, LNNS 654, pp. 26–39, 2023.
https://doi.org/10.1007/978-3-031-28451-9_3

1   Introduction

In recent years, the automation of industrial processes has taken a step forward towards a more fine-grained control and actuation with the widespread adoption of technologies, such as Industrial Internet of Things (IIoT) and Artificial Intelligence, that propel what is being called the fourth industrial revolution, or Industry 4.0 [1,2]. The main idea underlying Industry 4.0 is to collect large amounts of data at every stage of the production process, and to exploit them to make automated decisions as informed as possible, in order to reach the production goals in the most efficient way, while reducing or eliminating the need for human intervention. Such approach opens up countless new challenges in IIoT. Among these, how to optimally deploy sensors in a complex industrial machinery, in order to detect variations in the state of the system and enable targeted, proactive interventions and maintenance; how to transmit IIoT data in a reliable and energy-efficient way; how to effectively address potential security and privacy issues of cloud computing; how to implement reliable and real-time distributed decision making, moving the computing load to the edge of the network and within IIoT systems; how to efficiently process IIoT data streams with high variety, volume, and velocity; and how to flexibly support a heterogeneous set of applications, services, prediction models and visualization tools that provide information to stakeholders. The sheer amount and heterogeneity of data available in large IIoT systems amplify these challenges, in terms of scalability and information integration.

Some of these issues can be addressed by introducing computing nodes physically close to where data is produced [3,4]. These devices, which form what is called the *edge* layer, allow shifting the computing load away from the cloud, reducing latency of computing tasks, relieving IIoT systems from much of the computing load due to data pre-processing, but also of more complex tasks such as anomaly detection, or training and execution of machine learning models.

The modularity of this approach and the distribution of the computing load has several advantages. Among these, it allows alleviating the burden on the centralized part of the infrastructure, in particular for time-sensitive applications. Moreover, it enables processing information closer to the source makes it possible to perform computations without transmitting sensitive information throughout the entire network. In general, moving the computation to the edge improves the computational performance and the communication latency and robustness [5]. Edge nodes can also add context to the data collected, thus enabling informed decisions pertaining to the part of the network to which they are connected.

However, this novel paradigm also introduces several new challenges. First, there is no clear consensus on how an heterogeneous edge-based architecture should be structured, in order to efficiently support the above mentioned services in an IIoT environment [6]. Edge nodes may be heterogeneous, and have limited resources, making scalability and efficient real time orchestration a key issue in real scenarios. The sensing, communication and computing infrastructure needs to be resilient to different types of faults and service disruptions. Thus, it must be designed and managed by taking into account reliability and service availability requirements, in order to deliver the target levels of service in case of hardware/software failures. New learning paradigms, and in general new decentralized algorithmic patterns, need to be developed and efficiently supported by the edge/cloud infrastructure to fully exploit the possibilities offered by the availability of large data streams. To the best of our knowledge, most IIoT architectures are not edge-based [7,8], and very few edge-based IoT architectures have been proposed so far. Debauche et al. propose an edge infrastructure to deploy microservices and AI applications at the edge layer, which is used for IoT applications in agriculture [9]. Guimarães et al. propose an edge-based IoT architecture to monitor industrial nodes [10]. These architectures are however tailored for a narrow, specific application domain, and though they demonstrate the potential of edge computing in IIoT, they do not specify how to generalize their approaches to other domains and applications.

To achieve the goal of designing a general edge-based IoT architecture, in this work we outline ABIDI, a framework for context-aware and veracious data analytics with automated knowledge discovery and reasoning for IIoT. The ABIDI framework encompasses the entire IIoT stack, from the devices to the edge, and to the cloud or central infrastructure, where the application performs the desired computation. The goal of this framework is to enable the efficient and reliable collection of data and the development of AI applications that can be seamlessly deployed on a variety of IIoT scenarios. This is achieved by designing an IIoT architecture whose efficiency depends on both the integration between its modules and the optimization within each module.

In particular, we enable improvements of network performance and reliability by designing a methodology to select the best communication technologies in different contexts, and by proposing an IIoT network architecture which allows reducing latency and energy consumption while easing integration with upper layers. We propose an edge architecture that enables the AI-based IIoT systems, distributing the computation between the cloud/central infrastructure and edge nodes transparently to the application developers. We introduce new privacy-preserving, fully distributed and scalable learning schemes which do not need any parameter server and benefit from node mobility. We further develop visualization tools for data quality assessment that provide insight on the structure, contextual properties and dependencies present in the data streams and thus assist in the development of case dependent pre-processing methods, and we implement energy load prediction models for real world use cases.

The paper is organized as follows. In Sect. 2 we outline the architecture of our framework, we describe our approaches to the implementation of its main functional components. In Sect. 3 we present an application of our framework to a real world case. Finally, Sect. 4 discusses some of the key open research issues that our approach implies, and Sect. 5 presents our main conclusions.

2   The ABIDI Framework

A schematic representation of the overall architecture of the ABIDI framework is presented in Fig. 1. It is divided into three main layers: 1) the *IoT layer*, encompassing the IoT devices and the communication network; 2) the *edge layer*, providing low-latency decision making for IIoT devices and end user devices; and 3) the *cloud layer*, composed by a big data processing level, a data analysis level focusing on prediction of future events and patterns, and an application level.

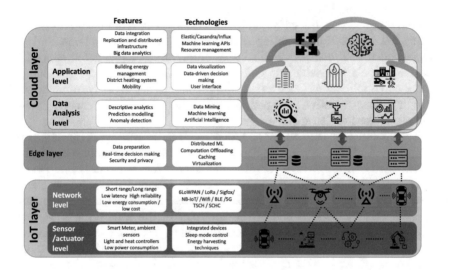

Fig. 1. A high-level representation of the architecture of the ABIDI framework.

IoT Layer. A first key component of the ABIDI framework at the IoT layer is a methodology for the selection of the most appropriate wireless communication technology (in terms of resource efficiency, but also of reliability and QoS support) for each use case or final application. Another important element of the ABIDI IoT layer is the use of energy harvesting techniques to power IIoT devices, taking advantage of the many energy sources typically available at industrial facilities.

Edge Layer. It is typically composed by an heterogeneous set of autonomous computing and communication devices, such as gateways, industry robots, wireless access points and cellular base stations. This layer is responsible for several functionalities related to *data quality*: 1) collection, aggregation and contextualization of the data coming from IIoT environments; 2) aggregation/real time monitoring and collection of metrics about data quality, such as data integrity, consistency, accuracy, completeness, validity, uniqueness and timeliness; 3) data creation (e.g. auto-filling values in forms, automatic extraction of data) and data enrichment; 4) data maintenance (reactive: data correction; proactive: business rules) and data unification (matching and deduplication); and 5) (in synergy with cloud) data protection (e.g. identification of sensitive data, detection of fraudulent behavior) and data retirement (end of life).

Given the tight resource constraints of IoT devices, another key role of the edge layer is the implementation of computation offloading services. Offloading computation (and power consumption) intensive tasks to the edge enables faster decision making of applications running at the edge (thus improving capability to handle latency-sensitive applications), and it saves energy in IoT devices, extending their lifetime.

With respect to the IIoT layer, edge devices implement IIoT and network coordination functions in a self-organized and autonomous manner. This includes enabling IoT integration by acting as gateways for local IoT systems, and joint management of IoT, network and edge resources. Such coordinated control has as its main goals to enable the delivery of the QoS required by the different verticals and applications (such as the support for latency-sensitive applications), and to implement reactive (and possibly proactive) schemes for ensuring service continuity in case of disruptions.

For what concerns the use of ML and data intensive strategies (for the implementation of ABIDI platform applications as well as for the management of the platform itself) the edge layer plays a double role. On one side, it implements mechanisms for model training which are close to data and thus resource efficient and context aware. In addition, it executes local machine learning prediction models. With this respect, one of the key roles of the edge layer is to enable the implementation of learning architectures which are able to provide high levels of data security and privacy preservation, of scalability (with respect to both participating systems and of applications) and of resiliency to infrastructure failures. Indeed these features are critical in present day IIoT scenarios in which data (as well as computing resources) are spread across an ever growing number of heterogeneous devices, and in which harnessing locally available devices, even in an opportunistic manner, is key to achieve high levels of QoS (e.g. in terms of latency of computing tasks) in a resource efficient manner.

To perform efficient inference and learning at the edge, the ABIDI architecture is designed to enable the communication not only of data, but also of models and computational tasks. This increases the overall efficiency of the infrastructure, by distributing the computation in an organic manner in the edge layer, and between the centralized infrastructure and the edge. For example, in a classical IoT network, sensors collect data and transmit it to the central server, which is in charge of all the computation. In an edge infrastructure, the intermediate layers can instead manage part of the operations, such as aggregating data or spotting malfunctioning devices, transmitting to the central server only the correct, aggregated information. Edge nodes can therefore relieve the central server of unnecessary operations, making local decisions. This paradigm brings clear advantages in terms of computational and transmission speed.

Cloud Layer. The data collected by the IIoT devices and potentially the results of the elaboration at the edge level are transferred to the cloud layer. The ABIDI infrastructure relies on suitable database technologies to collect the data. Different applications may require different databases, or a pre-existing infrastructure could be integrated in the ABIDI architecture. Regardless of how this infrastructure is defined and which hardware and software are used, the storage of the data collected remains a potential bottleneck in any data-centric pipeline. Therefore, the ABIDI architecture adopts a flexible data infrastructure that can be optimized for different tasks.

The data analysis level from the cloud layer includes automated and semi-automated data cleaning, data visualization tools, and machine learning for predictive and descriptive modeling. The successful operations of final applications, such as evidence-based decision support tools, depends on the quality of the data, such as their timeliness and reliability. In ABIDI architecture, automated data cleaning methods are applied to solve any data quality deficiencies that are relatively simple to treat, and to perform basic fault detection procedures. The adoption of automated methods, when they are reliable, allows minimizing human effort, which is crucial when operating with big data. As a solution for optimizing between reliability and human effort, semi-supervised methods are applied in cases that cannot be reliably solved using automated methods.

In the data analysis level, interactive exploratory data visualization tools are utilized to enable effortless monitoring and inspection of the big data and of the data quality. The visualization tool prepares the developers of automated data processing system to improve the quality of their data to meet the contextual requirements, to reflect the needs of decision-making process and to allow providing domain specific answers to the user. Through an effective visualization, the massive amount of data becomes accessible and understandable, which makes it possible to both ensure the appropriateness of the automated pre-processing steps and to add use case dependent methods above the automated ones. The combined application of these two approaches allows achieving high quality standards for IIoT data, particularly in those application contexts where it is often plagued by noise, or where it is often incomplete and inaccurate.

Machine learning regression is applied in the analysis level for descriptive modeling and for prediction of IIoT data streams. The descriptive models estimate the value of a data variable at a certain moment, and the estimations are useful for missing value imputation and anomaly detection. Predictive regression models differ from the descriptive ones in that they estimate values of the variables at a future moment. The predictions can offer substantial profits when utilized in decision support tools. The ABIDI architecture includes a full automated pipeline for creating baseline regression models for time series prediction.

Technology Selection in IIoT Network. Although in industrial environments, traditionally, assets have been connected using wired communication technologies (based on Field-bus or Industry Ethernet), recent advances on wireless communications have enabled the access to new elements and data, providing advantages in terms of flexibility, mobility, installation, and cost, among others [11]. While wireless sensor networks (WSNs) have been largely used in building automation, smart city or agriculture domains, the industrial environments differ from these due to their particular constraints, especially in terms of latency, environment, heterogeneity and mobility [12]. There are many wireless communication technologies and protocols that may be named as Industrial IoT networks [13]. Regarding existing literature that presents technical features, existing deployments, and future trends, the ABIDI framework considers following IoT network technologies as the most relevant: BLE, ZigBee, WiFi, WirelessHart, LoRaWAN, Sigfox, 6LoWPAN, NB-IoT, LTE Cat-M1, and 5G.

Although there are many surveys and reviews on IIoT networks, such as [14], few studies have considered factors beyond technical parameters, including the constraints of factories environments and its integration with the other layers of the IIoT architecture [15]. Through reviewing literature and technical specifications, Table 1, which summarizes the main parameters of each technology, has been created to assist technology selection. As it can be observed, the different IIoT network technologies have their strengths and weaknesses, and therefore cannot comply with all the requirements of every use case or application.

The ABIDI framework is based on a two-step procedure for selecting the appropriate communication technology for a specific use case or application, as follows.

1. Determine the essential use case specific requirements set by the final application. These requirements may be divided in the following categories:
   - **Technical factors:** They include technical characteristics such as the transmission capacity (data rate), the time taken from the instant the node transmits the message until it arrives to the final application (latency), the communication coverage (range), the bi-directionality (duplex) and the loss of messages (reliability).
   - **Implementation factors:** They integrate those factors especially relevant during the IoT network implementation phase. The most important one is cost, which is the sum of the cost of IoT devices and nodes plus the cost of network infrastructures (for those technologies that demand the deployment of private network elements, such as 6LoWPAN, Zigbee, WiFi or LoRa), or the cost of data plans (for those technologies that provide the network infrastructure, such as Sigfox, 5G, or NB-IoT).
   - **Functional factors:** They cover factors that affect everyday working of IoT applications, including the autonomy of the devices (energy consumption), which is determined by the time the IoT device is turned on and especially by the energy consumption during the communication process.
2. Compare these requirements with Table 1, and select the most suitable technology. This step is implemented via Machine Learning based algorithms, which recommend the best communication technology based on use case requirements, and on all system constraints.

Table 1. Summary of the main parameters considered in ABIDI methodology for IIoT technology selection.

Technology	Data rate[a]	Latency[b]	Range[c]	Duplex	Reliability	Consumption	Cost
BLE	Mbps	30 ms	100 m	half	low	low	low
ZigBee	kbps	40 ms	100 m	half	high	low	low
WiFi	Mbps	30 ms	100 m	half	med	med	low
WirelessHart	kbps	10 ms	200 m	half	high	med	high
LoRaWAN	kbps	300 ms	10 km	half	med	med	med
Sigfox	bps	4 s	50 km	limited[d]	high	high	med
6LoWPAN	kbps	20 ms	100 m	half	med	low	low
NB-IoT	kbps	2 s	10 km	half	high	high	high
LTE Cat-M1	kbps	2 s	10 km	half	high	high	high
5G	Gbps	10 ms	10 km	half	high	high	high

[a, b, c] Approximate values—in the order of magnitude.

[d] Sigfox provides limited bidirectional capacity: the IoT device can upload up to 140 12-byte messages a day, but it can only receive four 8-byte messages.

3   A Building Management Use Case

In order to assess the ABIDI framework, we implemented it in a smart building testbed at CEDINT-UPM in Madrid, Spain, a three-story construction that hosts offices, research labs, and other facilities. It is equipped with 30 IoT power meter devices are installed at panel boards, allowing specific energy consumption monitoring of 560 electrical lines; 40 IoT ambient sensor devices measuring temperature, luminosity, humidity and presence detection—apart from battery level; and 30 HVAC controllers, which provide set-point temperature, fan speed, working mode (cold/heat), state (on/off) and indoor temperature data. By means of an Elastic Stack-based IoT Platform, data collected were distributed and replicated to provide inputs for machine learning (ML) and visualization tools. The two main goals have been: i) optimizing energy consumption by context-aware data analytics of energy consumption patterns, taking into account energy measurements, ambient parameters and user behaviour; and ii) ensuring data reliability and veracity, by improving communications, and detecting and correcting missing or wrong measurements.

We applied the ABIDI framework methodology to select the optimal communication technology. The main technical requirements were low data rate, medium reliability, non-critical latency, variable sending frequency (30 s–15 min), and bi-directionality. To this end, we performed an experimental characterization for communication reliability and energy consumption.

Experimental results for reliability (latency and error rate) of the different technologies were obtained using an ad-hoc testbed (Table 2). Latency was measured considering an end-to-end trip, from the Industrial IoT node to the application server. For error rate, the same latency packets were used. Based on these results, 6LoWPAN outperformed alternative protocols with regards to communication latency.

Table 2. Experimental latency, error rate and consumption results

Technology	6LoWPAN	LoRaWAN	Sigfox	BLE	WiFi
Latency (ms)	20	290	3700	26	32
Error Rate (%)	0.01	0.6	0	0.03	0
Tx. Consumption (mAs)	0.8	6.3	804.8	1.0	3.4

As a second step, we experimentally measured the energy consumption of the transmission process at 5 V (Table 2). These measurements were taken using a Nordic Semiconductor Power Profiler Kit II. Then, the transmission current demand per se was integrated during the time of packet transmission. Regarding power consumption, 6LoWPAN outperformed again the other technologies, especially Sigfox, which was expected to have a greater consumption as its on-air time is much longer.

Bottom line, considering the number on IoT nodes (100) and area of deployment (50 m × 40 m), the variable sending frequency and the non-restrictive requirements in terms of latency an reliability, BLE and 6LoWPAN seemed to be the best choices. However, the features of 6LoWPAN mesh topology, which enables the utilization of a single network coordinator or access point for the entire use case (together with the fact that it implements IPv6 connectivity, allowing direct access from the Internet), made 6LoWPAN the final choice. To infer energy consumption patterns, we have combined temperature measurements of indoor ambient sensors and the HVAC energy consumption measured with BatMeter smart meters, as the latter data sources alone proved insufficient.

A baseline XGBoost regression model was built for short-term (one hour ahead) HVAC energy consumption prediction. Error metrics CV-RMSE, Rel-RMSE and MASE for the model were 0.292, 0.811, and 0.870, respectively. Rel-RMSE and MASE measures include a built-in comparison to a naïve time series prediction model, and the value being less than one indicates the model is performing better than the naïve model. This showed that the suggested baseline model is capable of providing useful outputs in short-term predictions.

In data pre-processing phase, it was possible to automatically detect a malfunctioning sensor in the monitored area by inspecting the rate of data packages sent by each sensor node. Semi-supervised methods, where IoT data streams were combined with relevant metadata, allowed imputing missing temperature data by utilizing peer sensors in the room. While a reasonably light approach was enough to meet the data quality requirements of the desired application, a more thorough visual interface was also developed for this setup (Fig. 2) for ease of inspection of data streams.

The layout of the final application is in Fig. 3. The quality of data was ensured for each of the five parameters utilized in the final application. Gradients of temperature data sources were inspected together with the gradient of the electricity consumption data, so as to label times of HVAC usage in every room. The application layout shows a figure of each data stream, highlights the times of HVAC usage in each temperature figure, visualizes the number of active HVAC

Table 3. Basic information of the UPM data set

Indoor ambient sensors	Number of sensors	34
	Sensors per space	1–4
	Parameters	Temperature, humidity, light, motion
Smart meters	Number	32
	Parameters	Power consumption (528 lines)
Time range of data collection	Start	In steps through 2018–2020
	End	Ongoing (12/2021)
Frequency of data collection		From seconds to an hour

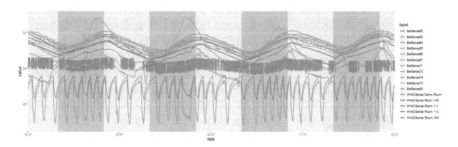

Fig. 2. Application screenshot, detecting HVAC usage in the UPM building, displaying all of the relevant data streams.

units as a function of time, and provides a summary of the estimated electricity consumption per room for a given period of time. These data allowed determine new opportunities for optimization of power consumption. In particular, the fact that the highest peaks in power consumption were caused by HVAC units being turned on simultaneously in multiple rooms suggest that smart scheduling of HVAC duty cycles could substantially reduce these peaks, and thus contribute to preventing outages (Table 3).

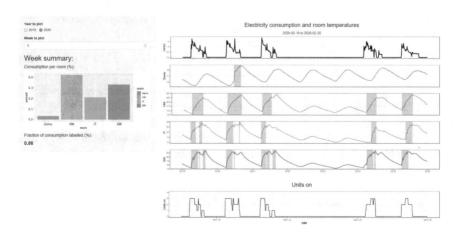

Fig. 3. Screenshot of the visualization tool, detecting HVAC usage in UPM building.

4   Key Research Challenges

We discuss some technical challenges for full implementation of the ABIDI approach, including IIoT architecture and protocols, energy harvesting, network optimization, Cloud infrastructure optimization, and Application layer. It is crucial to identify and analyze those challenges for seeking novel solutions.

Reliable IIoT Architecture and Protocols. In order to increase performance and reliability for IIoT networks, and meet with the most-demanding communication requirements (e.g. robot control), some leading-edge techniques must be implemented at various levels of the architecture. Specifically, MAC layer enhancements such as Time Slotted Channel Hopping (TSCH) could be used in 6LoWPAN. TSCH avoids packet losses and reduces latency by dynamically changing the carrier frequency in a globally synced mesh network among all the nodes in the network [16]. On the other hand, for other communication technologies, such as LoRaWAN or NB-IoT, the scheme of Static Context Header Compression (SCHC) could be implemented. SCHC allows compression of IPv6/UDP/CoAP packets, with the aim of making them suitable for transmission over their restricted links of these technologies and providing higher interoperability by using IPv6 connectivity [17].

Resource Optimization of ML Training at the Edge. One of the key open issues in gossip learning lies in the lack of understanding of the relationship between patterns of exchange of models and of movement of agents, and some of the primary performance parameters of the scheme. A key challenge concerns how to optimally tune model merging as a function of the context and of the specific problem. Different merging strategies have shown to perform very differently according to the specific model, but also as a function of the degree of dynamicity of the environment. New approaches need to be designed in order to improve their efficiency in heterogeneous settings, i.e. when applied to set of nodes with very diverse sensing and computing capabilities. Finally, strategies for improving the communication efficiency of these schemes have to be designed, and the trade-off between performance and resource efficiency has to be characterized.

Energy Harvesting. As already mentioned, the location of IoT devices within manufacturing equipment and processes means that they have to be battery-powered. Energy harvesting (EH) rises as a green, sustainable, and virtually infinite power supply to wireless devices, obtaining the available energy from the environment to reduce the need for storage components. Power generation density depends mainly on the real characteristics of the ambient energy availability for the IoT device location. Even if RF appears to be a common energy source provided by manufacturing equipment and existing wireless communications, its power density is small compared to other energy sources such as light or magnetic induction. A more in-depth analysis of the power density of the diverse energy harvesting techniques in factories is needed. Another deciding factor is the availability of the energy source, which may be steady (RF) or more unpredictable (light), affecting the power supply profile. Time variation of the energy sources should be characterized. Finally, a recent trend of study is the use of hybrid energy harvesting schemes, combining high-power-density techniques (PV) with more steady sources (RF).

Edge/Cloud Balance-Network Optimization. The flexibility of the ABIDI architecture at the edge layer allows it to adaptively distribute the computation. For this, the particular application deployed at the edge level will be containerized providing the architecture with more flexibility at the edge layer. This containerization provides the edge-layer with the option to dynamically adapt the computational resources by using an (intra-)edge layer load balancing mechanism such as Kubernetes. The flexibility of the architecture will then be complemented with an inter-layer load-balancing mechanism which allows the edge layer to offload tasks to the server infrastructure. To this end, appropriate load balancing mechanisms need to be designed, capable of efficiently cooperate with the data-offloading and task balancing processes.

Cloud Infrastructure Optimization. On the cloud side, for each use case the database used must be tailored to the specific needs to optimize its performance in terms of throughput. This is better done when the scenario characteristics in terms of data and operations are fully determined, to obtain the database configuration that best serves the application. This approach can be also applied to entire software pipelines, such as when Kafka is employed to transmit data from the edge nodes to the database on the cloud, to optimize every step of the data collection process. The configuration methodology remains the same, requiring only to define the interface between Irace and the desired database/pipeline [18].

Application Layer. Turning the current approach taken for improving data quality in cloud environment into a full, low-effort pipeline applicable to a wide range of use cases is a key challenge. A full pipeline from data to decision support tool visualization has currently only been implemented for time series regression models, and expanding to other kind of tasks, such as classification, is important to widen the spectrum of covered use cases. Utilizing Bayesian Estimation or some other suitable algorithm for hyperparameter tuning instead of using a grid search could also improve computational efficiency of the process.

5   Conclusions

Technologies such as artificial intelligence and Internet of Things are reshaping industrial processes to the point that relevant actors are calling this transformation the fourth industrial revolution. The combination of big data and automated decision making is helping companies in transitioning from general mass production to a smart production that uses information to increase efficiency and reduce waste and operational costs. This transformation does not come without challenges, since current approaches are limited in scope and application.

In this work we have presented the ABIDI architecture for Industrial Internet of Things. ABIDI is a general framework that can be instantiated to address different real world cases, making use of the most suitable technologies for each scenario. It encompasses the whole IIoT stack, from the sensors and network layer to the final application, combining the use of cloud architectures with an edge layer of computational nodes that can improve the performance and robustness of the final application, and can perform distributed AI tasks. We have discussed how the components of our architecture address the shortcomings of the current state of the art. Finally, we have reported a real world scenario where we instantiated our architecture, and we have outlined the steps necessary to reach the full vision of the ABIDI infrastructure.

Acknowledgment. This work has been supported by the CHIST-ERA project CHIST-ERA-17-BDSI-001 ABIDI "Context-aware and Veracious Big Data Analytics for Industrial IoT". This work has been partially supported by COST INTERACT, the FARI Institute, and by SNF Dymonet project. AF is supported by Service Public de Wallonie Recherche under grant n° 2010235 - ARIAC by DIGITALWALLONIA4.AI. Published with a contribution from 5 × 1000 IRPEF funds in favour of the University of Foggia, in memory of Gianluca Montel.

References

1. Tao, F., Qi, Q., Liu, A., Kusiak, A.: Data-driven smart manufacturing. J. Manuf. Syst. **48**, 157–169 (2018)
2. Sisinni, E., Saifullah, A., Han, S., Jennehag, U., Gidlund, M.: Industrial internet of things: challenges, opportunities, and directions. IEEE Trans. Ind. Inf. **14**(11), 4724–4734 (2018)
3. Shi, W., Cao, J., Zhang, Q., Li, Y., Xu, L.: Edge computing: vision and challenges. IEEE Internet Things J. **3**(5), 637–646 (2016)
4. Qiu, T., Chi, J., Zhou, X., Ning, Z., Atiquzzaman, M., Wu, D.O.: Edge computing in industrial internet of things: architecture, advances and challenges. IEEE Commun. Surv. Tutor. **22**(4), 2462–2488 (2020)
5. Ejaz, M., Kumar, T., Ylianttila, M., Harjula, E.: Performance and efficiency optimization of multi-layer IoT edge architecture. In: 2nd 6G Wireless Summit (6G SUMMIT), pp. 1–5. IEEE (2020)
6. Sittón-Candanedo, I., Alonso, R.S., Rodríguez-González, S., García Coria, J.A., De La Prieta, F.: Edge computing architectures in industry 4.0: a general survey and comparison. In: Martínez Álvarez, F., Troncoso Lora, A., Sáez Muñoz, J.A., Quintián, H., Corchado, E. (eds.) SOCO 2019. AISC, vol. 950, pp. 121–131. Springer, Cham (2020). https://doi.org/10.1007/978-3-030-20055-8_12
7. Boyes, H., Hallaq, B., Cunningham, J., Watson, T.: The industrial internet of things (IIoT): an analysis framework. Comput. Ind. **101**, 1–12 (2018)
8. Sobin, C.: A survey on architecture, protocols and challenges in IoT. Wirel. Pers. Commun. **112**(3), 1383–1429 (2020)
9. Debauche, O., Mahmoudi, S., Mahmoudi, S.A., Manneback, P., Lebeau, F.: A new edge architecture for AI-IoT services deployment. Procedia Comput. Sci. **175**, 10–19 (2020)
10. Guimarães, C.S.S., Jr., de Andrade, M., De Avila, F.R., Gomes, V.E.D.O., Nardelli, V.C.: IoT architecture for interoperability and monitoring of industrial nodes. Procedia Manuf. **52**, 313–318 (2020)

11. Gungor, V.C., Hancke, G.P.: Industrial wireless sensor networks: challenges, design principles, and technical approaches. IEEE Trans. Ind. Electron. **56**(10), 4258–4265 (2009)
12. Catenazzo, D., O'Flynn, B., Walsh, M.J.: On the use of wireless sensor networks in preventative maintenance for industry 4.0. In: 2018 12th International Conference on Sensing Technology (ICST), pp. 256–262 (2018)
13. Li, X., Li, D., Wan, J., Vasilakos, A.V., Lai, C.-F., Wang, S.: A review of industrial wireless networks in the context of industry 4.0. Wirel. Netw. **23**(1), 23–41 (2015)
14. Raza, S., Faheem, M., Guenes, M.: Industrial wireless sensor and actuator networks in industry 4.0: exploring requirements, protocols, and challenges-a MAC survey. Int. J. Commun. Syst. **32**(15), e4074 (2019)
15. Liu, Y., Kashef, M., Lee, K.B., Benmohamed, L., Candell, R.: Wireless network design for emerging IIoT applications: reference framework and use cases. Proc. IEEE **107**(6), 1166–1192 (2019)
16. Urke, A.R., Kure, Ø., Øvsthus, K.: A survey of 802.15.4 TSCH schedulers for a standardized industrial internet of things. Sensors **22**(1) (2022)
17. Sanchez-Gomez, J., Gallego-Madrid, J., Sanchez-Iborra, R., Santa, J., Skarmeta, A.F.: Impact of SCHC compression and fragmentation in LPWAN: a case study with LoRaWAN. Sensors **20**(1) (2020)
18. Silva-Muñoz, M., Franzin, A., Bersini, H.: Automatic configuration of the Cassandra database using Irace. PeerJ Comput. Sci. **7**, e634 (2021)

DASA: An Efficient Data Aggregation Algorithm for LoRa Enabled Fog Layer in Smart Agriculture

Mayank Vyas[1], Garv Anand[1], Ram Narayan Yadav[1],
and Sanjeet Kumar Nayak[2(✉)]

[1] Department of Electrical and Computer Science Engineering, Institute of
Infrastructure Technology, Research and Management (IITRAM), Ahmedabad, India
{mayank.vyas.20e,garv.anand.20e,ramnarayan}@iitram.ac.in
[2] Department of Computer Science and Engineering, Indian Institute of Information
Technology, Design and Manufacturing, Kancheepuram, Chennai, India
sanjeetn@iiitdm.ac.in

Abstract. In a smart agriculture system many resource-constrained sensors are installed near the crops as well as at some strategic locations in an agriculture field to collect relevant crop and environment data in real-time. This data is then used for both critical latency-sensitive decision making as well as for long-term planning. Nowadays, with the help of smart IoT systems, resolving the problems like irrigate fields, avoid animal intrusions, notify the farmer about the seasonal rainfall etc. becomes easier. The edge of the IoT networks regularly receives a huge amount of data generated by sensors that need to be delivered to the server present in the remote data centers/cloud for additional real time control or long term decision making. However, transmitting huge amount of these IoT data across the network toward the cloud imposes a high overhead in terms of bandwidth demand and latency on the IoT network. So, the key challenge in building a smart agriculture system include high communication latency and bandwidth consumption incurred with computing over the data on the cloud. Also, frequent Internet disconnections in rural areas may lead to improper latency sensitive decision making at cloud due unavailability of data. In this paper to resolve such issues of cloud based smart agriculture system, we present a LoRa-based three-tier smart agriculture system comprised of (i) Field layer, (ii) Fog computing layer, and (iii) Cloud computing layer. In particular, a data aggregation algorithm through a LoRa enabled fog computing layer for smart agriculture (DASA) is proposed to compress the total amount of IoT data to be uploaded to the cloud. We present the performance of our proposed scheme and compare with the existing frameworks for smart agriculture system in terms of compression ratio, compression time, compression power, and amount of data transmitted to cloud from fog computing layer. Comparison results show that the proposed algorithm significantly decreases the volume of data to be uploaded to the cloud platform and achieves highest compression ratio among other existing schemes. We also tested the performance of the proposed data aggregation algorithm on real testbed.

M. Vyas and G. Anand have equal contributions.

L. Barolli (Ed.): AINA 2023, LNNS 654, pp. 40–52, 2023.
https://doi.org/10.1007/978-3-031-28451-9_4

Keywords: IoT · Fog computing · Clustering · Compression · LoRa · Smart agriculture system

1   Introduction

The vast emerging domain of Internet of Things (IoT) has witnessed the ever-growing demand for ease in human behavior in various fields like smart homes, smart offices, and smart agriculture [6,18]. The aim of smart agriculture is to improve the productivity and reduce farmers efforts by utilizing IoT and advanced communication technology [5,15,20]. Various applications such as switching ON/OFF water sprinklers, detection of animal invasion, fire detection etc. related to smart agriculture application require huge amount of data transmission from agriculture fields to data centers/clouds for real-time decision making [19,22]. Because of continuous data transmission from sensors deployed in agriculture fields, the edge nodes of the IoT networks regularly receives a huge amount of data. However, transmitting these huge IoT data across the network toward the data center or the cloud impose a high overhead in terms of bandwidth demand and latency on the IoT network. Further, as per the available reports most of the farming lands in India are placed in remote areas and have very poor Internet connectivity. Hence, it prevents the deployment of various short and long range communication techniques such as cellular, WiFi, and NB-IoT [4,9]. To address this issue of communication requirements, we utilized the strength of Long Range (LoRa) communication technology [10] for smart agriculture which may help farmers in India. It helps in wireless communication up to long distances (up to few Kms) without using an Internet connection. So, to address the intermittent connectivity issue, in this paper, we present a LoRa-based three-tier smart agriculture framework comprised of (i) Field layer, (ii) Fog computing layer, and (iii) Cloud computing layer.

Fog computing layer brings the processing and decision-making power close to the end devices, thus in a way reducing the communication time. We can store the data locally and thus, helps in reducing the uploading cost to the cloud [11]. Therefore, with the use of fog computing layer, rapid response to the end devices can be made and moderate level of computation can also be performed at the edge of the network itself (more importantly, even in case of Intermittent Internet connection) [7]. When there is a poor Internet connection, the fog computing layer need to efficiently utilise the bandwidth. So, instead of sending all the data received from the field layer, if we can aggregate the data before communicating to the cloud computing layer, then the issue mentioned above can be addressed. In this paper, a data aggregation algorithm is developed for the LoRa enabled Fog Computing layer that efficiently compresses the data before sending it to the cloud. The proposed data aggregation algorithm for Fog Layer in Smart Agriculture (DASA) consists of four phases: (i) selecting proper attribute of the data for clustering using Euclidean Similarity (ii) determination of optimum number of cluster for the received data (iii) clustering on the received data, and (iv) compression on each cluster of data. We performed experiments with DASA on real dataset [1] and also compared with existing data compression schemes like LZW and Huffman. We also conduct experiments and show the performance of the proposed algorithm on real testbed. We found DASA provides high compression ratio for the data as compared to the existing schemes.

The remainder of the paper is organized as follows. Section 2 presents the existing related works. The proposed communication framework for LoRa enabled data transmission in smart agriculture is provided in Sect. 3. Section 4 presents the proposed efficient data aggregation framework. Section 5 presents various results in support of performance evaluation of DASA. Finally, Sect. 6 concludes the paper and discusses some future works.

2   Related Work

In [21], a decentralized multi competitive clustering protocol was proposed which can be used for wireless senor network assisted smart agriculture application. Degree of node, residual energy and distance were used for performing clustering. In [3], the authors showed that how wireless sensor network can be used to build clustering for wireless sensor network. They did the entire experiment in Egypt and it results in increased productivity and crop output. Similarly, an energy efficient static multi-hop based routing protocol is proposed in [8], which helps in finding the static path for data transfer in smart agricultural applications.

Recently, authors in [13], proposed a novel fuzzy criteria based clustering mechanism to find the optimum cluster head in smart agriculture application. This approach helps the users of the application by providing the optimum route to minimize the usage of energy. In [16], the authors proposed a Gateway clustering energy efficient centroid protocol where centroid is used determine the cluster head. Also, to reduce the overhead on cluster head, each cluster's gateway node were used to transmitting data to base station. The scheme discussed above does not provide considerable amount of data compression and cannot be extended for locations having intermittent Internet connection.

3   Proposed Communication Framework

The communication framework for efficient data collection and processing in smart agriculture is shown in Fig. 1. It consists of three layers, namely, field layer, fog computing layer, and cloud computing layer. They are linked by cross-layer upstream and downstream for data and control information flows.

Field Layer: It includes the actual sensors deployed in the agriculture field that measure the physical parameters of interest such as air temperature, air humidity, solar radiation, soil temperature at various depths, wind speed, rainfall, etc. Further, this layer can also accommodate intelligent actuators for irrigation, to execute commands send from higher layers such as fog or cloud level.

Fog Computing Layer: The fog nodes collect data from the sensors deployed in the field and execute the data processing algorithms for intelligent aggregation that will reduce network traffic and bandwidth demand for transmitting data to cloud. Various decision making algorithms that exploit the incoming data at this level can be deployed for local decisions, thus avoiding completely the increased cost and latency of the upper layer.

Cloud Computing Layer: Data is transferred towards a cloud platform for long-term storage and for performing analytical decisions. To implement this, we aim to use The Thing Network [2] platform in conjunction with the data aggregation algorithm deployed for the fog computer layer. Various decisions can be made in this layer in case of latency and intermittent internet connection issues. Also, the anomalous data can be stored in the cloud for future reference.

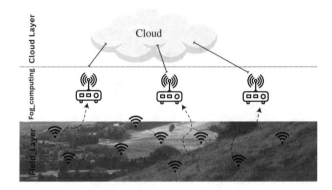

Fig. 1. Proposed communication framework using fog computing for smart agriculture

4   DASA: Proposed Data Aggregation Algorithm

In this section, the data aggregation algorithm (DASA) is discussed in detail for the communication framework provided in Sect. 3. DASA consists of four steps: (i) selecting proper attribute of the data for clustering using euclidean similarity (ii) determining the optimum number of cluster for the received data (iii) clustering on the received data, and (iv) compression on each cluster of data. The detailed algorithm is provided in Algorithm 1 and discussed below. The idea of applying clustering is to partition the whole data received (in various time interval) into groups based on similarity of data values. After clustering, each cluster has nearly similar data. So, if we apply compression in a cluster, then the compressed data will be smaller in length and finally, will results in higher compression ratio. Now, we discuss each steps of DASA in detail.

We have used a clustering method inspired from agglomerative hierarchical clustering, in which the data points are grouped into clusters using euclidean distance and later the clusters are merged together using Ward distance [17]. This clustering uses a bottom-up approach, where each data point starts in its own cluster. These clusters are then joined using a greedy approach, by taking the two most similar clusters together and merging them. This bottom-up approach treat each data point as single cluster and then, successively agglomerates pairs of clusters until all the clusters have been merged into a single cluster (which contains all the data).

Algorithm 1: DASA

Input: Input $= (x_1, y_1), (x_2, y_2), (x_3, y_3)\ldots..(x_n, y_n)$
Output: Compressed BIN File of Clustered Data
begin
    for $i \leftarrow 1$ *to* n **do**
      | $C_i \leftarrow (x_i, y_i)$
    end
    $C = \{C_1, C_2, C_3 \ldots C_n\}$
    $C \leftarrow algo\_to\_find\_optcluster()$
    $C_{\text{opt}} \leftarrow graph\_for\_optimal\_clusters$
    while $|C| > |C_{opt}|$ **do**
        $\arg_{min} dist(C_i, C_j) \leftarrow Wardist(C_i, C_j), \forall C_i, C_j \in C$
        Merge clusters $C_k \leftarrow \{C_i, C_j\}$
        Remove $C \leftarrow C \setminus \{C_i, C_j\}$
        add $C \leftarrow C \cup C_k$
    end
    $(C_1, C_2 \ldots C_{opt})$
    for *each C' in C* **do**
      | compute the frequency of each character in C'
    end
    $Node \leftarrow dict[character : frequency]$
    construct MaxHeap(frequency,character)
    while $Node.char \mathrel{!}= character$ **do**
        $encd_{file} \leftarrow 0 \leftarrow Node.left()$
        $encd_{file} \leftarrow 1 \leftarrow Node.right()$
        $list - encd - file \leftarrow encd_{file}$
        $encd_{file} = \text{``} \quad \text{''}$
    end
    Return $BIN(list - encd - file, Node)$
end

We have used two-dimensional clustering, in which the algorithm selects two suitable attributes from the received data. We used the Euclidean distance method to find the suitable attributes. After the selection of attributes, the clustering algorithm gets two attributes to begin clustering and it produces the optimal number of clusters based on the value it received from the $algo\_to\_find\_optcluster()$ method. This is used to find the optimal number of clusters for the received data point in the given time period [12,14].

Below, we provide how to compute the Euclidean distance between the data points of two attributes.

Euclidean Similarity

Let $x = \{x_1, x_2, \ldots x_p\}$ and $y = \{y_1, y_2, \ldots y_p\}$. Then, Euclidean metric is defined as Eq. (1).

$$d(x,y) = (\sum_{i=1}^{p} |x_i - y_i|)^m)^{1/m} \tag{1}$$

In DASA, we used $d(x,y)$ as Euclidean distance with $m = 2$.

Ward Linkage: Ward distance is a mathematical norm that is used to merge the cluster on the basis of a minimum error scheme. Instead of measuring the distance directly, it analyzes the variance of clusters. This method is said to be the most suitable method for quantitative variables. Ward's method says that the distance between two clusters, A and B, is the amount of sum of squares increase when we merge the clusters. The lesser the value of the function which calculates the increase in the error of the sum of squares, more the respective clusters will be merged. Mathematically, the Ward linkage is provided as Eq. (2).

$$\triangle(A,B) = \sum_{i \in A \cup B} ||\vec{z_i} - \overrightarrow{m_{A \cup B}}||^2 - \sum_{i \in A} ||\vec{z_i} - \overrightarrow{m_A}||^2 -$$
$$\sum_{i \in B} ||\vec{z_i} - \overrightarrow{m_B}||^2 = \frac{n_A n_B}{n_A + n_B} ||\overrightarrow{m_A} - \overrightarrow{m_B}||^2 \tag{2}$$

In Eq. (2), $\vec{z_i}$ and $\vec{m_j}$ denote the center of cluster i and j, and n_j is the number of data points in each cluster j. $\triangle$ is called the merging cost of combining the clusters A and B. With clustering, the sum of squares starts out at zero (because every point is in its own cluster) and then grows as we merge clusters. Ward's method keeps this growth as small as possible.

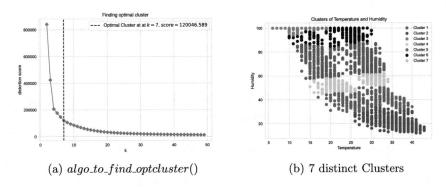

(a) *algo_to_find_optcluster()*   (b) 7 distinct Clusters

Fig. 2. Clustering of data

The *algo_to_find_optcluster()* method in Algorithm 1 generates the optimal number of clusters based on the numeric values of the attribute given to the clustering method. As shown in Fig. 2(a), $k = 7$ is the number of required clusters, and the 7 clusters are represented in Fig. 2(b).

Next, we discuss the data compression method i.e., huffman coding, which is used in the third phase (see Algorithm 1). In this phase, every data value is used for each cluster to construct a heap data structure. Using a while loop a tree is constructed with binary value 0 for the left child and binary value 1 for the right child. Subsequently, a binary tree is created for each cluster. Finally, the encoded values of each cluster are stored and compressed to a binary file, which results as an output of DASA.

5   Performance Evaluation

The aim of the proposed data aggregation technique is to decrease the communication cost by compressing the volume of data that need to be transmitted using the Internet to the cloud from the fog layer. After receiving the sensors' data, fog nodes perform preprocessing and remove the redundant data and noises from the collected data before applying DASA. After that, fog nodes transmit the compressed file to the cloud along with the dictionary required to decode the file to get the original data points.

In this section, we evaluate the proposed data aggregation approach that will be executed at the fog node. First, we use the available dataset [1] for preliminary experiments and performance evaluation. Then, we also evaluated the performance of the proposed DASA on our own testbed (see below). In this section, we compared our proposed DASA Algorithm with existing baseline compression algorithms like LZW lossless compression and Huffman coding (denoted as Huff).

To analyse the performance, We use different performance metrics, such as compression ratio (CR), compression and decompression time, compression power, and size of transmitted data.

– *Compression Ratio (CR in %):* Mathematical definition is provided in Eq. (3):

$$CR(\%) = (1 - \frac{D_{Compressed}}{D_{Original}}) * 100 \qquad (3)$$

where $D_{Compressed}$ refers to the data size after compression (Compressed Data) and $D_{Original}$ is the data size before compression (Original Data).
– *Compression Time:* It is defined as the total time required (measured in seconds) by the scheme to generate the compressed file from the original file.
– *Compression Power (CP):* It is typically defined as the ratio between the uncompressed volume of data to the volume of compressed volume of data. Mathematically defined as Eq. (4):

$$CP = \frac{D_{Original}}{D_{Compressed}} \qquad (4)$$

where $D_{Compressed}$ refers to the data size after compression (Compressed Data) and $D_{Original}$ is the data size before compression (Original Data).

In the dataset [1], there are temperature, humidity, and soil moisture at two different depth as attributes. First, we used Euclidean similarity method to select two attributes that will be used to cluster the data points. Using Euclidean similarity, we found that temperature and humidity are suitable attributes for clustering. Then, we perform clustering to group the data points. For that, we use the *algo_to_find_optclusters*() function to find the optimal number of clusters. Figure 2(a) shows the scores at the various number of clusters and optimal number of clusters are 7. The final clusters are shown in Fig. 2(b).

Test Bed Setup: Various sensors connected to the Arduino board to monitor the soil moisture, and environmental parameters such as temperature, humidity, etc. as shown in Fig. 3. The Arduino board programmed to acquire data from the sensors connected to it and send them to the gateway (fog node) via LoRa communication, the latest wireless technology that allows long distance communication, and low power consumption. It is more suitable for smart agriculture applications. We use a star topology to connect sensor nodes to the gateway, which is kept at a distance of 1000 mts and at a height of 40 ft. During the experiments, we found that the latency of the data is around 1.1 sec. In this way, we can have very few gateways to handle large agricultural lands. The characteristics of LoRa communication protocol and sensor's accuracy used in our implementation are presented in Table 1 and Table 2 respectively.

(a) Sensor Node   (b) Fog Node

Fig. 3. Test bed setup

Table 1. Preliminary information about LoRa communication protocol

Standard	LoRaWAN
Topology	Star
Communication Range	Upto 5 km
Data rate	Upto 50 kbps
Frequency band	868 MHz and 915 MHz

Characteristic of Fog Node: At the fog layer, we connected Raspberry Pi with 10 channels LoRaWAN GPS concentrator as shown in Fig. 3(b). We use the PG1301, a LoRa based multi-channel transmitter/receiver designed to simultaneously receive several LoRa packets using random spreading factors on random channels. Its goal is to enable robust connection between a central wireless data concentrator and a massive amount of wireless end-points spread over a very wide range of agriculture field. The PG1301 is based on SX1301 + 2 × 1257 IC.

Table 2. Different types of sensors and their accuracy

Sensors	Accuracy and range
Flame	100% and 1 m
Humidity	$+3\,\mathrm{g/m^3}$ *and* $0-100\%Rh$ *at* $80-190\,^{\circ}C$
Soil moisture	99% and 60

Figure 4(a) shows the compression ratio for various compression techniques including DASA using the dataset [1] and Testbed. The compression ratio is an important metric that determines the bandwidth demands to transmit the data in the network. Figure 4(a) shows that DASA algorithms perform better than the original LZW and Huffman compression schemes. The results indicate that DASA generates a better compression ratio in comparison with other approaches. DASA compresses the data size up to 57.39% (for the dataset [1]) and upto 38.39% (for the testbed). Figure 4(b) shows the compression time for various approaches. The result shows that our proposed DASA takes higher compression time as compared to LZW and Huffman (for the dataset [1] and testbed). Figure 4(c) shows the compression power (CP) for various compression techniques including DASA using both the datasets. The compression power is an important metric that determines the amount of data compressed. Figure 4(c) shows that DASA algorithm perform better than the lossless LZW and Huffman. The results indicate that DASA is having a better compression power in comparison with other approaches. DASA compresses the data with a CP of 2.35 whereas for LZW and Huffman it is 1.18 and 2.09 respectively(for the dataset [1]). Similarly, for the testbed dataset DASA achieves a CP of 1.62.

In the next experiment, we compare the performance of proposed scheme in terms of volume of transmitted data to the cloud. It is the size of the final compressed file that is to be uploaded to the cloud, i.e. the file size after applying the DASA/existing compression technique. We use The Things Network cloud in our experiment. Figure 5 and Fig. 6 show the size of transmitted data for the dataset [1] and our won data generated from testbed. The result shows that DASA approach can generate a lower size of data (to be sent) as compared to the lossless LZW and Huffman (in case of dataset [1]). The size of the transmitted file is larger in the case of Huff and LZW as compared DASA (see Fig. 5).

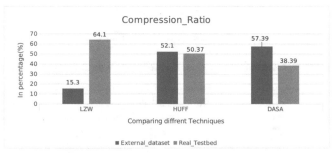

(a) Compression Ratio of various techniques

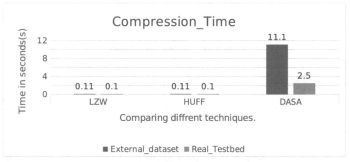

(b) Compression Time of various techniques

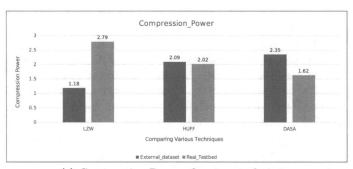

(c) Compression Power of various techniques

Fig. 4. Comparison of various schemes interms of several performance metric

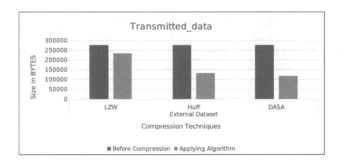

Fig. 5. Comparison of size of transmitted data (Dataset [1])

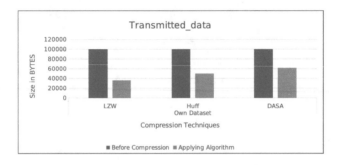

Fig. 6. Comparison of size of transmitted data (Testbed Dataset)

6   Conclusions and Future Work

In this paper, we proposed a data aggregation algorithm to reduce bandwidth consumption of LoRa enabled Fog layer in smart agriculture (DASA). The entire volume of sensor data is efficiently aggregated before uploading to the cloud. Initially, we performed experiments on existing data set using python programming language, and then we also conducted experiments and show the performance of DASA on real testbed. Comparison results show that the proposed algorithm significantly decreases the volume of data to be uploaded to the cloud. Thus, can be highly effective for performing data communication from fog layer to cloud where Internet is intermittent. In the future, we aim to find a representative (a member is called a representative if it is similar to most of the members and different from the members in other cluster) of each cluster and only the representative will be send to the cloud. We will also analyze the trade-off between the cost of information loss vs bandwidth demand and storage requirement of the cloud.

Acknowledgment. We acknowledge TIH-IoT at IIT Bombay for sponsoring this work through Technology Development Program via sanctioned letter no. TIH-IoT/21-22/ TDP/Acad/SL/004.

References

1. Smart Fasal-smart irrigation and fertilization system for precision agriculture using internet of things and cloud infrastructure (wheat dataset (Feb 2021 to April 2021)). http://smartfasal.in/ftp-dataset-portal/
2. The things network. http://swww.thethingsnetwork.org/
3. Abd El-kader, S.M., El-Basioni, B.M.M.: Precision farming solution in Egypt using the wireless sensor network technology. Egyptian Inform. J. **14**(3), 221–233 (2013)
4. Ahmed, N., De, D., Hussain, I.: Internet of things (IoT) for smart precision agriculture and farming in rural areas. IEEE Internet Things J. **5**(6), 4890–4899 (2018)
5. Atmaja, A.P., El Hakim, A., Wibowo, A.P.A., Pratama, L.A.: Communication systems of smart agriculture based on wireless sensor networks in IoT. J. Robot. Control (JRC) **2**(4), 297–301 (2021)
6. Babangida, L., Perumal, T., Mustapha, N., Yaakob, R.: Internet of things (IoT) based activity recognition strategies in smart homes: a review. IEEE Sens. J. (2022)
7. da Costa Bezerra, S.F., Filho, A.S., Delicato, F.C., da Rocha, A.R.: Processing complex events in fog-based internet of things systems for smart agriculture. Sensors **21**(21), 7226 (2021)
8. Dubey, A.K., Upadhyay, D., Thilagam, P.S.: An energy-efficient static multi-hop (ESM) routing protocol for wireless sensor network in agriculture. In: 2018 2nd International Conference on Micro-Electronics and Telecommunication Engineering (ICMETE), pp. 277–280. IEEE (2018)
9. Farooq, M.S., Riaz, S., Abid, A., Umer, T., Zikria, Y.B.: Role of IoT technology in agriculture: a systematic literature review. Electronics **9**(2), 319 (2020)
10. Gkotsiopoulos, P., Zorbas, D., Douligeris, C.: Performance determinants in LoRa networks: a literature review. IEEE Commun. Surv. Tutor. (2021)
11. Kalyani, Y., Collier, R.: A systematic survey on the role of cloud, fog, and edge computing combination in smart agriculture. Sensors **21**(17), 5922 (2021)
12. Kumar, U., Legendre, C.P., Lee, J.C., Zhao, L., Chao, B.F.: On analyzing GNSS displacement field variability of Taiwan: hierarchical agglomerative clustering based on dynamic time warping technique. Comput. Geosci. **169**, 105243 (2022)
13. Pandiyaraju, V., Logambigai, R., Ganapathy, S., Kannan, A.: An energy efficient routing algorithm for WSNs using intelligent fuzzy rules in precision agriculture. Wirel. Pers. Commun. **112**(1), 243–259 (2020)
14. Patel, P., Sivaiah, B., Patel, R.: Approaches for finding optimal number of clusters using k-means and agglomerative hierarchical clustering techniques. In: 2022 International Conference on Intelligent Controller and Computing for Smart Power (ICICCSP), pp. 1–6. IEEE (2022)
15. Qazi, S., Khawaja, B.A., Farooq, Q.U.: IoT-equipped and AI-enabled next generation smart agriculture: a critical review, current challenges and future trends. IEEE Access (2022)
16. Qureshi, K.N., Bashir, M.U., Lloret, J., Leon, A.: Optimized cluster-based dynamic energy-aware routing protocol for wireless sensor networks in agriculture precision. J. Sens. **2020** (2020)
17. Roux, M.: A comparative study of divisive and agglomerative hierarchical clustering algorithms. J. Classif. **35**(2), 345–366 (2018)
18. Sinha, B.B., Dhanalakshmi, R.: Recent advancements and challenges of internet of things in smart agriculture: a survey. Futur. Gener. Comput. Syst. **126**, 169–184 (2022)

19. Sungheetha, A., Sharma, R.: Real time monitoring and fire detection using internet of things and cloud based drones. J. Soft Comput. Paradigm (JSCP) **2**(03), 168–174 (2020)
20. Tao, W., Zhao, L., Wang, G., Liang, R.: Review of the internet of things communication technologies in smart agriculture and challenges. Comput. Electron. Agric. **189**, 106352 (2021)
21. Venkateshwar, A., Patil, V.C.: A decentralized multi competitive clustering in wireless sensor networks for the precision agriculture. In: 2017 International Conference on Current Trends in Computer, Electrical, Electronics and Communication (CTCEEC), pp. 284–288. IEEE (2017)
22. Zamora-Izquierdo, M.A., Santa, J., Martínez, J.A., Martínez, V., Skarmeta, A.F.: Smart farming IoT platform based on edge and cloud computing. Biosys. Eng. **177**, 4–17 (2019)

FSET: Fast Structure Embedding Technique for Self-reconfigurable Modular Robotic Systems

Aliah Majed[1], Hassan Harb[2(✉)], Abbass Nasser[1], and Benoit Clement[3]

[1] ICCS-Lab, Computer Science Department, American University of Culture
and Education (AUCE), Beirut, Lebanon
{aliah.majed,abbass.nasser}@auce.edu.lb
[2] College of Engineering and Technology, American University of the Middle East,
Egaila 54200, Kuwait
hassan.harb@aum.edu.kw
[3] Lab-STICC, UMR CNRS 6585, Ensta-Bretagne University, Brest, France
benoit.clement@ensta-bretagne.fr

Abstract. The rapid growth in communication technologies has lead to
a new generation of robotics called as Modular Robotic System (MRS).
The most crucial process in MRS is self-reconfiguration, which is regarded
as the major challenge for such technology. Indeed, creating new morphology and behaviors manually is a time-consuming and costly process,
especially when dealing with complex structures. In this paper, we have
proposed a fast self-reconfiguration technique called FSET, i.e. Fast SET,
dedicated to MRSs. Our proposed technique consists mainly in two stages:
root selection and morphology formation. The final goal of these stages is
to enhance the time cost to get new morphology of the traditional SET
algorithm thus, ensure fast self-reconfiguration. The root selection stage
selects a small number of modules in order to find the best tree roots that
effects the topological conditions that leads to successful of the embedding
process or not. The morphology formation stage uses the traditional SET
algorithm to calculate the embedding truth table where the initial roots
used are taken from the first stage. Finally, we show the efficiency of our
mechanism through simulations on real scenario using M-TRAN, in terms
providing a fast reconfiguration process in MRS and reducing the energy
consumption of modules thus, increasing its lifetime.

1 Introduction

Since the early 1980's, the robotics industry has seen an increase in the production and manufacture of millions of robots of different types and missions. These
robots are altering our everyday lives and assisting us in making our jobs more
efficient and successful. Furthermore, the rapid advancement of technology in the
new century has ushered in a new robotic era: Modular Robotic System (MRS).

The most important attribute of modular robots, regardless of their design,
is their ability to reconfigure their morphology, a mechanism known as self-
reconfiguration (See Fig. 1), which is regarded as the major challenge for such

© The Author(s), under exclusive license to Springer Nature Switzerland AG 2023
L. Barolli (Ed.): AINA 2023, LNNS 654, pp. 53–66, 2023.
https://doi.org/10.1007/978-3-031-28451-9_5

technology [1,2]. Self-reconfiguration is the mechanism by which a modular robot's initial arrangement of modules (and thus initial form, also known as configuration) is changed into a target configuration. Indeed, MRS use self-reconfiguration to get new morphology and new behavior to performs specified tasks. However, creating new morphology and behaviors manually is a time-consuming and costly process, especially when dealing with complex structures. As a result, designing an algorithm that creates new morphology and new behavior from already existing modular robotic structures, takes a great attention from researchers and communities and became an active research field nowadays.

Therefore, to avoid the above-mentioned challenge, Structure Embedding Technique (SET) for the self-reconfigurable modular robotic system has been introduced. SET decides if a given modular robot structure can be embedded into another structure in order to form new morphology. In this paper, we present a fast SET, abbreviated FSET, a technique for modular robotic systems to minimize the delay to get new morphology. The proposed technique consists of a two-stage algorithm and can highly outperform the traditional SET in terms of the time cost to get new morphology and energy consumption of modules. The first stage of our technique, called root selection, has an objective to find the best roots of the initial trees, by selecting a small number of modules instead of the whole sets. The second stage, which is called morphology formation, uses the first stage's root of trees, to calculate the embedding truth table between modules in order to check the embeddability of the two modular robotic designs, resulting in the formation of new morphologies. Consequently, the calculation time cost of our FSET will highly minimize that of traditional SET due to the small number of the training modules used in the first stage and the low number of iteration loops needed in the second stage.

The rest of the paper is organized as follows. In Sect. 2, we present related works in self-reconfiguration techniques used in MRS. Section 3 detail the SET mechanism. The system demonstration and the results are presented in Sect. 4. Finally, Sect. 5 concludes our paper and gives some perspectives.

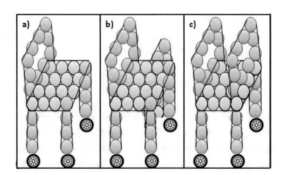

Fig. 1. Sample self-reconfiguration of about 52 3D modules from a chair into a stroller. (a) Chair initial configuration; (b) An intermediate configuration from the self-reconfiguration process; (c) Stroller goal configuration.

2   Related Work

Various methods for self-reconfiguration introduced for MRS can be found in the literature. Initially, researchers in the field of modular self-reconfigurable robotics based their focus on the hardware issue of creating metamorphic robots; then, research interest eventually arose in the generalized management of categories of these systems and various software frameworks were proposed; More recently, in the field of Self-Organizing Particle Systems, researchers have begun to show interest in what may be called a theoretical kind of metamorphic system.

In fact, three categories can be extracted from these three methods. We would then refer to self-reconfiguration algorithm, which Bottom-Up, Top-Down, and Theoretical, as well as [3]. Such methods vary by how they relate to their target execution platform, and therefore, by the nature of the constraints that make up the algorithm model used. We'll go through each of the three self-reconfiguration methods mentioned above, as well as the possible solutions.

From one hand, the authors of [4–9] have proposed Bottom-Up methods to self-reconfiguration in MRS. The Bottom-Up method involves a focus on modular robotic hardware at first. They've come up with a range of module models, ranging from UCMs like Telecube [4] and Crystalline [5] to hybrids like M-TRAN [6] and Roombot [7]. bi-partite models like the Robotic Molecule [9] and I-Cubes [10, 11], as well as self-reconfigurable structures like Fracta [8]. There are several other models in the literature, but these are the ones that are used in the algorithms under consideration.

Due to the complexities of the geometry of hardware modules or their motion capacities, this approach credibly provides a very complicated self-reconfiguration preparation. Non-holonomic motion constraints are typical in these systems, complicating the reconfiguration method. Motion constraints can either be local: caused by module geometry and blocking constraints; or they can be global: like the connectivity constraint which specifies that the whole system's graph must stay connected at all times. Several strategies for achieving holonomy at the expense of granularity have been devised, including using higher holonomy module aggregates (meta-modules) [10] or arranging the device into a porous structure [9] from which modules can flow unconstrainedly (a scaffold). Although the kinematics in the Bottom-Up method are typically more complicated, modules are more likely to expect a greater understanding of their surroundings. Sensor data about their orientation, location in the system, neighborhood, and other factors are used to produce these environmental information.

The authors in [9] have suggested a centralized solution for their bipartite Molecule robot, depending on a three-level hierarchical planner. The highest level of preparation is task-level planning, which chooses a configuration that is suitable for the task at hand. It then admits on a motion plan for Molecules to turn the initial configuration into the target configuration using configuration planning. The configuration planner uses trajectory planning to shift individual modules to their target positions at a lower stage. They also implemented the aforementioned scaffolding principle to ensure that Molecules reached into the target configuration, but this increased the granularity of their device dramatically since a single scaffold tile consisted of 54 modules. While these early works using centralized planners laid the foundations for most of the field and implemented useful problem simplification techniques, they lack the robustness, scalability, and autonomy that self-reconfiguration needs. As a result, researchers transformed to the decentralized self-reconfiguration process, as we'll see below.

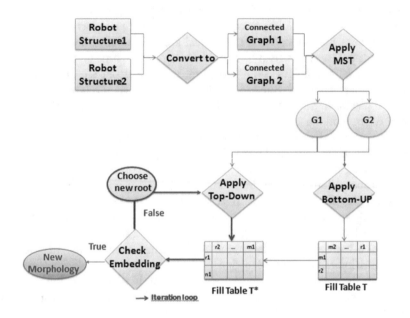

Fig. 2. SET algorithm flowchart.

On the other hand, a lot of Top-Down methods have proposed to self-reconfiguration in MRS [3,12–16]. The Top-Down method plays a critical role in constructing shape formulation techniques that aren't related to a particular hardware application and can be applicable to a variety of MSR in a generalized way.

The authors in [12] developed the Pixel meta-module framework for lattice-based modular robots, which could greatly simplify reconfiguration planning in large modular robots. The key idea is to split the reconfiguration problem into two tasks: planning and resource allocation. The former's job is to figure out the meta-module positions in the target configuration need to be filled next, while the latter's job is to figure out where the meta-modules that will fill that position should come from.

In [16] the authors presents a two-level hierarchical approach to completely generic algorithms (for any architecture), in which the planning problem is formulated as a distributed Markov Decision Process (MDP). An MDP is defined by the four-tuple S, A, T, R, where S is the set of states, represented by open positions to be filled by modules, and is equal to the number of faces of the modules; A is the set of actions, represented by the disconnection of a connector from a neighbor module and the reconnection to another, potentially using a different connector; T is a deterministic or stochastic transition function that determines the next action to take; R is the estimated reward, which is set to 1 to reduce the number of moves.

The authors overcome this MDP by introducing dynamic programming in a distributed environment using message passing. The MDP works at the planner's higher levels, deciding for each moving module which other modules and connectors should link to during the next time phase. The low-level planner then calculates the sequence of individual module movements that the mobile module can take to detach from its current neighbor and reconnect at its new anchor point. Modules look at the structure to make sure they aren't an articulation point in the system's graph, then determine if they are mobile or not and lock a portion of it during their motion if they are. Since several modules can lock the same part of the structure, they can shift in parallel, speeding up the reconfiguration phase. Designing an effective kinematic planner to serve as the transition function T is the most difficult part of this scenario.

The authors in [17] developed their PacMan self-reconfiguration algorithm for two-dimensional unit-compressible modules to three-dimensional structures. As compared to surface moving modules, one advantage of UCM is that they can migrate through the volume of the system, theoretically benefiting from a higher number of parallel motions and a shorter distance to their destination. The authors use a technique known as virtual relocation to transfer modules from one end of the configuration to the other, switching their identities as they compress and decompress along the path to their target point.

PacMan depends on a two-stage distributed planning algorithm in which: (1) modules locally calculate the difference between the current and target shapes to determine which modules should move; and (2) A suitable search (depth-first search) for a mobile module is carried out from the desired locations, with pellets dropped along the way to mark the route that the selected module should follow. Our proposed method consists of two phases, where it applies one of the bottom-up algorithms in its first phase, while in the second phase it applies one of the top-down algorithms. Recognizing if two complete configurations are the same [18], detecting graph automorphisms [19], and recognizing similar substructures for efficient reconfiguration are all examples of existing work in graph representations of modular robots. Our work stands out by incorporating task implications on configurations and specifying criteria for replicating a design's capabilities by replicating its design.

3   FSET Technique

In the literature, one can find a huge number of self reconfiguration algorithms like SET, PacMan, scaffold-based etc. However, SET is one the most popular algorithms used in self reconfiguration. Unfortunately, traditional SET suffer from its huge calculation time cost needed to obtain the new morphology. In order to overcome this problem, we propose a new version of SET called FSET, Fast SET, which highly enhances the time cost of traditional SET. Our FSET consists of two stages, root selection and morphology formation stages, and calculate the embedding truth table according to the topological embedding condition. In the next sections, we first recall the traditional SET and its topological embedding conditions then we detail the two stages of our technique.

3.1 Recall of SET Algorithm

SET is one of self reconfiguration algorithm that decides if a given modular robot structure can be embedded into another structure to form new morphology (Fig. 2.). The process of SET starts by taking the two robotic structures and convert them into two connected graph. Then, it applies MST to them and randomly selects the initial roots for the trees.

We consider that the robotic system is modeled as a connected graph G (M, L): M denotes the set of nodes representing the modules, and $L \subseteq M \times M$ denotes the set of links that connecting such modules.

SET maintain a $|M_1| \times |M_2|$ truth table T, where $T[m_1, m_2]$ is true, under a specified rooting $r_1 \in M_1$, $r_2 \in M_2$. At the end of the algorithm, $T[r_1, r_2]$ answers whether structure (S_1) embeds in Structure (S_2) under r_1 and r_2; if the answer is negative, it repeat the process for a new rooting until it either get a positive answer or we exhaust all possible rootings, in which case we conclude that S_1 does not embeds in S_2.

SET is a two pass algorithm. At first, all entries of the truth table are false. After that, it start proceed bottom-up, starting from the leaves of S_1, and keep going gradually towards r_1. As a starting point, it consider a leaf $m_1 \in M_1$ and check whether it embeds in the leaves of S_2, to calculate the truth table that give us the new morphology according to the topological condition (see [20] for details). Basically, it fill the truth table by traversing G_1 in reverse pre-order, where at each step of the traversal process the nodes of G_2 in reverse pre-order.

After first pass, the second pass is lunched, where it involves a top-down message passing. It iterate top-down, starting from the roots of S_1, and progressively down to v_1. Then it compute a new table called T^*, based on the topological condition in top-down and the preceding truth table in bottom-up pass. However, $T^*[r_1, n] = true$ iff $T_n[r_1, n] = true$, $\forall n \in M_2$. It is then not hard to see that S_1 embeds in S_2 iff at least one entry of the $r_1 - th$ row of T^* is true. If the answer is negative, This process is repeated until we either get a positive answer or we exhaust all possible rootings.

Indeed, it is proved that the loop process generated in the algorithm will always end. Subsequently, SETis highly dependent on the randomly initial tree roots. SET algorithm is one of the simple method in the self reconfiguration approach that has been used in wide range of domains (Fig. 3).

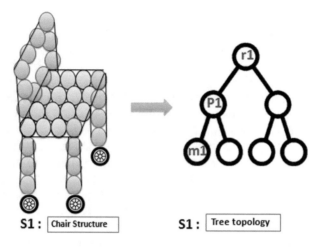

Fig. 3. Convert MRS to connected graph.

3.2   FSET: Fast SET Algorithm

3.2.1   Root Selection Stage

Mostly, the efficiency and performance of the SET algorithm are greatly affected by initial tree roots as different initial tree roots often lead to different embedding or failure in the embedding and thus, failure to obtain the new morphology. Therefore, the calculation time cost for the embedding truth table between the modules of the two robotic structures to form new morphology will be high. Hence, the selection of the initial root trees is becoming a challenge for the SET algorithm.

The first stage of our adapted SET is called root selection and aims to solve the above problem. We propose to select a subset/training from the two sets of modules G_1 and G_2 of the connected graph that represents the two modular robotic structures, in order to find the approximate final tree roots $R = r_i, r_j$ that generate the final morphology. Our intuition is to reduce the number of iterations needed in the traditional SET to obtain the final morphology as fast as possible, thus enhancing the processing time of the SET.

Obviously, the efficiency of the selection root stage is highly related to the percentage, represented by Ts (i.e. training size), of a training set of modules. Subsequently, increasing the value of Ts leads to an increase in the calculation time of FSET so no profit will be noticed compared to traditional SET. On the other hand, the lowest the value of Ts is the better the processing time, but the error in the final obtained morphology will increase. Therefore, selecting the appropriate value of Ts is very essential in the first stage of our technique. Indeed, we believe that Ts should be determined by the decision-makers or experts depending on the application requirements.

3.2.2   Morphology Formation Stage

After having the approximate initial root of trees R, the second stage is lunched which aims to reduce the number of iteration loops in SET. So, it take the obtained roots in the first stage and the whole sets of modules G , and then it apply SET over G in order to get the final morphology.

Algorithm 1 describes the procedure of the second stage of our technique. First, we determine the number of modules needed to find the tree roots in the first stage of our technique (line 2). Based on this number, we randomly select the training sets among the whole sets of modules G1 and G2 (lines 3–6). The modules in the training set represent now the approximate roots of the trees. Then, we calculate the T* embedding truth table. This process is repeated until we either get a positive answer or we exhaust all possible rootings (lines 14–20). At this moment, the first stage is accomplished and the initial roots are determined (line 21). After that, the second stage is running where the process starts by considering the roots obtained in the first stage as the initial roots of the trees. Then, we calculate the T* embedding truth table based on the obtained roots from the first stage. Again, the loop is repeated until we either get a positive answer or we exhaust all possible rootings that give us the embedding sequence that form the new morphology (lines 21–30).

Algorithm 1.   FSET Algorithm.

Require: Two sets of modules: $G_1 = \{m_1, m_2, \ldots, m_k\}$; $G_2 = \{n_1, n_2, \ldots, n_q\}$;
   Percentage of training set: T_s.
Ensure: Embedding truth table T* that generate the new morphology.
   1: $G_s \leftarrow \emptyset$
   2: $N_s \leftarrow [(T_s \times k)/100]$
   3: **for** $i \leftarrow 1$ to N_s **do**
   4:     // randomly selects the training set of modules among G_1 and G_2
   5:     $G_s \leftarrow G_s U G_i$
   6: **end for**
   7: **for** $i \leftarrow 1$ to 2 **do**
   8:     $T_i \leftarrow \emptyset$
   9:     // randomly choose roots r_i among G_s belongs T_i
  10: **end for**
  11: // Initially, all entries are false
  12: (i.e $T[m_i, n_j]$=false)
  13: trace the two tree in a bottom-up to calculate the embedding truth table T
      between modules
  14: **repeat**
  15:     **for** $i \leftarrow 1$ to s **do**
  16:         **for** $j \leftarrow 1$ to s **do**
  17:             // involves a top-down message passing to calculate embedding truth
                  table $T^*[m_i, n_j]$ between modules
  18:         **end for**
  19:     **end for**

20: **until** $r_t h$ row of T^* contain at least one true value or exhaust all possible rootings

21: extract the roots r_i and r_j from the previous T*

22: //use the whole set of modules G_1 and G_2

23: **repeat**

24: **for** $i \leftarrow 1$ to k **do**

25:   **for** $j \leftarrow 1$ to q **do**

26:    // involves a top-down message passing to calculate embedding truth table $T^*[r_i, r_j]$ between modules

27:   **end for**

28: **end for**

29: **until** $r_t h$ row of T^* contain at least one true value or exhaust all possible rootings

30: **return** T^* that generate the new morphology

---

4   Simulation and Results

In order to evaluate its efficiency, we tested our mechanism on one of the most used MRS proposed in recent years, e.g. M-TRAN [6]. Indeed, M-TRAN is a three-dimensional modular robotic system, with characteristics of both lattice and chain (linear) types of modular robot. Each M-TRAN module is made up of two semi-cylindrical pieces that can rotate 180° around their axis and have an independent battery, two degrees of freedom motion, six surface connections, and intelligence with inter-module communication. The M-TRAN system can perform flexible and adaptive locomotion in various configurations using coordination control based on a CPG [21]. Figure 4 presents the components of a M-TRAN module.

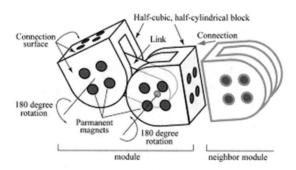

Fig. 4. M-TRAN components.

In our simulations, we used two robotic that have chair and wall design and are made out of M-TRAN modules. We obtain a new design with stroller morphology quickly after using the FSET method (Fig. 5).

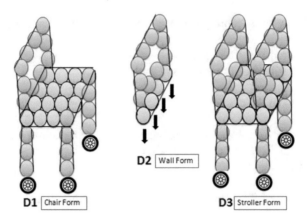

Fig. 5. Morphologies adapted in our simulation.

The objective of our simulations was to confirm that our technique can successfully achieve intended results for reducing the delay to obtain new morphology and reducing the energy consumption in modules that leads to extend the MRS lifetime. In order to evaluate the performance, we compare our results to the traditional SET. In our simulations, we evaluated the performance using the following parameters:

- The Number of Modules for substructure n_b, takes the following values: 100, 200, 300, 1000 and 2000.
- the Number of Modules for superstructure n_p, takes the following values: 500, 950, 2000 and 4000.
- the percentage of module chosen, T_s, takes the following values: 5, 10, 15 and 20.

4.1   Execution Time

Sometimes, getting new morphology fast time as possible to the end-user is a crucial operation especially in e-health and military applications. Figure 6, shows the execution time for both FSET and SET when varying the number of modules (for both the substructure and the superstructure respectively). The results show that FSET can optimize the execution time, comparing always to the SET, from 10% (while varying number of module from (100, 500) to (300, 2k) module) to 37% (while varying number of module from (500, 2k) to (2k, 4k) module).

Obviously, the execution time of FSET will be highly affected by the selection of the tree roots as well as the number of iteration loops to obtain the final morphology. Therefore, FSET outperforms the SET where the processing time to get new morphology is twice accelerated when using FSET, compared to SET Algorithm.

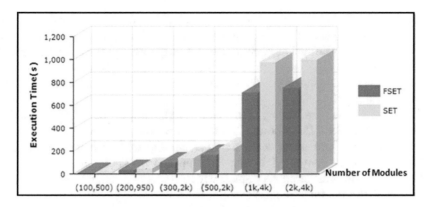

Fig. 6. Processing time for FSET and SET.

4.2 Iteration Loop

One of the factor that can delay the obtain of new morphology is the number of iterations. In Fig. 7, we show how many iterations are generated by the two robotic structures to find the final morphology for both FSET and the SET. It is important to know that a high number of iterations can increase the complexity of the proposed algorithm. The obtained results show that, The number of iterations is reduced by at least 30% as shown in these figure when applying FSET on the SuperBot modules. Therefore, FSET minimize the morphology delay by reducing the number of iterations.

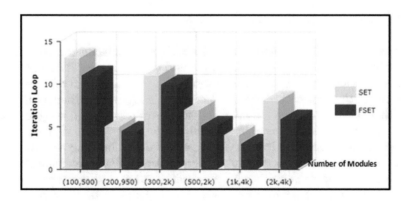

Fig. 7. Iteration loop number for FSET and SET.

4.3 Energy Consumption

Energy consumption is a crucial parameter to assess in MRS since it impacts the overall system's functionality. Indeed, the data transmission or activity done during the transition consumes the majority of a module's limited energy. In our simulation, we implemented the same energy model that used in [3] to calculate the energy consumption in SuperBot modules, assuming that each module has fixed energy units, based on the MRS size, then we considered that each message transmission consumes 0:2 unit and each successful embedding modules consumes 0:8 unit.

Figure 8, shows the energy consumed in the SuperBot robot depending on modules number. The obtained results show that the energy consumption increases with the increasing of the modules number while it is optimized, using FSET, up to 68% compared to the SET approach. Therefore, our proposed technique can be considered very efficiently in terms of reducing the energy consumption of the modular robotic system, thus, increasing its lifetime.

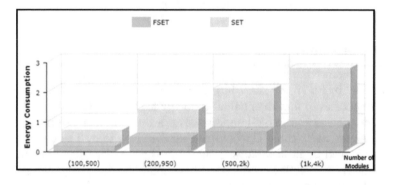

Fig. 8. Energy consumption for FSET and SET.

5   Conclusion and Future Work

In this paper, we have proposed a fast self reconfiguration technique called FSET, i.e. Fast SET, dedicated to MRSs. Our proposed technique consists mainly in two stages: root selection and morphology formation. The final goal of these stages is to enhance the time cost of creating new morphology of traditional SET algorithm thus, ensure fast self reconfiguration. The root selection stage selects a small number of modules in order to find the best tree roots that effects the topological conditions that leads to successful of the embedding process or not. The morphology formation stage uses the traditional SET algorithm to calculate the embedding truth table where the initial roots used are taken from the first stage. Finally, we demonstrated FSET's efficiency in terms of complexity, optimality (lowest number of steps), and time efficiency by simulating its performance on a real robot called SuperBot.

As a future work, we will study how to handle designs with a small number of cycles and how to reduce the kinematic checking runtime. We will go from detecting embeddability to design synthesis in the long run. We consider our embedding method is a good place to start for this line of research, and preliminary results are promising.

References

1. Thalamy, P., Piranda, B., Bourgeois, J.: Distributed self-reconfiguration using a deterministic autonomous scaffolding structure. Ph.D. dissertation, UBFC (2019)
2. Thakker, R., Kamat, A., Bharambe, S., Chiddarwar, S., Bhurchandi, K.: Rebis-reconfigurable bipedal snake robot. In: 2014 IEEE RSJ International Conference on Intelligent Robots and Systems, Chicago, USA, pp. 309–314 (2014)
3. Thalamy, P., Piranda, B., Bourgeois, J.: A survey of autonomous self-reconfiguration methods for robot-based programmable matter. Robot. Auton. Syst. **120**, 103242 (2019)
4. Vassilvitskii, S., Yim, M., Suh, J.: A complete, local and parallel reconfiguration algorithm for cube style modular robots. In: Proceedings 2002 IEEE International Conference on Robotics and Automation (Cat. No. 02CH37292), vol. 1, pp. 117–122. IEEE (2002)
5. Murata, S., Yoshida, E., Tomita, K., Kurokawa, H., Kamimura, A., Kokaji, S.: Hardware design of modular robotic system. In: Proceedings of the 2000 IEEE/RSJ International Conference on Intelligent Robots and Systems (IROS 2000) (Cat. No. 00CH37113), vol. 3, pp. 2210–2217. IEEE (2000)
6. Murata, S., Yoshida, E., Kamimura, A., Kurokawa, H., Tomita, K., Kokaji, S.: M-TRAN: self-reconfigurable modular robotic system. IEEE/ASME Trans. Mechatron. **7**(4), 431–441 (2002)
7. Sproewitz, A., et al.: Roombots-towards decentralized reconfiguration with self-reconfiguring modular robotic metamodules. In: 2010 IEEE/RSJ International Conference on Intelligent Robots and Systems, pp. 1126–1132. IEEE (2010)
8. Yoshida, E., Murata, S., Kurokawa, H., Tomita, K., Kokaji, S.: A distributed method for reconfiguration of a three-dimensional homogeneous structure. Adv. Robot. **13**(4), 363–379 (1998)
9. Kotay, K.D., Rus, D.L.: Algorithms for self-reconfiguring molecule motion planning. In: Proceedings. 2000 IEEE/RSJ International Conference on Intelligent Robots and Systems (IROS 2000) (Cat. No. 00CH37113), vol. 3, pp. 2184–2193. IEEE (2000)
10. Unsal, C., Khosla, P.K.: A multi-layered planner for self-reconfiguration of a uniform group of I-cube modules. In: Proceedings 2001 IEEE/RSJ International Conference on Intelligent Robots and Systems. Expanding the Societal Role of Robotics in the the Next Millennium (Cat. No. 01CH37180), vol. 1, pp. 598–605. IEEE (2001)
11. Ünsal, C., Kiliççöte, H., Khosla, P.K.: A modular self-reconfigurable bipartite robotic system: implementation and motion planning. Auton. Robot. **10**(1), 23–40 (2001)
12. Dewey, D.J., et al.: Generalizing metamodules to simplify planning in modular robotic systems. In: 2008 IEEE/RSJ International Conference on Intelligent Robots and Systems, pp. 1338–1345. IEEE (2008)
13. Yim, M., Zhang, Y., Lamping, J., Mao, E.: Distributed control for 3D metamorphosis. Auton. Robot. **10**(1), 41–56 (2001)
14. Fitch, R., Butler, Z., Rus, D.: Reconfiguration planning for heterogeneous self-reconfiguring robots. In: Proceedings 2003 IEEE/RSJ International Conference on Intelligent Robots and Systems (IROS 2003) (Cat. No. 03CH37453), vol. 3, pp. 2460–2467. IEEE (2003)
15. Fitch, R., Butler, Z., Rus, D.: In-place distributed heterogeneous reconfiguration planning. In: Alami, R., Chatila, R., Asama, H. (eds.) Distributed Autonomous Robotic Systems 6, pp. 159–168. Springer, Tokyo (2007). https://doi.org/10.1007/978-4-431-35873-2_16

16. Fitch, R., McAllister, R.: Hierarchical planning for self-reconfiguring robots using module kinematics. In: Martinoli, A., et al. (eds.) Distributed Autonomous Robotic Systems. Springer Tracts in Advanced Robotics, vol. 83, pp. 477–490. Springer, Heidelberg (2013). https://doi.org/10.1007/978-3-642-32723-0_34
17. Butler, Z., Rus, D.: Distributed planning and control for modular robots with unit-compressible modules. Int. J. Robot. Res. **22**(9), 699–715 (2003)
18. Park, M., Chitta, S., Teichman, A., Yim, M.: Automatic configuration recognition methods in modular robots. Int. J. Robot. Res. **27**(3–4), 403–421 (2008)
19. McKay, B.: Nauty user's guide (v2. 4). Computer Science Dept., Australian National University (2007)
20. Mantzouratos, Y., Tosun, T., Khanna, S., Yim, M.: On embeddability of modular robot designs. In: 2015 IEEE International Conference on Robotics and Automation (ICRA), pp. 1911–1918. IEEE (2015)
21. Kurokawa, H., Tomita, K., Kamimura, A., Kokaji, S., Hasuo, T., Murata, S.: Self-reconfigurable modular robot M-TRAN: distributed control and communication. In: Proceedings of the 1st international conference on Robot communication and coordination, pp. 1–7 (2007)

A Blockchain Based Authentication Mechanism for IoT in Agriculture 4.0

Oumayma Jouini[1(✉)] and Kaouthar Sethom[2]

[1] Innov'COM Laboratory, University of Carthage, SUP'COM,
University of Tunis El Manar, ENIT, Ariana, Tunis, Tunisia
oumayma.jouini@enit.utm.tn
[2] Innov'COM Laboratory, University of Carthage, SUP'COM, ENICarthage,
Ariana, Tunisia

Abstract. Modern agriculture is increasingly oriented toward the integration of TIC technologies such as Robots, drones, and big data analytics, to combine resource protection and economic, social, and environmental needs. The rising domain of "Agritech", brings together internet-connected devices which are helping to monitor the health of crops and livestock and increase livestock productivity. However, IoT devices are resource-constrained devices, incapable of securing and defending themselves, and can be easily hacked and compromised. In this paper, we propose a new authentication mechanism for IoT devices based on blockchain smart contract integration in smart farming

Keywords: Smart farming · Internet of Things · Blockchain · Smart contracts · Wireless sensor networks · Smart agriculture

1 Introduction

Promoting sustainable agriculture, combating climate change and land degradation, halting biodiversity loss, ending world hunger and ensuring food security, and reducing social inequalities at the same time is becoming the global challenge in which contemporary agriculture plays a crucial role.

In this scenario, technological innovation has proven to be the decisive ally of the sector, in the production and distribution of agricultural products and, even more, in the development of knowledge. Contemporary agriculture borrows the logic of Industry 4.0 by developing a synergy between the typical technologies of Interconnected Agriculture and Precision Agriculture, such as the Internet of Things, the Internet of Agriculture, Big Data Analysis, and blockchain.

Thanks to this approach, Intelligent Agriculture can bring significant improvements to the sector and create a strong economic, environmental, and social impact. Examples include the more efficient use of plant protection products or the reduction of diesel consumption, with benefits in terms of cost reduction, reduction of CO_2 emissions into the atmosphere, and significant repercussions on the reputation and the image of farmers, given the growing consumer awareness of these issues.

© The Author(s), under exclusive license to Springer Nature Switzerland AG 2023
L. Barolli (Ed.): AINA 2023, LNNS 654, pp. 67–76, 2023.
https://doi.org/10.1007/978-3-031-28451-9_6

This new form of agriculture which engages ICT and IoT technologies bring also new risks and dangers in terms of ICT security. Even though many research efforts have concentrated on designing network security protocols, cryptography solutions, and device security, several challenges remain, especially with respect to data integrity, service trustworthiness, and the lack of metrics for device security. These challenges are often hard to deal with effectively, due to their range of possible cyber and physical security threats. It is in this same perspective that we position our contribution described in this paper. We propose to combine blockchain trust with IoT power to secure our agriculture 4.0 framework.

The remainder of the paper is as follows: Sect. 2 is dedicated to the literature review, in Sect. 3 we discuss our proposed architecture describing the blockchain authentication, the smart contract, and the performance evaluation. We conclude our paper in Sect. 4.

2 Literature Review

2.1 IoT Vulnerabilities and Threats in Agriculture 4.0

Connected objects are now restructuring the entire agricultural value chain. They allow the prediction and prevention of crises without even the physical presence of the farmer in his crops. The data collected by these sensors are stored in clouds accessible in real-time by farmers. Even if this data-driven technological infrastructure is designed to strengthen the analytical capacity and strategic vision of actors in the agricultural sector, some thread needs to be addressed.

The presence of malware and data theft is a risk in virtually all types of connected systems, and smart farming is no exception.

As the number of IoT devices actively used in agriculture grows, so does the number of entry points for malicious third-party programs. There is an urgent need for security policies for the agricultural IoT.

For all the advantages that precision agriculture offers, the potential for mass disruption is also enormous. The likelihood of unsecured devices being located and compromised is certain. The impact is potentially catastrophic. Without proper cybersecurity on the IoT interfaces for the physical devices and sensors precision agriculture uses, data could be easily lost or stolen, the food supply could be disrupted, or human lives could be placed at risk [1].

2.2 Blockchain Technology

A blockchain is based on a shared ledger distributed among network nodes. The ledger is a sequential chain of cryptographic hash-linked blocks (Fig. 1), recording the history of the transactions that take place between the peers in the network [2].

Instead of relying on a centralized third party, such as a bank, to mediate transactions, nodes in a blockchain network use a consensus protocol to agree on ledger content, and cryptographic hashes and digital signatures to ensure the integrity of transactions [3].

The core aspect of each Blockchain network is the consensus protocol. It ensures that the shared ledgers are exact replicas and reduces the likelihood of fraudulent transactions because tampering would have to take place simultaneously in numerous locations. Digital signatures guarantee that transactions originated from senders (signed with private keys) and not forgers. The decentralized peer-to-peer blockchain network precludes any one member or group of participants from having complete control over the system's supporting infrastructure or undermining it. Participants in the network are all equal and adhere to the same protocols. They could consist of one or more of the following: private individuals, government agencies, organizations, or a mix of all of these.

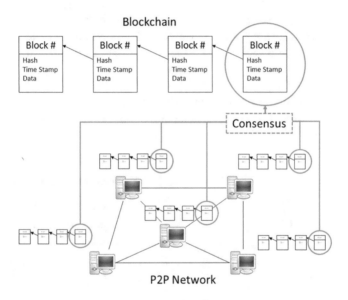

Fig. 1. Key elements of blockchain systems [4]

2.3   Blockchain Solutions in IoT Agriculture

the benefits of blockchain technology include secure data storage, anonymity, and transparency. Although the user's public key and transaction data can be seen in the public blockchain, the user's identity and private key will not be made public. IoT applications using blockchain technology were investigated by researchers [5–7].

The authors [8] developed a smart contract based on climate and soil condition monitoring metrics in smart agriculture. A detailed smart-contract implementation, however, is not provided. Furthermore, no real-time experiments detecting agricultural conditions and testing the proposed smart-contract-based metric monitoring are carried out.

AgriBlockIoT, a blockchain-based solution for agriculture food supply-chain management, was proposed by Caro et al. [9]. However, there is no end-to-end implementation of the Agriculture blockchain, including the ability for sensors to send data in real-time. Furthermore, the message network latency for updating blockchain transactions is higher. We addressed these issues and implemented a more realistic blockchain-based solution for sending sensor alert data as a blockchain transaction.

3   Proposed Architecture

In this section, we will describe the main components of the proposed architecture (Fig. 2). Our testbed is implemented to monitor the behavior of dairy cows. Each cow is equipped with a collar. Equipped with sensors, the smart object detects the behavior parameters of the animal and transmits them via a relay device. The data is analyzed by specific algorithms. The breeder is informed instantly. This monitoring system needs to be installed in local and remote locations of farms that will assist the concerned farmers in monitoring their cattle activities from diverse locations for the whole day.

The cow monitoring system is composed of hardware components, the edge node, the Blockchain system, and the end-user application. We hereafter present a reliable, scalable, and authentic decentralized end-user and IoT device authentication technique that utilizes a graphical user interface with connectivity to

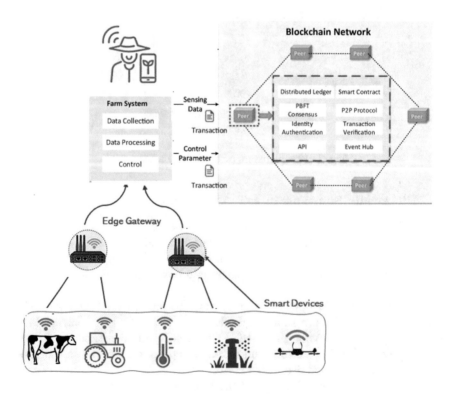

Fig. 2. Proposed blockchain enabled architecture

blockchain algorithms. These algorithms consist of the logic that authenticates end-user access to IoT devices and identifies the devices that are accessible by the end user. Through these, issues of a centralized third party and the double dependency problem can be removed.

3.1   Iot Devices

Tests were conducted in collaboration with two farms located in Mateur an agricultural region in the North of Tunisia. Cows are equipped with connected collars [10].

The collars offer four types of services: heat detection, feed monitoring, animal activity, and calving detection. Neck-mounted accelerometers are used to classify walking, standing, lying, panting, feeding, and ruminating, as well as grazing behaviour.

The information is visible on an application accessible by computer, tablet or smartphone (Fig. 3).

A box with an antenna collects information from sensors up to 10 km in radius. They are then transmitted via the internet to a remote server which processes them and returns them to the Farmer mobile application, in the form of alerts and graphics.

Long-distance communication (LORAWAN) [12] is interesting because it allows the farmer to monitor animals spread over several buildings and the cows grazing around the buildings. To effectively manage the two key periods (heat and calving) of suckler cow production, we developed a mobile application for farmers based on the information collected from the collar (Fig. 4). The collar uses a unique accelerometer to identify cow movement patterns and detect pathologies in advance. It helps farmers improve the pregnancy rate by identifying which cows need to be inseminated and when.

Fig. 3. Cow connected collar [11]

Fig. 4. Mobile application for cow monitoring

3.2   Edge Gateway

An IoT physical gateway bridges IoT devices, sensors, equipment, systems, and the cloud to facilitate their communication. It recovers the data captured by the various IoT devices, to transmit them in a reliable, coordinated, and efficient manner. The gateway is thus essential for edge computing functions. Indeed, the sensors will produce disparate data, with different formats from each other. Sending all of this data directly to the remote server would be economic, energy, and strategic nonsense. The IoT gateway aggregates and sorts this mass of data, to transmit only those really useful for the end use of the solution.

The communication between the gateway and IoT devices uses the MQTT protocol.

3.3   Blockchain for Authentication

This section discusses the technology used to build the backend part of the architecture. We will use the Blockchain for authentication purposes and for IoT data storage. The main challenge that we face is the expensive cost of the blockchain writing process and the high volume generated by IoT devices.

Some works [10,13,14] use an Off-chain database for storing information. Off-chain data is any non-transactional data such as MongoDB or Casandra. We will explain our solution in sections A and B.

A Smart Contract for Authentication

Modern blockchains allow the definition of smart contracts (SCs). An SC is a computer protocol designed to digitally ease, verify, or enforce the terms of the agreement between users. We will use the SC for user authentication in our case.

We implement a blockchain algorithm that allows the authenticated user access to the authenticated IoT device. Registration of end-users, smart devices, access control, authentication, and functionalities are deployed in the smart contract through the blockchain. The main idea here is that the node/user that requires some sensor data needs first to be authenticated to access that sensor data. This will be the responsibility of the blockchain.

Fig. 5. Smart contract functions

- **Step 1:** During this step, each farmer who joins the blockchain network, registers its public and private key pair. Then, he is allowed to register all his own IoT devices on the system through a simple interface with a mobile app. He must indicate the list of allowed users or nodes to interact with his sensors. The system will add the farmer devices to the Blockchain using the smart contract function Add device (Fig. 5). IoT devices of any organization are registered in a smart contract via a web interface by the owner of the organization i.e., the farmer.
- **Step 2:** Any communication between device A and user B must first go into the authentication process. An authentication request is sent by A current gateway to the Blockchain with the id of B. The smart contract verifies if B is allowed to communicate with device A, The system uses the public key of B stored in the blockchain to verify B's signature on the B request sent initially to A. If the platform validates B's identity; it sends the confirmation response to the gateway of A else the authentication process fails.
- **Step 3:** After successful authentication, the gateway allows user B to communicate with sensor A.

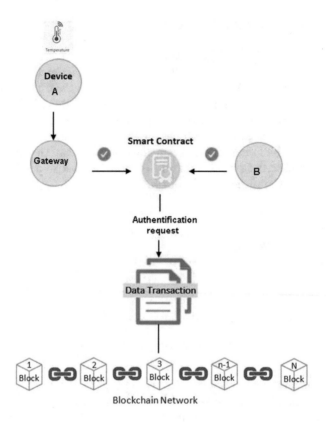

Fig. 6. Blockchain authentication process

B Information Reseizing

Instead of storing information from several IoT devices in the blockchain with each periodic update, we will just store the root of the Merkel tree [13] relating to this update. We do this to optimize the costs of our solution because writing in the blockchain is very expensive, and is proportional to the amount of information to be stored inside the blockchain. We will therefore take advantage of the Merkle tree which will help us to condense the information (initially long) to only store its root. Here is an example of building a Merkle tree: let's take a system composed of 3 sensors a, b and c.

- **h0** = hash(a) # where hash(a) = SHA256(SHA256(a))
- **h1** = hash(b)
- **h2** = hash(c)
- **h3** = hash(c) # since we have an odd number of transactions (3), we duplicate the last one, c. This duplication makes it possible to have an even number of leaves on the tree, even if the number of transactions is odd.

- **h01** = hash(h0+h1) # + means here the concatenation of character strings h1 and h2
- **h23** = hash(h2+h3)
- **hroot**= hash(h01+h23) # is the Merkle root of the block that we store in the Blockchain.

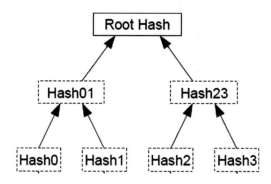

Fig. 7. Merkle tree example

4   Conclusion

IoT devices are more and more used in smart farming for more precise and auto-mated processes. In such cases, it's important to protect data communication and secure the system. Wireless sensors have limited computation capacity and power source which makes it essential to design authentication protocols that are both secure as well as lightweight. In this article, we externalize the authentica-tion protocol inside the blockchain through the IoT gateway. We thus designed a new smart contract-based blockchain authentication mechanism for IoT devices identification in a smart farming environment.

References

1. Khan, M.A., Algarni, F., Quasim, M.T.: Decentralised internet of things. In: Khan, M.A., Quasim, M.T., Algarni, F., Alharthi, A. (eds.) Decentralised Internet of Things. SBD, vol. 71, pp. 3–20. Springer, Cham (2020). https://doi.org/10.1007/978-3-030-38677-1_1
2. Cai, W., Wang, Z., Ernst, J.B., Hong, Z., Feng, C., Leung, V.C.: Decentralized applications: the blockchain-empowered software system. IEEE Access **6**, 53 019–53 033 (2018)
3. Quasim, M.T., Khan, M.A., Algarni, F., Alharthy, A., Alshmrani, G.M.M.: Blockchain frameworks. In: Khan, M.A., Quasim, M.T., Algarni, F., Alharthi, A. (eds.) Decentralised Internet of Things. SBD, vol. 71, pp. 75–89. Springer, Cham (2020). https://doi.org/10.1007/978-3-030-38677-1_4

4. https://softwareengineeringdaily.com/2021/05/18/ethereum-2-0-whats-the-big-deal

5. Wang, H., et al.: IoT based clinical sensor data management and transfer using blockchain technology. J. ISMAC **2**(03), 154–159 (2020)

6. Rathee, G., Sharma, A., Saini, H., Kumar, R., Iqbal, R.: A hybrid framework for multimedia data processing in IoT-healthcare using blockchain technology. Multimed. Tools Appl. **79**(15), 9711–9733 (2020)

7. Deebak, B., Memon, F.H., Dev, K., Khowaja, S.A., Wang, W., Qureshi, N.M.F.: TAB-SAPP: a trust-aware blockchain-based seamless authentication for massive IoT-enabled industrial applications. IEEE Trans. Industr. Inform. **19**, 243–250 (2022)

8. Voutos, Y., Drakopoulos, G., Mylonas, P.: Smart agriculture: an open field for smart contracts. In: 2019 4th South-East Europe Design Automation, Computer Engineering, Computer Networks and Social Media Conference (SEEDA-CECNSM), pp. 1–6. IEEE (2019)

9. Caro, M.P., Ali, M.S., Vecchio, M., Giaffreda, R.: Blockchain-based traceability in agri-food supply chain management: a practical implementation. In: 2018 IoT Vertical and Topical Summit on Agriculture-Tuscany (IOT Tuscany), pp. 1–4. IEEE (2018)

10. Thakur, D., Kumar, Y., Kumar, A., Singh, P.K.: Applicability of wireless sensor networks in precision agriculture: a review. Wireless Pers. Commun. **107**(1), 471–512 (2019)

11. https://www.iotglobalnetwork.com/products/single/id/3197/lorawan-gps-cattle-collar

12. Haxhibeqiri, J., De Poorter, E., Moerman, I., Hoebeke, J.: A survey of LoRaWAN for IoT: from technology to application. Sensors **18**(11), 3995 (2018)

13. Pinter, P.J., Jr., Hatfield, J.L., Schepers, J.S., Barnes, E.M., Moran, M.S., Daughtry, C.S., Upchurch, D.R.: Remote sensing for crop management. Photogram. Eng. Remote Sens. **69**(6), 647–664 (2003)

14. Kamble, P., Shirsath, D., Mane, R., More, R.: IoT based smart greenhouse automation using Arduino. Int. J. Innov. Res. Comput. Sci. Technol. (IJIRCST) 2347–5552 (2017). ISSN

Performance Evaluation of FC-RDVM and LDIWM Router Placement Methods for WMNs by WMN-PSODGA Hybrid Simulation System Considering Load Balancing and Chi-Square Distribution of Mesh Clients

Admir Barolli[1], Shinji Sakamoto[2(✉)], Elis Kulla[3], Leonard Barolli[4], and Makoto Takizawa[5]

[1] Department of Information Technology, Aleksander Moisiu University of Durres, L.1, Rruga e Currilave, Durres, Albania
admirbarolli@uamd.edu.al
[2] Department of Information and Computer Science, Kanazawa Institute of Technology, 7-1 Ohgigaoka, Nonoichi, Ishikawa 921-8501, Japan
shinji.sakamoto@ieee.org
[3] Department of System Management, Fukuoka Institute of Technology, 3-30-1 Wajiro-Higashi, Higashi-Ku, Fukuoka 811-0295, Japan
kulla@fit.ac.jp
[4] Department of Information and Communication Engineering, Fukuoka Institute of Technology, 3-30-1 Wajiro-Higashi, Higashi-Ku, Fukuoka 811-0295, Japan
barolli@fit.ac.jp
[5] Department of Advanced Sciences, Faculty of Science and Engineering, Hosei University, 3-7-2, Kajino-machi, Koganei-shi, Tokyo 184-8584, Japan
makoto.takizawa@computer.org

Abstract. In this paper, we present WMN-PSODGA hybrid simulation system for optimization of mesh routers in Wireless Mesh Networks (WMNs). We consider Chi-square distribution of mesh clients and compare the results of a Fast Convergence Rational Decrement of Vmax Method (FC-RDVM) with Linearly Decreacing Inertia Weight Method (LDIWM). The simulation results show that FC-RDVM has better load balancing than LDIWM.

1 Introduction

Wireless Mesh Networks (WMNs) are a good choice for last-mile networks since they have cost-effective scalability, fault-tolerance, and load distribution capabilities. WMNs are particularly well-suited for linking edge devices, where data is generated, due to their ability to cooperate for sharing data and computation, which results in enhanced analytics capabilities, improved response times, better performance, and increased security and privacy. One of the critical requirements

L. Barolli (Ed.): AINA 2023, LNNS 654, pp. 77–85, 2023.
https://doi.org/10.1007/978-3-031-28451-9_7

of the Internet of Things (IoT) is scalability, and centralized networks need more bandwidth to satisfy this requirement. In this way, edge mesh networks are based on distributed intelligence, eliminating bottlenecks and supporting large-scale networks.

Designing and engineering WMNs is a significant challenge, as it involves complex relationships of various parameters. The optimization process is a trade-off among different parameters including mesh router connectivity, mesh client coverage, Quality of Service (QoS), and network cost. The optimization is challenging as it requires balancing multiple objectives and constraints that often have conflict.

In the mesh router node placement problem, a grid area is given where a specified number of mesh router nodes must be deployed. A fixed number of mesh client nodes with arbitrary positions within the grid area are also given. The objective is to assign the mesh router nodes such locations in the grid area that achieve optimal network connectivity and client coverage while also balancing the load of mesh routers. Network connectivity is measured by the Size of Giant Component (SGC) of the resulting WMN graph. The client coverage is determined by the Number of Covered Mesh Clients (NCMC), which is simply the number of mesh client nodes that fall within the radio coverage of at least one mesh router node. To ensure load balancing, a new parameter called the Number of Covered Mesh Clients per Router (NCMCpR) is included in the fitness function.

Mesh node placement in WMNs can be seen as a family of problems, which is shown (through graph theoretic approaches or placement problems, e.g., [2,6]) to be computationally hard to solve for most of the formulations [13]. Node placement problems are known to be computationally hard to solve [4,5,14]. In previous works, some intelligent algorithms have been investigated for the node placement problem [1,3,7,8].

In our previous work [10], we implemented a Particle Swarm Optimization (PSO) based simulation system called WMN-PSO. Also, we implemented another simulation system based on Genetic Algorithms (GA), called WMN-GA [9], for solving the node placement problem in WMNs. Then, we designed and implemented a hybrid simulation system based on PSO and Distributed GA (DGA). We call this system WMN-PSODGA.

In this paper, we compare the simulation results of the Fast Convergence Rational Decrement of Vmax Method (FC-RDVM) with Linearly Decreasing Inertia Weight Method (LDIWM) considering Chi-square distribution of mesh clients.

The rest of the paper is organized as follows. We introduce intelligent algorithms in Sect. 2. Section 3 presents the implemented hybrid simulation system. The simulation results are given in Sect. 4. Finally, we give conclusions and future work in Sect. 5.

2   Intelligent Algorithms for Proposed Hybrid Simulation System

2.1   Particle Swarm Optimization

PSO is a local search algorithm that uses a group of particles to find solutions in the search space of a problem. The movement of these particles is inspired by the collective behaviour of different species, such as how they move together in their habitat to find food or migrate to better environments. The optimal solution is achieved through the interaction between the particles in the flock, with each member benefiting from the experiences of others. Each particle determines its movement towards the optimal solution based on its own best-known location, the best location achieved by other flock members (best fitness), and random perturbations. In other words, each particle communicates with some other particles and is influenced by the best point found by any member of its topological neighbourhood. Eventually, all members of the flock converge on a solution.

2.2   Distributed Genetic Algorithm

Genetic Algorithms (GA) is a robust search technique that searches through a population of individuals and can operate on various representations. However, the design of the representation of operators can be challenging. GAs have been successfully applied to various optimization problems in different domains. However, one of the main drawbacks of this technique is that it is computationally expensive and time-consuming.

The GA has the ability to avoid falling prematurely into local optima and can eventually escape from them during the search process. Distributed GA (DGA) has an additional mechanism to escape from local optima by considering multiple islands. Each island computes its own solution and then migrate them as shown in Fig. 1. The main GA operators are Selection, Crossover and Mutation. We have implemented different operators in our system.

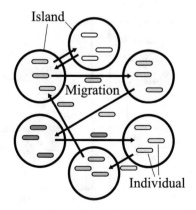

Fig. 1. Model of migration in DGA.

3   Description of WMN-PSODGA Hybrid Intelligent Simulation System

In this section, we present the proposed WMN-PSODGA hybrid intelligent simulation system. The system includes these components: initialization, particle-pattern, gene coding, fitness function and replacement methods.

The pseudo-code of our system is shown in Algorithm 1 and the Migration function is shown in Fig. 2.

Our proposed system starts by generating an initial solution randomly by *ad hoc* methods [15]. We determine the velocity of particles through a random process that considers the area size. For example, when the area size is $W \times H$, the velocity is randomly chosen between $-\sqrt{W^2 + H^2}$ and $\sqrt{W^2 + H^2}$.

Algorithm 1. Pseudo code of WMN-PSODGA system.

Simulation time:= T_{max}, $t := 0$;
Initial solutions: $\boldsymbol{P}$;
Initial global solutions: $\boldsymbol{G}$;
/* begin PSODGA */
while $t < T_{max}$ **do**
    sub_process(PSO);
    sub_process(DGA);
    waiting_sub_processes();
    evaluation($\boldsymbol{G}^t, \boldsymbol{P}^t$);
    /* Migration() moves solutions (see Fig. 2). */
    migration();
    $t = t + 1$;
end while
updating_solutions($\boldsymbol{G}^t, \boldsymbol{P}^t$);
return Best found solution;

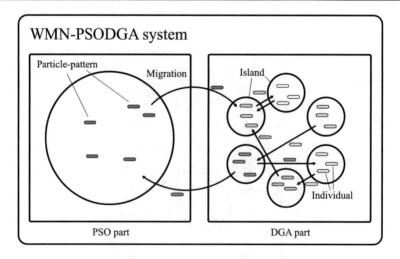

Fig. 2. Model of WMN-PSODGA migration.

A particle is a mesh router. The fitness value of a particle-pattern is computed by combination of mesh routers and mesh clients positions. Each particle-pattern is a solution as shown is Fig. 3. We represent a WMN by a gene and each individual in the population is a combination of mesh routers.

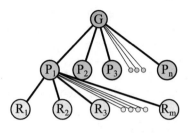

G: Global Solution
P: Particle-pattern
R: Mesh Router
n: Number of Particle-patterns
m: Number of Mesh Routers

Fig. 3. Relationship among global solution, particle-patterns, and mesh routers in PSO part.

In WMN-PSODGA, we use the following fitness function:

$$Fitness = \alpha \times SGC(\boldsymbol{x}_{ij}, \boldsymbol{y}_{ij}) + \beta \times NCMC(\boldsymbol{x}_{ij}, \boldsymbol{y}_{ij}) + \gamma \times NCMCpR(\boldsymbol{x}_{ij}, \boldsymbol{y}_{ij}).$$

This function uses these indicators:

SGC (Size of Giant Component)
    The SGC is the maximum number of connected routers.
NCMC (Number of Covered Mesh Clients)
    The NCMC is the number of covered mesh clients by mesh routers.
NCMCpR (Number of Covered Mesh Clients per Router)
    The NCMCpR is the number of clients covered by each router. The NCMCpR indicator is used for load balancing.

The weight coefficients of the fitness function are α, β, and γ for SGC, NCMC, and NCMCpR, respectively, where $\alpha + \beta + \gamma = 1$.

Table 1. The common parameters for each simulation.

Parameters	Values
Distribution of Mesh Clients	Chi-square
Number of Mesh Clients	48
Number of Mesh Routers	16
Radius of a Mesh Router	4.0–4.5
Number of GA Islands	16
Number of Migrations	200
Evolution Steps	9
Selection Method	Random
Crossover Method	BLX-α
Mutation Method	Uniform
Crossover Rate	0.8
Mutation Rate	0.2
Replacement Method	FC-RDVM, LDIWM
Area Size	32.0 × 32.0

The mesh routers are moved according to their velocities. There exist many techniques for replacing mesh routers. In this paper, we consider LDIWM and FC-RDVM.

Linearly Decreasing Inertia Weight Method (LDIWM)
   In LDIWM, C_1 and C_2 are set to 2.0, constantly. On the other hand, the ω parameter is changed linearly from unstable region ($\omega = 0.9$) to stable region ($\omega = 0.4$) with increasing of iterations of computations [12].
Fast Convergence Rational Decrement of Vmax Method (FC-RDVM)
   In FC-RDVM [11], a value of V_{max} which is maximum velocity of particles is considered. The V_{max} decreases with the increasing of iterations as shown in Eq. (1).

$$V_{max}(k) = \sqrt{W^2 + H^2} \times \frac{T - k}{T + \delta k} \tag{1}$$

Where W and H are the width and the height of the considered area, respectively. Also, T and k are the total number of iterations and a current number of iteration, respectively. The k is a variable varying from 1 to T, which is increased by increasing the iterations. The δ is a curvature parameter.

4   Simulation Results

In this section, we present the simulation results for FC-RDVM and LDIWM. The coefficients of the fitness function are set as $\alpha = 0.6$, $\beta = 0.3$, and $\gamma = 0.1$. The simulation parameters are summarized in Table 1.

   The visualization results are shown in Fig. 4, with Fig. 4(a) and Fig. 4(b) illustrating the results for LDIWM and FC-RDVM, respectively. Figure 5 presents

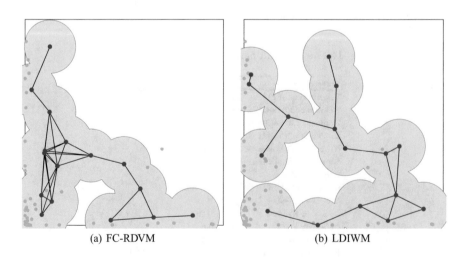

(a) FC-RDVM   (b) LDIWM

Fig. 4. Visualization results after optimization.

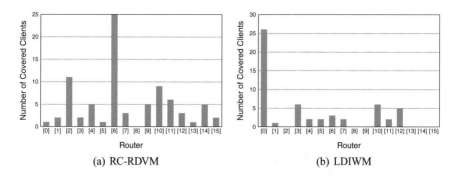

Fig. 5. Number of covered clients by each router after optimization.

the number of mesh clients covered by each router for each replacement method, whereas Fig. 6 depicts the standard deviation, with r being the correlation coefficient.

As shown in Fig. 4 and Fig. 5, both router replacement methods achieve full coverage and high connectivity. Nevertheless, the router replacement methods show different performance on load balancing.

We show the indicator of load balancing in Fig. 6(a) and Fig. 6(b). A decrease in the standard deviation indicates a tendency towards a better load balancing among routers. The standard deviation for LDIWM is an increasing line, while FC-RDVM is a decreasing line. Thus, the FC-RDVM performs better than LDIWM in terms of load balancing.

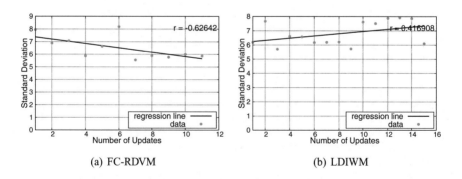

Fig. 6. Transition of standard deviations.

5   Conclusions

In this work, we evaluated the performance of WMNs using a hybrid simulation system based on PSO and DGA (called WMN-PSODGA). We compared the simulation results of FC-RDVM and LDIWM router replacement methods for the Chi-square distribution of mesh clients.

The simulation results show that both methods achieve good coverage and high connectivity, but they do not obtain the same results for load balancing, with FC-RDVM outperforming LDIWM.

In future work, we will consider the implementation of different crossover, mutation and other router replacement methods.

References

1. Barolli, A., Sakamoto, S., Ozera, K., Barolli, L., Kulla, E., Takizawa, M.: Design and implementation of a hybrid intelligent system based on particle swarm optimization and distributed genetic algorithm. In: Barolli, L., Xhafa, F., Javaid, N., Spaho, E., Kolici, V. (eds.) EIDWT 2018. LNDECT, vol. 17, pp. 79–93. Springer, Cham (2018). https://doi.org/10.1007/978-3-319-75928-9_7
2. Franklin, A.A., Murthy, C.S.R.: Node placement algorithm for deployment of two-tier wireless mesh networks. In: Proceedings of the Global Telecommunications Conference, pp. 4823–4827 (2007)
3. Girgis, M.R., Mahmoud, T.M., Abdullatif, B.A., Rabie, A.M.: Solving the wireless mesh network design problem using genetic algorithm and simulated annealing optimization methods. Int. J. Comput. Appl. **96**(11), 1–10 (2014)
4. Lim, A., Rodrigues, B., Wang, F., Xu, Z.: k-Center problems with minimum coverage. Theor. Comput. Sci. **332**(1–3), 1–17 (2005)
5. Maolin, T., et al.: Gateways placement in backbone wireless mesh networks. Int. J. Commun. Netw. Syst. Sci. **2**(1), 44–50 (2009)
6. Muthaiah, S.N., Rosenberg, C.P.: Single gateway placement in wireless mesh networks. In: Proceedings of the 8th International IEEE Symposium on Computer Networks, pp. 4754–4759 (2008)
7. Naka, S., Genji, T., Yura, T., Fukuyama, Y.: A hybrid particle swarm optimization for distribution state estimation. IEEE Trans. Power Syst. **18**(1), 60–68 (2003)
8. Sakamoto, S., Kulla, E., Oda, T., Ikeda, M., Barolli, L., Xhafa, F.: A comparison study of simulated annealing and genetic algorithm for node placement problem in wireless mesh networks. J. Mob. Multimed. **9**(1–2), 101–110 (2013)
9. Sakamoto, S., Kulla, E., Oda, T., Ikeda, M., Barolli, L., Xhafa, F.: A comparison study of hill climbing, simulated annealing and genetic algorithm for node placement problem in WMNs. J. High Speed Netw. **20**(1), 55–66 (2014)
10. Sakamoto, S., Oda, T., Ikeda, M., Barolli, L., Xhafa, F.: Implementation and evaluation of a simulation system based on particle swarm optimisation for node placement problem in wireless mesh networks. Int. J. Commun. Netw. Distrib. Syst. **17**(1), 1–13 (2016)
11. Sakamoto, S., Barolli, A., Liu, Y., Kulla, E., Barolli, L., Takizawa, M.: A fast convergence RDVM for router placement in WMNs: performance comparison of FC-RDVM with RDVM by WMN-PSOHC hybrid intelligent system. In: Barolli, L. (eds.) Complex, Intelligent and Software Intensive Systems. CISIS 2022. LNNS, vol. 497, pp. 17–25. Springer, Cham (2022). https://doi.org/10.1007/978-3-031-08812-4_3
12. Shi, Y., Eberhart, R.C.: Parameter selection in particle swarm optimization. In: Porto, V.W., Saravanan, N., Waagen, D., Eiben, A.E. (eds.) Evolutionary Programming VII. EP 1998. LNCS, vol. 1447, pp. 591–600. Springer, Berlin, Heidelberg (1998). https://doi.org/10.1007/BFb0040810

13. Vanhatupa, T., Hannikainen, M., Hamalainen, T.: Genetic algorithm to optimize node placement and configuration for WLAN planning. In: Proceedings of the 4th IEEE International Symposium on Wireless Communication Systems, pp. 612–616 (2007)
14. Wang, J., Xie, B., Cai, K., Agrawal, D.P.: Efficient mesh router placement in wireless mesh networks. In: Proceedings of the IEEE Internatonal Conference on Mobile Adhoc and Sensor Systems (MASS-2007), pp. 1–9 (2007)
15. Xhafa, F., Sanchez, C., Barolli, L.: Ad hoc and neighborhood search methods for placement of mesh routers in wireless mesh networks. In: Proceedings of the 29th IEEE International Conference on Distributed Computing Systems Workshops (ICDCS-2009), pp. 400–405 (2009)

Uncertainty Handling with Type-2 Interval-Valued Fuzzy Logic in IoT Resource Classification

Renato Dilli[1]([⊠]), Renata Reiser[1], Adenauer Yamin[1], Hélida Santos[2], and Giancarlo Lucca[2]

[1] UFPEL - Federal University of Pelotas, Pelotas, RS, Brazil
{renato.dilli,reiser,adenauer}@inf.ufpel.edu.br
[2] FURG - Federal University of Rio Grande, Rio Grande, RS, Brazil
{helida,giancarlo.lucca}@furg.br

Abstract. The growing supply of Internet-connected resources, often providing more than one service, add complexity to the procedures for discovering, classifying, and selecting the most appropriate resources to meet client demands. The specification of client preferences can lead to inaccuracies and uncertainties, as it depends on prior knowledge and experience for the correct details of parameters such as minimum, maximum, and measurement scales. This paper aims to address uncertainties in specifying and processing client preferences when classifying a set of discovered IoT (Internet of Things) resources. We propose a software architecture for resource discovery and classification in IoT called EXEHDA-Resource Ranking. The proposal stands out in IoT resource classification, exploring three approaches: (i) initial selection of resources with MCDA algorithm; (ii) pre-classification of newly discovered resources with machine learning; and (iii) treatment of uncertainty in preference processing using Type-2 Interval-valued Fuzzy Logic. In addition, one scenario containing resource request simulations applying different client preferences can be demonstrated in EXEHDA-RR features.

1 Introduction

Currently, around 27 billion things are connected to the Internet and this number is expected to reach 125 billion by 2030 [1]. Moreover, each of these connected things can provide more than one service, selecting resources whose services best meet a client's requirements among a large number of capable resources. Due to the dynamic aspects of IoT, we cannot anticipate all eventualities [2] and the resource discovery and classification activities can become a complex task demanding new technologies to deal with uncertainty. Thus, the classification of services may be affected by distinct sources of uncertainty, including vagueness of user preferences about Quality of Service (QoS) attributes and client uncertainty about the adequacy of the max and min attribute values [3].

L. Barolli (Ed.): AINA 2023, LNNS 654, pp. 86–98, 2023.
https://doi.org/10.1007/978-3-031-28451-9_8

Client preferences define the optimality criterion which is applied in the service selection from a set of service candidates. These preferences generally refer to user-relevant QoS properties [4]. Preference descriptions are evaluated in order to classify previously discovered candidate services as an optimization problem. This classification process gets the best service according to established preferences [5]. In general, client preferences in QoS requirements have a diffuse nature. In addition, QoS requirements are client dependent, even if the functional requirements remain the same [6].

The central motivation of this work emerges from a high scalable scenario and the uncertainties of the client in specifying their preferences, classifying and selecting the most appropriate resources among a ratio to fit IoT resources.

The use of Type-1 Fuzzy Logic (T1FL) is due to its ability to deal with uncertainties, which is not possible to reach by only considering the classical approach. The Interval-valued Type-2 Fuzzy Logic (IT2FL) is considered to be potentially more suitable for modelling not only the uncertainty related to client preferences, but also the imprecision related to the Fuzzy Classifier performed on EXEHDA-RR (Resource Ranking). The additional complexity arises from the inclusion of a Footprint Of Uncertainty (FOU), offering extra degrees of freedom to T2FL in comparison to T1FL and reducing the number of rules in the rule bases [7].

This paper mainly discusses the flexible proposal named EXEHDA-RR, which aims to list the most appropriate resources among a list of fit resources in a high scalability scenario typical of IoT. This proposal, an extended version of a previous paper published at the 2019 IEEE International Conference on Fuzzy Systems [8], considers client preferences exploring not only MCDA (Multi-Criteria Decision Analysis) algorithm for the best resource selections, but it also takes into account machine learning strategies, employing decision trees for pre-classification of new resources and handling uncertainties in applying client preferences with T2FL.

To characterize the functionalities of the proposal, a usage scenario was created with different client preferences. The results obtained reached the objectives, selecting the most appropriate resource, according to the client's interest.

The rest of the paper is structured as follows. Section 2 provides an overview of the EXEHDA-RR software architecture. In addition, components and modules responsible for discovery and classification of IoT resources are also presented. In Sect. 3, EXEHDA-RR proposal validation is demonstrated. The scenario simulates client requests with different preferences. Section 4 presents the related work. And, finally, Sect. 5 presents the conclusions.

2   EXEHDA-RR: Software Architecture

EXEHDA [9] is the middleware, focus of this work, responsible for providing the basic computational infrastructure for IoT. This computational environment is made up of execution cells in which computational devices are distributed.

EXEHDA-RR is a new resource discovery and classification service proposed for EXEHDA middleware. The software architecture was modeled considering the dynamics in which resources enter and leave the environment, thus controlling the presence of devices, including low computing power, connected to gateways.

In the resource classification process, which aims to satisfy the client requests, the evaluation of QoS attributes is a challenging step. The definition of the importance of QoS attributes by the client is an activity that depends on the experience and knowledge of each individual. The uncertainties caused by the specification and processing of client preferences are dealt with using Interval Type-2 Fuzzy Logic. Client preferences are translated into rules for the fuzzy inference engine, changing the final classification of resources.

Clients can set their preferences for each of the QoS attributes. Relevance for each of them can be Low, Medium or High. According to the selected preference, additional rules will be activated in the fuzzy rule base.

In EXEHDA-RR, the QoS attributes of resources are normalized and submitted as input to the Type-2 Fuzzy system, and fuzzified using trapezoidal and triangular membership functions. The use of Type-2 Fuzzy membership functions in the input of fuzzy controllers can result in fewer rules when compared to a conventional fuzzy system [10]. The Fuzzy Classifier consists of the following steps:

Resources rated by MCDA Classifier or by ML (Machine Learning) Pre-Classifier as best, i.e. "1" are normalized before being processed by the Fuzzy Classifier. In the normalization process, the best QoS attribute value is normalized to 100. This is done for each of the attributes. Attributes which already have their values ranging from 0 to 100 are not normalized, such as the AV - Availability attribute.

It is a process turning a crisp entry into a fuzzy value of a Type-1 fuzzy set (FS) and/or a Type-2 fuzzy set (FS2), which indicates the degree of pertinence of the element $x \in U$ in the fuzzy set A, often given by a triangular form (A = (a, b, c)) or a trapezoidal form ($\tilde{A} = (a, b, c, d)$), whose corresponding analytical expressions are given by Eq. (1) and Eq. (2) as follows:

$$
\mu_A(x) = \begin{cases} 0, & \text{if } x \leq a; \\ \frac{x-a}{b-a}, & \text{if } a < x \leq b; \\ \frac{c-x}{c-b}, & \text{if } b \leq x < c; \\ 0, & \text{if } x \geq c. \end{cases} \tag{1}
$$

$$
\mu_A(x) = \begin{cases} \frac{x-a}{b-a}, & \text{if } a \leq x < b; \\ 1, & \text{if } b \leq x \leq c; \\ \frac{d-x}{d-c}, & \text{if } c < x \leq d; \\ 0, & \text{otherwise.} \end{cases} \tag{2}
$$

For each QoS attribute, an entry in the Type-2 Fuzzy interval system is added. The system was modeled with a single output representing the weight of importance for the evaluated resource. The QoS attribute is identified by a linguistic variable. Table 1 shows the linguistic variables for for the following attributes: RT (Response Time), AV (Availability), and TH (Throughput). The first column shows the linguistic terms. The HighRT Term was modeled with a triangular membership function and the lower and upper bounds are presented as the Upper MF and Lower MF columns. And, the MediumRT and LowRT Linguistic Terms were modeled with trapezoidal membership functions.

Table 1. Linguistic variables - RT, AV and TH

Linguistic Term	Upper MF	Lower MF
H - HighRT	(56,100,100)	(68,100,100)
M - MediumRT	(31,46,68,83)	(43,58,56,71)
L - LowRT	(0,0,38,58)	(0,0,26,46)

The output linguistic terms related to the variable Weight were modeled using trapezoidal membership functions. Linguistic terms and their values are both presented in Table 2.

Table 2. Linguistic variable - weight

Linguistic Term	Upper MF	Lower MF
H - HighWeight	(46,76,100,100)	(58,88,100,100)
M - MediumWeight	(6,36,63,93)	(18,48,51,81)
L - LowWeight	(0,0,28,53)	(0,0,16,41)

Fuzzy rule-based systems run faster than conventional rule-based systems because they are easier to understand, read, add and modify services. In addition, they also have natural knowledge in uniform representation including separate knowledge from processing and a reduced time complexity [11].

The rule base describes how language variables are processed linguistically in the inference machine. In the developed controller, the Mamdani [12] method was applied. And, the decision-making was based on "If X Then Y" rules, in which both the antecedent and consequent values of the linguistic variables are expressed through fuzzy sets.

The fuzzy operators mainly used are the union and intersection fuzzy aggregators. The Mamdani method aggregates the rules through the OR operator (union), which is modeled by the maximum operator, and in each rule, the logical operator AND is modeled by the minimum operator (intersection).

The Type-2 Fuzzy system supports the analogous inference rule structures of the Type-1 Fuzzy system, and the difference is related to the nature of the membership functions [13].

EXEHDA-RR processes client preferences through fuzzy rules. Client preference is reported for three QoS attributes. The system allows the use of more attributes, but the rule base must be increased to cover all conditions. QoS attributes are represented by linguistic variables in the fuzzy system input, with three linguistic terms each, resulting in 27 rules. These rules are always processed even if the client does not wish to specify preferences, meaning that all quality attributes are important. The attributes: RT - Response Time, AV - Availability, and TH - Throughput will have the same weight of importance in the final classification.

The logic applied in building the rules is that the smallest value of the linguistic terms among the input language variables is used in the output variable. See, for instance, Rule 08 which can be described as follows: If RT = Low and AV = High and TH = Medium then Weight = Low. The input variable RT has the lowest linguistic term (Low), this term is the same as the output variable (Weight).

Table 3 contains additional rules for applying client preference, where it is specified as follows: RT - Low, AV - High, and TH - Medium. To the set of 27 rule bases, two new rules have been added: rule 28 checking if the AV resource attribute is High. If it is verified, the output variable Weight is set High; Rule 29 checks whether the TH attribute is Medium, if so, the Medium value is assigned to the Weight output variable. These rules aim to bring the fuzzy relevance of the input variables to the output variable, thus ensuring the preference set by the client.

Table 3. LHM client preference

Rule	Antecedent	Consequent
28	HighAV and MediumTH	MediumWeight
29	MediumTH and HighTH	HighWeight

Table 4 contains additional rules for applying client preference, where it is specified as follows: RT - Low, AV - High, and TH - Low. The following rules were added to the set of 27 base rules: rule 28, which checks whether the value of the AV attribute of the resource is High and, if so, the value High is assigned to the Weight output variable; rule 29, checks if the value of the resource's RT attribute is Medium and, if so, the value Medium is assigned to the Weight output variable; and rule 30 that checks whether the value of the resource's TH attribute is Medium and, if so, the value Medium is assigned to the Weight output variable.

Table 4. LHL client preference

Rule	Antecedent	Consequent
28	HighAV	HighWeight
29	MediumRT	MediumWeight
30	MediumTH	MediumWeight

Table 5 contains additional rules for applying client preference, where it is specified as follows: RT - Medium, AV - High, and TH - Medium. To the set of 27 base rules, two new rules have been added: rule 28 checks if the RT resource attribute value is Medium. If so, the Medium value is assigned to the output variable Weight; Rule 29 checks whether the value of the AV attribute is High, if so, such high value is assigned to the Weight output variable.

Table 5. MHM client preference

Rule	Antecedent	Consequent
28	MediumRT and MediumTH	MediumWeight
29	HighAV	HighWeight

Table 6 contains additional rules for applying client preference, where it is specified as follows: RT - High, AV - Low, and TH - Low. To the set of 27 base rules, the new rule has been added: rule 28 checking if the value of the RT resource attribute is High. In such case, the value High is assigned to the output variable Weight; rule 29 checks if the value of the feature's AV attribute is Medium, if so, the Medium value is assigned to the output variable Weight; Rule 30, checks if the value of the TH resource attribute is Medium, if so, such medium value is assigned to the Weight output variable.

Table 6. HLL client preference

Rule	Antecedent	Consequent
28	HighRT	HighWeight
29	MediumAV	MediumWeight
30	MediumTH	MediumWeight

Table 7 contains the additional rule for applying client preference, specified as follows: RT - High, AV - High, and TH - Low. Again, to the set of 27 base rules, a new rule has been added: rule 28 checks if the RT resource attribute is High and the AV attribute is High, if the condition is true, the lowest relevance is assigned between the two input variables. Weight output variable.

Table 7. HHL client preference

Rule	Antecedent	Consequent
28	HighRT and HighAV	HighWeight

Table 8 also contains additional rules for applying client preference specified as follows: RT - Medium, AV - Low and TH - High. So, to the set of 27 base rules, two new rules have been added: rule 28 verifying the RT resource attribute as Medium and the TH attribute as High. In such case, if the condition is true, the lowest relevance is assigned between the two input variables. Thus, Rule 29 checks if the TH attribute is High and, if High is assigned to the Weight output variable.

Table 8. MLH client preference

Rule	Antecedent	Consequent
28	MediumRT and HighTH	MediumWeight
29	HighTH	HighWeight

Finally, it follows the specification of Table 9 containing additional rules for applying client preference: RT - High, AV - Low, and TH - Medium. Two new rules were added to support the set of 27 rule bases: Rule 28 checking if the RT resource attribute request is High. If High, its output variable Weight is obtained; Additionally, Rule 29 checks if the TH attribute is Medium, if Medium, it is assigned to the Weight output variable.

Table 9. HLM client preference

Rule	Antecedent	Consequent
28	HighRT	HighWeight
29	MediumTH	MediumWeight

Providing the output of each rule-based input built by composition operators and fuzzy logic (FL) connectors, the fuzzy inference machine processes Mandani type rules, considering the logical operator "AND". Thus, the output linguistic variable "Weight" will be assigned to the least relevance among the input linguistic variables (RT, AV and TH).

Considering the best choice of granularity, a Type-2 Fuzzy set is transformed into a Type-1 Fuzzy set, which is called type-reduced process. The type reduction block aims to use the Karnik-Mendel (KM) algorithm determining the minimum (y_L) and maximum (y_R) of Type-1 fuzzy set centroids.

Turning a fuzzy number into a real number, this step produces a quantifiable result in FL. Among many different methods of defuzzification producing quantifying results after the fuzzification step, this work considers the Centroid method as proposed by Karnik-Mendel [13], as the average between two values: the former calculating the left endpoint of the interval (C_L), and the latter providing its right end (C_R) given as follows:

$$C_L = \min_{L \in \mathbb{N}} \mathbf{c}(T(L)); C_R = \max_{R \in \mathbb{N}} \mathbf{c}(T(R)),$$

when $x_1 < x_2 < \cdots < x_N$ and $L, R \in \mathbb{N}$ are points switching from $\overline{\mu}_{\tilde{A}}$ to $\underline{\mu}_{\tilde{A}}$, and analogously, from $\underline{\mu}_{\tilde{A}}$ to $\overline{\mu}_{\tilde{A}}$, respectively:

$$\mathbf{c}(T(L)) = \frac{\sum_{i=1}^{L} x_i \overline{\mu}_{\tilde{A}}(x_i) + \sum_{i=L+1}^{N} x_i \underline{\mu}_{\tilde{A}}(x_i)}{\sum_{i=1}^{L} \overline{\mu}_{\tilde{A}}(x_i) + \sum_{i=L+1}^{N} \underline{\mu}_{\tilde{A}}(x_i)}; \tag{3}$$

$$\mathbf{c}(T(R)) = \frac{\sum_{i=1}^{R} x_i \underline{\mu}_{\tilde{A}}(x_i) + \sum_{i=R+1}^{N} x_i \overline{\mu}_{\tilde{A}}(x_i)}{\sum_{i=1}^{R} \underline{\mu}_{\tilde{A}}(x_i) + \sum_{i=R+1}^{N} \overline{\mu}_{\tilde{A}}(x_i)}. \tag{4}$$

Concluding the defuzzification step, since many elements in a list of generated intervals in the IVFL (interval-valued fuzzy logic) approach may not be comparable by the usual partial order, a total order for intervals is considered based on the admissible orders, as proposed in [14].

Thus, in this proposal the development of applications were not restricted to the use of aggregation operators (as the arithmetic mean) performed over upper and lower bound limits of interval data in the final procedure of the defuzzification step.

Such new strategy of interval data preserving enables us to apply metrics (see for instance, diameter) extending the analysis of interval results by considering not only the uncertainty associated to input data but also the imprecision related to the classification process calculations. This proposal models both uncertainties, the possible indecision of client preferences along with imprecision of computational calculations. In addition, we remain capable to guarantee the comparison of all output interval data.

So, this paper considers Xu and Yager's order defined, for all $X, Y \in \mathbb{U}$, as follows:

$$[\underline{X}, \overline{X}] \leq_{XY} [\underline{Y}, \overline{Y}] \Leftrightarrow \begin{cases} \underline{X} + \overline{X} < \underline{Y} + \overline{Y}; \text{or} \\ \underline{X} + \overline{X} = \underline{Y} + \overline{Y} \text{ and } \overline{X} - \underline{X} \leq \overline{Y} - \underline{Y}. \end{cases} \tag{5}$$

Resources are classified in descending order of defuzzified intervals by the admissible order proposed by [14].

3 EXEHDA-RR Proposal Validation

In order to evaluate the potential of the EXEHDA-RR Classifier, one scenario was defined. The Fuzzy Classifier was developed in Java, using the Juzzy framework, as proposed by [10].

The scenario simulates resource requests, with different client-defined preferences. The presence of IoT resources was induced by inserting resources from the QWS dataset version 2.0 available at [15] with 2,505 features. This dataset contains thirteen attributes, from which five attributes were selected to be inserted into the Component Directory.

In this scenario, the Client Component sends eight requests by the "AnalysisWSAppLabImplService" resource, each one having a different preference for the QoS RT, AV and TH attributes.

The Directory Component processes the client request and the query result for "AnalysisWSAppLabImplService" returns 102 resources which can fulfill the client request. These resources are rated by the MCDA Classifier by analyzing the five quality attributes: RT, AV, TH, RE, and LA. The best resources are rated "1" and the worst features are "4". The MCDA Classifier returns 13 resources (Table 10) as the best ones. So, these resources are sent to the Fuzzy Classifier to apply client preferences.

In the fuzzy classifier, attributes are normalized before being processed by the Juzzy framework. Table 10 presents the 13 resources resulting from the MCDA classifier, with their original and normalized QoS attribute values. Resource URIs have been replaced with IDs to make it easier to see data in the tables.

In Tables 18 and 19 described below, we present the interval data obtained in the defuzzification step via IVFL.

Table 10. Resources selected by MCDA classifier

ID	Original			Normalized		
	RT	AV	TH	RT	AV	TH
01	221.48	90	10.9	99.77	90	98.2
02	239.22	85	7.7	92.37	85	69.37
03	261.09	85	5.4	84.63	85	48.65
04	256.39	85	5.1	86.18	85	45.95
05	220.96	87	11.1	100	87	100
06	243.91	85	7.1	90.59	85	63.96
07	223.83	92	10.8	98.72	92	97.3
08	250.74	86	7	88.12	86	63.06
09	246.43	85	9.7	89.66	85	87.39
10	277.35	86	6.3	79.67	86	56.76
11	280.39	86	6.4	78.8	86	57.66
12	275.87	85	6.5	80.1	85	58.56
13	267.43	86	5.7	82.62	86	51.35

Figure 1 shows the rankings obtained by each resource, according to the preference established by the client. The resources are represented by the IDs from 1 to 13 in the horizontal axis of the figure, and the classification obtained by the resource is presented in the vertical axis.

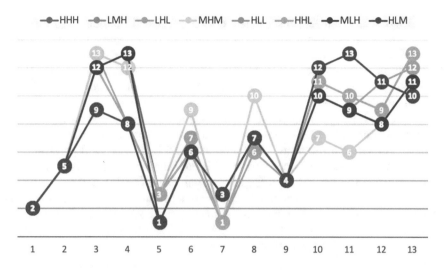

Fig. 1. Scenario 2 - fuzzy classifier ranking considering client preferences

Note that preference sets with the Availability attribute with H (High), that is, high importance in client preference rated resource "1" as the best resource of ID 07 (preferences HHH, LHM, LHL, MHM and HHL).

4   Related Works

Current proposals for resource classification address uncertainties in specifying client preferences, but many are not concerned with scaling IoT resources or adjusting relevance functions to the operational profile of the client application. Belouaar et al. [16] is a fuzzy logic-based model to help web service consumers select the most appropriate service based on their preferences. The authors employed fuzzy logic to represent vague or inaccurate data, with triangular membership functions and used the Gravity Center method for defuzzification.

In Rangarajan's work [17] machine learning is used to predict QoS properties. The QWS dataset is used for ML algorithm training and testing. The proposed model can use up to five QoS attributes and reaches over 90% accuracy.

Suchithra [18] proposes a fuzzy multi-attribute decision making algorithm for service selection. The proposed method periodically collects user feedback to update the service QoS rating. Classification logic uses a strategy of dynamically updating the QoS constraints of the services involved.

Kumar et al. [19] propose a multicriteria decision making framework incorporating fuzzy TOPSIS and AHP to solve QoS-based service selection problems. The weights obtained using the AHP methodology are included in the decision-making process and are utilized by the fuzzy TOPSIS computations for determining the ranking order of the available cloud service alternatives.

Patiniotakis et al. [20] discuss the cLoud Service Recommender (PuLSaR) which uses a multicriteria decision approach. PuLSaR enables unified assessment of cloud services, taking into account accurate and inaccurate metrics and addressing their lack of expression. The proposal addresses linguistically expressed preferences and characteristics of cloud services that do not have a fixed or precise values.

Gohar et al. [21] propose a fuzzy-based approach for web service discovery. It models the ranking of QoS-aware web services as a fuzzy multi-criteria decision-making problem. To represent the functional and imprecise QoS preferences of both the web service consumer and provider in linguistics term, fuzzy rule base is created with the help of Java Expert System Shell (JESS) API.

All Related Works consider client preferences and QoS attributes in their proposals. Fuzzy logic is discussed in the works [16, 19, 20] and [21]. None of them apply Type-2 Interval-valued Fuzzy Logic to solve uncertainties in specifying preferences or optimal QoS attribute parameters.

EXEHDA-RR differs from other Related Works by resolving possible uncertainties arising from client preferences through Type-2 Fuzzy Logic and by being able to work with a high resource scalability typical of the Internet of Things.

5   Conclusions

The EXEHDA-RR software architecture for resource discovery and classification at IoT was introduced. EXEHDA-RR considers the dynamism of the computing infrastructure provided by IoT, with a large number of clients and resource providers. Through the assessment of functional and non-functional requirements, resources are classified and the highest evaluated are made available to clients. Many uncertainties can occur in the client's definition of preferences, either due to the lack of knowledge of the scales of each QoS attribute or the percent importance of each QoS attribute.

The main contributions of this work are as follows:

- Handling uncertainty with Type-2 Fuzzy Logic by modeling membership functions for each QoS attribute, enabling domain experts to define the optimal conditions for each attribute. Specification of client preferences through linguistic variables, reducing misconceptions due to ignorance of the ideal scales of attributes.
- Client preference is implemented in the fuzzy system by adding mandani-type rules, increasing the weight of the most important client specified attributes.

The Fuzzy Classifier receives the client-defined preferences and activates the appropriate rules for the fuzzy inference engine by aggregating the rules for defuzzifying the final weight of the resource.

In a simulation scenario, it was possible to verify the change in the classification of resources already considered better, usually containing close QoS values. Client preference is applied and resources are categorized to meet client needs by delivering the most appropriate resources among many that can meet client request.

In the research continuity, we highlight two important goals: (i) employment of consensus measures aimed at the participation of various specialists in defining membership functions; and (ii) promotion of new computational effort assessment metrics as well as accuracy.

Acknowledgements. This study was partially financed by CAPES - Coordenação de Aperfeiçoamento de Pessoal de Nível Superior - Brasil, Finance Code 001.

References

1. Graham, L.: The Internet of Things: a movement, not a market, IHS Markit Ltd
2. Salah, N.B., Saadi, I.B.: Fuzzy AHP for learning service selection in context-aware ubiquitous learning systems. In: 2016 International IEEE Conferences on Ubiquitous Intelligence & Computing, Advanced and Trusted Computing, Scalable Computing and Communications, Cloud and Big Data Computing, Internet of People, and Smart World Congress, pp. 171–179 (2016)
3. Platenius, M.C., von Detten, M., Becker, S., Schafer, W., Engels, G.: A survey of fuzzy service matching approaches in the context of on-the-fly computing. In: CBSE 2013 - Proceedings of the 16th ACM SIGSOFT Symposium on Component Based Software Engineering (April 2017), pp. 143–152 (2013)
4. Liu, F.G., Xiao, F., Lin, Y.D.: Combining experts' opinion with consumers' preference in web service QoS selection. In: Proceedings - International Conference on Machine Learning and Cybernetics, vol. 4, pp. 1740–1746 (2013)
5. Wang, H., Olhofer, M., Jin, Y.: A mini-review on preference modeling and articulation in multi-objective optimization: current status and challenges. Complex Intell. Syst. **3**(4), 233–245 (2017). https://doi.org/10.1007/s40747-017-0053-9
6. Tripathy, A.K., Tripathy, P.K.: Fuzzy QoS requirement-aware dynamic service discovery and adaptation. Appl. Soft Comput. J. **68**(November), 136–146 (2018)
7. Wu, D., Mendel, J.M.: Uncertainty measures for interval type-2 fuzzy sets. Inf. Sci. **177**(23), 5378–5393 (2007)
8. Argou, A., Dilli, R., Reiser, R., Yamin, R.: Exploring type-2 fuzzy logic with dynamic rules in IoT resources classification. In: IEEE International Conference on Fuzzy Systems, vol. 2019-June (2019). https://dx.doi.org/10.1109/FUZZ-IEEE.2019.8858944
9. Lopes, J., et al.: A middleware architecture for dynamic adaptation in ubiquitous computing. J-Jucs **20**(9), 1327–1351 (2014)
10. Wagner, C.: Juzzy - a java based toolkit for type-2 fuzzy logic. In: Proceedings of the 2013 IEEE Symposium on Advances in Type-2 Fuzzy Logic Systems, T2FUZZ 2013–2013 IEEE Symposium Series on Computational Intelligence, SSCI 2013, 1 April 2013, pp. 45–52 (2013)
11. Priya, N.H., Chandramathi, S.: QoS based optimal selection of web services using fuzzy logic. J. Emerg. Technol. Web Intell. **6**(3), 331–339 (2014)
12. Mamdani, E.H., Assilian, S.: An experiment in linguistic synthesis with a fuzzy logic controller. Int. J. Man-Mach. Stud.
13. Karnik, N.N., Mendel, J.M.: Centroid of a type-2 fuzzy set. Inf. Sci. **132**(1–4), 195–220 (2001)
14. Xu, Z., Yager, R.R.: Some geometric aggregation operators based on intuitionistic fuzzy sets. Int. J. Gen. Syst.

15. Al-Masri, E., Mahmoud, Q.H.: QoS-based discovery and ranking of web services. In: Proceedings - International Conference on Computer Communications and Networks, ICCCN, Honolulu, HI, USA, 2007, pp. 529–534 (2007)
16. Belouaar, H., Kazar, O., Kabachi, N.: A new model for web services selection based on fuzzy logic. Courrier du Savoir **1**(26), 393–400 (2018)
17. Rangarajan, S.: Qos-based web service discovery and selection using machine learning. EAI Endorsed Trans. Scalable Inf. Syst. **5**(17)
18. Suchithra, M., Ramakrishnan, M.: Non functional QoS criterion based web service ranking. In: Proceedings of the International Conference on Soft Computing Systems, ICSCS
19. Kumar, R.R., Mishra, S., Kumar, C.: Prioritizing the solution of cloud service selection using integrated MCDM methods under Fuzzy environment. J. Supercomput. **73**(11), 4652–4682 (2017). https://doi.org/10.1007/s11227-017-2039-1
20. Patiniotakis, I., Verginadis, Y., Mentzas, G.: PuLSaR: preference-based cloud service selection for cloud service brokers. J. Internet Serv. Appl. **6**(1), 1–14 (2015). https://doi.org/10.1186/s13174-015-0042-4
21. Gohar, P., Purohit, L.: Discovery and prioritization of web services based on fuzzy user preferences for QoS. In: IEEE International Conference on Computer Communication and Control (IC4) (2015)

An Analysis of Representative Works of Japanese Literature Based on Emotions and Topics

Miki Amano[1], Kenshin Tsumuraya[2], Minoru Uehara[2(✉)], and Yoshihiro Adachi[3]

[1] Department of Information Science and Arts, Toyo University, Saitama 3508585, Japan
s1B101901966@toyo.jp
[2] Graduate School of Information Sciences and Arts, Toyo University, Saitama 3508585, Japan
{s3B102200130,uehara}@toyo.jp
[3] RIIT Toyo University, Saitama 3508585, Japan
adachi@toyo.jp

Abstract. In literary research, subjective analysis by researchers and critics using artisanal analog methods is still the mainstream approach. In contrast, text-processing techniques that make full use of machine learning are already being actively applied to the analysis of Internet public opinion and various types of reviews. In this study, we analyzed the works of representative writers of modern Japanese literature in terms of emotions and topics. For emotion-based analysis, we combined a dictionary-based emotional analysis technique with the RoBERTa machine learning model. For topic-based analysis, we used a method for clustering distributed representations generated by Sentence-BERT. The method proposed in this study for analyzing the characteristics of literary works using machine learning techniques represents one of the most advanced and objective digital approaches to literary research.

1 Introduction

Although some studies of modern Japanese literature make use of digital techniques such as self-organizing maps [1] and fuzzy cluster analysis [2], most current literary studies adopt an analog approach based on researchers' and critics' subjective skills. Meanwhile, in the fields of natural language processing and AI, research on text analysis techniques using collective knowledge is progressing. For example, there is a text analysis technique that uses a machine learning model that has learned all pages of Japanese Wikipedia [3, 4] and a text analysis technique that uses an emotional word dictionary containing emotional words collected from a large number of review sentences and novels [5–7]. Research is also being conducted on how to classify and retrieve Japanese sentences based on topic using machine learning technology [8].

In this study, we propose an innovative digital approach to literary research based on collective intelligence that is not biased toward individual subjectivity. Our approach uses two novel analytical techniques to characterize each author and each literary work: one is a hybrid emotion-based analysis technique (abbr. HEAT) for Japanese texts that

L. Barolli (Ed.): AINA 2023, LNNS 654, pp. 99–112, 2023.
https://doi.org/10.1007/978-3-031-28451-9_9

combines results obtained using the Robustly Optimized BERT Pretraining Approach (RoBERTa) [9] and the Emotional Expression Analysis System (EEAS) [5]; the other is a topic-based analysis technique for Japanese texts (abbr. TBAT), which uses a distributed representation generated by a Japanese version of Sentence-BERT [10]. The contributions of our paper are as follows:

- The Japanese RobERTa model (abbr. JRoBERTa) was constructed by fine-tuning the Rinna RoBERTa model [11] using a teacher-labeled corpus to classify Japanese sentences into 11 emotional categories.
- By using HEAT with JRoBERTa, we analyzed the emotions of the works of representative writers of modern Japanese literature and clarified the characteristics of each writer and work based on these emotions.
- A Japanese Sentence-BERT (abbr. JSBERT) using the National Institute of Information and Communications Technology (abbr. NICT) BERT Japanese Pre-trained model [12] was fine-tuned using the Japanese translation of the Stanford Natural Language Inference (JSNLI) corpus [13].
- By using TBAT with JSBERT, we analyzed the topics of works by representative writers of modern Japanese literature and extracted the characteristics of each work based on the topics.

In Sect. 2, we describe the emotion- and topic-based analytical techniques used in this study. In Sect. 3, we explain the results of the emotion-based analysis of modern Japanese literary works. In Sect. 4, we describe the results of the topic-based analysis of modern Japanese literary works. Section 5 summarizes the study.

2 Analytical Techniques

In this study, HEAT was used for the emotional analysis of Japanese literary works, and TBAT was used for topic analysis. These techniques use a dictionary-based emotional analysis technique that we developed and the machine learning models JRoBERTa and Sentence-BERT. These are described in this section.

2.1 Heat

EEAS. EEAS is an emotional expression analysis system that uses an emotional word dictionary (abbr. EWD) and performs high-speed emotional analysis of Japanese sentences [5, 6]. The EWD contains approximately 7,800 emotional words labeled according to 11 emotional categories: joy, anger, sadness, fear, shame, like, dislike, excitement, peacefulness, surprise, and request. The EEAS can perform an emotional analysis of 10,000 Japanese sentences in approximately 1.3 s.

JRoBERTa. Bidirectional Encoder Representations from Transformers (abbr. BERT) is a language representation model with a deep bidirectional transformer structure. As of September 2018, it had achieved state-of-the-art results on 11 natural language processing tasks [14]. RoBERTa is a model that modifies the pretraining precedence of BERT and improves the performance of natural language processing [9]. Rinna RoBERTa [11] is a model pre-trained on Japanese Wikipedia [15] and Japanese CC-100 [16] for analyzing Japanese sentences.

JRoBERTa is a Japanese RoBERTa for the classification of 11 emotions that has been fine-tuned from Rinna RoBERTa using a corpus of about 13,000 sentences with 11 emotional category teacher labels. When evaluated on the same test data extracted from the corpus, JRoBERTa outperforms EEAS in emotion classification accuracy for emotional categories other than sadness, fear, and shame, which have very few teacher-labeled training sentences.

HEAT. HEAT is an emotional analysis technique that uses the results of analysis by either JRoBERTa or EEAS, whichever is more accurate, for each of the 11 emotional categories. Table 1 shows HEAT's accuracy of analysis for 11 emotional categories.

Table 1. HEAT's accuracy of analysis for 11 emotional categories.

Emotion category	Technique	Precision	Recall	F1-score
Joy	JRoBERTa	0.819	0.785	0.801
Anger	JRoBERTa	0.833	0.333	0.476
Sadness	EEAS	0.230	0.898	0.367
Fear	EEAS	0.365	0.836	0.508
Shame	EEAS	0.429	1.000	0.600
Like	JRoBERTa	0.714	0.753	0.733
Dislike	JRoBERTa	0.726	0.737	0.732
Excitement	JRoBERTa	0.333	0.583	0.424
Peacefulness	JRoBERTa	0.900	0.600	0.720
Surprise	JRoBERTa	0.696	0.667	0.681
Request	JRoBERTa	0.715	0.823	0.765
Average		0.615	0.729	0.619

2.2 TBAT

Procedure. In a previous study [8], we built JSBERT, taking Tohoku Japanese BERT [17] as a pre-trained model and fine-tuning it using the JSNLI corpus. In this study, we used the NICT BERT Japanese Pre-trained model [12] as a pre-trained model with a mode that does not split words by "wordpiece" so that we could label the resulting clusters with appropriate words. The processing steps performed by TBAT are as follows: first, JSBERT generates a distributed representation for each sentence in the target sentence data set. Next, this distributed representation set is clustered using the hierarchical clustering Ward method, and the target sentence set is also clustered correspondingly. Cluster labels are then automatically attached to each cluster in the clustering results.

Analysis Example. The Test2385 corpus is a dataset of 2,385 sentences with 12 teacher labels that we collected from online news and reviews for this study. This corpus comprises 192 sentences labeled "IT (information technology)," 194 sentences labeled "games," 223 sentences labeled "sports," 187 sentences labeled "illness," 391 sentences labeled "study," 180 sentences labeled "animals," 158 sentences labeled "weather," 135 sentences labeled "novels," 171 sentences labeled "politics," 229 sentences labeled "cooking," 118 sentences labeled "disasters," and 207 sentences labeled "clothes."

The accuracy of topic-based clustering using distributed representations generated by JSBERT from the Test2385 corpus is shown in Table 2. In this study, each cluster in the clustering result is labeled with the teacher label given most often to the sentences that make up the cluster. In Table 2, the JSBERT-normal column indicates the accuracy of analysis when using the distributed representation generated from all the words composing the input sentence, and the JSBERT-nouns column indicates the accuracy of analysis when using the distributed representation generated only from the noun words composing the input sentence. Table 2 shows that topic-based clustering of sentences can be performed with good accuracy by TBAT.

Table 2. Accuracy of topic-based clustering for Test2385.

	JSBERT-normal			JSBERT-nouns		
	Precision	Recall	F1-score	Precision	Recall	F1-score
IT	0.889	0.922	0.905	0.903	0.870	0.886
Games	0.955	0.758	0.845	0.955	0.773	0.855
Sports	0.876	0.883	0.879	0.734	0.915	0.814
Illness	0.942	0.861	0.899	0.926	0.807	0.863
Study	0.886	0.936	0.910	0.896	0.944	0.919
Animals	0.944	0.944	0.944	0.871	0.978	0.921
Weather	0.906	0.728	0.807	0.813	0.962	0.881
Novels	0.908	0.881	0.895	0.923	0.889	0.906
Politics	0.981	0.895	0.936	0.981	0.930	0.955
Cooking	0.972	0.926	0.949	0.960	0.952	0.956
Disasters	0.667	0.949	0.783	0.932	0.814	0.869
Clothes	0.794	0.932	0.858	0.994	0.860	0.922
Average	0.893	0.885	0.884	0.907	0.891	0.896

Cluster Labeling. For the cluster labels of each cluster, the list of words (cluster-label candidates) appearing in the cluster is sorted by a label-word evaluation value, defined as $\alpha \times \text{n}\mathit{TfIdf} + (1 - \alpha) \times \text{n}\mathit{Cosim}$ ($0 \leq \alpha \leq 1$), and the top n words of the sorted cluster-label candidates are adopted. Here, $\text{n}\mathit{TfIdf}$ is the cluster-based TFIDF value normalized between 0 and 1, and $\text{n}\mathit{Cosim}$ represents the cosine similarity value between the distributed representation of the word appearing in the cluster and the cluster centroid normalized between 0 and 1. We show part of the cluster corresponding to "clothes" in Table 3 and the labels automatically attached to that cluster in Table 4. When $\alpha = 0.5$ for all 12 clusters, including the "clothes" cluster in Table 4, the appropriate words for each cluster label were included in the top 10 words of the sorted cluster-label candidates.

Table 3. Part of the cluster corresponding to "clothes."

Sentence	English translation
ティアードワンピースは、スカート部分がギャザーやフリルで段々になっている服の表現です。	A tiered dress is a piece of clothing in which the skirt is tiered with gathers and frills
衣服をリサイクルするメリットは、部屋が整理されるだけではなく、ゆくゆくは環境保護につながることもあります。	The benefit of recycling clothes not only helps keep your room tidy but can ultimately help protect the environment
欲しかった服が買えて嬉しいです。	I am happy that I could buy the clothes I wanted
今のトレンドは白ブラウスと黒タイトスカートです。	The current trend is white blouse and black tight skirt
靴の手入れの基本はまず汚れや水分をきれいな布などで拭き取る事で、靴が濡れている場合はシューズキーパーや新聞紙などで形を整え、湿気の少ない場所で乾かします。	The basics of shoe care are to first wipe off dirt and moisture with a clean cloth, etc., and if the shoes are wet, shape them with a shoe keeper or newspaper, and dry them in a dry place
今年は肩を出したトップレスがトレンドだ。	Tops with exposed shoulders are trending this year
厚底ブーツを履く人が増えている。	More and more people are wearing thick-soled boots
戦後のファッション界の先駆けとして、国際的な評価を得て活躍したファッションリーダーの森英恵さんが、東京都内の自宅で死去していたことが発表された。	It was announced that Hanae Mori, an internationally acclaimed fashion designer who was active as a pioneer in the post-wat fashion world, had died at her home in Tokyo

Table 4. Top-10 cluster labels automatically attached to the "clothes" cluster according to the label-word evaluation value when α = 0.5.

Sorted by n*Cosim*	n*Cosim*	Sorted by n*TFIDF*	n*TFIDF*	α = 0.5
clothes[服]	1.00000	suit[スーツ]	1.00000	material[素材]
inner[インナー]	0.92488	material[素材]	0.75000	coordination[コーディネート]
material[素材]	0.91043	shirt[シャツ]	0.66667	fashion[ファッション]
wearing[着]	0.90665	jacket[ジャケット]	0.58333	tops[トップス]
clothes[服装]	0.90342	sweater[セーター]	0.50000	clothes[服装]
coordination[コーディネート]	0.90332	coordination[コーディネート]	0.50000	item[アイテム]
item[アイテム]	0.90313	tops[トップス]	0.50000	trend[トレンド]
clothing[衣類]	0.87842	dress[ワンピース]	0.50000	silhouette[シルエット]
fashion[ファッション]	0.85108	fashion[ファッション]	0.50000	clothing[衣類]
tops[トップス]	0.84679	skirt[スカート]	0.50000	material[生地]

3 Analysis of Literary Works Based on Emotions

In this section, we describe the results of the emotional analysis of modern Japanese literary works.

3.1 Target Literary Works

The subjects of emotional analysis were 8 novels and 6 essays by Haruki Murakami [村上春樹], 4 novels by Jun Ikeido [池井戸潤], 4 novels by Keigo Higashino [東野圭吾], 3 novels and 2 essays by Naoki Matayoshi [又吉直樹], and 5 novels by Kanae Minato [湊かなえ], for a total of 32 works. Table 5 provides information about some of the works that were the subjects of emotion analysis in this study.

3.2 Emotional Analysis Results

Using HEAT, an emotion analysis of the above 32 works was performed. The appearance rate of emotional words in a literary work is defined as a percentage value obtained by dividing the number of emotional words appearing in the work by the total number of words in the work. Table 6 shows the average rates of appearance of emotional words in the works analyzed for each author.

Table 5. Some of the titles of each writer's works used for the emotion analysis.

Writer	Type	Year	Title
Haruki Murakami	Novel	1987	*Norwegian Wood* [ノルウェーの森]
Haruki Murakami	Novel	2002	*Kafka on the Shore* [海辺のカフカ]
Haruki Murakami	Novel	2009–2010	*1Q84*
Haruki Murakami	Essay	2007	*What I Talk About When I Talk About Running* [走ることについて語るときに僕の語ること]
Jun Ikeido	Novel	2012	*The Lost Generation Strikes Back* [ロスジェネの逆襲]
Jun Ikeido	Novel	2014	*Icarus-Flying on Silver Wings* [銀翼のイカロス]
Keigo Higashino	Novel	2006	*The Devotion of Suspect X* [容疑者Xの献身]
Naoki Matayoshi	Novel	2015	*Spark* [花火]
Naoki Matayoshi	Novel	2017	*Theater* [劇場]
Naoki Matayoshi	Essay	2016	*Overcome the Night* [夜を乗り越える]
Kanae Minato	Novel	2008	*Confessions* [告白]
Kanae Minato	Novel	2009	*Expiation* [贖罪]
Kanae Minato	Novel	2012	*The Snow White Murder Case* [白ゆき姫殺人事件]

Table 6. Average rates of appearance of emotional words in each author's works.

Writer	Haruki Murakami		Jun Ikeido	Keigo Higashino	Naoki Matayoshi		Kanae Minato
Type	Novel	Essay	Novel	Novel	Novel	Essay	Novel
Rate	8	13	11	8	18	18	14

The rate of appearance of emotional words is fairly constant for each author, with little difference between works by each author. As shown in Table 5, the works of Naoki Matayoshi and Kanae Minato have high rates of appearance of emotional words, and the rate of appearance of emotional words appearing in novels and essays by Haruki Murakami is different.

Table 7 shows the rates of appearance of emotional categories in works by Haruki Murakami and by Naoki Matayoshi. The rate of appearance of an emotional category in a work is the percentage of the number of emotion words belonging to the emotional category divided by the total number of emotional words in the work.

Table 7. Rates of appearance of emotional categories in works by Haruki Murakami (a) and Naoki Matayoshi (b).

	(a) Haruki Murakami			
category	Norwegian Wood	Kafka on the Shore	1Q84	What I Talk About When I Talk About Running
joy	6%	5%	4%	10%
anger	1%	1%	1%	1%
sadness	33%	29%	28%	24%
fear	8%	18%	20%	11%
shame	2%	2%	1%	4%
like	13%	12%	9%	10%
dislike	21%	19%	21%	31%
excitement	4%	2%	2%	3%
peacefulness	1%	2%	1%	2%
surprise	4%	2%	3%	1%
request	7%	8%	9%	3%

	(b) Naoki Matayoshi		
category	Spark	Theater	Overcome the Night
joy	7%	4%	7%
anger	1%	1%	1%
sadness	25%	26%	15%
fear	20%	23%	12%
shame	6%	8%	4%
like	6%	3%	25%
dislike	21%	24%	22%
excitement	5%	5%	3%
peacefulness	1%	1%	0%
surprise	1%	3%	3%
request	5%	2%	7%

Figure 1 shows the results of visualizing the rates of appearance of the emotional categories in Table 7 using an emotional radar chart (ERC) [5]. An ERC is a chart in which each emotional category is arranged on the axis of a regular polygon so that the characteristics of the rate of appearance (or appearance frequency) of each emotional category in the entire novel can be understood at a glance. Positive emotions such as "joy," "like," and "peacefulness" are placed on the upper side of the ERC; negative emotions such as "anger," "sadness," "fear," "shame," and "dislike" are placed on the lower side; and neutral emotions such as "excitement" and "surprise" are placed in-between. Opposite emotional categories, such as "like" and "dislike," "joy" and "sadness," and "peacefulness" and "fear," are placed on opposite sides of the ERC axis.

From Table 7 and Fig. 1, it can be seen that the rates of appearance of emotional categories have similar characteristics in novels by the same author and that the characteristics differ from author to author. In addition, it can be seen that novels and essays have different characteristics in terms of the rates of appearance of emotional categories.

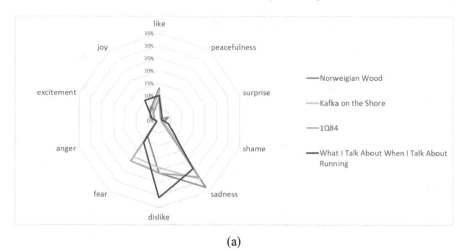

(a)

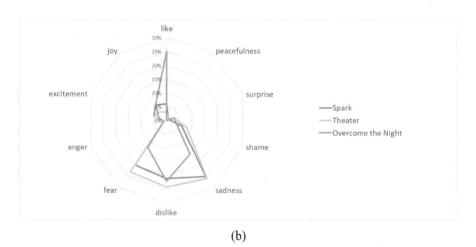

(b)

Fig. 1. ERCs of works by Haruki Murakami (a) and Naoki Matayoshi (b).

Because we found that the rates of appearance of emotional categories have different characteristics for each author, we created Table 8, showing the average rates of appearance of emotional categories in the works of each author, and visualized the ERC based on Table 8 shown in Fig. 2.

Table 8. Average rates of appearance of emotional categories in works by the five writers.

	Haruki Murakami		Jun Ikeido	Keigo Higashino	Naoki Matayoshi		Kanae Minato
	Novel	Essay	Novel	Novel	Novel	Essay	Novel
Joy	5%	10%	2%	4%	5%	7%	6%
Anger	1%	1%	5%	2%	1%	1%	2%
Sadness	27%	24%	29%	23%	25%	16%	26%
Fear	17%	9%	17%	21%	22%	17%	16%
Shame	1%	3%	4%	5%	8%	7%	4%
Like	13%	17%	2%	3%	4%	19%	7%
Dislike	20%	25%	19%	21%	23%	22%	18%
Excitement	2%	2%	4%	2%	4%	3%	7%
Peacefulness	1%	1%	1%	1%	1%	1%	1%
Surprise	3%	3%	5%	5%	2%	3%	5%
Request	8%	5%	13%	14%	4%	5%	9%

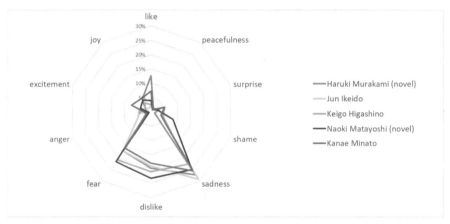

Fig. 2. ERC of the average rates of appearance of emotional categories in works by the five writers.

Table 8 and Fig. 2 show that the "like" emotional category appears more frequently in the works of Haruki Murakami than in those of other authors, the "excitement" emotional category occurs more frequently in Kanae Minato's works, and the "shame" emotional category occur more frequently in Naoki Matayoshi's works. Table 8 also shows that the "like" emotional category appears frequently in Naoki Matayoshi's essays. In addition, Table 8 shows that the works of Jun Ikeido and Keigo Higashino have a high rate of appearance of the "request" emotion category.

An emotional sequence diagram (ESD) is a diagram that visualizes how emotional expressions change over the course of literary works. In an ESD, the horizontal axis is the time or the order of appearance of each sentence in the work, and the vertical axis is the frequency of appearance of each emotion. Each emotional category sequence in the ESD corresponds to 11 emotional categories (joy, like, peacefulness, surprise, excitement, request, shame, sadness, anger, fear, and dislike) in order from the top. That is, the upper series represents positive emotions, and the lower series represents negative emotions. An ESD makes it possible to visualize changes in emotions in a novel in an easy-to-understand manner. Figure 3 shows the ESD of the middle section of Haruki Murakami's work "*1Q84*." Around the middle of this figure, the frequency of appearance of the "excitement" and "sadness" emotional categories is high. This part is an important point in the story because it is the part where "the main character's emotions are greatly shaken by the death of a close friend."

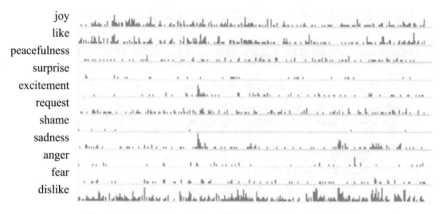

Fig. 3. ESD for the middle of Haruki Murakami's "*1Q84*."

4 Analysis of Literary Works Based on Topics

Topic analysis of the above 32 works was performed using TBAT. Table 9 shows the results of a topic analysis using TBAT of Haruki Murakami's essay, "*What I Talk About When I Talk About Running*." This table shows some of the cluster information generated by clustering into 20 clusters using a distributed representation of the sentences in the essay, along with the corresponding cluster labels.

In Table 9, the "rank" column is the rank of each cluster when sorted in descending order by the number of sentences contained in each cluster. The "sentence rate" column is the percentage of the number of sentences belonging to each cluster divided by the total number of sentences (2,994). As shown in Table 9, topic-based analysis using TBAT enables a detailed analysis of the topical features of each work.

Table 10 summarizes the characteristics of the topics common to the works of each writer.

Table 9. Some clustering results and corresponding cluster labels.

Rank	Top five cluster labels	Number of sentences	Sentence rate
1	race [レース], distance [距離], pace [ペース], runner [ランナー], time [タイム]	320	11%
2	marathon [マラソン], kilo [キロ], runner [ランナー], race [レース], distance [距離]	264	9%
5	novel [小説], writer [作家], the work [作品], literature [文学], sentence [文章]	226	8%
6	myself [自分], human [人間], consciousness [意識], personality [性格], body [肉体]	192	6%
10	the sun [太陽], season [季節], temperature [気温], weather [天候], feeling [気持ち]	129	4%
11	life [人生], age [年齢], human [人間], life [生活], myself [自分]	108	4%
13	hydration [水分], water supply [給水], supply [補給], dehydration [脱水], training [練習]	103	3%
16	crawl [クロール], swimming [水泳], form [フォーム], race [レース], training [練習]	83	3%
17	triathlon [トライアスロン], bicycle [自転車], race [レース], competition [競技], training [練習]	80	3%
19	music [音楽], record [レコード], rhythm [リズム], playing [演奏], jazz [ジャズ]	69	2%

Table 10. The characteristics of topics common to the works of each writer.

Writer	Type	Frequent topics	Comment
Haruki Murakami	novel	she [彼女], woman [女性], sex [性]	All the novels analyzed have clusters with three labels in the top ten. All the works have clusters with 'room' label in the top ten
Haruki Murakami	essay	novel [小説]	All the novels analyzed have clusters with this label in the top ten
Jun Ikeido	novel	company [会社], bank [銀行]	All the works analyzed contain 'company' with in the top 10 clusters. 'Bank' is the top cluster label in the three novels

(*continued*)

Table 10. (*continued*)

Writer	Type	Frequent topics	Comment
Keigo Higashino	novel	Topics related to characters and behavior [登場人物の様子や行動に関する話題]	Keigo Higashino's works emphasize the visible parts, such as the behavior and appearance of the characters
Naoki Matayoshi	novel, essay	entertainment [芸能], room [部屋]	Haruki Murakami has different topics in his novels and essays, but Naoki Matayoshi has many topics in common. Four of five works contain 'company' within the top 10 clusters
Kanae Minato	novel	myself [自分], idea [考え], feeling [気持ち]	There were clusters with 'myself' as the top label for 4 of the 5 works

5   Conclusion

In this paper, we proposed techniques for literary analysis based on emotions and topics using machine learning. Then, using these techniques, we reported the results of our analysis of representative works of modern Japanese literature. The analysis of literary works based on emotion using HEAT revealed that each writer had significantly different characteristics in terms of emotional expression. In the topic-based analysis of literary works using TBAT, we were able to analyze each work's characteristics in terms of the topics it addresses in detail.

As a future task, first, we need to improve the accuracy of emotion analysis and topic analysis processing. We wish to improve the accuracy of EEAS's emotion analysis by adding emotion words and reviewing the assigned emotional categories. We also wish to collect a large corpus with teacher labels for emotional classification and use it to train the JRoBERTa emotion classification model to improve its accuracy. In addition, we wish to improve the accuracy of topic-based clustering by increasing the training data of Sentence-BERT, for instance, by adding MultiNLI [18] translated into Japanese. It would also be invaluable to have literary researchers and critics examine the analytical results obtained by the technique of this paper, evaluate the validity of the results, and give us feedback on how to improve the technique.

The emotion- and topic-based literary analysis techniques proposed in this paper can be applied not only to Japanese but also to other languages. These techniques can provide objective information about literary works. We believe that this information will be useful to literary scholars and critics.

References

1. Yoshida, T., Kobayashi, I.: An approach to a bird's-eye analysis on novels based on emotional expressions. DEIM Forum, A8–6 (2011). (in Japanese)

2. Kikuchi, Y., Kato, C., Maeshiro, H.: An exploration of the fiction of Murakami Haruki using fuzzy cluster analysis. J. Visual. Soc. Japan **31**(Suppl. 1), 141–146 (2011). (in Japanese)
3. Adachi, Y., Negishi, T.: Development and evaluation of a real-time analysis method for free-description questionnaire responses. In: Proceedings of the 15th IEEE International Conference on Computer Science and Education (2020)
4. Tsumuraya, K., Takahashi, H., Adachi, Y.: Emotion analysis and topic analysis of Japanese sentences by BERT. In: The 84th National Convention of IPSJ (2022). (in Japanese)
5. Adachi, Y., Kondo, T., Kobayashi, T., Etani, N., Ishii, K.: Emotion analysis of Japanese sentences using an emotion-word dictionary. J. Visual. Soc. Japan **41**(161), 21–27 (2021). (in Japanese)
6. Seyama, Y., Astremo, A., Adachi, Y.: Extraction of "request" intentions for review analysis on social media and e-commerce sites. In: FIT2021 (2021). (in Japanese)
7. Yonghui, H., Tsumuraya, K., Uehara, M., Adachi Y.: High-speed emotional analysis of free-descriptive sentences using an emotional-word dictionary. In: IEEE ICCSE2022 (2022)
8. Tsumuraya, K., Amano, M., Uehara, M., Adachi Y.: Topic-Based Clustering of Japanese Sentences Using Sentence-BERT. CANDAR2022 (2022)
9. Liu, Y., et al.: RoBERTa: A robustly optimized BERT pretraining approach. arXiv:1907. 11692 (2019)
10. Reimers, N., Gurevych, I.: Sentence-BERT: sentence embeddings using siamese bert-networks. In: Proceedings of the 2019 Conference on Empirical Methods in Natural Language Processing, Association for Computational Linguistics, (2019). https://arxiv.org/abs/1908.10084
11. rinna Co., Ltd.: japanese-roberta-base (2022). https://huggingface.co/rinna/japanese-roberta-base. Accessed 29 oct 2022
12. National Institute of Information and Communications Technology. NICT BERT Japanese Pre-trained models (2022). https://alaginrc.nict.go.jp/nict-bert/index.html. Accessed 29 oct 2022
13. Yoshikoshi, T., Kawahara, D., Kurohashi, S.: Multilingualization of a Natural Language Inference Dataset Using Machine Translation, SIG Technical Reports, vol. 2020-NL-244 No.6 (2020). (in Japanese)
14. Devlin, J., Chang M. W., Lee, K., Toutanova, K.: BERT: Pre-training of Deep Bidirectional Transformers for Language Understanding (2018). arXiv:1810.04805
15. Japanese Wikipedia (2022). https://dumps.wikimedia.org/jawiki/. Accessed 29 oct 2022
16. CC-100: Monolingual Datasets from Web Crawl Data (2022). https://data.statmt.org/cc-100/. Accessed 29 oct 2022
17. Tohoku NLP Group.: Pretrained Japanese BERT models (2022). https://github.com/cl-tohoku/bert-japanese.31 oct 2022
18. Williams, A., Nangia, N., Bowman, S.: A broad-coverage challenge corpus for sentence understanding through inference. In Proceedings of the 2018 Conference of the North American Chapter of the Association for Computational Linguistics: Human Language Technologies, vol. 1, 1112–1122 (Long Papers) (2018)

Analysis and Comparison of Machine Learning Models for Glucose Forecasting

Théodore Simon$^{(\boxtimes)}$, Jianfei Zhang, and Shengrui Wang

Université de Sherbrooke, Sherbrooke, Canada
{theodore.simon,jianfei.zhang,shengrui.wang}@usherbrooke.ca

Abstract. Continuous blood glucose monitoring (CGM) is a central aspect of the modern study of diabetes. It is also a way of improving the quality of life of patients. To make appropriate decisions for patients with diabetes, it needs an effective tool to monitor these levels in order regarding insulin administration and food intake to keep blood glucose levels within the range target. Efficient and accurate prediction of future blood sugar levels repeatedly benefits the diabetic patient by helping them to reduce the risk of blood sugar level extremes, including hypoglycemia and hyperglycemia. In this study, we implemented several time-series models, including statistical and machine-learning-based models, using two direct and recursive strategies, to forecast glucose levels in patients. We applied these models to data collected from 171 patients in a clinical study. For the 30-min prediction horizon, the average of mean absolute percentage errors (MAPEs) and root mean squared errors (RMSEs) for each model respectively shows that ARIMA, XGBoost, and TCN can yield more accurate forecasts. We also highlight the difference between statistical and machine-learning-based models, where statistical models perform effectively in predicting CGM levels, although they cannot perceive changes in variation, like neural-network-based models.

1 Introduction

Diabetes mellitus (DM) [1] is a metabolic disease, involving inappropriately elevated blood glucose levels. DM has several categories, including type 1, type 2, maturity-onset diabetes of the young (MODY), gestational diabetes, neonatal diabetes, and secondary causes due to endocrinopathies, steroid use, etc. The main subtypes of DM are Type 1 diabetes mellitus (T1DM) and Type 2 diabetes mellitus (T2DM), which classically result from defective insulin secretion (T1DM) and/or action (T2DM) [6]. Insulin is an endocrine hormone released from the pancreas that facilitates uptake of glucose by a variety of cells. In the long term, hyperglycemia can lead to the development of certain complications, particularly in the eyes, kidneys, nerves, heart, and blood vessels. Diabetes is a chronic disease that cannot be cured, but can be controlled. In 2021, there have been more than 530 million adults living with diabetes around the world – e.g., 51 million in North America and 61 million in Europe; this number is projected to reach 640 million by 2030 and 780 million by 2045 [9]. In addition, 541 million people are estimated to have impaired glucose tolerance in 2021. It is also estimated that over 6.7 million deaths aged 20–79 died from diabetes-related causes in 2021 (see IDF Diabetes Atlas 10th edition 2021. https://diabetesatlas.org/en).

Of the diabetes cases, about 5–10% are type 1 (which is formerly called insulin-dependent diabetes or juvenile-onset diabetes) – *i.e.*, an autoimmune condition requiring external administration of insulin for the regulation of blood glucose [7]. Although type-1 diabetes is hard to be cured or prevented, we can manage blood glucose through subcutaneous administration of insulin either by injection or continuous infusion. That's why type-1 diabetes patients must monitor their BG levels throughout the day and take necessary actions to prevent hypoglycemia (low BG levels) and hyperglycemia (high BG levels). A patient needs to measure his BG concentration several times throughout the day and night using the finger-stick test or a sensor connected to a smartphone. Accordingly, in recent years, there has been increased use of flash glucose monitoring and continuous glucose monitoring (CGM) for recording an individual's BG levels continuously (e.g., 288 measurements per day at 5-min intervals). Using these recorded historical CGM to accurately forecast future BG levels would improve diabetes treatment by enabling proactive treatment and help patients prevent both low and high blood sugar levels and allow them to meet glycemic targets and decrease the risk of long-term complications.

Forecasting the future glucose levels based on the past observed data is a typical time-series prediction problem – i.e., predicting the continuation of these series. These predictions in advance (e.g., 30-min-ahead forecasts) can make glucose level management possible and aid decisions accordingly [5]. Hence, we focus on statistical and machine learning models that have been well-established and extensively used for time-series forecasting, such as autoregressive integrated moving average (ARIMA), extreme gradient boosting (XGBoost), Prophet, long short-term memory networks (LSTM) and temporal convolutional network (TCN). These methods have shown very promising results in the context of time-series forecasting. Those using deep-learning take longer to train but are more likely to detect variations while linear models are faster but less accurate. How well these models perform glucose forecasting is the concern of this work.

In this paper, we propose a comparative empirical evaluation of different statistical and ML-based time series forecasting models with different forecasting strategies. To validate these techniques, we applied them to two different CGM datasets regarding the blood glucose levels of Type-1 patients. We designed the particular CGM data preprocessing method that transforms the irregular data into time series. In the evaluation part, we made various 5, 10, 15, 20, 25, and 30-min-ahead predictions of the blood glucose for the next 24 h. These predictions were made based on different historical data, via different forecasting strategies. We investigated the models' settings in every forecasting task – i.e., which of past glucose observations would be more useful for the forecast and at what time the glucose level can be accurately predicted by various machine learning models. Overall, we can summarize the contributions of this work as follows:

- Analysis and comparison of the performance of statistical and ML-based models with various forecasting strategies;
- Investigating various historical data to be used for optimizing the predictions;
- Analysis on a dataset from a clinical study of 171 patients with diabetes.

2 Methodology

Given a univariate time series $y_{t-K+1}, \ldots, y_t$ comprising K observed glucose levels, our **forecasting task** is to predict the next H glucose levels $\hat{y}_{t+1}, \hat{y}_{t+2}, \ldots, \hat{y}_{t+H}$, at time $t+1, t+2, \ldots, t+H$. These glucose levels can be inferred by the forecasting model f with parameter Θ, given the K observed glucose levels, *i.e.*,

$$\hat{y}_{t+1}, \hat{y}_{t+2}, \ldots, \hat{y}_{t+H} = f(y_{t-K}, \ldots, y_{t-1}; \Theta). \tag{1}$$

In what follows, we will present our forecasting strategies, models used for performing forecasting, and evaluation method.

2.1 Forecasting Strategies

To implement the forecasting task shown in Eq. 1, we used two different forecasting approaches: direct and recursive strategies.

- To predict the next H glucose levels, with the direct strategy, suppose we have H trained models, say $f_1, f_2, \ldots, f_H$, each for a particular forecasting horizon h. Given the time series $y_{t-K}, \ldots, y_{t-1}$, we have

$$\hat{y}_{t+1} = f_1(y_{t-K}, \ldots, y_{t-1}; \theta_1)$$
$$\hat{y}_{t+2} = f_2(y_{t-K}, \ldots, y_{t-1}; \theta_2)$$
$$\vdots$$
$$\hat{y}_{t+H} = f_H(y_{t-K}, \ldots, y_{t-1}; \theta_H)$$

where the optimal parameter $\theta_h (h = 1, 2, \ldots, H)$ can be estimated as follows

$$\theta_h^* = \underset{\theta_h}{\arg\min} \sum_t [y_{t+h} - f_h(y_{t-K}, \ldots, y_{t-1}; \theta_h)]^2.$$

- For recursive forecasting, the glucose level at $t+1$, $\hat{y}_{t+1}$, is predicted by the trained model f based the previous K values, $y_{t-K+1}, \ldots, y_t$, as follows:

$$\hat{y}_{t+1} = f(y_{t-K+1}, \ldots, y_t; \theta_1),$$

Then, y_{t+1} will be replaced by $\hat{y}_{t+1}$ for predicting the next time $\hat{y}_{t+2}$ via the model f, as follow:

$$\hat{y}_{t+2} = f(y_{t-K+2}, \ldots, \hat{y}_{t+1}; \theta_2).$$

By applying this rolling strategy consecutively, we can obtain the glucose level at $t+H$:

$$\hat{y}_{t+H} = f(\hat{y}_{t-K+H}, \ldots, \hat{y}_{t+H-1}; \theta_H).$$

An advantage of using the recursive strategy is that only one model is required. saving significant computational time, especially when a large number of time series and forecast horizons are involved. The strategy also ensures that the fitted model matches the assumed data generating process as closely as possible. On the other hand, the recursive forecasts are not equal to the conditional mean, even when the model is exactly equivalent to the data generating process.

2.2 Forecasting Models

In this section, we will review five commonly used time-series forecasting models, including ARIMA, XGBoost, LSTM, TCN, and Kalman Filter. They will perform glucose forecasting in this study.

- The autoregressive moving average (ARMA) can only be used for stationary time series data. In practice, however, many time series may show non-stationary behavior. To address this, the autoregressive integrated moving average (ARIMA) is proposed [3], which is a generalization of an ARMA model to include the case of non-stationarity as well. In ARIMA models, a non-stationary time series is made stationary by applying finite differencing of the data points.
- Extreme Gradient boosting (XGBoost) [4] merges hundreds of low-accuracy prediction models into a single high-accuracy model through a quantifiable boosting algorithm. These models must frequently be integrated to obtain good prediction accuracy under tolerable parameter values. It may need to be iterated or repeated multiple times or more to attain sufficient accuracy if the data collection is vast or complicated.
- As a typical recurrent neural network (RNN), the long short-term memory (LSTM) network [8] has received considerable attention since it can handle long-term dependencies well and effectively outperform most previously proposed time series forecasting approaches.
- In order to address the memorizing and time-consuming issues, temporal Convolution Neural Networks (TCNs) [2] over LSTMs for many time series prediction tasks that involve long-term memories. Temporal convolutional networks (TCN) were introduced for video-action segmentation tasks to overcome the issues faced by the RNN-based approaches. TCN can be seen as the combination of 1D convolutional neural network with dilated and causal convolutions. TCN has dilated convolutional network architecture with two hidden layers.
- Prophet [10] is a procedure for forecasting time series data based on an additive model where non-linear trends are fit with yearly, weekly, and daily seasonality, plus holiday effects. It works best with time series that have strong seasonal effects and several seasons of historical data. Prophet is robust to missing data and shifts in the trend, and typically handles outliers well. In our case, the data have already been pre-processed and are not subject to seasonal events.

2.3 Walk-Forward Evaluation

We employ the expanding walk-forward evaluation [12] to evaluate the validity and reliability of the models. For each patient, we forecast the CGM levels 48 times (once every 30 min) through the period of 24 h on day-3. For each l-ahead forecast made for time t, we use all known CGM readings as of $t - l$ for training the model. For example, the 10-min-ahead forecasting model used for predicting the CGM levels at 18:00:00 will be trained based on all known CGM readings as of 17:50:00. To measure the performances of the models, we use the mean absolute percentage error (MAPE) and the root mean squared error (RMSE) for each lead time across all forecasting times. Let $\hat{y}_t$ and y_t be the forecasted and true CGM level at time t, respectively. Given T timestamps, the evaluation metrics can be computed as follows:

$$\text{MAPE} = \frac{1}{T} \sum_{t=1}^{T} \left| \frac{y_t - \hat{y}_t}{y_t} \right| \times 100\%. \tag{2}$$

$$\text{RMSE} = \sqrt{\frac{1}{T} \sum_{t=1}^{T} (y_t - \hat{y}_t)^2}. \tag{3}$$

These two measures are commonly used when assessing the performance of a technique on time series forecasting. Their advantage is that the average forecast error of a model is expressed in the same unit of the variable to be predicted. MAPE represents the average of the absolute percentage of each entry in a dataset, showing, on average, how accurate the forecasted quantities were in comparison with the actual quantities. MAPE is one of the most widely used measures of forecast accuracy, due to its advantages of scale-independent and interpretability. RMSE takes a value in the range $[0, 1]$; the lower the better for our forecasts. RMSE assigns high penalties to large errors since the prediction errors are squared; this is useful when we abstain from large forecasting errors.

3 Experiments

3.1 Data and Pre-processing

The original CGM data have been provided by the International Diabetes Closed Loop (iDCL) trial of JAEB CENTER FOR HEALTH RESEARCH (2018). Eligible participants started a run-in phase of 2 to 8 weeks that have been customized based on whether the participant is already a pump for CGM users or not. Study participants have been recruited from 7 clinical centers in the United States without regard to gender, race, or ethnicity. At least 50 participants with HbA1c above 7.5% and 50 participants with HbA1c below 7.5% (people with an HbA1c level above 6.5% are considered diabetic). At least 50 participants aged between 14 and 25, and 50 participants aged 26 and older.

The glucose level for each patient has been read approximately every 5 min for at least two weeks. However, unexact recording times make the timescale of glucose readings unaligned. We particularly amend the recording times – *e.g.*, round off 14:29:43 to 14:30:00 and 14:36:14 to 14:35:00. Additionally, there are a fair number of missing records due to technical errors (e.g., failure of wearable devices). We fill in missing values via the regression method [11]. Figure 1 (left) shows an example of a patient's CGM readings with a bunch of missing values; Fig. 1 (right) presents the CGM data preprocessed by data imputation.

3.2   Results

3.2.1   Performance

Table 1 shows the performance, in terms of MAPE (top of the cell) and RMSE (bottom of the cell), of various models. Each MAPE and RMSE result is the average error of the (171 patients × 48 forecasts/patient) respectively made 5, 10, 15, 20, 25, and 30 mins in advance across the next 24 h (N.B.: we make a forecast every 30 min for each patient for the next 24 h). For each model (exclude Prohet), we have various settings about the lookback steps (*i.e.*, how many previous CGM readings fed into the model) and predicting steps (*i.e.*, how many CGM values predicted for the future in each loop). It can be seen from the figure that TCN model yields accurate 5-min-ahead forecasts, which are better than 30-min-ahead ones by LSTM with 2 pred-steps. The results that we obtain in Table 1 show that when we compare the MAPE scores between the prediction at 5 min and that at 30 min with a look back of 4 and a predstep of 1, we have a deterioration of the MAPE of almost 10%. The results show that LSTM model can be improved when using 6 past values to predict the next 2 values in a loop and TCN using the 4 past values to forecast the next 2 values. Additionally, forecasting the next 2 values once can achieve a 2% improvement, compared to producing one value for the future. We notice that the TCN model is 2% more precise on all prediction levels. The best models for near predictions are linear models such as ARIMA and

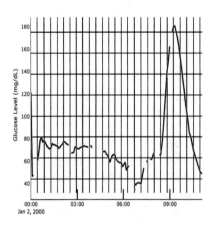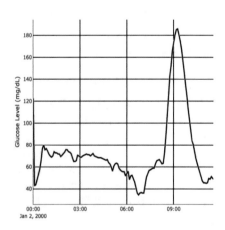

Fig. 1. Raw CGM readings (left) and CGM data (right) preprocessed by data imputation, for a 60-year-old woman.

Table 1. MAPE of the forecasts for the next 24 h

Model	Lookback	PredSteps	MAPE (top) and RMSE (bottom)					
			5	10	15	20	25	30
LSTM	4	2	7.27	9.1	10.95	12.8	14.76	16.69
			10.25	12.55	14.97	17.39	20.01	22.59
		1	7.34	9.31	11.28	13.28	15.44	17.56
			9.66	**11.94**	**14.27**	**16.64**	**19.19**	**21.77**
	6	1	7.12	8.95	10.83	12.78	14.87	16.99
			10.6	12.99	15.5	18.07	20.86	23.65
		2	**7.02**	**8.76**	**10.57**	**12.39**	**14.43**	**16.42**
			10.93	13.31	15.88	18.44	21.28	24.04
TCN	4	1	5.51	6.52	7.94	9.76	12.09	14.77
			7.04	**8.05**	**9.62**	**11.45**	**14.01**	**16.74**
		2	**5.49**	**6.33**	**7.58**	**8.91**	**10.83**	**12.81**
			7.72	8.72	10.35	12.13	14.76	17.43
	6	2	5.57	6.4	7.62	8.93	10.9	12.88
			8.97	10.06	11.85	13.83	16.85	19.89
		1	5.57	6.45	7.77	9.33	11.51	13.88
			8.44	9.52	11.32	13.42	16.42	19.55
ARIMA	1	6	1.84	4	5.97	7.9	9.48	10.88
			3.86	8.24	12.11	16.22	19.17	21.93
	2	6	**1.36**	**2.98**	**4.72**	**6.66**	**8.35**	**9.96**
			2.9	**6.14**	**9.54**	**13.58**	**16.57**	**19.49**
XGBoost	4	2	1.84	3.88	6.18	8.5	10.49	12.38
			2.61	**5.4**	**8.61**	11.78	14.47	17.09
		1	**1.84**	**3.87**	**6.16**	**8.48**	**10.45**	**12.33**
			2.62	5.42	8.65	11.85	14.56	17.2
	6	1	1.86	3.91	6.21	8.54	10.55	12.47
			2.83	5.86	9.4	12.84	15.74	18.57
		2	1.85	3.88	6.19	8.5	10.53	12.47
			2.92	6.05	9.76	13.3	16.29	19.21
Prophet	6	1	28.03	28.53	29.21	30.09	30.71	31.15
			38.64	39.4	40.31	41.35	42.17	42.82

XGBoost. They are respectively 5.66 and 5.18% more efficient than the LSTM model and 3.13 and 3.63% more efficient than the TCN model for 5-min predictions. The least efficient model is the prophet model created by Méta which allows us to give a prediction interval but is not very precise when asked to predict the exact sequence of our time series. For the three short-horizon (*i.e.*, 5-, 10, and 15-min-ahead) forecasts, for all models; this demonstrates that forecasting for the

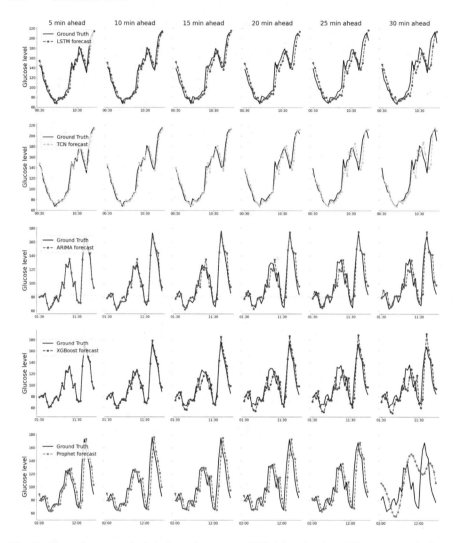

Fig. 2. True glucose and their best forecasts yielded by the five different models for a patient

far future is more challenging than for the near future. The results, in Table 1, show that linear models (*e.g.*, ARIMA and XGBoost) using recursive strategy perform better, in comparison with other models. For all forecasting tasks, they can yield forecasts with less difference to the true values, throughout the forecasting period. The prophet model is the least efficient on our dataset. This model performs better on data following a seasonal recurrence.

3.2.2   Visualization
Figure 2 shows the true (solid lines) and forecasted (dotted lines) CGM levels for a 46-year-old patient with type 1 diabetes. For each model, the 6 subfigures

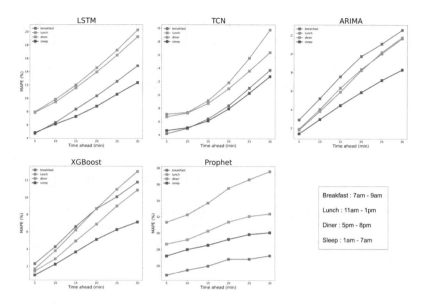

Fig. 3. MAPE by period

present the forecasts made 5, 10, 15, 20, 25, and 30 min in advance, respectively. All these forecasts are made by the models with the optimal settings that lead to the lowest MAPE and RMSE (as shown in Table 1). Note that the more we have predicted the value of CGM far in the future the bigger the difference between the forecasts and true values. For example, the 5-min-ahead LSTM's forecasts are extremely close to the ground truth and when you go further in the forecast, 30 min ahead, meaning that the model is much more accurate for short-term forecasts. Comparing the 30-min-ahead forecasts between LSTM and ARIMA, we see that the former has a kind of lag while the latter yields the forecasts closer to the true value except for those turning points. Models such as LSTM and TCN are more comfortable in detecting a change in the trend of glucose levels. It is particularly noticeable in the predictions at 25 and 30 min, the points are certainly far from the true values but have anticipated an increase or a decrease in the glucose level.

The 30-min-ahead forecasts of the TCN model follow the sharp drop in glucose levels observed at 2pm. Linear models simply extend a trend. For example, the 25-min-ahead and 30-min-ahead XGBoost models prolong the movements of the glucose level but do not anticipate the glucose level decrease at 1pm. To verify the effectiveness of these different models, we tested them on data from the 171 patients over 4 d, 3 training days and 1 test day. We have therefore noticed that linear models are more efficient in general and when the volatility of the time series remains low. However, models using neural networks are slower but make it possible to predict with more precision a potential drop in glucose levels, which is an advantage in the context of CGM. The prophet model has great difficulty analyzing our dataset and identifying paternal effects that it could replicate.

Figure 3 shows the MAPE of the forecasts made for specific periods of a day: breakfast time (between 7am–9am) for 20 patients, lunchtime (between 11am–1pm) for 10 patients, dinner time (5pm–8pm), and time for sleep (1am–7am). For the majority of the models, we can observe that our predictions are better during the night period and more difficult during the full day at mealtime and especially during lunch.

4 Conclusions

In this work, we compared 5 models, statistical and ML-based models to make 6 steps ahead forecasts with 5 min intervals. Data from 171 diabetic patients followed for 14 d and with measurements every 5 min. We cleaned and completed the data and then we managed to transform the date times to correspond and start by Jan 1, 2000. With different methods, we predicted the evolution of glucose levels over the 6 stages and for 24 h in these people. We also divided the day into periods to analyze the accuracy of our forecasts over these periods. The RMSE and MAPE measurements showed us the differences between the different models for our data set. Our predictions are also more accurate during quiet periods such as the night and more difficult during meals, the lunch period being for most models the least conducive to predictions. To conclude, we highlighted the difference between statistical and learning models, indeed statistical models are very effective in predicting the CGM 30 min ahead but have difficulty in perceiving changes in variation as well as neural network models.

References

1. Ali, J.B., Hamdi, T., Fnaiech, N., Di Costanzo, V., Fnaiech, F., Ginoux, J.M.: Continuous blood glucose level prediction of type 1 diabetes based on artificial neural network. Biocybernetics Biomed. Eng. **38**(4), 828–840 (2018)
2. Bai, S., Kolter, J.Z., Koltun, V.: An empirical evaluation of generic convolutional and recurrent networks for sequence modeling (2018)
3. Box, G.E., Jenkins, G.M., Reinsel, G.C., Ljung, G.M.: Time Series Analysis: Forecasting and Control. John Wiley & Sons, Hoboken (2015)
4. Chen, T., Guestrin, C.: XGBoost: a scalable tree boosting system. In: KDD, pp. 785–794. Association for Computing Machinery (2016)
5. Contreras, I., Vehi, J., et al.: Artificial intelligence for diabetes management and decision support: literature review. J. Med. Internet Res. **20**(5), e10775 (2018)
6. Eren-Oruklu, M., Cinar, A., Quinn, L., Smith, D.: Adaptive control strategy for regulation of blood glucose levels in patients with type 1 diabetes. J. Process Control **19**(8), 1333–1346 (2009)
7. Foster, N.C., Miller, K.M., Tamborlane, W.V., Bergenstal, R.M., Beck, R.W.: Continuous glucose monitoring in patients with type 1 diabetes using insulin injections. Diab. Care **39**(6), e81–e82 (2016)
8. Hochreiter, S., Schmidhuber, J.: Long short-term memory. Neural Comput. **9**(8), 1735–1780 (1997)

9. Jaloli, M., Cescon, M.: Long-term prediction of blood glucose levels in type 1 diabetes using a cnn-lstm-based deep neural network. J. Diab. Sci. Technol. 19322968221092785 (2021)
10. Taylor, S.J., Letham, B.: Forecasting at scale. PeerJ Preprints 5:e3190v2 (2017)
11. Yuan, Y.C.: Multiple Imputation for Missing Data: Concepts and New Development (Version 9.0). SAS Institute Inc., Rockville (2010)
12. Zhang, J., Pathak, H.S., Snowdon, A., Greiner, R.: Learning models for forecasting hospital resource utilization for COVID-19 patients in Canada. Sci. Rep. **12**(8751) (2022). https://doi.org/10.1038/s41598-022-12491-z

On Enhancing Network Slicing Life-Cycle Through an AI-Native Orchestration Architecture

Rodrigo Moreira[1]([✉]), Joberto S. B. Martins[2], Tereza C. M. B. Carvalho[3], and Flávio de Oliveira Silva[4]

[1] Institute of Exact and Technological Sciences, Federal University of Viçosa, Rio Paranaíba 38810-000, Brazil
rodrigo@ufv.br

[2] UNIFACS, Salvador 41820-020, Brazil

[3] USP, São Paulo 05508-010, Brazil
terezacarvalho@usp.br

[4] Faculty of Computing, Federal University of Uberlândia, Uberlândia 38400-902, Brazil
flavio@ufu.br

Abstract. Legacy experimental network infrastructures can still host innovative services through novel network slicing orchestration architectures. Network slicing orchestration architectures available in state-of-the-art have building blocks that structurally change depending on the problem they are trying to solve. In these orchestrators, life-cycle functions of network slices experience advances on numerous fronts, such as combinatorial methods and Artificial Intelligence (AI). However, many of the state-of-the-art slicing architectures are not AI-native, making heterogeneity and the coexistence and use of machine learning paradigms for network slicing orchestration hard. Also, using AI in a non-native way makes network slice management a challenger and shallow. Hence, this paper proposes and evaluates a distributed AI-native slicing orchestration architecture that delivers machine learning capabilities in all life cycles of a network slice. Carried experiments suggest lower error using distributed machine learning models to predict Radio Access Network (RAN) resource consumption in slicing deployed over different target domains.

1 Introduction

Artificial Intelligence (AI) has been edge-cutting in many science areas, especially in sharing network resources. After emerging technologies such as Software-defined Networking (SDN) and Network Function Virtualization (NFV) share network resources to deliver them to end users as network slicing is straightforward. Furthermore, cloud computing has allowed AI techniques to evolve

L. Barolli (Ed.): AINA 2023, LNNS 654, pp. 124–136, 2023.
https://doi.org/10.1007/978-3-031-28451-9_11

quickly. In mobile networks, similar evolution happened supported by technological enablers such as AI, specifically machine learning used to perform and support the network slicing requirements of users and modern applications. These requirements drove the standardization and evolution of mobile networks, leading to the creation of experimental architectures and testbeds [1, 2].

Many network slicing architectures arose, especially in the 5G and B5G mobile networks fields. These architectures used and evolved technological enablers capable of meeting the demands of the life cycle of a network slice. The artificial intelligence techniques predominantly used in network slicing are Reinforcement Learning (RL), Deep Reinforcement Learning (DRL), and classification. They enhance the network slicing life cycle, guarantee Quality of Service (QoS), admission control, resource allocation, resource scheduling, and others [3].

In our state-of-the-art survey, we noticed that Slicing Orchestration Architectures encompass machine learning techniques as a use case to handle specific tasks on the network slice life-cycle [4]. Hence, the usage of Machine Leaning (ML) technique happens on an ad hoc basis, coming across as challenging to evolve the machine learning paradigms on the Orchestration Architectures to address specific slice orchestration requirements and the end user. Hence, this paper sheds light on those problems by proposing and evaluating an AI-native slicing orchestration architecture with distributed learning agents spread over target domains to manage different slices in detail in their various life stages [5].

Among our main contributions include an AI-native architecture reference for network slicing orchestration. A distributed learning evaluation in three different RAN slicing services hosted in other target domains. A qualitative assessment of federated learning with agents in these target domains regarding the performance and generalizability of the global model.

The remainder of this work is organized as follows: Sect. 2 brings the works that walked towards constructing an AI-native Orchestration slicing architecture. Section 3 qualitatively describes the Orchestration architecture proposed in this article. In Sect. 3, we present the experimental scenario we used to evaluate the proposed architecture. In Sect. 5, we bring and discuss experimental results, while in Sect. 6, we draw conclusions and point out emerging challenges.

2   Related Work

In this section, we bring some prior works highlighting aspects related to ours. Many state-of-the-art efforts bring AI capabilities on the network slicing life-cycle as functionality that could be deployed and evolve after the final build-block orchestration architecture [6]. However, we advocate that an AI-native orchestration architecture enhances the network slicing life-cycle, enabling fine-grained management tasks over the network slicing. Different from ours, they used a centralized training model and storing.

Casetti et al. [7] realizes the Machine Learning as a Service (MLaaS) concept applied to a 5G Network Slicing Architecture. In this work, using already-trained algorithms, an Application Programming Interface (API) offers to Management and Orchestration (MANO) entities some insights about the network slice life-cycle. They tried their approach using two relevant scenarios. While Kukliński et al. [8] proposed a reference architecture for Zero-touch network and service management where the MLaaS concept feeds the Slice Provider that uses ML algorithm in the commissioning and deployment phase of network slicing. Their solutions do not consider the ML capabilities in other slicing phases. They have a centralized structure considering training and MLaaS architecture.

Mason et al. [9] bring a distributed approach for resource allocation in a network slicing orchestration. In their approach, there is the Training Manager, which enforces agents with embedded RL and carries the training using local information, avoiding interactions with the central managers. Further, using transfer learning, they combine the agents to enhance the model's generality. Different from ours, they built native Reinforcement Learning architecture. Our framework goes beyond, with a complete life-cycle management of the machine learning models that help the network slice orchestration in its entire life-cycle.

On the 5G mainstream, Garrido et al. [10] proposed a management and orchestration solution capable of predicting traffic in the 5G network when slicing is running, and prediction is used in the admission control mechanism. Their orchestration solution has a distributed agent who runs a RL algorithms in different domains. With this structure, they achieved a method for context-aware for traffic prediction and admission control. Our work goes beyond with a machine learning native structure that distributes AI capabilities among all orchestration life-cycles and target domains empowering.

In addition, Theodorou et al. [11] proposed and evaluated a 5GZorro project which builds a cross-domain network slicing with an AI-drive closed-loop automation orchestration. Among the domains, the paper describes the usage of Distributed Ledger Technology (DLT) to support inter-domain SLA. In their approach, the Intelligent Slice layer is placed on the closed loop of the network slicing life-cycle. The orchestration mechanism proposed by the authors has a distributed MANO architecture for network slicing orchestration. Their machine learning capabilities embedded on MANO does not take into account a distributed structure to support different machine learning capabilities for all network slicing life-cycle.

Bega et al. [12] proposed and evaluated an AI-based network slice manage-ment framework in which AI plays a significant role in the network slicing life cycle. The proposed framework employs AI to support admission control, net-work resource orchestration, including instantiation and runtime, and resource scheduling. It does this intelligent slicing management using two machine learn-ing paradigms: unsupervised and reinforcement learning. Unlike our works, the network slicing orchestration mechanism does not consider distributed and other tenant domains, and we go further, proposing a full-stack machine learning framework for all network slicing life-cycle and capable of handling distributed machine learning over the slicing architecture.

Similarly, D'Oro et al. [13] presented and evaluated an orchestration tool for RAN in the 5G mainstream. They conducted experiments using a relevant testbed, seven base stations, and 42 user equipment. The rationale behind their orchestration relies on a functional build-block orchestration of Radio Access Network (RAN) elements to support real-time and non-real-time slice-as-a-service. The proposed orchestrator has blocks to address requests of a network slice. The slicing description feeds the Request Collector. Further, it feeds the Orchestration Engine, which has an ML/AI Catalog with models capable of fitting the slicing service requirement.

Similarly, Chergui et al. [14] built and evaluated a framework with AI-driven blocks for managing and orchestrating massive network slicing. The framework aimed to address scalability and zero-touch. To do so, they have a Monitoring System block to gather data from tenant infrastructure, Analytical Engine to process data collected from Monitoring System, and a Decision Engine which is a decision-making element that considers a local ML to interact with NAMF to employ resources guarantee to the network slice.

Our work goes beyond, and we propose an AI-native block that supports the network slicing life-cycle and handles agent model distribution over the target tenants' domains hosting the network slices. Also, our approach for MLaaS encompasses all slicing life-cycle phases.

3 SFI2 AI-Native Architecture

Slicing Future Internet Infrastructures (SFI2) is an ongoing joint project in Brazil that enrolls numerous institutions that seek to integrate different experimental facilities in Brazil [15,16]. Here, we idealized a slicing orchestration for the SFI2 reference architecture following the MLaaS rationale based on the assumption of the existence of hybrid domains over which network slices are deployed. SFI2 integrates resources and services from legacy experimental infrastructures by defining a new reference architecture and functionalities, allowing network slicing instantiation and deployment on heterogeneous domains.

The SFI2 reference architecture (Fig. 1) [16] contains functional building blocks with predefined roles. It follows the technical specification Slice Management of 3rd Generation Partnership Project (3GPP) [17]: preparation, commissioning, operation, and decommissioning phase. For the network slice to be deployed on SFI2 domains, the functional blocks of the architecture exchange messages for a data plane, intelligence plane, data control, and management planes, and the slice monitoring plane network, service, and infrastructure.

The upper left block (Tenant/Experimenter Slice Description) represents the SFI2 interface that receives input from users whose content is the specification of the network slice to be deployed. The slice specification in the SFI2 architecture contains the syntax and semantics necessary for the SFI2 architecture to implement the network slice on the target domains it operates on. The user must be authorized by Identity and Access Management (IAM) to submit the network slice deployment request.

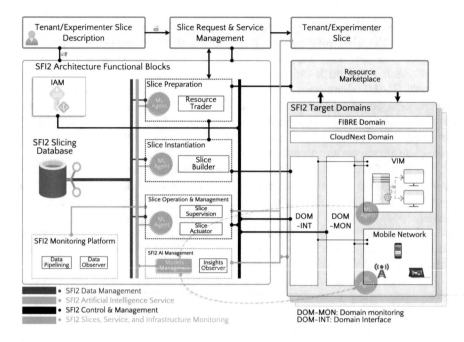

Fig. 1. AI-Native orchestration architecture.

The upper block (Slice Request & Service Management) has the dual role of making the slice specification processor and deploying and managing the service for the requested network slice. In this block, the services of the internal block of the SFI2 architecture are claimed for establishing the network slice. When the network slice deployment ends, the user has access to the network slice service and the network slice management through the upper right block (Tenant/Experimenter Slice).

Within the SFI2 architecture, the Slice Preparation component, namely Resource Trader, lists and organizes the resources necessary for establishing the network slice required by the user tenant; for this purpose, this block interacts with other blocks, such as the Resource Marketplace. In the SFI2 Architecture, the Marketplace is responsible for the dynamic discovery of resources that domains voluntarily advertise to the SFI2 Architecture. These domains are composed of heterogeneous technologies that are the SFI2 Target Domains such as FIBRE [18], CloudNext [19], and others.

Through the Slice Builder, the Slice Instantiation Block instantiates the network slice in the target domains. In addition, in Slice Instantiation, we have the role of Slice Resource Orchestration that aggregates the slice parts constituting a single network slice as a service, in addition to offering the user the run-time lifecycle of their network slice. All network slice deployment phases are registered in the SFI2 Slicing Database.

Additionally, the Slice Operation & Management block contains functions defined as Slice Supervision that correspond to monitoring the service deployed in the domains on which the network slice is deployed. The Slice Supervision service that the SFI2 Architecture contains exposes services and functionality to the user for the upper block (Tenant/Experimenter Slice) through the monitoring interface. Network slice monitoring works in conjunction with the SFI2 Monitoring Platform block, which is full-stack monitoring of the SFI2 Architecture.

The SFI2 Monitoring Platform has Monitoring Agents (MAGs) acting as node exporters throughout the SFI2 Architecture, including Targets Domains and network slices that, when deployed, export statistics to the monitoring service. By receiving heterogeneous monitoring data, service, infrastructure, and slices, the block has a Data Pipelining component through which they treat data that arrives from different sources asynchronously.

The SFI2 AI Management block is the shift-cutting paradigm block of the SFI2 architecture. This block interacts with the other SFI2 blocks through the Control and Management interface. The functional rationale of this block is of a distributed nature, where each block of the SIF2 architecture has the ML Agent component, which interacts with the Models API through the Artificial Intelligence Service interface. Each ML Agent can download from the Models API the specific models for each part of the network slice lifecycle and offer specialized support to this phase through predictions, regressions, and clustering. For example, in the Slice Preparation phase, the Resource Trader can search the ML Agent for the SFI2 Target Domain that is most likely to support the network slice specification considering historical data already learned by the learning model machine.

Similarly, the Slice Operation & Management can make use of the ML Agent can query its local machine learning model about resource consumption by an already deployed network slice and be able to make decisions through the Slice Actuator about it. This distributed MLaaS model in network slicing architecture is a significant innovation, especially for the orchestration of Future Internet infrastructure resources.

We detail the operation of the SFI2 AI Management block, in Fig. 2 we present the details of our ML framework that makes the architecture of SFI2 ML-oriented. This architectural block has three interfaces through which the machine learning service is organized. The SFI2 Artificial Intelligence Service Interface refers to the control of distributed learning mechanisms and the distribution of data from machine learning models that can be trained to support the lifecycle of a network slice in the SFI2 Architecture.

The SFI2 Control & Management channel refers to the exchange of messages to remote procedure calls to model manager sub-modules for the tasks of training, validating, storing, and distributing trained ML models to be used throughout the lifecycle of a network slice deployed over SFI2 Infrastructures. The SFI2 Slices, Services, and Infrastructure Monitors interface is the interface that the AI Management block has with the SFI2 Architecture monitoring service.

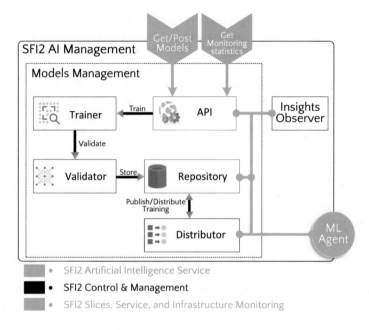

Fig. 2. SFI2 MLaaS reference architecture.

Between the Distributor and the ML Agent, there are remote procedure calls and data distribution for distributed training. The distribution of these data will depend on the specification of the learning model. It can be *Non*-Independent and Identically Distributed (IID) and IID. The *Non*-IID distribution refers to the dataset distributed across the training agents not being identically distributed across the agents and to different data samples on which the model will conduct the training. The IID data distribution refers to equally and identically distributed samples over the training agents, with a high probability of different agents being exposed to similar data in the training phase.

4   Experimental Setup

According to Fig. 3, we built an experimental scenario assuming three different target domains using distributed ML Agents that train locally on resource consumption in the RAN. These training agents operate on the federated learning paradigm and report the weights of the local models being trained to the central model, hosted by SFI2 architecture with proceeds with FedAvg [20] algorithm. This algorithm aims to average the weights of the machine learning models reported to the central model.

The quality criterion for the distributed training that takes place in the SFI2 architecture is the Loss considering the Mean Squared Error (MSE) metric, which is a measure of the quality of an estimator. In addition, we use a Learning Rate of 0.001, a value widely used in the literature for tuning parameters in an

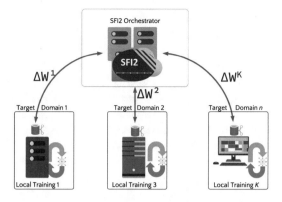

Fig. 3. Experimental Setup.

optimization algorithm, and 0.9 for momentum, which is the extension to the gradient descent optimization algorithm.

A single dataset in a time-series way with 11 attributes concerning usage and resource consumption of a 5G RAN. In our experiments, we used the Long Term Evolution (LTE) Physical Downlink Control Channel (PDCCH) dataset, which contains different Base Stations (BSs) consumption records from three different locations in Barcelona city [21]. The first location El Born has 5,421 samples. The second location, Les Corts, with 8,615 samples, and the third one, PobleSec, with 19,909 samples collected. We summarize the 11 features in just one column by summing all the features and then normalizing the training feature to have a smooth graph. For reproducibility, we let our distributed learning method available as open-source in a GitHub repository https://github.com/romoreira/5G-nodeB-TrafficPrediction/tree/combinedts.

We carried out experiments aiming to answer the following questions.

- How viable is the realization AI-Native network slicing orchestration in our specific network slicing architecture?
- What is the performance in distributed training for resource consumption prediction in virtualized RAN?
- Is it possible to predict RAN resource consumption using distributed machine learning?

5 Results and Discussion

To answer question one, we idealized a native distributed learning architecture and an experimental scenario, as shown in Fig. 3, where each Local Training is a SFI2 Target domain, denoting virtualized RAN resources in Future Internet Infrastructures. Thus, we consider different target domains managed by SFI2 architecture where each target domain has RAN service.

As a result, our training agents worked distributedly in each target domain, sending information at each training round to the central model managed in the SFI2 architecture. After the end of the training phase, the model hosted by SFI2 architecture has global knowledge about the training agents and can perform consumption forecasting abound specific RAN.

In order to answer question two, we train the Convolutional Neural Network (CNN) Omni-Scale CNN [22], achieving higher performance for time-series forecasting. Each ML-agent individually performed the training process for RAN consumption forecasting and sent model weights to SFI2 global model. Figures 4-A, 4-B, and 4-C bring the split of the dataset for training and those reserved for testing, using the division of the data set 80% for training and 20% for testing.

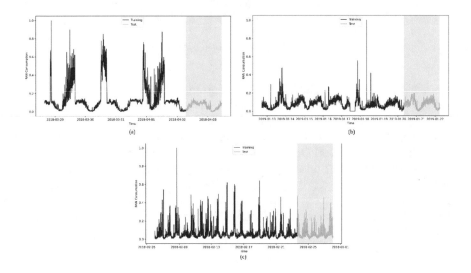

Fig. 4. Target domains - locally train and test dataset split.

At the end of the training cycle, consisting of 30 epochs, we reached a Distributed Loss of 0.0003619, following the Losses of each ML-agent for Target Domain 1 of 0.0028681 and Target Domain 2 of 0.0003704, and Target Domain 3 0.0005889. The metric for calculating the Loss was the MSE that brings the sum of the differences between the real and the predicted in the training and testing process.

In addition, we verified the generalization capability of the models of each distributed ML-Agent. We represented the Loss rate over the training epochs as shown in Fig. 5-A, 4-B, and 4-C. Despite the noise during the training process, there is a generalization in the learning of the local training agent models.

These noises are due to the distribution of aggregated weights in Avg method by the server to the local agents. This causes the local agent to receive weights from other Convolutional Neural Networks (CNNs), trained with unseen data, weights trained by other agents distributed along the target domains of the SFI2 Architecture. This training process is typical Non-Independent and Identically

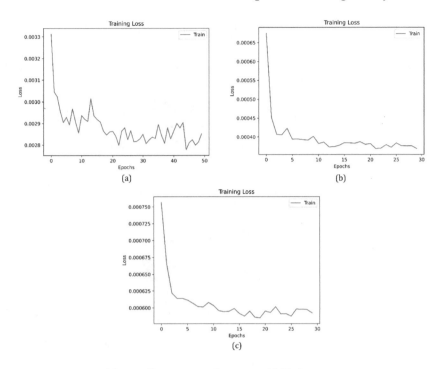

Fig. 5. Convergence learning of ML–Agents.

Distributed (Non-IID); each training agent has a different data set in quantities and distribution.

Answering question three, after the distributed training rounds, we aimed to measure the generalization capacity of the SFI2 global model, which aggregated the weights of each training agent spread over the virtualized infrastructure. The global model forecasted the RAN consumption for the three locations of the target domains. Figure 6-A represents the generalization capability of the global model hosted by SFI2 Architecture for forecasting the RAN consumption of target domain 1.

Following, we evaluate the ability of the global model to forecast the RAN consumption of the Les Corts location (target domain 2) by the Fig. 6-B, and we find that the forecasted RAN consumption gradually follow the natural tendency leading us to admit that distributed learning, especially federated learning, is suitable to perform distributed forecasting for AI-native Orchestration Architectures. Followed by the PobleSec location (target domain 3), with the same behavior in the learning and forecasting task fin Fig. 6-C.

Finally, an AI native for network slicing orchestration architecture fills the state-of-the-art gaps and potentially can drive the construction of new orchestration architectures for network slicing.

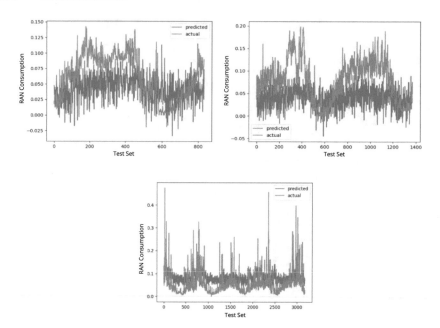

Fig. 6. Real *versus* predict in global SFI2 model - three target domains.

6   Concluding Remarks

This paper proposes and evaluates an AI-native approach for network slicing orchestration considering the SFI2 architecture. After diving into the state-of-the-art, we found that predominantly existing orchestration techniques adopt machine learning as a separate architecture feature for the build blocks. The evaluation results allow us to hypothesize that AI-native slicing architectures support more precise orchestration for slicing over heterogeneous infrastructures.

Thus, our empirical results suggest that federated learning can be used natively in the build-block of network slicing orchestration architectures to allow AI capabilities enhances the life-cycle orchestration of the entire network slicing life-cycle. In our experimental evaluation, we verified that the centralized model in SFI2 Architecture is capable of handling precisely RAN consumption forecasting, leading us to admit that AI-native Architecture brings versatility in the management of network slices. Additionally, we endorse that our rationale could be evolved and generalized in other slicing architectures.

We intend to carry out further measurements in future work, such as the co-existence of different machine learning models hosted by SFI2 Orchestration Architecture to support different management and orchestration of the network slicing in different phases on its life cycle. In addition, it is worth evaluating other machine learning technologies such as Recurrent Neural Network (RNN) and other divisions of the dataset.

Acknowledgements. The author thanks the National Council for Scientific and Technological Development (CNPq) under grant number 421944/2021-8 (call CNPq/MCTI/FNDCT 18/2021), and the Research Support Foundation of the State of São Paulo (FAPESP) grant number 2018/23097-3, for the thematic project Slicing Future Internet Infrastructures (SFI2).

References

1. Esmaeily, A., Kralevska, K.: Small-scale 5G testbeds for network slicing deployment: a systematic review. Wireless Commun. Mobile Comput. **2021**, e6655216 (2021)
2. Silva, A.P., et al.: 5GinFIRE: an end-to-end open5G vertical network function ecosystem. Ad Hoc Netw. **93**, 101895 (2019)
3. Ssengonzi, C., Kogeda, O.P., Olwal, T.O.: A survey of deep reinforcement learning application in 5G and beyond network slicing and virtualization. Array **14**, 100142 (2022)
4. Khan, L.U., Yaqoob, I., Tran, N.H., Han, Z., Hong, C.S.: Network slicing: recent advances, taxonomy, requirements, and open research challenges. IEEE Access **8**, 36009–36028 (2020)
5. 3GPP, 3rd Generation Partnership Project; Technical Specification Group Services and System Aspects; Management and orchestration; Concepts, use cases and requirements (Release 17) (2022)
6. Zhang, S.: An overview of network slicing for 5g. IEEE Wirel. Commun. **26**(3), 111–117 (2019)
7. Casetti, C., et al.: Ml-driven provisioning and management of vertical services in automated cellular networks. IEEE Trans. Netw. Serv. Manage. **19**(3), 2017–2033 (2022)
8. Kukliński, S., et al.: A novel architectural approach for the provision of scalable and automated network slice management, in 5G and beyond. In: Maglogiannis, I., Macintyre, J., Iliadis, L. (eds.) AIAI 2021. IAICT, vol. 628, pp. 39–51. Springer, Cham (2021). https://doi.org/10.1007/978-3-030-79157-5_4
9. Mason, F., Nencioni, G., Zanella, A.: Using distributed reinforcement learning for resource orchestration in a network slicing scenario. IEEE/ACM Trans. Netw. 1–15 (2022)
10. Garrido, L.A., Dalgkitsis, A., Ramantas, K., Verikoukis, C.: Machine learning for network slicing in future mobile networks: design and implementation. In: 2021 IEEE International Mediterranean Conference on Communications and Networking (MeditCom), pp. 23–28 (2021)
11. Theodorou, V., et al.: Blockchain-based zero touch service assurance in cross-domain network slicing. In: 2021 Joint European Conference on Networks and Communications & 6G Summit (EuCNC/6G Summit), pp. 395–400 (2021)
12. Bega, D., Gramaglia, M., Garcia-Saavedra, A., Fiore, M., Banchs, A., Costa-Perez, X.: Network slicing meets artificial intelligence: an AI-based framework for slice management. IEEE Commun. Mag. **58**(6), 32–38 (2020)
13. D'Oro, S., Bonati, L., Polese, M., Melodia, T.: Orchestran: Network automation through orchestrated intelligence in the open ran. In: IEEE INFOCOM 2022 - IEEE Conference on Computer Communications, pp. 270–279 (2022)
14. Chergui, H., Ksentini, A., Blanco, L., Verikoukis, C.: Toward zero-touch management and orchestration of massive deployment of network slices in 6g. IEEE Wirel. Commun. **29**(1), 86–93 (2022)

136   R. Moreira et al.

15. de Oliveira Silva, F., de Brito Carvalho, T.C., Martins, J.S.B., Both, C.B., Macedo, D.F.: SFI2 Technical Report - TR01/2021 SFI2 - Slicing Future Internet Infrastructures Round Table. Technical report TR01/2021, SBRC/WPEIF 2021, São Paulo, Brazil (2021)

16. Martins, J.S.B., Carvalho, T.C., Flavio, S., Moreira, R.: SFI2 network slicing reference architecture. Technical report TR03/2022, SFI2 Technical report (2022)

17. Ferrús, R., Sallent, O., Pérez-Romero, J., Agusti, R.: On the automation of ran slicing provisioning and cell planning in NG-RAN. In: 2018 European Conference on Networks and Communications (EuCNC), pp. 37–42 (2018)

18. Salmito, T.: FIBRE - an international testbed for future internet experimentation. In: Simpósio Brasileiro de Redes de Computadores e Sistemas Distribuídos - SBRC 2014, (Florianopolis, Brazil), pp. 969 (2014)

19. Brasileiro, F., Brito, A., Blanquer, I.: Atmosphere: adaptive, trustworthy, manageable, orchestrated, secure, privacy-assuring, hybrid ecosystem for resilient cloud computing. In: 2018 48th Annual IEEE/IFIP International Conference on Dependable Systems and Networks Workshops (DSN-W), pp. 51–52 (2018)

20. Nilsson, A., Smith, S., Ulm, G., Gustavsson, E., Jirstrand, M.: A performance evaluation of federated learning algorithms. In: Proceedings of the 2nd Workshop on Distributed Infrastructures for Deep Learning, DIDL'18, (New York, NY, USA), pp. 1–8. Association for Computing Machinery (2018)

21. Trinh, H.D., Fernández Gambín, N., Giupponi, L., Rossi, M., Dini, P.: Mobile traffic classification through physical control channel fingerprinting: a deep learning approach. IEEE Trans. Netw. Serv. Manag. **18**(2), 1946–1961 (2021)

22. Tang, W., Long, G., Liu, L., Zhou, T., Blumenstein, M., Jiang, J.: Omni-scale CNNs: a simple and effective kernel size configuration for time series classification. In: International Conference on Learning Representations (2022)

Adaptive Inference on Reconfigurable SmartNICs for Traffic Classification

Julio Costella Vicenzi[1]([⊠]), Guilherme Korol[1], Michael Guilherme Jordan[1], Mateus Beck Rutzig[2], and Antonio Carlos Schneider Beck Filho[1]

[1] Universidade Federal do Rio Grande do Sul, Porto Alegre, Brazil
{julio.vicenzi,guilherme.korol,michael.jordan,caco}@inf.ufrgs.br
[2] Universidade Federal de Santa Maria, Santa Maria, Brazil
mateus@inf.ufsm.br

Abstract. Traffic classification is an essential part of network management and monitoring, and has been widely used to help improve security and detect anomalies. Convolutional neural networks (CNN) have shown good performance for many of these traffic classification tasks, at the cost of a heavy computational burden. To efficiently classify incoming packets, hardware accelerators such as FPGAs, which are integrated with smart network cards, can provide the required parallelism with energy efficiency to execute such CNNs. Optimization techniques such as pruning can further improve inference speeds at the cost of some accuracy, but are usually static (i.e., cannot change after deployment). Given that, this work implements an adaptive framework to process different traffic classification tasks efficiently. It works by generating multiple pruned CNN hardware models during design time; and at run-time exploits the accuracy vs throughput trade-off of pruning by dynamically and automatically switching between the pre-generated models according to the number of incoming packets and the different classification tasks at a given moment. Compared to the regular (static) solution, our implementation improves the Quality of Experience up to 1.14×, executing up to 1.51× more inferences and improving energy efficiency per inference up to 1.35×.

1 Introduction

In computer network monitoring, traffic classification is a crucial task that extracts relevant features, such as the type of application used, protocols, and encryption. These features provide an understanding of how the network resources are used, and can also aid in security monitoring for anomaly and intrusion detection, providing information on potential network attacks and improper use of network resources. Deep neural networks (DNNs) have been extensively employed for these classification problems, initially with Multi-Layer Perceptron (MLP) networks, and more recently with Convolutional Neural Networks (CNNs), providing highly accurate classifications at the cost of a heavy computational burden. However, internet speeds are growing each year, with newer network infrastructure supporting port speeds of over 100 Gbps. This speed increase leads to ever greater number of packets flowing through the network

L. Barolli (Ed.): AINA 2023, LNNS 654, pp. 137–148, 2023.
https://doi.org/10.1007/978-3-031-28451-9_12

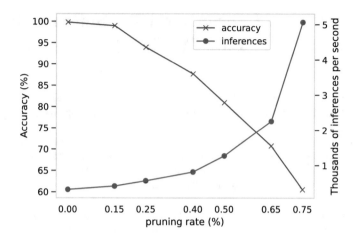

Fig. 1. Accuracy (blue) and inferences per second (red) w.r.t. pruning rate.

that require classification. Therefore, computation offloading is essential to handle CNN's highly parallel workload amidst this challenge.

SmartNICs provide a platform that help overcoming these challenges. Through the use of Software-Defined Networks (SDN) and integrated accelerators such as Field-Programmable Gate Arrays (FPGA) [13], SmartNICs allow for the use of fully custom and flexible accelerators. Especially for such *reconfigurable* SmartNICs, the flexibility provided by reconfiguring the FPGA with new accelerators allows for multiple custom designs to be explored on-demand.

Even though reconfigurable SmartNICs can improve the efficiency of CNN processing, relying exclusively on the hardware cannot sustain performance and energy gains indefinitely - especially in a heavy-load context of large CNNs and high traffic volumes. In this context, the CNNs must be optimized as well. Pruning is a popular technique to reduce the CNN size and computation requirements at the cost of accuracy. Pruning removes entire parts of a CNN, from neurons to entire filters, reducing the number of parameters required to be stored in memory and the amount of calculations performed in each inference. Figure 1 shows an example of the performance *versus* accuracy trade-off created by pruning. As the pruning rate increases, the accuracy drops while the throughput greatly increases. Thus, pruning fits well with the network's traffic fluctuations. By reconfiguring the SmartNIC, it is possible to achieve faster inferences when the network is overloaded or higher accuracy levels whenever the network is experiencing a slower traffic. On top of that, such flexibility also allows switching among multiple traffic classification tasks at runtime.

In summary, multiple heavy-load CNNs are required to process traffic data in very dynamic network environments. At the same time, reconfigurable SmartNICs are getting more popular and are part of many network infrastructures. In this work, we propose a Framework for adapting the inference processing to the network's traffic influx dynamically. To the best of our knowledge, this is the first work

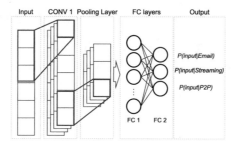

Fig. 2. A sample CNN.

to explore CNN pruning in the context of network traffic classification over FPGAs. Particularly, the framework leverages this already-available reconfigurability to switch at runtime the inference processing between (i) multiple traffic classification tasks, and (ii) multiple pruned versions of these CNNs. The accuracy-throughput trade-off is exploited via a library of different accelerator models created at design-time that can be smartly selected at runtime.

The paper is organized as follows: Sect. 2 presents the background and related work for CNNs, traffic classification, and hardware acceleration. Section 3 details the proposed framework, Sect. 4 the methodology, and Sect. 5 evaluates the results. Finally, Sect. 6 concludes the work.

2   Background and Related Work

CNNs and Network Traffic Classification: CNNs are used for classification tasks, initially employed for image classification problems, they have found great success for traffic classification problems as well. Figure 2 shows an example, where the network receives a traffic sample (e.g., a packet) as input and predicts what category, from a given set of problem-defined classes (or labels), it fits best. The output is a list with the probabilities of the input belonging to the respective class. In this example, three classes are shown, representing how the traffic is generated: Email, Streaming or P2P applications.

To build a CNN, it is first necessary to define a topology, which defines the number and types of layers and their parameters. The three main building blocks of CNNs are convolutional (CONV), fully connected (FC), and pooling layers. Both CONV and FC layers contain learnable parameters, in the form of weights and biases, that are adjusted during training. Once the training is complete, the model is ready to predict unknown inputs (i.e., packets that were not used during training). This second phase is called inference and is the focus of this work. During inference, a CONV layer receives a multidimensional array with multiple channels called feature maps. The parameters of a CONV layer form filters, that are slid over the feature map (shown as blue squares in Fig. 2). In practice, once trained, filters have learned which relevant features characterize the input to differentiate between classes (e.g., what makes an Email packet different from a P2P one).

The CONV layer output is usually down sampled using a pooling layer. After the last CONV layer, the feature maps are flattened into 1d arrays and input into FC layers. These perform matrix-vector multiplication between its input and weights, and correlate the previously extracted features to the possible classes. Batch normalization layers are a common way of accelerating the training of DNNs, being added after CONV or FC layers. The layer acts as a regularization step of the intermediate inputs between the other layers of the DNN, normalizing the inputs by their mean and standard deviation.

CNN Optimizations and Pruning: alongside the implementations of hardware accelerators, many works have speedup the inference by optimizing the CNN model via compression methods such as structured [12] and non-structured pruning [8], adaptive inference [10,11], and quantization [2]. The use of pruning is able to reduce both the memory footprint of the model by reducing the number of parameters and also the number of computations during inference [4]. Additionally, quantization can be used alongside pruning, further reducing the used memory by using fewer bits to represent parameters.

Modern CNN models contain millions of parameters and are often redundant, and although this can lead to benefits during training, the over-parametrization can lead to inefficient inference times. Pruning explores this redundancy, by reducing the number of parameters in a DNN, whilst achieving a higher accuracy than a model with the same number of parameters trained from scratch [12]. Theoretically, a pruning algorithm is capable of removing parameters at no accuracy cost, however this is an NP-hard problem. Additionally, unstructured pruning might create sparse matrices that can actually cause slowdowns. Filter pruning is the method used in this work, which avoids sparsity, by removing entire filters channels [12]. This also decreases the number of channels in the output feature map, since each channel corresponds to a filter, leading to roughly a quadratic reduction in memory footprint and computation.

The exploration of the resource and accuracy trade-off introduced with pruning was implemented by frameworks such as NestDNN [7], ReForm [17] and DMS [9]. NestDNN's focus on multi-tenant applications in smartphones, dynamically scheduling a variant from a multi-capacity models generated using pruning, based on runtime information. ReForm provides a resource-aware inference mechanism in mobile devices. DMS optimizes Quality-of-Service (QoS), by adding and removing filters at runtime based on QoS goals the number of users changes.

FPGA-Based CNN Acceleration: FINN is a tool used in this work that is capable of mapping a CNN topology to an FPGA accelerator. It implements modules using High-Level Synthesis (HLS) to configure each layer according to the CNN topology. These accelerators generated by FINN are called dataflow since they are connected in pipeline architecture, where different modules are dedicated to each CNN layer.

Figure 3 (a) displays an example of a CNN to a simplified FINN accelerator. The CONV layers are mapped into a Sliding Window Unit (SWU), responsible for handling the input feature map to be processed, and a Matrix-Vector

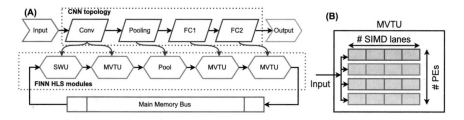

Fig. 3. A sample CNN-dataflow mapping (A) and an overview of parallelism in MVTU modules (B).

Threshold Unit (MVTU) module, that performs the convolution operation. The components in a MVTU are shown in Fig. 3 (b). Two parameters can be configured by the user to adjust the module parallelism: the number of processing elements (PE) and SIMD lanes. However, the choice for PE/SIMD are somewhat constrained by the CNN topology (i.e., the number and size of convolutional channels and kernels).

FPGAs can be dynamically reconfigured to process different tasks at runtime. This process involves loading the configuration file (bitfile) into the device, changing the hardware's function. This process is not instant, and takes longer depending on the bitfile's size, the system's memory configuration, and the FPGA device itself.

FPGA-Based Traffic Classification: FPGAs have been used to accelerate machine learning tasks ranging from network intrusion to traffic classification. They provide high throughput, low power consumption and ensure low latency, enabling more computationally intensive algorithms to be employed [6,15]. Additionally, FPGAs have been adopted into the network infrastructure solutions, such as programmable network interface controllers [13]. Many hardware accelerators for network-based machine learning algorithms have been proposed, such as [15] The work shows a decision tree algorithm C4.5 and discretization algorithm Minimum Description Length, increasing classifications per second from 75–150 million when running on a multicore CPU to 7500 million when using an FPGA. [1,3,18].

Wrap-Up and Our Contributions: CNNs are able to provide high accuracy for network classification problems, and can make use of pruning to reduce the memory footprint and speedup inferences, at the cost of some accuracy. This optimization fits well with FPGA accelerators, as they can be reconfigured at runtime with more aggressive pruning rates for higher throughput or lower pruning rates for higher accuracy. Different from works focused on building custom accelerators [6,15], we instead focus on leveraging the adaptive properties of FPGAs, using pruning to optimize the employed CNN, creating a design space with an accuracy vs. throughput trade-off. The adaptability also enables the exploration of multiple classification tasks as they are requested by the network manager, allowing for multiple types of traffic classification analysis to be performed over time.

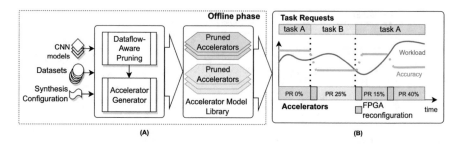

Fig. 4. Overview of the proposed framework. (A) shows the offline phase, where a library of accelerators is generated. (B) shows a runtime example of the framework in action, reconfiguring the FPGA according to the requested task and workload.

3 Framework

The proposed framework is based on a hardware system composed of an FPGA-enabled smart network interface card (SmartNIC). SmartNICs can process packets locally (without host support) to implement various tasks, ranging from intrusion detection to traffic classification, our case-study. These multiple tasks, requested one at a time, combined with the intrinsically variable network conditions create a dynamic environment in which the platform is required to adapt. In this context, as each task uses a different CNN, the configurability of the FPGA can be exploited to offload computations with custom hardware accelerator solutions, enabling the required adaptivity and efficiency. Precisely, the framework exploits the throughput-accuracy trade-off through different versions of accelerators that are generated at design-time. Based on the design space generated, an algorithm selects at runtime the best design according to both the requested task (i.e., specific CNN) and network condition (i.e., traffic volume).

Figure 4 shows an overview of the framework. It uses a two-step solution. First, the offline design-time phase generates the library of accelerators for each task. Next, the runtime phase continually monitors the network conditions to select the currently most appropriate accelerator. Thus, this runtime search may reconfigure the FPGA whenever needed (change of either task or pruning rate).

Offline Step and Design-Time: to generate a set of accelerators with varied accuracy-performance profiles for each task, a pruning method is employed, considering both the network topology and the constraints of the accelerator. Initially, the design-time phase will train the not-pruned CNNs for each task. These CNNs follow a user-defined topology and are trained using the quantization-aware library Brevitas [14] on the provided datasets. Once trained, an accelerator can be generated and added to the library by exporting the CNN to an ONNX file and using the FINN workflow.

For FINN, the number of output channels in a convolutional layer (or neurons, in the case of a FC layer) must be a multiple of the number of PEs, and the number of input channels must be a multiple of the number of SIMD lanes. These requirements ensure the complete use of all processing elements available.

Given this restriction, we leverage the Dataflow-Aware pruning method presented in [11] that takes the original CNN and SIMD and PE configuration and a pruning rate (PR). For each convolutional layer, the method attempts to remove $k = Ch_{original} \, pr$ channels, and tests the restriction of $(Ch^i_{out} - k) \bmod PE_i = 0$ and $(CH^i_{out} - k) \bmod SIMD_{i+i} = 0$, where PE_i and $SIMD_{i+1}$ are the number of PE and SIMD lanes of the current layer i and following layer $i+1$, and Ch^i_{out} is the number of output channels in the current layer (notice that $Ch^i_{out} = Ch^{i+1}_{in}$). In the case of a convolutional layer followed by an FC layer, the SIMD condition changes to $((CH^i_{out} - k)FM) \bmod SIMD_{i+1} = 0$ where FM is the size of the feature map entering the FC layer. If the value of k does not meet these conditions, k is decreased by one and the process continues until they are met or $k = 0$. With the number of pruned channels defined, the criteria for which channels to prune follows [12], where the relative importance of a filter in each channel is calculated by the sum of the absolute weight values (ℓ_1-norm). The k filters with the lowest sum are pruned.

Once all convolutional layers are pruned, the new CNN model is retrained, exported as an ONNX file, synthesized, and added to the accelerator library together with information on its test accuracy (extracted after training) and performance (extracted after synthesis).

Runtime Step: at runtime, all traffic passing through the network is considered for classification. If, however, the traffic volume (i.e., number of requested inferences) surpasses the accelerator's processing capability, those inferences are considered lost and the whole application undermined. To cope with such high-demand applications, the Framework has to continually adapt the SmartNIC inference processing to the current network conditions. This search happens (i) when the requested task differs from the one currently running or (ii) when the incoming traffic workload changes by more than 25%. In any case, when a new accelerator is selected, it is reconfigured into the FPGA by the Framework.

The algorithm for accelerator selection takes as input the current number of incoming inferences (workload) and the requested task currently being processed. The criteria for selection is calculated based on the current workload and the achievable quality of experience of each accelerator in the library for that task, following the formula in Eq. 1, where W_{model} indicates the maximum throughput of the accelerator, W_{input} the current workload and α the model's accuracy. If the accelerator is capable of processing more frames than the current workload, then we consider the maximum percentage of processed frames of 1 (100%), represented by $min(\cdot)$. This formula is evaluated on each accelerator. Then, the one with the highest value is selected for FPGA configuration and execution.

$$QoE_m = min(\frac{W_{model}}{W_{input}}, 1)\alpha \qquad (1)$$

As an illustrative example of the framework in action, Fig. 4 (B) shows different tasks requested for traffic classification over time. The blue curve represents the number of incoming traffic for inference as workload, and the green curve the accuracy of the current employed accelerator. The accelerator selection algo-

Table 1. CNN topology of the evaluated models. The final layer differs in number of outputs for each task. All CONV and FC layers implement bias.

Layer	Input	Kernel size	Stride	Padding	Output
1D Convolution + BatchNorm + ReLU	784 * 1	25	1	Same	784 * 32
1D Max Pool	784 * 32	4	4	Same	196 * 32
1D Convolution + BatchNorm + ReLU	196 * 32	25	1	Same	196 * 64
1D Max Pool	196 * 64	4	4	Same	49 * 64
Fully Connected + BatchNorm + ReLU	49 * 64	–	–	None	1024
Fully Connected + BatchNorm + SoftMax	1024	–	–	None	2/6/12

rithm adapts to the request type, selecting the model from the library with the best possible QoE for each scenario. Here, initially an accelerator with 0% pruning is selected for task A, and when task B is requested, the FPGA is reconfigured with a 25% pruning accelerator. Notice that FPGA reconfiguration is not instantaneous, and will cause a reduction in performance if it is performed too frequently. During the second processing of task A, the workload changes drastically, causing the framework to select an accelerator capable of processing more inferences per second, with a higher pruning rate but lower accuracy.

4 Methodology

The encrypted VPN traffic classification problem from the ISCX VPN-nonVPN traffic dataset [5] is used for the evaluation of the framework. The dataset offers four representations of the packets and Session+All is selected since it presents the best accuracy results. The samples are 784 bytes long and labeled into 12 classes, differentiating between regular (non VPN) and protocol (VPN) encrypted traffic and by their traffic type: email, chat, streaming, file transfer, VoIP or P2P. This leads to four possible distinct classification tasks: **2 classes**: VPN and non VPN; **6 classes VPN**: classify exclusively VPN packets by traffic type; **6 classes nonVPN**: exclusively nonVPN traffic; **12 classes**: classify by both protocol and traffic type.

Based on the 1d CNNs proposed in [16], four CNNs are trained, with three modifications to fit into FINN's dataflow design. First: inputs, weights, bias and activations are quantized. Second: Batch normalization layers are added as they work well with the implemented quantization technique and can be absorbed by the quantization layer. Third: the size of the max pool kernels is increased to 4. All tasks are implemented using a similar topology, varying the number of neurons on the output layer (2, 6 or 12). Table 1 details the CNN models.

Training was done with the quantization-aware library Brevitas [14]. The CNNs use 2-bit inputs and 4-bit weights, bias, and activations. Models were trained for 500 epochs with a learning rate of 0.01, using stochastic gradient descent with a minibatch size of 50. Pruned CNNs are retrained using the same hyperparameters for 150 epochs. The TOP-1 test accuracy results are reported.

All accelerators used for the experiments were generated using Xilinx's FINN [2] and synthesized using Xilinx Vivado for resource usage and power information, and Verilator is used for RTL simulations of performance. The accelerators are synthesized to an Alveo U200 Data Center Acceleration card at 100 MHz, which is commonly used as smartNIC and network accelerator card. The time for FPGA reconfiguration is 300ms. FINN provides flexibility when selecting the parallelism of design via PE and SIMD configuration. To fairly evaluate the pruned accelerator's processing capabilities, all implementations were constrained equally. Each accelerator is composed of four MVTU's, with PEs fixed at 4, 8, 32, with the last layer using the same number of classes: 2, 6 or 12, while SIMD is fixed at 5, 4, 8, and 32. The Dataflow-Aware pruning allows for six unique pruning accelerators under this constraint: 15%, 25%, 40%, 50%, 65% and 75%.

To evaluate the framework, we propose four scenarios that vary the in terms of workload and the frequency new tasks are requested:

- **Stable Workload (S):** varying the number of incoming packets by 12.5% every 8 s;
- **Variable Workload (V):** varying the number of incoming packets by 75% every 2 s;
- **Low Switching rate (L):** a different task is requested every 15 s;
- **High Switching rate (H):** tasks change every 2 s.

These characteristics are combined, forming the scenarios SH, SL, VH and VL, to cover the different possible network conditions. The four scenarios run for 360 s. As a baseline in our experiments, the unmodified original FINN hardware models with no pruning are used (Table 1), and each time a new task is requested the FPGA is reconfigured.

Quality of Experience (QoE) (as shown in Eq. 1, performance, and energy per inference are used to evaluate the framework.

5   Results

Offline Step Evaluation: Initially, the design space created by the library pruned accelerators generated at the offline phase is analyzed. Figure 5 (A) shows the impact pruning has on each classifier, with the x-axis showing the percentage of channels pruned and the y-axis showing the achieved accuracy for post training evaluation. At the leftmost side of the graph the pruning rate is zero, with the dashed lines representing the accuracies of the state-of-the-art implementation shown in [16]. It is clear that even with quantization, our unpruned classifiers achieve accuracy levels close to the original models, with accuracy drops of 0.20, 1.17, 6.37 and 3.30% for 2, 6 vpn, 6 nonVPN and 12 classes tasks, respectively. Some accuracy reduction is expected when using quantization, due to the reduced precision of calculations.

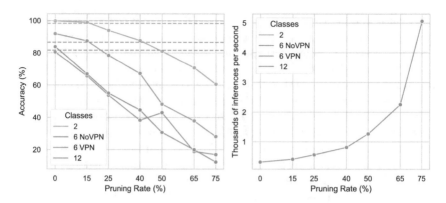

Fig. 5. Accuracy vs. pruning rate (A) and inferences per second vs. pruning rate (B) for each accelerator model (the different lines overlap each other, making only the 12 classes visible).

We can also notice that the impact of pruning differs for each task. Noticeably, 6 nonVPN and 12 classes tasks have a steep accuracy reduction, even with only 15% pruning, whereas 2 and 6 VPN tasks maintain a smoother decrease in accuracy for up to 40% pruning. This is due to the fact that 6 nonVPN and 12 classes original models already achieve lower accuracy compared to 2 and 6 VPN, requiring more parameters to achieve great accuracy, making the pruning of parameters affect accuracy more harshly. The inferences per second is similar across tasks as they overlap each other in the graph, and they increase exponentially with the pruning rate, as shown in Fig. 5 (B). On the 2 classes task, our Framework is capable of increasing the inferences per second from 318, with no pruning, to 565, at 25% pruning, at a 6.58% accuracy drop. On the 12 classes task, a more significant accuracy drop (28.82%) is observed for the same performance increase. This leads to a diversity of optimization opportunities produced by the trade-off of accuracy and number of inferences in each task, which will be exploited by the dynamic part of the framework.

Runtime Step Evaluation: Table 2 shows the average QoE, increase in the number of processed inferences, and energy per inference for each scenario (Sect. 4) considering the framework and their comparison to the baseline. The framework is able to adapt to the changing network requests, with an increase of up to 1.14× in QoE, enabled by the framework's model selection algorithm, adapting the running accelerator based on each task's trade-off in both stable and variable workloads for all evaluated scenarios. The use of pruned accelerators with higher processing capability allows the processing of greater workloads while maintaining a good accuracy level, resulting in a better QoE when compared to the baseline implementation.

The number of reconfigurations reflects the optimization opportunities found by our approach. For the stable workload with a low task switching scenario (SL), it only presented 2 additional reconfigurations compared to the base-

Table 2. Results comparing the baseline to the proposed framework.

Scenario	SH		SL		VH		VL	
Algorithm	Baseline	Framework	Baseline	Framework	Baseline	Framework	Baseline	Framework
Average QoE (%)	61.87	70.11	69.19	77.16	60.52	69.10	68.97	74.57
QoE increase	1.00	**1.13**	1.00	**1.12**	1.00	**1.14**	1.00	**1.08**
Increase in processed inferences	1.00	**1.27**	1.00	**1.34**	1.00	**1.51**	1.00	**1.31**
Energy per inference (mJ)	12.89	9.82	11.62	8.37	12.89	8.33	11.62	8.70
Increase in energy efficiency	1.00	**1.24**	1.00	**1.28**	1.00	**1.35**	1.00	**1.25**
FPGA reconfigurations	135	149	18	20	135	166	18	114

line, mostly changing hardware configurations according to the requested task. Still, our dynamic QoE algorithm could select the best accelerator even with few opportunities for optimization. Considering the variable workload and low switching (VL) scenario, the baseline performs only 18 reconfigurations (one for each task switch), whereas the framework performs 114. The workload variance offers many more opportunities to leverage pruning rates for the same task, using higher pruning rates for greater workloads and more accurate accelerators for smaller ones. Even considering the time spent reconfiguring the FPGA, the framework can process more inferences than the baseline model.

The framework's effectiveness is highlighted in the variable workload high task switching rate (VH) scenario. Compared to the baseline, it resulted in 31 additional reconfigurations. Still, our approach could accurately find several opportunities for optimization within the same task, even in a scenario where the tasks are often switched. On the other hand, SH stable workload scenario offers almost as many chances for reconfiguration, but since the workload variance is smaller, changing accelerators has less impact on the total processed frames.

Overall, the gains in energy efficiency show that even though the framework uses an algorithm based on maximizing QoE, it also leads to energy savings, due to the higher number of processed frames enabled by the use of pruning.

6   Conclusion

This work proposed a framework for processing multiple traffic classification tasks, while dynamically adapting the inference processing on reconfigurable SmartNICs. The framework achieved up to $1.14\times$ improvement on QoE and up to $1.51\times$ more inferences were processed at $1.35\times$ better energy efficiency when compared to a state-of-the-art baseline.

Acknowledgements. This study was financed in part by Coordenação de Aperfeiçoamento de Pessoal de Nível Superior - Brasil (CAPES) - Brazil - Finance Code 001, São Paulo Research Foundation (FAPESP) grant #2021/06825-8, FAPERGS and CNPq.

References

1. Aceto, G., Ciuonzo, D., Montieri, A., Pescapè, A.: Mobile encrypted traffic classification using deep learning: experimental evaluation, lessons learned, and challenges. IEEE Trans. Netw. Serv. Manag. **16**(2), 445–458 (2019)
2. Blott, M., et al.: FINN-R: an end-to-end deep-learning framework for fast exploration of quantized neural networks. ACM Trans. Reconfigurable Technol. Syst. **11**(3), 1–23 (2018)
3. Boutaba, R., et al.: A comprehensive survey on machine learning for networking: evolution, applications and research opportunities. J. Internet Serv. Appl. **9**(1), 1–99 (2018). https://doi.org/10.1186/s13174-018-0087-2
4. Choudhary, T., Mishra, V., Goswami, A., Sarangapani, J.: A comprehensive survey on model compression and acceleration. Artif. Intell. Rev. **53**(7), 5113–5155 (2020). https://doi.org/10.1007/s10462-020-09816-7
5. Draper-Gil, G., Lashkari, A.H., Mamun, M.S.I., Ghorbani, A.: Characterization of encrypted and VPN traffic using time-related features. In: Proceedings of the 2nd International Conference on Information Systems Security and Privacy, pp. 407–414. SCITEPRESS - Science and Technology Publications, Rome, Italy (2016)
6. Elnawawy, M., Sagahyroon, A., Shanableh, T.: FPGA-based network traffic classification using machine learning. IEEE Access **8**, 175637–175650 (2020)
7. Fang, B., Zeng, X., Zhang, M.: NestDNN: resource-aware multi-tenant on-device deep learning for continuous mobile vision. In: MobiCom, pp. 115–127 (2018)
8. Han, S., Pool, J., Tran, J., Dally, W.J.: Learning both weights and connections for efficient neural networks. In: NIPS, pp. 1135–1143 (2015)
9. Kang, W., Kim, D., Park, J.: DMS: dynamic model scaling for quality-aware deep learning inference in mobile and embedded devices. IEEE Access **7**, 168048–168059 (2019)
10. Korol, G., Jordan, M.G., Rutzig, M.B., Beck, A.C.S.: Synergistically exploiting CNN pruning and HLS versioning for adaptive inference on multi-FPGAs at the edge. ACM Trans. Embed. Comput. Syst. **20**(5s), 1–26 (2021)
11. Korol, G., Jordan, M.G., Rutzig, M.B., Beck, A.C.S.: AdaFlow: a framework for adaptive dataflow CNN acceleration on FPGAs. In: DATE, pp. 244–249. IEEE (2022)
12. Li, H., Kadav, A., Durdanovic, I., Samet, H., Graf, H.P.: Pruning filters for efficient convnets. In: ICLR (2017)
13. Lockwood, J.W., McKeown, N., Watson, G., et al.: NetFPGA–an open platform for gigabit-rate network switching and routing. In: MSE, pp. 160–161. IEEE Computer Society (2007)
14. Pappalardo, A.: Xilinx/Brevitas (2021). https://doi.org/10.5281/zenodo.3333552
15. Tong, D., Qu, Y.R., Prasanna, V.K.: Accelerating decision tree based traffic classification on FPGA and multicore platforms. IEEE Trans. Parallel Distrib. Syst. **28**(11), 3046–3059 (2017)
16. Wang, W., Zhu, M., Wang, J., Zeng, X., Yang, Z.: End-to-end encrypted traffic classification with one-dimensional convolution neural networks. In: 2017 IEEE International Conference on Intelligence and Security Informatics (ISI), pp. 43–48. IEEE, Beijing, China (2017)
17. Xu, Z., Yu, F., Liu, C., Chen, X.: Reform: static and dynamic resource-aware DNN reconfiguration framework for mobile device. In: DAC, pp. 1–6 (2019)
18. Zhang, W., Wang, J., Chen, S., Qi, H., Li, K.: A framework for resource-aware online traffic classification using CNN. In: CFI, pp. 5:1–5:6. ACM (2019)

eDeepRFID-IPS: Enhanced RFID Indoor Positioning with Deep Learning for Internet of Things

Belal Alsinglawi[1(✉)] and Khaled Rabie[2]

[1] Western Sydney University, Parramatta, Sydney, NSW, Australia
B.Alsinglawi@westernsydney.edu.au
[2] Manchester Metropolitan University, Manchester, UK
k.rabie@mmu.ac.uk

Abstract. In smart environments, indoor positioning systems provide several options for smart computing users, businesses, and industries, thereby dramatically enhancing human well-being and productivity. Smart homes and smart indoor environments are prominent emerging technologies in the Internet of Things era and future communications, with applications such as providing personalized healthcare to the elderly and those with impairments by connecting them to the world via high-speed wireless communication infrastructure. This study offers a new approach to real-time indoor positioning using passive RFID technology to estimate the real-time location of smart home users based on their movements in smart environment space. An experimental indoor positioning system technique intends to improve assisted living and identify daily activities in a smart environment. To demonstrate this, we conducted a case study on indoor positioning using RFID technology. The experimental investigation is based on a location-based system that leverages the creation of deep learning algorithms in conjunction with radio signal strength indicator (RSSI) measurements of passive RFID-tagged devices. The proposed architecture encourages more precise identification of smart home objects and the ability to precisely locate users in real-time with good measured precision while minimizing technical and technological barriers to the adoption of location-based technologies in the daily lives of smart environment inhabitants. This will eventually facilitate the realization of location-based Internet of Things (IoT) systems.

1 Introduction

Internet of Things (IoT) presents a new type of home automation, communication with other devices, intelligence, and connectivity to everyday devices [1]. The capabilities of IoT are incorporated into smart home devices and sensors, which bring greater benefits. Smart home devices can be stationary or mobile, measuring and sensing the condition of an object and its interactions with its users, acting as actuators to perform activities (opening doors, turning off appliances), or combining both functions. IoT systems enable these devices to be queried by other platforms, controller devices, or IoT applications and services that coordinate many items without human intervention. In addition, the

© The Author(s), under exclusive license to Springer Nature Switzerland AG 2023
L. Barolli (Ed.): AINA 2023, LNNS 654, pp. 149–158, 2023.
https://doi.org/10.1007/978-3-031-28451-9_13

data acquired from IoT smart home devices at the perception layer can be sent to the intermediate data layers of data gathering, processing, and analytics on the middleware layer (known as the edge and fog layers) using appropriate IoT communication protocols. The processed data or data attributes are then streamed and shared in a centralized or decentralized computing paradigm (e.g., federated learning platforms) to serve users or other IoT services connected via IoT microservices or IoT-end service points. This generates services with added value for users who benefit from IoT peculiarities. These services are based on the development of IoT architecture layers [2]. Therefore, consumers of smart homes benefit from IoT services, modeling, and the interweaving of information that enables smart home systems to make more intelligent decisions based on the users' preferences or to offer more tailored and optimal services. This cannot be achieved without integrating context-aware technology into IoT.

The indoor positioning system (IPS) is a network of sensors used to locate individuals or objects in indoor environments where GPS and other satellite technologies lack precision or fail entirely in indoor spaces [1]. Contextual data in IoT systems is primarily utilized to provide customized services, enhance the quality and precision of the information, find adjacent services, and finally facilitate implicit user interactions [3]. Consequently, indoor positioning - also known as indoor localization or location-based systems- is considered a crucial enabler for this technology in the era of IoT and high-speed wireless technologies such as 5G. In recent years, research on indoor placement in smart homes, also referred to as indoor localization, has been one of the primary focuses of the IoT community. In the context of location-aware systems, numerous techniques have been developed to improve the precision of tracking humans' positions or objects in smart homes and localizing fixed or mobile objects. These approaches use many IoT technologies, such as radio frequency-based technologies (RFID, Bluetooth, Zigbee, UWB, WLAN), ultrasonic sensors [1], light-based location, and many more wireless light-based techniques [3], to find the target items within sensing accuracies. These technologies play a crucial role in a variety of smart indoor applications, such as healthcare monitoring systems for the elderly and disabled [5, 6].

Typically, they try to improve these individuals' daily activities [6], including their well-being and their ability to perform daily domestic tasks. Nonetheless, these technologies also serve as the fundamental components of smart grid communication systems. RFID is a desirable technology for indoor localization in IoT smart homes. In RFID-based systems, the RFID reader sends energy to an antenna, which converts the energy into an RF wave that is sent to a read zone, once the tag is within the read zone. Once the RFID tag has been read within the read zone, its internal antenna receives power from the RF waves. The energy flows from the antenna of the tag to the IC and powers the chip, which then generates a signal and transmits it back to the RF system. This process is called "backscatter technique." [7].

Figure 1 depicts a typical RFID system with antennas attached to an RFID reader. These antennas return captured data from sensed tags (e.g., passive tags) to the reader for location processing. Active (battery-powered) RFID tags can also be semi-active or passive (without a built-in battery). RFID tracking systems may employ active, passive, or semi-active methods. RFID tracking applications include in-hospital patient tracking, smart cities and smart grids, asset tracking, supply chain management, security, medical and healthcare asset tracking, industry 4.0, and perhaps industry 5.0, as well as several more.

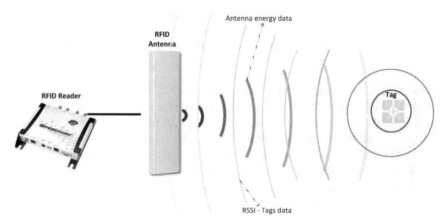

Fig. 1. Illustrates the communication paradigm for the passive RFID localization system [1].

Current studies in literature such as [8–10] have examined passive RFID for RFID location-based indoor systems by installing many tags on the floor or attaching them to smart home devices to properly identify them. Users must wear multiple tags and deploy them in indoor environments where some may be invisible to readers. Furthermore, some of these methods are restricted to certain tagged objects that users must carry or attach to their mobile objects in order to achieve their full potential. These methods are ineffective; hence, users either employ many tags, which require constant maintenance, or alter the positions of fixed tags periodically. Moreover, users of smart homes must measure their locations and input them into the indoor positioning system (IPS). Therefore, this is impractical, particularly in expansive or dynamic contexts where tags must be modified in the presence of several sensed objects with non-static locations. To overcome this issue, we build on prior research that identified residents of smart houses using a single passive RFID tag based on passive RFID backscattering principles [5]. Using the geometric properties of the RFID 3 antenna, this technique converts the RSSI into distance relative to the target tag (passive RFID tag). In this paper, we present a deep learning prediction approach for indoor localization RFID systems in smart homes to improve the locations of smart home users, aiming for a more precise, cost-effective, and operationally efficient system to identify and estimate the users' location for better assistive living in smart home objects.

This paper is structured as follows: the background section provides some context for RFID systems. The second section examines the literature review and related studies on RFID localization systems. The proposed RFID localization and predictive system for smart homes on the IoT are introduced in Sect. 3. The fourth section examines and describes the evaluation investigation of the proposed system on a real indoor positioning dataset. Section 5 concludes the paper and presents recommendations for the future.

2 Related Work

In the literature, various RFID indoor localization methods have been presented. The literature research finds three prevalent detection and position estimation algorithms using RFID technology. IPS approaches are based on geometry (distance estimation), Scene Analysis, fingerprinting, proximity and machine learning (ML) related methods [1].

Estimating Distance-Related Methods: The geometrical characteristics of triangles are utilized by distance estimation techniques to determine the positions of targets (RFID-tagged targets). Trilateration and multilateration are the two most common approaches for estimating distance. Time-based approaches, including time of arrival (TOA) and time difference of arrival (TDOA), are also frequently employed in indoor localization [1]. Some studies that utilized distance estimation reported approaches in smart home indoor environments where the distance estimation methods "trilateration", "multrilateration" and "RSSI" were extensively used [11–13].

RFID Scene Analysis Associated Methodologies: The scene analysis approach captures the features (fingerprints) of a scene and then estimates the position of the tagged objects by matching the online measurements to detect the location fingerprints that are the closest match. Certain probabilistic scene analysis algorithms, such as those described in [14], determine the location of an RFID tag.

Proximity-Related Approaches: In the proximity approach, the position is determined by the relative placement of a dense array of antennae. When a mobile target enters the radio signal coverage of a single antenna, the antenna interprets the target as a collocate item throughout the duration of its coverage. If multiple antennas identify the same target, only the antenna with the strongest signal will gather it. The cell of origin (COO) [1] determines the location of the mobile target, which is within the cell's coverage limitations. The localization approach is straightforward and does not demand intensive implementation. However, accuracy is dependent on the number of antennas and the signal's range. This also implies that the approximate position of the tagged object at a given time is utilized.

RSSI-Related Methodologies: The RSSI calculates the distance between two RFID readers by measuring the signal strength at the receiving end [15]. According to numerous studies, indoor positioning with RSSI is advantageous because RSSI is slow to respond to abrupt changes and is a highly averaged indicator for distance approximations in inside location systems [16]. The works described in [11–13] are some of the most current RFID-based localization strategies that employ RSSI for indoor locations.

Approaches Related to ML: ML models facilitate research in numerous domains and applications. Due to developments in data science and artificial intelligent (AI)-optimization techniques, using ML to detect or predict the location of an RFID tag of interest has become a prominent technique among researchers in recent years. ML models, for instance, make use of probabilistic approaches, regression-based or even classification-based models, and deep neural networks when applicable within the context of IPS. RFID tag forecasting is contingent upon the nature of the prediction and the types of input information. Although there are genuine attempts in the field to increase RFID indoor positioning accuracy for tracking fixed and mobile items, the majority of these initiatives rely on tag density on the smart home floor to accomplish the desired outcomes. K-nearest-neighbor (kNN), support vector regression (SVR) and artificial neural networks (ANN), convolutional neural network (CNN), and long short-term memory networks (LSTM) [17] are examples of these typical works that predict the target position of RFID tags using machine learning-based methods.

Other Indoor Localization Methods: Indoor Positioning in 3D: As demonstrated in [18], the 3D RFID indoor-based localization technique relies on the RFID tag array to supply 3D location information explicitly. In IPS systems, the utilization of angle of arrival (AoA)-based localization is an additional notable technique. 3D Angle of Arrival (AoA) localization, as described in [19], is an effective technique that uses the signal arriving at the antenna (i.e., RFID) to estimate the location of the sender (i.e., RFID tag). Linear Quadratic Estimation (LQE) techniques, such as the Kalman filter as described in [20], estimate unknown variables by calculating a joint probability distribution over the variables for each time frame. The Kalman filter estimates the real state of the target object (e.g., an RFID tag) given noisy input in RFID systems (input with some inaccuracy). The Kalman filter estimates the true position (e.g., the actual RFID tag), and then predicts the future state of the RFID tag's position based on previous readings (RFID tag prior placements). Recent RFID-based localization studies, such as [20], estimate the target RFID tag using the Kalman filter.

The literature has not reported a unanimous method surpassing other localization methods in RFID-IPS based systems, particularly the passive RFID systems. Passive RFID localization is currently unattainable due to a number of obstacles, such as the technological challenge of passive RFID systems, whose generated signals are susceptible to a number of environmental challenges, causing RSSI values to fluctuate and making indoor positioning inherently difficult. However, advancements in signal-filtering techniques have enabled passive RFID to be trusted in indoor positioning due to its cost-effectiveness and ease of use, and it has permeated many aspects of daily life, from booming gate access gates, access buildings to RFID applications in big corporations, such as supply chain management. This study is driven by research results and advancements in the literature to work towards an enhanced localization approach utilizing passive RFID technology to advance the core research for many ongoing projects, including assistive homes, smart cities, industry 4.0, and many others that are based on the fundamental IoT context-aware indoor positioning paradigm.

3   The proposed eDeepRFID-IPS System

Figure 2 shows the eDeepRFID-IPS system architecture. The eDeepRFID-IPS works in three stages, namely, the sensing stage, data preprocessing stage, and prediction stage.

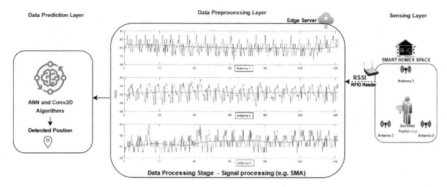

Fig. 2. eDeepRFID-IPS indoor positioning deep learning prediction architecture in smart homes passive RFID systems.

Sensing Stage (Perception Layer): This layer typically consists of the RFID system's primary physical components. The user is connected to the smart home using an RFID passive tag carried by the user (e.g., attached to a walking stick), and the identifiable users are interacting with other home objects (e.g., home appliances) to define an activity that the user is conducting at a given time. There is a spatial-temporal relationship between the tagged subject and the tagged smart home device. Consequently, the tag uniquely identifies each individual or object, and the tag transmits backscatter signals to the RFID reader via the RFID antenna's passive mode. RFID antennas read the tag's RSSI values, which are generated by passive RFID tags and calculated as the antenna's gain in real-time (dBi). The RFID reader emits less energy to detect tags. In passive tracking, RFID readers require more energy to localize RFID tags with no power supply and rely only on the reader's signals. RFID reader is connected to the LAN network, which is interfaced with the local server, and sends data to the local machine through the Simple Network Management Protocol (SNMP) network monitoring protocol [3]. The local server acts as an edge device for further data processing sage. In this work, we used the RFID indoor positioning dataset that was used in the previous work [3].

The dataset contains the RFID tag locations, which were tracked in a stationary and real-time experimental testbed area with a size of 2.75 m × 3.0 m The dataset consists of the RFID RSSI values of the tags, and since the RFID reader reports back to the local server 4 RSSI readings per second for the corresponding location of the tag at a time (t). The RSSI of the tags is reported by the three antennas that are connected to the RFID reader. The tag locations in real-time, the antenna locations (reference points), and the RSSI values of the tag in relation to each antenna are obtained using the Friss equation and the Trilateration geometrical algorithm [5] to obtain the estimated actual locations of the tag, which are measured by distance (meter). The dataset also contains the actual location of the tag in the real-time indoor positioning scenario. This dataset used the following scenario: "A user holding a walking stick walks for five minutes within the testbed area".

Data Preprocessing Stage: The proposed approach is designed to collect streaming data from the RFID reader for each antenna connected to the reader during the data processing phase. Multiple elements, such as environmental interference (walls, magnetic fields, etc.) and the human body, typically contribute to signal noise in RFID readings (RSSI) for RFID signals. Before moving on to the next phase of the data prediction stage, it is vital to apply the right filtering technique, such as the simple moving average filter (SMA), to smooth RSSI values and eliminate noise from the RFID signal. The SMA is a signal smoothing technique that reduces the offbeat RSSI readings [3], reducing the impact of environmental interference (noise in RSSI readings) or reducing the odd RSSI values due to this problem. Noting that this problem involves a time series, capturing the RFID tag's mobility within the testbed region is crucial. This was obtained in a single second by calculating the average values for each time. Therefore, we have two points that define points before and after the prediction point, and we can split the dataset using a validation split that predicts at least 20 percent of the unseen data.

Prediction Stage (Indoor Positioning Prediction Layer): The processed, filtered, and manipulated RSSI data from the previous layer becomes the input to the RFID prediction layer. In this layer, we build and evaluate two neural networks). The ANN [17] model is built with five layers (1 input layer with three hidden layers and an output layer). The second model, the Convolutional Neural Networks (CNN), considers the time-based data collected from the RFID sensors. In both models, RSSI values are real-time input mapped to their corresponding real-world locations (ground truth) in the floor grid of the smart home test bed. The RSSI, the locations of the antennas, and the ground truth values become the input neurons to the deep neural network architecture to learn from complex location-based patterns.

Neural Network Architecture (DNN) with Conv2D: a supervised deep learning model was created with fully-connected neural networks that learn with CNN (Conv2D). Using a semi-standard [21] CNN to capture the spatial-temporal understanding and dependencies between the reference points (RFID antennas) and the target passive RFID tag by converting the long (x, y) vectors input into an image so that we could allow the replacement of sensors or transferring this CNN setup to a new building with zero knowledge, then the system will transfer values into the new settings (knowledge transfer) as long as we have a similar setup (testbed) to behave in a similar manner. The x and y location values of 3X RFID antennae were converted into an input of 5×5 box pixels, allowing us to feed CNN with this information.

Layer input is convolved with the 2D convolution layer (Conv2D) to produce a tensor of outputs. A ReLU activation function was used, and padding (valid) kept the filter window inside the input image. Also, the kernel size parameter is used to set the height and width of the 2D convolution. Consequently, we have spatial data, MaxPooling2D layers, which downsample the input along its spatial dimensions (height and width) and extract the maximum value throughout an input window for each input channel. Using Root Mean Square Error (RMSE) [22] is the residuals' standard deviation (prediction errors). Residuals are a measure of how far data points are from the regression line; RMSE is a measure of how dispersed these residuals are. In other words, it indicates the degree of data concentration around the line of best fit. The RMSE is used to penalize the far error heavily.

4   Results and Discussion

Figure 3 describes the results of the real-time positioning of the eDeepRFID-IPS system based on the utilization of the Conv2D-CNN algorithm and the dataset in offline positioning. System performance evaluation is measured by the Root Mean Squared Error (RMSE) of the predicted values (estimated) to the actual ground truth locations (collected data) from the testbed. The ANN reported an RMSE of 3.67, whereas the Conv2D-CNN results indicate a good evaluation of the system's performance, with a 3.6. It is worth noting that the larger the RMSE, the weaker the predictions are.

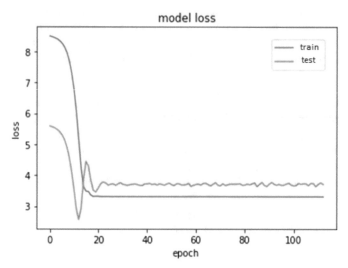

Fig. 3. Model loss (Conv2D) for eDeepRFID-IPS IIndoor positioning

The performance of the eDeepRFID-IPS model loss evaluation during training and validation is depicted in Fig. 3. The graph demonstrates that learning performance fluctuates slightly over time. This demonstrates the viability of the Conv2D-CNN algorithm for passive RFID indoor positioning systems. The eDeepRFID-IPS has multiple applications, and in addition to indoor positioning and localization, it is possible for activities of daily living (ADL) in elderly smart homes [23]. For example, a Fitbit can be attached to walking sticks or wheelchairs so that people (such as the elderly) can determine their location in relation to other household objects in real time. In addition, it may be used to define basic ADLs and some instrumental ADLs for smart home users; hence, the system's ability to correlate users' locations to their ongoing activity and detect the type of activity performed by the user at a particular time. This aids in a variety of domestic activities, including monitoring the health of the elderly in real-time and supporting people with disabilities. It can be applied effectively to prevent wandering off, particularly in nursing homes and with dementia patients who may wander out. With the simplicity of deploying eDeepRFID-IPS in real-world settings, it benefits the users; consequently, it offers a number of services with cost-effective, non-invasive, and extremely lightweight passive RFID tags that can be attached to almost any object in homes or carried by users, weighing almost 0.3 g.

eDeepRFID-IPS has some shortcomings, such as the fact that the model and system do not account for predicting RFID tag positions using unlabeled data; consequently, future research should feed the neural network model with unlabeled data. In addition, passive RFID tags are susceptible to environmental interference, and some of the readings are unreliable, which may influence the location accuracy or define a particular user activity. This difficulty is associated with passive RFID tags; however, implementing advanced or sophisticated techniques in combination with other accurate, cost-effective IoT sensors and technologies can overcome these obstacles and, as a result, unlock the tremendous positioning potential of RFID systems.

5 Conclusion

This research developed a reliable eDeepRFID-IPS positioning system and a deep learning prediction framework for the challenge of RFID-based indoor positioning systems in smart homes connected to IoT. eDeepRFID-IPS trains a DNN architecture (Conv2D) utilizing spatial data collected from a previous dataset that used passive RFID tags and an RFID system to position and track passive RFID tags in real-time while taking advantage of the geometrical relationship between the passive RFID tags in relation to their sensing antennas and the RSSI values generated by them. The ANN and Conv2D evaluations demonstrated that the system is capable of passively tracking in real-time using an offline dataset. In future work, we will evaluate the model's real-time and online position estimation capabilities while exploring moderately advanced deep neural network approaches. In addition, we intend to leverage eDeepRFID-IPS's capabilities in ADL applications and other important newly emerging IoT applications, such as smart manufacturing.

References

1. Alsinglawi, B., Elkhodr, M., Nguyen, Q.V., Gunawardana, U., Maeder, A., Simoff, S.: RFID localisation for Internet of Things smart homes: a survey. arXiv preprint arXiv:1702.02311 (2017)
2. Al-Qaseemi, S.A., Almulhim, H.A., Almulhim, M.F., Chaudhry, S.R.: IoT architecture challenges and issues: Lack of standardization. In: 2016 Future Technologies Conference (FTC), pp. 731–738. IEEE (2016)
3. Jiang, H., Li, J., Zhao, P., Zeng, F., Xiao, Z., Iyengar, A.: Location privacy-preserving mechanisms in location-based services: a comprehensive survey. ACM Comput. Surv. (CSUR) **54**, 1–36 (2021)
4. Sheikh, S.M., Asif, H.M., Raahemifar, K., Al-Turjman, F.: Time difference of arrival based indoor positioning system using visible light communication. IEEE Access **9**, 52113–52124 (2021)
5. Alsinglawi, B., Liu, T., Nguyen, Q.V., Gunawardana, U., Maeder, A., Simoff, S.J.: Passive RFID localisation framework in smart homes healthcare settings. In: The Promise of New Technologies in an Age of New Health Challenges: Selected Papers from 5th Global Telehealth Conference 2016, Auckland, New Zealand, 1–2 November 2016, pp. 1–8 (2016)

6. Qi, J., Yang, P., Newcombe, L., Peng, X., Yang, Y., Zhao, Z.: An overview of data fusion techniques for Internet of Things enabled physical activity recognition and measure. Inf. Fusion **55**, 269–280 (2020)
7. Tian, C., Ma, Y., Wang, B.: Cooperative localization for passive RFID backscatter networks and theoretical analysis of performance limit. IEEE Trans. Wireless Commun. (2022)
8. Vena, A., Illanes, I., Alidieres, L., Sorli, B., Perea, F.: RFID based Indoor Localization System to Analyze Visitor Behavior in a Museum. In: 2021 IEEE International Conference on RFID Technology and Applications (RFID-TA), pp. 183–186. IEEE (2021)
9. Zhu, C., Zhao, S., Xia, Y., Li, L.: An improved three-point localization method based on RSS for transceiver separation RFID systems. Measurement **187**, 110283 (2022)
10. Zeng, Y., Liao, Y., Chen, X., Tan, H.: UHF RFID indoor localization based on phase difference. IETE J. Res. 1–7 (2021)
11. Sheikh, S.M., Asif, H.M., Raahemifar, K., Kausar, F., Rodrigues, J.J., Mumtaz, S.: RSSI based implementation of indoor positioning visible light communication system in NS-3. In: ICC 2021-IEEE International Conference on Communications, pp. 1–6. IEEE (2021)
12. De Oliveira, L.S., Rayel, O.K., Leitao, P.: Low-Cost Indoor Localization System Combining Multilateration and Kalman Filter. In: 2021 IEEE 30th International Symposium on Industrial Electronics (ISIE), pp. 1–6. IEEE (2021)
13. Strzoda, A., Marjasz, R., Grochla, K.: How accurate is LoRa positioning in realistic conditions? In: Proceedings of the 12th ACM International Symposium on Design and Analysis of Intelligent Vehicular Networks and Applications, pp. 31–35 (2022)
14. Sequeira, J.S., Gameiro, D.: A probabilistic approach to RFID-based localization for human-robot interaction in social robotics. Electronics **6**, 32 (2017)
15. Avanzato, R., Beritelli, F., Capizzi, G., Napoli, C., Rametta, C.: Soil moisture estimation based on the RSSI of RFID modules. In: 2021 11th IEEE International Conference on Intelligent Data Acquisition and Advanced Computing Systems: Technology and Applications (IDAACS), pp. 647–652. IEEE (2021)
16. Jovalekic, N., Drndarevic, V., Pietrosemoli, E., Darby, I., Zennaro, M.: Experimental study of LoRa transmission over seawater. Sensors **18**, 2853 (2018)
17. Yang, T., Cabani, A., Chafouk, H.: A survey of recent indoor localization scenarios and methodologies. Sensors **21**, 8086 (2021)
18. Maheepala, M., Joordens, M.A., Kouzani, A.Z.: A low-power connected 3D indoor positioning device. IEEE Internet Things J. **9**, 9002–9011 (2021)
19. Alma'aitah, A., Alsaify, B., Bani-Hani, R.: Three-dimensional empirical AoA localization technique for indoor applications. Sensors **19**, 5544 (2019)
20. Soni, N., Mishra, A.: Intelligent RFID-based localization and tracking model design using unscented Kalman filter and convolution neural network. In: 2022 4th International Conference on Smart Systems and Inventive Technology (ICSSIT), pp. 1101–1107. IEEE (2022)
21. Karadayı, Y., Aydin, M.N., Öğrenci, A.S.: A hybrid deep learning framework for unsupervised anomaly detection in multivariate spatio-temporal data. Appl. Sci. **10**, 5191 (2020)
22. Chai, T., Draxler, R.R.: Root mean square error (RMSE) or mean absolute error (MAE). Geosci. Model Dev. Discuss. 7, 1525–1534 (2014)
23. Zhang, T., Zhao, D., Yang, J., Wang, S., Liu, H.: A smart home based on multi-heterogeneous robots and sensor networks for elderly care. In: Intelligent Robotics and Applications. ICIRA 2022 Lecture Notes in Computer Science, vol. 13455, pp. 98–104. Springer, Cham (2022). https://doi.org/10.1007/978-3-031-13844-7_10

A Model for Artificial Conscience to Control Artificial Intelligence

Davinder Kaur, Suleyman Uslu, and Arjan Durresi[✉]

Indiana University-Purdue University Indianapolis, Indianapolis, IN, USA
{davikaur,suslu}@iu.edu, adurresi@iupui.edu

Abstract. We propose *"Artificial Conscience - Control Module"* framework to control the AI systems and to make them adaptable based on the user requirements. AI users can have different needs from the same AI system, and these systems must adjust their output based on these requirements. The proposed framework enables users to provide context to the AI system by assigning weights to different evaluation metrics. Based on these weights, AI metrics-agents negotiate with each other using our trust engine to output a solution with maximum *"Artificial Feeling."* This framework can be easily implemented in any AI system where multiple metrics are involved. We have illustrated the proposed framework using an AI system for classifying people based on income.

1 Introduction

Artificial Intelligence (AI) systems have transformed our lives. Nowadays, almost every task is either guided or done by algorithms. The rapid development and growing use of these systems have raised many concerns. These systems have become complex and do not always yield safe and reliable results. Taking proper measures to design, develop, test and oversee these systems becomes very important. Different researchers have proposed various ways to accomplish this.

Some government and private research organizations have proposed different ethical guidelines and frameworks to make AI safe, reliable, and trustworthy. One of the primary requirements proposed by most of these frameworks is the involvement of the human agency to control AI [14]. Researchers have proposed the concept of bounded optimality [17,19] to control AI. Another line of researchers has proposed the concept of artificial conscience, which deals with replicating some aspects of the human consciousness in machines [4]. Baasr, a cognitive neuroscientist [1] proposed the Global Workspace Theory of brain consciousness using a theatre analogy. Solms [20,21] proposed that consciousness is related to feelings. Blum proposed the theory of consciousness through a computer science perspective [3], in which different processors compete with each other to get their information broadcasted to other processors.

In this paper, we have combined both lines of research on controlling AI and artificial conscience. We developed a model of an artificial conscience-control module, which can be used to make AI systems adapt according to the users'

L. Barolli (Ed.): AINA 2023, LNNS 654, pp. 159–170, 2023.
https://doi.org/10.1007/978-3-031-28451-9_14

requirements using the concept of "Artificial Feeling." This control module provides controllability of AI system decisions, making them safe, reliable, and trustworthy and hence increasing their acceptance in society. This paper is organized as follows. Section 2 presents the background and related work in the field of AI, artificial conscience, and trust. Section 3 describes our proposed framework, "Artificial Conscience Control Module." Section 4 illustrates our framework using an AI system for classifying people based on their income, and in Sect. 5, we conclude our paper.

2   Background and Related Work

This section presents background and related work for the need to control AI, artificial conscience, and the role of trust.

2.1   Need to Control AI

The wide adoption of AI systems does not imply that they are always safe and reliable [9]. It becomes essential to control and oversee these systems to prevent any harm caused by them to the users or society. Different government and private research organizations have proposed various guidelines and frameworks to make them safe, reliable, and trustworthy. One of the main requirements presented by these agencies is the involvement or control of humans in AI decision-making [14]. Different researchers have proposed various ways to involve humans in the AI lifecycle. European Union (EU) [6] suggested the involvement of humans in three phases: designing, developing, and overseeing. Other researchers have proposed the involvement of humans based on the risk associated with using AI. International Organization for Standardization (ISO) [7] also suggested the involvement of humans by integrating control points in the AI life cycle to increase the trust and adoption of AI systems. The proposed frameworks by all these influential organizations show the importance of human involvement in controlling AI.

2.2   Artificial Conscience and Controlling AI

Artificial conscience, also known as machine conscience, is a way to implement some aspects of human cognition that comes from the phenomenon of consciousness [4]. Different researchers have proposed other goals that can be achieved by artificial conscience. Some of them are autonomy, resilience, self-motivation, and information integration. To achieve these goals, there is a need to design conscious machines that can replicate some features of the conscious experience.

To replicate some features of the conscious experience, it's imperative to understand what consciousness is. Baars, a cognitive neuroscientist [1], proposed the Global Workspace Theory (GWT) of the brain and explained consciousness through the theater analogy as the activity of actors in a play performing on the stage of Working Memory, their performance under observation by a vast audience of unconscious processors sitting in the dark. Another set of researchers proposed the theory of consciousness through the perspective of theoretical computer science known as Conscious Turing Machine (CTM) or Conscious AI [3]. This theory is influenced by Alan Turing's powerful model of computation known as the Turing Machine and by the global theory of consciousness GWT. Based on this theory, different processors compete with each other to get their information on the stage/short-term memory so that it can be broadcasted to other processors. Solms [20,21] proposed that consciousness is endogenous and is related to feelings. However, some other researchers argued that the classic notion of rationality is unattainable for real agents [18]. They proposed the concept of bounded optimality [17,19], which deals with optimizing not the actions taken, but the algorithm used to select the action. These types of agents trade off between efficiency and error. All the work by different researchers deals with understanding the human conscience and how some aspects can be implemented in AI.

2.3   Role of Trust

Trust is a complex phenomenon and is a context-dependent concept. Different disciplines define trust differently. In general, trust is defined as "the confidence one entity has in another entity that it will behave as anticipated" [7]. Trust information is highly influential in decision-making when multiple entities are involved [22]. Different researchers have proposed various ways to calculate and manage trust information. Ruan et al. [16] proposed a trust management framework to quantify trust between entities based on the measurement theory. Because of the flexibility of this framework, it has been used in various decision-making applications like healthcare [10,11], social networks [8,13], crime detection [12], and the food-energy sector [23–28]. The use of trust in all these applications validates its potential to help capture negotiations between different metrics of AI system.

3   Trust Based Artificial Conscience - A Control Module for AI Systems

This section introduces the trust-based artificial conscience control module to control the AI systems based on the user's needs. This control module assumes that different users can have other requirements from the AI system. Based on their requirement, each user assigns weights to the evaluation metrics of the AI system. Based on these weights, these metrics negotiate with each other over a list of solutions to output the agreed solution. The negotiation between the metrics is controlled by the trust and trust sensitivity of the metrics. Section 3.1 explains the trust engine, Sect. 3.2 explains the concept of trust pressure and sensitivity, and Sect. 3.3 describes the framework of the control module.

3.1   Trust Engine

Our trust engine [16] calculates the trust between the entities based on their past interactions. In a decision-making problem, agents interact with each other several times, proposing and rating each other's solutions. The rating provided by an agent to another agent's solution measures the agent's impression of the other agent. The ratings are considered to be between [0,1], where 0 is the lowest rating and 1 is the highest rating. The impression of agent A^X toward agent A^Y, denoted by $m^{X:Y}$, is the mean of the ratings, where $r^{X:Y}$ is the rating of A^X to A^Y given over N number of rounds. Another essential component of trust is confidence, denoted by c. It is used to capture the consistency of the impression. Confidence is inversely related to the standard error of the mean. The formula for impression and confidence is given in Eq. 1.

$$m^{X:Y} = \frac{\sum_{i=1}^{N} r_i^{X:Y}}{N} \quad \text{and} \quad c = 1 - 2 * \sqrt{\frac{\sum_{i=1}^{N} (m^{X:Y} - r_i^{X:Y})^2}{N * (N-1)}} \qquad (1)$$

The value of the impression and confidence is used to calculate the trust between two agents. This trust framework can also be utilized when two agents are not directly connected using the trust propagation and aggregation methods [16].

3.2   Trust Pressure and Trust Sensitivity

Trust pressure and sensitivity are based on the concept of social psychology, which studies the influence of others on the behavior of individuals [27]. Our framework uses trust pressure and sensitivity to capture agents' changing behavior. The source of the trust pressure, denoted as P, is the difference between the target trust level of the agent, T_{target}, and the agent's trust level from other agents, $T_{current}$. How much the trust pressure affects the agent's behavior, namely effective trust pressure, P_e, depends upon his trust sensitivity, S_T as shown in Eq. 2. If the trust sensitivity of an agent is high, it will alter his behavior much faster than an agent with less trust sensitivity. In our framework, trust sensitivity is introduced using the weights of the metrics assigned by the users. These weights are translated into trust sensitivities. Both of them are inversely proportional to each other.

$$P = T_{target} - T_{current}, \quad P_e = P \times S_T \qquad (2)$$

3.3   Artificial Conscience - Control Module

We have proposed an artificial conscience control module to control the AI systems based on the user requirements. Users can have different expectations from the AI system based on their needs. The involvement of humans provides meaning to the working of the AI system, which is easily missed by the algorithms

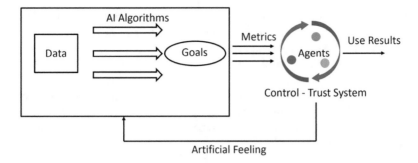

Fig. 1. Artificial conscience - control module to control AI

alone. For our framework, we assume that there is a decision-making task that needs to be performed by the AI system. Different machine-learning algorithms are deployed for that task, and multiple metrics are used to evaluate the solution. Users assign weights to these metrics based on their requirements to provide context and meaning to the decision-making task. The users' weights are used to calculate the trust sensitivities for the metrics. These metrics, which we call "agents," negotiate with each other based on the ratings/trust they get from other agents and their trust sensitivities to calculate "Artificial feeling" (AF) as a weighted average among agents. This framework consists of the following steps:

- At the beginning of the negotiation, each agent is given the goal of achieving the best option they can get among the pre-computed solutions. Different agents can have different importance on the parameters for their goals, leading to negotiation.
- Agents can negotiate for n number of rounds to reach a solution. n is application or user dependent. In the first round, each agent proposes the best solution based on their needs.
- If the agent receives an adequate level of trust which reflects the other agents' approval, it does not need to alter his solution in the subsequent rounds. Otherwise, it selects the next best solution, which would bring in more trust at the expense of a decline in benefits.
- The solutions are compared in terms of the benefit they provide using a distance metric. The distance between a proposed solution p and a goal g is calculated using the euclidean distance formula as shown in Eq. 3 where i is a parameter of a solution among d selected parameters for distance calculation.

$$dist(p, g) = \sqrt{\sum_{i=1}^{d} (p^i - g^i)^2} \tag{3}$$

- In each round, agents rate other agents' solutions based on their distance. For example, assume that there are two agents, namely A_A and A_B. If agent A_A proposes a solution closer to its goal but far away from the agent A_B's goal, then the agent A_A receives a low rating from the agent A_B. Based on the rating provided by the other agents, the trust of the agent is calculated using the framework described in Sect. 3.1.

- In short, each agent tries to minimize the distance from their proposed solution to their goal, considering the trust they receive depending on how much they are sensitive to trust. Their trust is determined by the distance of their proposed solution to other agents' goals. The amount of sacrifice is governed by trust sensitivity. The higher the trust sensitivity, the more the agent sacrifices to raise its rating.

- After n rounds, for each solution, the artificial feeling (AF) is calculated as the weighted average among agents. AF is used for comparing and selecting solutions of different AI algorithms.

This framework enables the control of the AI system based on the user's needs. Figure 1 shows the block diagram of our framework.

4   Implementation

This section describes the dataset used for the experiment, its implementation, and the results using our framework.

4.1   Data

We have performed our experiments on the real-world dataset to consider the accuracy and fairness aspects of the AI algorithm decision-making. US adult income dataset [5] is used for our experiments. The dataset contains 32,561 instances and 14 attributes. It includes two sensitive attributes: race and sex. It is pre-split into training and testing sets for the machine learning prediction task of predicting whether an individual makes more or less than $50,000$ per year. For the dataset, oversampling is performed on the minority target variable, and the categorical attributes are converted into numerical vectors using one-hot encoding to be used for training. In our study, we have considered "White" and "Male" as the privileged class for race and sex attributes.

4.2   Experimentation Setup and Results

In our study, the decision-making task that needs to be accomplished by the AI system is to classify the people according to their income. Two supervised machine learning models, namely support vector machine (SVM) and logistic regression (LR), are used to build the prediction model. A post-processing fairness technique, Calibrated Equalized-Odds, is used. This technique optimally adjusts the learned predictors to remove discrimination based on the equalized odds objective [15]. In our experiment, we used two types of metrics. One is the performance metric, i.e., accuracy (A), to quantify how well the system predicts true labels. Another metric type is the fairness metric to evaluate the system's

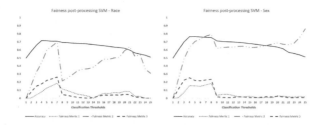

Fig. 2. Accuracy and fairness metrics trade-off for post-processing fairness enhancing SVM on Race and Sex sensitive attributes.

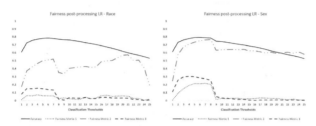

Fig. 3. Accuracy and fairness metrics trade-off for post-processing fairness enhancing LR on Race and Sex sensitive attributes

fairness, which is how unbiased the system predictions are concerning sensitive attributes. The three fairness metrics we use are Equal Opportunity Difference (F1), Disparate Impact (F2), and Average Odds Difference (F3) [2]. All these metrics are used to evaluate and compare different algorithms.

We have computed these metrics for sensitive attributes at 25 evenly distributed classification thresholds between (0.01–0.99). We labeled these 25 classification thresholds as solutions. In other words, there will be 25 solutions to choose from at each round of decision-making. The trade-off between the accuracy and fairness metrics for the 25 solutions is presented in Fig. 2 for post-processing Support Vector Machine (SVM) and in Fig. 3 for post-processing Logistic Regression (LR). The trade-off between the accuracy and fairness metrics at different classification thresholds shows that one solution/classification threshold cannot satisfy all the accuracy and fairness constraints for different users. There is a need for a mechanism that can prioritize one constraint over the other based on the user requirements.

In our proposed mechanism, all the metrics (A, F1, F2, F3) act as agents and negotiate with each other based on the ratings they get from other agents and their trust sensitivities. All these agents have a goal. The accuracy agent has the goal of $A = 1$, and the fairness agents have the goal of $F1 = 0$, $F2 = 0$, and $F3 = 0$. In our scenarios, the distance is one-dimensional, but the distance metric can be used for multi-dimensional scenarios, as shown in Eq. 3. Each agent aims to minimize their distance from the goal. For each solution, agents measure their distance from their goal. They rank the solutions from best to worst based on

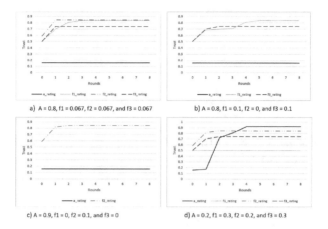

Fig. 4. Negotiation between different agents based on the weights and associated trust sensitivities for race attribute using post-processing LR algorithm

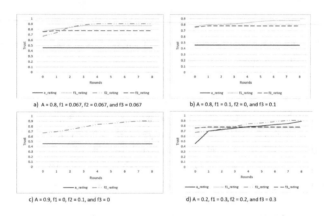

Fig. 5. Negotiation between different agents based on the weights and associated trust sensitivities for race attribute using post-processing SVM algorithm.

the distance to the goal. The best solution for an agent will be the solution with the minimum distance to the goal, and the worst solution will be the solution with the maximum distance from the goal. Each agent starts with their best solution and rates other agents' solutions based on their distance.

In the following rounds, each agent could (i) stay at the same solution or (ii) move to a new solution by increasing its distance from its goal and decreasing other agents' distance from its goals. And how much the agent sacrifices its own distance from its goal and increases other agents' distance to their goals depends on the rating it gets from other agents and its trust sensitivity. All the agents propose and rate each other solutions for multiple rounds.

To test our framework, we simulated eight rounds. In our experiments, we have considered trust sensitivity only for simplification purposes. If the trust sensitivity is below 0.7, the agent does not move. If the trust sensitivity is between (0.7–0.9), the agent moves to its next best solution leading to an increase in the rating, and if the trust sensitivity is between (0.9–1.0), which means the agent is highly sensitive, the agent moves to the next second-best solution to increase its ratings quickly. We simulated different trust sensitivities, which are associated with the weights of the metrics. Figures 4 and 5 show the movement of the agents across rounds based on the weights assigned to them and their associated trust sensitivities for the race attribute. Figure 4 shows the graphs for the post-processing logistic regression algorithm, and Fig. 5 shows the charts for the post-processing support vector machine algorithm. Each figure has four graphs for different metric weights. As seen in the figures, the movement of the accuracy agent in graph a) and graph d) is completely different. In graph a), the accuracy agent is not sensitive given its higher weight hence not moving from its best solution. However, in graph d), the accuracy agent is negotiating and moving away from its best solution given its less weight. This shows how agents with different weights and sensitivities behave differently and lead to other solutions. Table 1 summarizes different solutions/classification thresholds reached by each agent after eight negotiation rounds for different metric weights for race and sex attributes.

Table 1. Solution reached for each agent after eight rounds of negotiation based on user-defined weights for post-processing fairness SVM and post-processing LR

Metrics weights	Race post-processing SVM	Race post-processing LR	Sex post-processing SVM	Sex post-processing LR
a = 0.8, f1 = 0.0667, f2 = 0.0667, f3 = 0.0667	4,8,14,14	6,9,8,8	5,8,13,2	6,10,9,9
a = 0.8, f1 = 0.1, f2 = 0.0, f3 = 0.1	4,8,14	6,9,8	5,9,2	6,10,9
a = 0.9, f1 = 0.0, f2 = 0.1, f3 = 0.0	4,14	6,8	5,13	6,9
a = 0.2, f1 = 0.3, f2 = 0.2, f3 = 0.3	2,8,14,14	25,8,9,9	2,9,13,2	14,10,9,9

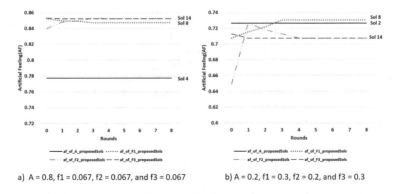

a) A = 0.8, f1 = 0.067, f2 = 0.067, and f3 = 0.067   b) A = 0.2, f1 = 0.3, f2 = 0.2, and f3 = 0.3

Fig. 6. Artificial Feeling (AF) during the rounds of negotiation for the proposed solutions of different agents.

Over the rounds of negotiation, we have calculated the "artificial feeling" (AF) based on the user weights for each selected solution. Figure 6 illustrates how the artificial feeling (AF) changes for each agent in the negotiation rounds based on the solution proposal and user-defined weights. At the end of the negotiation, the solution with the highest Artificial Feeling (AF) is selected. As seen in Fig. 6 graph a), Solution No. 14 is chosen for the given weight distribution because the AF of solution number 14 proposed by agents F2 and F3 is the highest, and based on graph b), Solution No. 8 is selected, which has the maximum AF value and was proposed by agent F1. This difference in solution numbers shows that the AF can capture the context of the AI system based on the user's defined weights. We have also compared the AF for the same weight distribution but different algorithms. For a weight distribution of (A = 0.8, f1 = 0.067, f2 = 0.067, f3 = 0.067), the maximum AF using SVM is 0.852, whereas the maximum AF using LR is 0.78. This shows that SVM performed better than LR for the same user requirement. So, our experiments show that the AF can capture different user contexts and be used as selection criteria for selecting the appropriate solution and algorithm for a given user context.

5   Conclusion

We have presented *"Artificial Conscience - Control Module"* framework to control the working of the AI system based on the user requirements. Based on the user-assigned weights, AI metrics-agents negotiate with each other using our trust engine and solves with maximum "Artificial Feeling." This type of framework can be applied to any AI system where multiple evaluation metrics are involved and when different users have different requirements from the AI system. Our framework uses the "Artificial Feeling" concept to select the best solution and algorithm for a given user requirement.

References

1. Baars, B.J.: In the theatre of consciousness. Global workspace theory, a rigorous scientific theory of consciousness. J. Conscious. Stud. 4(4), 292–309 (1997)
2. Bellamy, R.K., et al.: AI fairness 360: an extensible toolkit for detecting and mitigating algorithmic bias. IBM J. Res. Dev. 63(4/5), 1–15 (2019)
3. Blum, L., Blum, M.: A theory of consciousness from a theoretical computer science perspective: insights from the conscious turing machine. Proc. Natl. Acad. Sci. 119(21), e2115934119 (2022)
4. Chella, A., Manzotti, R.: Artificial Consciousness. Andrews UK Limited (2013)
5. Dua, D., Graff, C.: UCI machine learning repository. School of Information and Computer Science, University of California, Irvine, CA (2019)
6. EC: Ethics guidelines for trustworthy AI (2018). https://ec.europa.eu/digital-single-market/en/news/ethics-guidelines-trustworthy-ai
7. Information Technology – Artificial Intelligence – Overview of trustworthiness in artificial intelligence. Standard, International Organization for Standardization (2020)

8. Kaur, D., Uslu, S., Durresi, A.: Trust-based security mechanism for detecting clusters of fake users in social networks. In: Barolli, L., Takizawa, M., Xhafa, F., Enokido, T. (eds.) WAINA 2019. AISC, vol. 927, pp. 641–650. Springer, Cham (2019). https://doi.org/10.1007/978-3-030-15035-8_62

9. Kaur, D., Uslu, S., Durresi, A.: Requirements for trustworthy artificial intelligence – a review. In: Barolli, L., Li, K.F., Enokido, T., Takizawa, M. (eds.) NBiS 2020. AISC, vol. 1264, pp. 105–115. Springer, Cham (2021). https://doi.org/10.1007/978-3-030-57811-4_11

10. Kaur, D., Uslu, S., Durresi, A.: Trustworthy AI explanations as an interface in medical diagnostic systems. In: Barolli, L., Miwa, H., Enokido, T. (eds.) Advances in Network-Based Information Systems (NBiS 2022). LNNS, vol. 526, pp. 119–130. Springer, Cham (2022). https://doi.org/10.1007/978-3-031-14314-4_12

11. Kaur, D., Uslu, S., Durresi, A., Badve, S., Dundar, M.: Trustworthy explainability acceptance: a new metric to measure the trustworthiness of interpretable AI medical diagnostic systems. In: Barolli, L., Yim, K., Enokido, T. (eds.) CISIS 2021. LNNS, vol. 278, pp. 35–46. Springer, Cham (2021). https://doi.org/10.1007/978-3-030-79725-6_4

12. Kaur, D., Uslu, S., Durresi, A., Mohler, G., Carter, J.G.: Trust-based human-machine collaboration mechanism for predicting crimes. In: Barolli, L., Amato, F., Moscato, F., Enokido, T., Takizawa, M. (eds.) AINA 2020. AISC, vol. 1151, pp. 603–616. Springer, Cham (2020). https://doi.org/10.1007/978-3-030-44041-1_54

13. Kaur, D., Uslu, S., Durresi, M., Durresi, A.: A geo-location and trust-based framework with community detection algorithms to filter attackers in 5G social networks. Wireless Netw. 1–9 (2022). https://doi.org/10.1007/s11276-022-03073-y

14. Kaur, D., Uslu, S., Rittichier, K.J., Durresi, A.: Trustworthy artificial intelligence: a review. ACM Comput. Surv. **55**(2), 1–38 (2022)

15. Pleiss, G., Raghavan, M., Wu, F., Kleinberg, J., Weinberger, K.Q.: On fairness and calibration. Adv. Neural Inf. Process. Syst. **30** (2017)

16. Ruan, Y., Zhang, P., Alfantoukh, L., Durresi, A.: Measurement theory-based trust management framework for online social communities. ACM Trans. Internet Technol. **17**(2), 1–24 (2017)

17. Russell, S.: Human Compatible: Artificial Intelligence and the Problem of Control. VIKING (2019)

18. Russell, S.J.: Rationality and intelligence. Artif. Intell. **94**(1–2), 57–77 (1997)

19. Russell, S.J., Subramanian, D.: Provably bounded-optimal agents. J. Artif. Intell. Res. **2**, 575–609 (1994)

20. Solms, M.: The Hidden Spring: A Journey to the Source of Consciousness. W. W. Norton & Company (2019)

21. Solms, M., Friston, K.: How and why consciousness arises: some considerations from physics and physiology. J. Conscious. Stud. **25**(5–6), 202–238 (2018)

22. Sutcliffe, A.G., Wang, D., Dunbar, R.I.: Modelling the role of trust in social relationships. ACM Trans. Internet Technol. **15**(4), 1–24 (2015)

23. Uslu, S., Kaur, D., Rivera, S.J., Durresi, A., Babbar-Sebens, M.: Decision support system using trust planning among food-energy-water actors. In: Barolli, L., Takizawa, M., Xhafa, F., Enokido, T. (eds.) AINA 2019. AISC, vol. 926, pp. 1169–1180. Springer, Cham (2020). https://doi.org/10.1007/978-3-030-15032-7_98

24. Uslu, S., Kaur, D., Rivera, S.J., Durresi, A., Babbar-Sebens, M.: Trust-based game-theoretical decision making for food-energy-water management. In: Barolli, L., Hellinckx, P., Enokido, T. (eds.) BWCCA 2019. LNNS, vol. 97, pp. 125–136. Springer, Cham (2020). https://doi.org/10.1007/978-3-030-33506-9_12

25. Uslu, S., Kaur, D., Rivera, S.J., Durresi, A., Babbar-Sebens, M.: Trust-based decision making for food-energy-water actors. In: Barolli, L., Amato, F., Moscato, F., Enokido, T., Takizawa, M. (eds.) AINA 2020. AISC, vol. 1151, pp. 591–602. Springer, Cham (2020). https://doi.org/10.1007/978-3-030-44041-1_53

26. Uslu, S., Kaur, D., Rivera, S.J., Durresi, A., Babbar-Sebens, M., Tilt, J.H.: Control theoretical modeling of trust-based decision making in food-energy-water management. In: Barolli, L., Poniszewska-Maranda, A., Enokido, T. (eds.) CISIS 2020. AISC, vol. 1194, pp. 97–107. Springer, Cham (2021). https://doi.org/10.1007/978-3-030-50454-0_10

27. Uslu, S., Kaur, D., Rivera, S.J., Durresi, A., Babbar-Sebens, M., Tilt, J.H.: A trustworthy human-machine framework for collective decision making in food-energy-water management: the role of trust sensitivity. Knowl.-Based Syst. **213**, 106683 (2021)

28. Uslu, S., Kaur, D., Rivera, S.J., Durresi, A., Durresi, M., Babbar-Sebens, M.: Trustworthy acceptance: a new metric for trustworthy artificial intelligence used in decision making in food–energy–water sectors. In: Barolli, L., Woungang, I., Enokido, T. (eds.) AINA 2021. LNNS, vol. 225, pp. 208–219. Springer, Cham (2021). https://doi.org/10.1007/978-3-030-75100-5_19

A Vulnerability Risk Assessment Methodology Using Active Learning

Francisco R. P. da Ponte[✉], Emanuel B. Rodrigues, and César L. C. Mattos

Federal University of Ceará (UFC), Fortaleza, Brazil
fco.rparente@gmail.com, {emanuel,cesarlincoln}@dc.ufc.br

Abstract. Inadequate information security practices, such as using single metrics in Vulnerability Management (VM), can cause analysts to underestimate the likelihood and impact of vulnerability exploitation. Ideally, vulnerability, threat intelligence, and context information should be used in this task. Nonetheless, the lack of specialized tools makes this activity impractical since analysts have to manually correlate data from various security sources to identify the most critical vulnerabilities among thousands of organization assets. Although Machine Learning (ML) can assist in this process, its application has been little explored in the literature. Thus, we present a methodology based on Active Learning (AL) to create a supervised model capable of emulating the experience of experts in the Risk Assessment (RA) of vulnerabilities. Our experiments indicated that the proposed solution performed similarly to that of the analysts and achieved an average accuracy of 88% for critical vulnerabilities.

1 Introduction

Information Security (IS) generates constant concern for companies of all segments and sizes, since cybercriminals use increasingly sophisticated techniques to exploit flaws and attack their targets [1]. That happens because organizations often use poor practices to maintain the security of their networks.

Among those practices is the use of classic Vulnerability Management (VM) tools, which only consider the Common Vulnerability Scoring System (CVSS) to prioritize vulnerability remediation. Such methodology is inadequate since CVSS communicates the severity of vulnerabilities and not their risk [2]. Furthermore, as discussed in [3], remediation policies based on a single metric are not ideal.

Given this scenario, the research and consulting company Gartner drew the attention of security professionals to the need to prioritize the correction of vulnerabilities considering their risk [4]. In the context of IS, experts define risk as the likelihood that an attacker will exploit a vulnerability and its resulting impact on the organization [5].

Thus, to assess risk, analysts should analyze the vulnerability characteristics, threat intelligence, and context information. Threat intelligence is any information that can be used to identify threats and malicious users that can harm computer networks [6]. Context is any information used to categorize the criticality of assets within the organization.

L. Barolli (Ed.): AINA 2023, LNNS 654, pp. 171–182, 2023.
https://doi.org/10.1007/978-3-031-28451-9_15

In practice, to perform the Risk Assessment (RA), analysts use their professional experience to correlate information from various security feeds with the context of the organization's network assets. This activity is difficult and time-consuming, given the complexity of computer networks and the lack of specialized labor [7]. In addition, as it is a process performed manually by analysts, it is subject to errors, which can be caused, for example, by fatigue [8].

Researchers have successfully applied Machine Learning (ML) to solve some security problems, achieving high accuracy, lower response time [9], and helping to face the challenge of lacking skilled professionals [10]. The contributions presented in the literature range from intrusion detection systems to spam and malware detection. However, the use of ML in RA has not been thoroughly explored.

Therefore, the contribution of this work is the proposal of a ML-based methodology capable of classifying the risk of exploiting vulnerabilities, equivalent to the intuition and experience of cybersecurity professionals by using vulnerability, threat intelligence and context information.

Given the lack of labeled data and the high cost of labeling it, we use a ML technique called Active Learning (AL). This technique is capable of interactively querying the user to obtain the labels of the most challenging instances. Thus, the ML solution can achieve better results faster and with less data [11].

The rest of the paper is organized as follows. Section 2 presents the main related works that address RA. Section 3 describes our proposal to use AL to tackle the RA problem. Section 4 discusses the results obtained in our experiments, and finally, Sect. 5 draws the main conclusions and presents future work.

2   Related Work

This section describes the main works related to vulnerability risk assessment. Table 1 summarizes existing research, highlighting differences from our proposal.

Elbaz et al. used AL to train a ML classifier capable of identifying vulnerabilities of interest to security analysts [8]. The authors extracted the Common Platform Enumeration (CPE) information from newly published vulnerability descriptions and used it as input to the model. Their solution can classify vulnerabilities and choose whether to alert the analyst or not. This solution is limited since it only considers the CPE information. Besides, it disregards important information about vulnerability, threat intelligence, and context.

Kure et al. developed an approach to risk assessment that has two parts [12]. First, the assets are classified according to their criticality using fuzzy logic. Then, a ML model is used to predict which type of cyberattack malicious users can launch to compromise the assets. The parts work independently and do not communicate with each other. In addition, the main focus of the solution is to identify the risk of exploitation of some specific attacks, e.g., denial-of-service.

Walkowski et al. proposed a method to assess the risk by combining the CVSS score of the vulnerabilities with the impact that the exploitation of the asset would have on the confidentiality, integrity, and availability of organizational systems [13]. Wang *et al.* calculated the risk of exploiting assets by considering: (*i*) the vulnerabilities that affect the assets; and (*ii*) the vulnerable assets connected to them [14]. Experiments showed that it was possible to identify the assets with a higher risk of being exploited. However, both solutions fail to not consider threat intelligence information.

Gonzalez-Granadillo *et al.* developed a formula to calculate the risk of a vulnerability considering its root cause, the impact of its exploitation, and the organization's ability to recover from the attack [15]. Similarly, Chawla *et al.* calculated the risk considering the affected operating system (e.g., Linux or Windows) and the flaw used in the exploit (e.g., buffer overflow, unvalidated input, etc.) [16]. Both works disregard the global cybersecurity scenario (by not using threat intelligence information) and context. Thus, they cannot assess the real risk of exploiting vulnerabilities.

There are also proprietary solutions, such as Tenable's Vulnerability Priority Rating (VPR) [17], Rapid7's Real Risk Score (RRS) [18], and Kenna's Risk Score (RS) [19], whose authors claim to use threat intelligence and sophisticated machine learning algorithms. Nonetheless, these solutions do not disclose which attributes they considered in the RA. And although they claim to reduce the list of critical vulnerabilities by 97%, users do not have studies that indicate that their methodology is correct or efficient.

Table 1. Comparison between related work. Only our solution uses all three sets of information for classifying vulnerability risk, which makes the assessment more accurate.

Reference	CVSS	Threat Intelligence	Context	Tools
[8]	×			ML
[12]		×	×	Fuzzy + ML
[13]	×		×	Static model
[14]	×		×	Static model
[15]	×			Static model
[16]	×			Static model
This Work	×	×	×	ML

The bibliographic research showed a lot of works that developed solutions to assess the risk of exploiting vulnerabilities. However, none simultaneously considers vulnerability, threat intelligence, and context information. Thus, this work proposes using this set of data and the AL technique to capture the cybersecurity analyst experience and develop a ML model that can classify vulnerabilities regarding their risk.

3   Proposal and Methodology

In this section, we present the proposal and methodology used in the development of this work. A detailed description of the proposal is given in Sect. 3.1. Next we describe the dataset (Sect. 3.2), the labels used to classify the risk of vulnerabilities (Sect. 3.3) and the labeling strategy (Sect. 3.4). Finally, in Sect. 3.5, we detail the approach followed to train the ML model.

3.1   ML-Based Risk Assessment

The risk assessment is a security mechanism employed by experts to protect the assets and networks of a given organization. This process involves the correlation between vulnerability, threat, and asset information [5]. This data is available through open-source security feeds, social networks, Internet traffic analysis, vulnerability scanners, etc. When such data is augmented with context information and the analysts' personal experience, it provides the necessary knowledge to identify critical vulnerabilities [20].

However, without a specialized tool to help specialists in this activity, they have to manually classify the risk for each vulnerability in the organization's assets. Such a task is usually impracticable, given the sheer volume of security flaws [21] and the often small team of security analysts [7]. A possible solution to this problem is to use ML to assist in the classification process.

Nevertheless, we cannot simply apply a supervised learning technique, since in most cases there is no labeled data available. The logical step would be to label this data manually. That would be an expensive activity, even more for this specific problem, where we have few qualified experts to label large amounts of vulnerabilities [22].

To solve such a problem, we propose to use active learning, a technique that aims to improve the performance of the training process of ML models by selecting which item should be labeled by the specialists [11]. AL reduces the number of unnecessary entries labeled and consequently lessens the cost associated with this human-assisted task.

3.2   The Dataset

The dataset used in this work contains 23 attributes that help analysts classify the risk of vulnerabilities, and it was created from the work of Ponte *et al.* [23]. The attributes can be grouped into three sets: (*i*) 8 characteristics of the vulnerability, which help to identify its severity; (*ii*) 9 intelligence characteristics, which provide the information necessary to understand the global threat landscape; and finally, (*iii*) 6 characteristics of the environment, which help to assess the impact that exploiting the vulnerability will bring to the organization's systems.

It is important to note that although we are using a particular dataset, the proposed methodology is general enough and can be applied to any data. However, to correctly assess the risk of vulnerability exploitation, the chosen data must have information that can characterize vulnerabilities, threats, and context.

3.3   The Vulnerability Risk Labels

We have chosen the labels based on the severity rating made by Adobe [24] and Microsoft [25]. This classification is a valuable alternative to the one used in CVSS because it is familiar and prevents analysts from rating the vulnerability considering only the CVSS base score. Thus, the labels used to classify the risk of vulnerabilities in this work are: LOW, MODERATE, IMPORTANT and CRITICAL. A CRITICAL vulnerability will have a higher risk than a vulnerability rated IMPORTANT, and so forth.

3.4   Labeling Strategy via Active Learning

It is a common practice to label randomly chosen instances of a given dataset. However, in such a naive strategy, instances with redundant and insignificant characteristics have the same probability of being chosen as the most representative ones. Alternatively, the active learning technique considers an interactive algorithm to select which instances should be labeled at each iteration. Thus, the model can learn faster and without wasting resources, including human labelers.

Our process of active learning begins with a small amount of labeled examples and a large set of unlabeled data, following a pool-based strategy [11]. The labeled data is then used to train a ML model. Afterwards, the model analyzes the unlabeled instances and selects the one whose label prediction is the most uncertain, i.e., a query-by-uncertainty strategy. This technique uses the entropy to measure the average uncertainty inherent from all the possible outcomes of the analyzed instance. The entropy value is calculated by

$$H = -\sum_{k=1}^{4} P(y_k) \log_2 P(y_k), \tag{1}$$

where $P(y_k)$ represents the probability of the label y_k being applied to the analyzed instance. Thus, the model chooses the vulnerability with the highest value of entropy H.

The expert must then choose the most appropriate label for the selected instance, which will, in turn, be used to retrain the ML model. This process is iterative and will be repeated several times. Usually, the learning process stops after a certain number of elements are labeled or when the model achieves the expected accuracy value [11].

It is important to note that the trained model will emulate the knowledge of the experts. Thus, to avoid bias in the risk classification, the labeling process can be carried out by more than one expert, who form a committee. If there is a conflict between the applied label, they can discuss and reach a consensus or follow a majority vote.

3.5   Learning Strategy

After finishing the AL process, we can use a supervised learning technique with the labeled dataset to train a classifier to assess the risk of vulnerability exploitation. Supervised learning is a subcategory of ML algorithms which use labeled data to train regression or classification models [26]. Alternatively, we could use a semi-supervised learning technique, which receives as input a dataset containing few labeled data and many unlabeled data to train the ML model [26].

The proposed methodology is not specific to any given ML algorithm. However, we decided to use the same model for both the active learning process and the training of the final classifier. Thus, the models must have support for predicting probabilities, as we need to be able to estimate, at least approximately, the uncertainty of the predictions performed by the chosen ML algorithm.

4   Results

We have run several experiments to evaluate the performance of the supervised and semi-supervised learning when using active learning compared to a benchmark solution (simple labeling strategy by random selection). In the following subsections, we describe the setup of our experiments and discuss the obtained results.

4.1   Experiments

We used the data collected by Ponte *et al.* [23] on more than 200,000 vulnerabilities present in NIST[1] to create a subset of 260 items containing information about vulnerabilities, threat intelligence and context. Then, we asked three experts from a cybersecurity company, one from the Security Operations Center (SOC), one from the Threat Intelligence team, and one from the Vulnerability Management team, to perform the risk classification of all 260 selected vulnerabilities. Each professional manually classified all data. That was done independently due to conflicts in their agendas and took an average of 5 and a half hours.

We then analyzed the labels chosen by the experts and removed all vulnerabilities where it was not possible to form a majority vote, i.e., we removed each instance that was voted with 3 different labels. Such vulnerabilities are considered a source of label noise, which could worsen the classifier's performance. As a result, 52 vulnerabilities were removed. The remaining 208 were labeled as follows: 60 as LOW ($\approx 28\%$); 45 as MODERATE ($\approx 22\%$); 45 as IMPORTANT ($\approx 22\%$); and 58 as CRITICAL ($\approx 28\%$).

We designed our experiments to evaluate how an active learning approach can be an efficient and low-cost component of a risk assessment solution when compared to random labeling. Thus, we developed a "simulated oracle", i.e., a script that simulates the user interaction with the labelling strategy. The oracle provides the corresponding label to the selected instance by querying the labels previously given by the experts to the dataset. The active learning was compared with a random strategy, where a script randomly selects an instance to be labeled.

---

[1] National Institute of Standards and Technology: https://nvd.nist.gov/.

We also evaluated how different ML techniques (namely, supervised and semi-supervised learning) can impact the results. Thus, we run four scenarios: (i) active learning with supervised learning; (ii) active learning with semi-supervised learning; (iii) random labeling with supervised learning; and finally, (iv) random labeling with semi-supervised learning.

For each experiment, we evaluate the following ML algorithms: Random Forest (RF), Gradient Boosting (GB), Logistic Regression (LR), Support Vector Classification (SVC), and Multilayer Perceptron (MLP). Such a choice of models aim to include distinct learning strategies, from linear classifiers to artificial neural networks. Each scenario was repeated 100 times to compute statistics of interest.

4.2 Data Preprocessing and Evaluation Metrics

The attributes present in the dataset are continuous (4), discrete (1), Boolean (6), and categorical (12). We transform the categorical data into numerical values using the one-hot-encoding technique, since the elements do not have any hierarchy between them. Additionally, when necessary, we normalize the data with zero mean and unitary standard deviation.

To evaluate the performance of the models, we analyze the values of accuracy, precision, recall, and f1-score. It is important to note that, as our task is a multiclass problem, it was necessary to consider a weighted average to calculate the values of precision, recall, and f1-score. Thus, we first calculate the metrics for each label, then we find the average weighted by the support, i.e., the number of true instances for each label.

4.3 Discussion of Results

Table 2 presents the mean accuracy and standard deviation values after 100 queries from the AL for the four methodologies analyzed. The use of AL showed better results than labeling via random selection, except for the MLP algorithm, which performed equally.

Table 2. Mean accuracy (μ) and standard deviation (σ) values obtained for the different labeling and learning strategies. GB with Active + Supervised learning obtained the best results.

	Active + Semi	Random + Semi	Active + Super	Random + Super
	$\mu \pm \sigma$	$\mu \pm \sigma$	$\mu \pm \sigma$	$\mu \pm \sigma$
RF	0.72 ± 0.07	0.69 ± 0.07	0.72 ± 0.07	0.71 ± 0.08
GB	0.73 ± 0.07	0.71 ± 0.09	0.74 ± 0.08	0.70 ± 0.08
LR	0.65 ± 0.08	0.64 ± 0.07	0.66 ± 0.07	0.66 ± 0.07
SVC	0.69 ± 0.07	0.68 ± 0.08	0.70 ± 0.09	0.69 ± 0.08
MLP	0.63 ± 0.07	0.63 ± 0.08	0.65 ± 0.08	0.65 ± 0.06

We already expected that the MLP would not perform well, but we included it in the tests for the sake of completeness. MLP is a type of Artificial Neural Network (ANN) composed of at least one hidden nonlinear layer. ANNs are trained through a sequence of feedforward and backpropagation iterations, where the data is used to adjust the weights of the network. Thus, an ANN model requires much more data to properly train its parameters. The core of our problem involves the absence of large amounts of labeled data and the high cost related to the activity of manually acquiring new labeled instances. So, one could expect an ANN would not be the most suitable model, as it would suffer from overfitting, which happens when the model cannot generalize well beyond the training data.

In order to statistically evaluate the differences between the classifiers, we carried the Friedman rank test at significance level $\alpha = 0.05$, with accuracy chosen as the performance metric. Figure 1 shows the Critical Difference (CD) diagram for a Nemenyi post-hoc test of the AL combined with each supervised learning strategy. As one can see, GB outperforms all other algorithms, being the most accurate classifier, near rank 1, while MLP was the least accurate, near rank 5, which corroborates with Table 2. The critical distance value was 0.610, thus, we conclude that there were no significant differences between SVC and LR algorithms, as indicated by the thick horizontal line connecting them in Fig. 1.

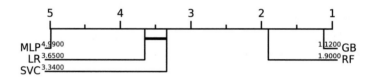

Fig. 1. Critical difference diagram for the Nemenyi test, performed on the average accuracy values for each active learning + supervised technique scenario. GB algorithm was the best, near rank 1.

Therefore, for the sake of brevity, we will focus the additional analysis on GB, which presented the highest average accuracy (Table 2) and outperformed all the other algorithms (Fig. 1). Figure 2 shows the evolution of accuracy values throughout the training process of the GB algorithm for the four different compared methodologies. As can be observed, the technique that combines active learning with supervised learning achieved the best results.

We can see from the curves in Fig. 2 that all methodologies start with similar accuracy values. However, those using active learning grow faster than the ones that randomly selects instances to label. Another way of interpreting these results is by comparing the Area Under the Curve (AUC) values. The methodologies that use active learning presented higher values of AUC than those that do not, as seen in the legend of Fig. 2. These results confirm the importance of carefully choosing which example should be labeled to provide greater accuracy improvements.

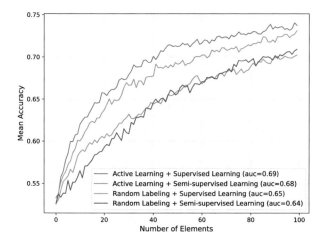

Fig. 2. Mean accuracy comparison of GB among all techniques. As can be seen, Active + Supervised Learning using GB grows faster and performed better than other techniques.

It is important to note that the considered pool of instances is small and almost uniformly divided between labels, as mentioned in Sect. 4.1. Thus, the probability of randomly choosing an entry with relevant characteristics was quite high, which favored the random selection technique. Even so, the active learning quickly surpass the random method (see Fig. 2). The difference between the curves should be more pronounced in a real-life scenario, with a pool of thousands of vulnerabilities.

The precision, recall, and f1-score observed were respectively 0.76 ± 0.08, 0.74 ± 0.08, and 0.73 ± 0.08 for the active learning technique combined with the supervised GB. Those values were high (above 70%), which shows that we have few False Positives (FP) and False Negatives (FN). This implies that our solution does not underestimate or overestimate the risk of vulnerabilities.

Figure 3 presents the confusion matrix for the best scenario combining the active learning with the supervised GB algorithm. We can see that despite the average accuracy being 74%, the model successfully classifies approximately 88% of all CRITICAL vulnerabilities, i.e., those whose risk and impact of exploitation is higher, correctly. In addition, almost all CRITICAL vulnerabilities that were incorrectly classified were said to be IMPORTANT, which appears just behind CRITICAL on the risk scale (Sect. 3.3).

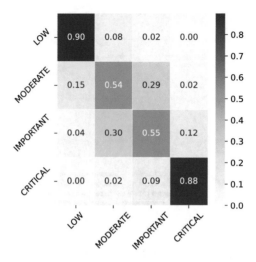

Fig. 3. Confusion matrix for the best case of Active + Supervised Learning using GB.

Finally, when we verify how each analyst performed individually when compared to the majority vote, we can see that they were correct in approximately 75% of the cases. This result is similar to the one achieved by the proposed AL strategy coupled with the supervised GB algorithm.

5   Conclusion and Future Work

In order to correctly assess the risk of exploiting security flaws, it is necessary to correlate information about vulnerabilities, threat intelligence and context. Most of the time, specialists do this manually and without the help of specialized tools, which makes this task complex and error-prone. Thus, we propose to combine the active learning technique with supervised learning algorithms to create a ML model capable of emulating the experience of experts in the risk assessment task.

The performed experiments indicated that the proposed methodology outperformed the random data labeling technique, even in an unfavorable scenario. Active learning combined with supervised learning using the GB algorithm was the best methodology. The obtained results presented an average overall accuracy of 74%, similar to how the analysts performed compared to the majority vote. Additionally, when we analyzed the accuracy for the CRITICAL label, we noticed that the model correctly classified 88% of the critical vulnerabilities.

In future work, we intend to analyze the performance of the proposed methodology in a real-life scenario, with experts manually labeling vulnerabilities queried in an online fashion by AL from a larger pool of samples.

Acknowledgment. The authors would like to thank CAPES for the financial support and the National Center of High Performance Processing (CENAPAD-UFC) of the Federal University of Ceará for the availability of the computational resources used in the experiments.

References

1. Afaq, S.A., Husain, M.S., Bello, A., Sadia, H.: A critical analysis of cyber threats and their global impact. In: Computational Intelligent Security in Wireless Communications, pp. 201–220. CRC Press (2022)
2. Spring, J., Hatleback, E., Manion, A., Shic, D.: Towards improving CVSS. Software Engineering Institute Carnegie Mellon University (2018)
3. Dey, D., Lahiri, A., Zhang, G.: Optimal policies for security patch management. INFORMS J. Comput. **27**(3), 462–477 (2015)
4. Lawson, C., Schneider, M., Bhajanka, P., Gardner, D.: Market Guide for Vulnerability Assessment (2019). https://www.gartner.com/en/documents/3975388. Accessed 19 May 2022
5. Andress, J.: Foundations of Information Security: A Straightforward Introduction. No Starch Press, San Francisco (2019)
6. Trifonov, R., Nakov, O., Mladenov, V.: Artificial intelligence in cyber threats intelligence. In: 2018 International Conference on Intelligent and Innovative Computing Applications (ICONIC), pp. 1–4. IEEE (2018)
7. Furnell, S., Fischer, P., Finch, A.: Can't get the staff? The growing need for cybersecurity skills. Comput. Fraud Secur. **2017**(2), 5–10 (2017)
8. Elbaz, C., Rilling, L., Morin, C.: Automated risk analysis of a vulnerability disclosure using active learning. In: Proceedings of the 28th Computer & Electronics Security Application Rendezvous (2021)
9. Geluvaraj, B., Satwik, P.M., Ashok Kumar, T.A.: The future of cybersecurity: major role of artificial intelligence, machine learning, and deep learning in cyberspace. In: Smys, S., Bestak, R., Chen, J.I.-Z., Kotuliak, I. (eds.) International Conference on Computer Networks and Communication Technologies. LNDECT, vol. 15, pp. 739–747. Springer, Singapore (2019). https://doi.org/10.1007/978-981-10-8681-6_67
10. Shaukat, K., Luo, S., Varadharajan, V., Hameed, I.A., Xu, M.: A survey on machine learning techniques for cyber security in the last decade. IEEE Access **8**, 222310–222354 (2020)
11. Settles, B.: Active learning literature survey [White paper]. University of Wisconsin-Madison Department of Computer Sciences (2009)
12. Kure, H.I., Islam, S., Ghazanfar, M., Raza, A., Pasha, M.: Asset criticality and risk prediction for an effective cybersecurity risk management of cyber-physical system. Neural Comput. Appl. **34**(1), 493–514 (2022). https://doi.org/10.1007/s00521-021-06400-0
13. Walkowski, M., Krakowiak, M., Jaroszewski, M., Oko, J., Sujecki, S.: Automatic CVSS-based vulnerability prioritization and response with context information. In: 2021 International Conference on Software, Telecommunications and Computer Networks (SoftCOM), pp. 1–6. IEEE (2021)
14. Wang, W., Shi, F., Zhang, M., Xu, C., Zheng, J.: A vulnerability risk assessment method based on heterogeneous information network. IEEE Access **8**, 148315–148330 (2020)

15. Gonzalez-Granadillo, G., Diaz, R., Veroni, E., Xenakis, C.: A Multi-factor Assessment Mechanism to Define Priorities on Vulnerabilities affecting Healthcare Organizations (2021)
16. Chawla, G., Sharma, N., Rawal, N.: IVSEV: improved vulnerability scoring mechanism with environment representative and vulnerability type. Int. J. Sci. Technol. Res. **8**(10), 1043–1047 (2019)
17. Tenable, Inc.: Whitepaper: Focus on the 3% of vulnerabilities likely to be exploited [White paper] (2020). https://lookbook.tenable.com/predictive-prioritization/technical-whitepaper-predictive-prioritization. Accessed 20 June 2022
18. Rapid7, Inc.: Rapid7 whitepaper: The four pillars of modern vulnerability management [White paper] (2021). https://www.rapid7.com/info/whitepaper-the-four-pillars-of-modern-vulnerability-management/. Accessed 20 June 2022
19. Kenna Security, Inc.: Understanding the Kenna Risk Score Prioritizing Vulnerabilities with Data Science [White paper] (2020). https://www.vmware.com/content/dam/digitalmarketing/vmware/en/pdf/docs/vmwcb-whitepaper-understanding-the-kenna-security-vulnerability-risk-score.pdf. Accessed 20 June 2022
20. Bromander, S.: Understanding Cyber Threat Intelligence: Towards Automation [Doctoral's Thesis, University of Oslo]. The University of Oslo Institutt for informatikk (2021). https://www.duo.uio.no/handle/10852/84713
21. Kenna Security Inc., Cyentia Institute.: Winning the Remediation Race [White paper] (2019). https://website.kennasecurity.com/wp-content/uploads/2020/09/Kenna_Prioritization_to_Prediction_Vol3.pdf. Accessed 20 June 2022
22. Miller, B., Linder, F., Mebane, W.R.: Active learning approaches for labeling text: review and assessment of the performance of active learning approaches. Polit. Anal. **28**(4), 532–551 (2020)
23. Ponte, F.R.P., Rodrigues, E.B., Mattos, C.L.: CVEjoin: An Information Security Vulnerability and Threat Intelligence Dataset. figshare. Dataset (2022). https://doi.org/10.6084/m9.figshare.21586923.v3
24. Adobe, Inc.: Adobe: Severity ratings (2022). https://helpx.adobe.com/security/severity-ratings.html. Accessed 16 Aug 2022
25. Microsoft, Inc.: Microsoft: Security update severity rating system (2022). https://www.microsoft.com/en-us/msrc/security-update-severity-rating-system. Accessed 16 Aug 2022
26. Murphy, K.: Probabilistic Machine Learning: An Introduction. MIT Press, Cambridge (2022)

A Fuzzy Inference and Posture Detection Based Soldering Motion Monitoring System

Kyohei Toyoshima[1], Chihiro Yukawa[1], Yuki Nagai[1], Genki Moriya[2], Sora Asada[2], Tetsuya Oda[2(✉)], and Leonard Barolli[3]

[1] Graduate School of Engineering, Okayama University of Science (OUS), 1-1 Ridaicho, Kita-ku, Okayama 700 0005, Japan
{t22jm24jd,t22jm19st,t22jm23rv}@ous.jp

[2] Department of Information and Computer Engineering, Okayama University of Science (OUS), 1-1 Ridaicho, Kita-ku, Okayama 700 0005, Japan
{t18j088mg,t19j008as}@ous.jp, oda@ous.ac.jp

[3] Department of Information and Communication Engineering, Fukuoka Institute of Technology, 3-30-1 Wajiro-Higashi-ku, Fukuoka 811-0295, Japan
barolli@fit.ac.jp

Abstract. Recently, governments are promoting the employment of persons with disabilities, and employers must ensure safety in the workplace. On the other hand, a single manager has difficulty in preventing accidents and injuries caused by human error. In addition, in order to learn soldering techniques, persons with disabilities need to repeat the same process. In this paper, we present a web system for depth camera based soldering motion monitoring. Experimental results show the danger level of soldering motion based on fuzzy inference. We also implement the web system and show the visualization results of the soldering worker state. From experimental results, the proposed system is able to monitor the soldering workers.

1 Introduction

The safety is the highest priority in an electronics manufacturing plant. In an industrial accident, especially one caused by safety, the cost in terms of human casualties, property damage and product damage is extremely high. In addition, employers must ensure workplace safety as the government promotes the employment of physically, intellectually, and mentally handicapped workers. On the other hand, it is difficult for a single manager to prevent accidents and injuries caused by human error, such as workers lack of concentration or cutting corners due to simple tasks.

Some factories instruct persons with disabilities to solder electronic components. However, it takes a long time for the disabled to master soldering skills, as they must repeat the same tasks repeatedly. Therefore, instructors need to monitor the soldering of workers, and there is a requirement to ensure the safety

L. Barolli (Ed.): AINA 2023, LNNS 654, pp. 183–191, 2023.
https://doi.org/10.1007/978-3-031-28451-9_16

of beginners and persons with disabilities during soldering work at production sites and to reduce the workload of instructors [1–8].

In this paper, we propose a web system for depth camera based soldering motion monitoring. It also considers the time duration of each dozing, left or right body orientation, and forward leaning posture based on fuzzy inference to give an indication. We analyze soldering motion based on object detection [9–14] and posture estimation [15–19]. It is also possible to centrally control and check the status of the worker soldering.

The structure of the paper is as follows. In Sect. 2, we present the proposed system. In Sect. 3, we describe the experimental results. Finally, conclusions and future work are given in Sect. 4.

2   Proposed System

In this section, we present the proposed system. Figure 1 shows an overview of the proposed system and Fig. 2 shows the system environment. In addition, the system detects dangerous motions of the soldering worker based on the recognized body parts and notifies the soldering worker using a smart speaker.

By using color images acquired with a depth camera, the proposed system detects the dozing state and estimates the posture of the upper body. The dozing state decides the left and right Eye Aspect Ratio (EAR) [20] from the vertical and horizontal Euclidean distances based on the 2-dimensional landmarks of both eyes. Then, the eyelids are considered closing when the sum of EAR is less than 0.4 (threshold value). In case the eyelids are closed for less than 5 [sec.], the soldering worker is considered awake and in case the eyelids are closed for more than 5 [sec.], the soldering worker is considered dozing state.

The posture during soldering [21] is estimated considering three dimensions based on the key points (x, y) obtained by skeletal estimation using MediaPipe [22–26] and the distance z data to each key point from the depth image of the depth camera [27]. The safe posture is when the body is parallel to the worktable during soldering. The angle from the perpendicular line connecting the reference line shown in Fig. 3 and the left and right shoulders is decided as the body orientation. The attention body orientation is when the body orientation is tilted more than 30 [deg.] to the left or right and the attention posture is when the chin is below the line connecting both shoulders.

The inputs for fuzzy inference are the Dozing State Time (DST), the Body Orientation (BO) and the Attention Posture Time (APT). The Danger Motion Level (DML) is decided between 0.0 to 10.0 and output. Figure 4(a), Fig. 4(b), Fig. 4(c) and Fig. 4(d) show the input and output membership functions. The fuzzy rulebase is also shown in Table 1. APT is set to H when the value of BO is M, the value of APT is L and the value of DML is H. The danger level during soldering works by the proposed system is based on fuzzy inference, which considers the workers status and changes the contents of the instructions. Therefore, the proposed system can effectively prevent the danger associated with soldering work. Google AIY VoiceKit V2 outputs voice based on Table 2 and Table 3.

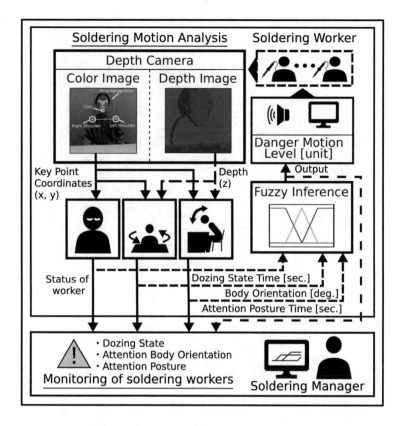

Fig. 1. Overview of the proposed system.

For monitoring the state of the soldering worker, the proposed system displays the occurrence of a dozing state, attention body orientation and attention posture in the form of a waveform. Transmission from two clients to one management server at 1.0 [*sec.*] intervals. The transmitted data are device number, current time, doze state time, attention body orientation time, attention posture time and danger motion level. The management server saves the received data in a JSON format file and displays the elapsed time of each workers state on a browser.

3 Experimental Results

In this section, we present the experimental results. We implemented the proposed Web-based system in two Jetson Nano: No. 1 Jetson Nano and No. 2 Jetson Nano. In No. 1 Jetson Nano scenario, the implemented system performed

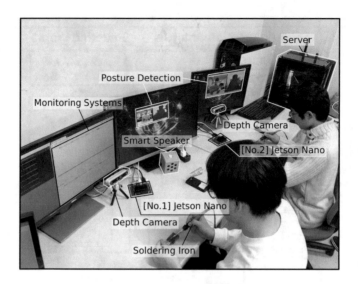

Fig. 2. System environment.

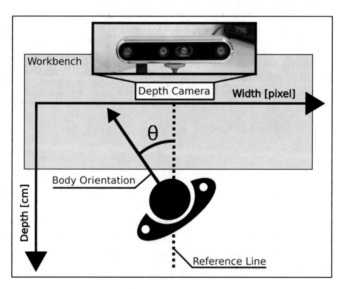

Fig. 3. Measuring of body orientation by depth camera.

the dangerous motion in this order: dozing state for 5 [*sec.*], attention body orientation for 5 [*sec.*] and attention posture for 5 [*sec.*]. While in No. 2 Jetson Nano scenario, the dangerous motion order is: attention posture for 5 [*sec.*], dozing state for 5 [*sec.*] and attention body orientation for 5 [*sec.*].

Figure 5(a) shows the output result of fuzzy inference when *BO* is 45.5 [*deg.*]. In the case of a high value of *DST* and a high value of *APT*, the value of *DML*

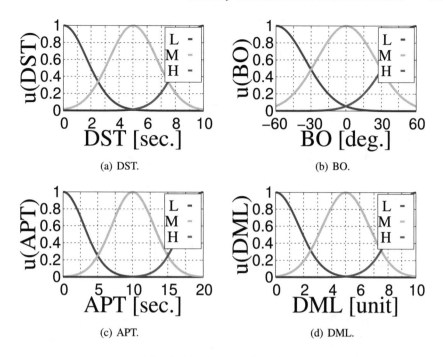

Fig. 4. Membership functions.

Table 1. Fuzzy rule-base.

DST	BO	APT	DML	DST	BO	APT	DML	DST	BO	APT	DML
H	H	H	H	M	H	H	H	L	H	H	H
H	H	M	H	M	H	M	M	L	H	M	M
H	H	L	H	M	H	L	M	L	H	L	L
H	M	H	H	M	M	H	M	L	M	H	M
H	M	M	H	M	M	M	M	L	M	M	M
H	M	L	H	M	M	L	L	L	M	L	L
H	L	H	H	M	L	H	H	L	L	H	H
H	L	M	H	M	L	M	M	L	L	M	M
H	L	L	H	M	L	L	M	L	L	L	L

is high. Also, in the case of a low value of *DST* and a low value of *APT*, the value of *DML* is low. Figure 5(b) shows the output result of fuzzy inference when *APT* is 8 [*sec.*]. In this case, it can be seen that the value of *DML* is high when the value of *BO* is more than ±30 [*deg.*] or the value of *DST* is high. This experimental result shows that the proposed system can decide and provide the current danger level of the soldering worker based on fuzzy inference.

Table 2. Status of work on the DML.

Range of DML [unit]	Work state in process
$0 \leq DML < 2.5$	Safety
$2.5 < DML < 5$	Dangerous
$5 < DML < 7.5$	Extremely Dangerous
$7.5 < DML \leq 10$	Stop Work

Table 3. Examples of voice output for danger detection.

Conditions	Content of Output
Angle between the reference line and the body is more than ±30 [deg.]	"Please turn the body to the front."
Chin is lower than the shoulders	"Please correct your posture."
Eyelids closed for more than 5 [sec.]	"Please wake up."

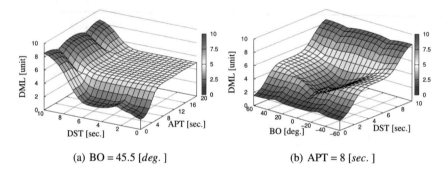

(a) BO = 45.5 [*deg.*]   (b) APT = 8 [*sec.*]

Fig. 5. Output results of fuzzy inference.

Figure 6 shows the visualization results of the web system. Figure 6(a) shows the safety state and Fig. 6(b) shows the dangerous state. The administrator accesses the web system, which displays the data stored in the management server. From Fig. 6(b), it can be seen that each state of the worker is displayed in a waveform, in the same pattern as in the experimental scenario. From the experimental results, we conclude that the proposed system is an effective method for monitoring workers performing soldering.

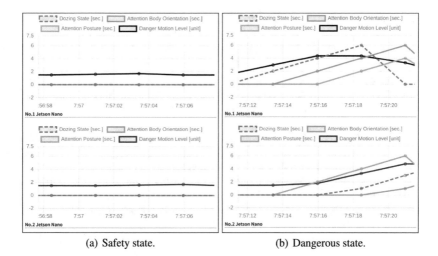

(a) Safety state.   (b) Dangerous state.

Fig. 6. Visualization results of the web system.

4   Conclusions

In this paper, we presented a soldering motion analysis system based on a depth camera. From the experimental results, we showed that the proposed system can decide and provide the current danger level of a soldering worker based on fuzzy inference. Also, we showed that the proposed system can monitor the soldering workers by displaying their conditions in the waveforms.

In the future, we would like to make improvements to the proposed system and consider different scenarios.

Acknowledgement. This work was supported by JSPS KAKENHI Grant Number JP20K19793.

References

1. Yasunaga, T., et al.: Object detection and pose estimation approaches for soldering danger detection. In: Proceedings of The IEEE 10th Global Conference on Consumer Electronics, pp. 776–777 (2021)
2. Yasunaga, T., et al.: A soldering motion analysis system for danger detection considering object detection and attitude estimation. In: Proceedings of The 10th International Conference on Emerging Internet, Data & Web Technologies, pp. 301–307 (2022)
3. Toyoshima, K., et al.: Proposal of a haptics and LSTM based soldering motion analysis system. In: Proceedings of The IEEE 10th Global Conference on Consumer Electronics, pp. 1–2 (2021)
4. Toyoshima, K., et al.: Design and implementation of a haptics based soldering education system. In: Barolli, L. (ed.) Innovative Mobile and Internet Services in Ubiquitous Computing, IMIS 2022. Lecture Notes in Networks and Systems, vol.

496, pp. 54–64. Springer, Cham (2022). https://doi.org/10.1007/978-3-031-08819-3_6

5. Toyoshima, K., et al.: Experimental results of a haptics based soldering education system: a comparison study of RNN and LSTM for detection of dangerous movements. In: Barolli, L., Miwa, H. (eds.) Advances in Intelligent Networking and Collaborative Systems, INCoS 2022. Lecture Notes in Networks and Systems, vol. 527, pp. 212–223. Springer, Cham (2022). https://doi.org/10.1007/978-3-031-14627-5_20

6. Toyoshima, K., et al.: Analysis of a soldering motion for dozing state and attention posture detection. In: Barolli, L. (ed.) Advances on P2P, Parallel, Grid, Cloud and Internet Computing, 3PGCIC 2022. Lecture Notes in Networks and Systems, vol. 571, pp. 146–153. Springer, Cham (2022). https://doi.org/10.1007/978-3-031-19945-5_14

7. Oda, T., et al.: Design and implementation of an IoT-based e-learning testbed. Int. J. Web Grid Serv. **13**(2), 228–241 (2017)

8. Liu, Y., et al.: Design and implementation of testbed using IoT and P2P technologies: improving reliability by a fuzzy-based approach. Int. J. Commun. Netw. Distrib. Sys. **19**(3), 312–337 (2017)

9. Papageorgiou, C., et al.: A general framework for object detection. In: The IEEE 6th International Conference on Computer Vision, pp. 555–562 (1998)

10. Felzenszwalb, P., et al.: Object detection with discriminatively trained part-based models. IEEE Trans. Pattern Anal. Mach. Intell. **32**(9), 1627–1645 (2009)

11. Obukata, R., et al.: Design and evaluation of an ambient intelligence testbed for improving quality of life. Int. J. Space-Based Situated Comput. **7**(1), 8–15 (2017)

12. Oda, T., Ueda, C., Ozaki, R., Katayama, K.: Design of a deep Q-network based simulation system for actuation decision in ambient intelligence. In: Barolli, L., Takizawa, M., Xhafa, F., Enokido, T. (eds.) WAINA 2019. AISC, vol. 927, pp. 362–370. Springer, Cham (2019). https://doi.org/10.1007/978-3-030-15035-8_34

13. Obukata, R., et al.: Performance evaluation of an AmI testbed for improving QoL: evaluation using clustering approach considering distributed concurrent processing. In: Proceedings of IEEE AINA-2017, pp. 271–275 (2017)

14. Yamada, M., et al.: Evaluation of an IoT-based e-learning testbed: performance of OLSR protocol in a NLoS environment and mean-shift clustering approach considering electroencephalogram data. Int. J. Web Inf. Sys. **13**(1), 2–13 (2017)

15. Toshev, A., Szegedy, C.: DeepPose: human pose estimation via deep neural networks. In: Proceedings of The 27th IEEE/CVF Conference on Computer Vision and Pattern Recognition (IEEE/CVF CVPR-2014), pp. 1653–1660 (2014)

16. Haralick, R., et al.: Pose estimation from corresponding point data. IEEE Trans. Sys. **19**(6), 1426–1446 (1989)

17. Fang, H., et al.: RMPE: regional multi-person pose estimation. In: Proceedings of the IEEE International Conference on Computer Vision, pp. 2334–2343 (2017)

18. Xiao, B., et al.: Simple baselines for human pose estimation and tracking. In: Proceedings of the European Conference On Computer Vision (ECCV), pp. 466–481 (2018)

19. Martinez, J., et al.: A simple yet effective baseline for 3D human pose estimation. In: Proceedings of the IEEE International Conference on Computer Vision, pp. 2640–2649 (2017)

20. Soukupova, T., et al.: Real-time eye blink detection using facial landmarks. In: Proceedings of the 21st Computer Vision Winter Workshop, Rimske Toplice, Slovenia (2016)

21. Micilotta, A.S., Ong, E.-J., Bowden, R.: Real-time upper body detection and 3D pose estimation in monoscopic images. In: Leonardis, A., Bischof, H., Pinz, A. (eds.) ECCV 2006. LNCS, vol. 3953, pp. 139–150. Springer, Heidelberg (2006). https://doi.org/10.1007/11744078_11

22. Zhang, F., et al.: MediaPipe hands: on-device real-time hand tracking. arXiv preprint arXiv:2006.10214 (2020)

23. Shin, J., et al.: American sign language alphabet recognition by extracting feature from hand pose estimation. Sensors **21**(17), 5856 (2021)

24. Hirota, Y., et al.: Proposal and experimental results of a DNN based real-time recognition method for ohsone style fingerspelling in static characters environment. In: Proceedings of The IEEE 9th Global Conference on Consumer Electronics, pp. 476–477 (2020)

25. Erol, A., et al.: Vision-based hand pose estimation: a review. Comput. Vis. Image Underst. **108**, 52–73 (2007)

26. Lugaresi, C., et al.: MediaPipe: a framework for building perception pipelines. arXiv preprint arXiv:1906.08172 (2019)

27. Andriyanov, N., et al.: Intelligent system for estimation of the spatial position of apples based on YOLOv3 and real sense depth camera D415. Symmetry **14**(1), 148 (2022)

An Anomaly Detection System
for Intelligent Robot Vision Using LSTM

Chihiro Yukawa[1], Kyohei Toyoshima[1], Yuki Nagai[1], Masahiro Niihara[2],
Yuma Yamashita[2], Tetsuya Oda[2(✉)], and Leonard Barolli[3]

[1] Graduate School of Engineering, Okayama University of Science (OUS),
1-1 Ridaicho, Kita-ku, Okayama 700-0005, Japan
`{t22jm19st,t22jm24jd,t22jm23rv}@ous.jp`
[2] Department of Information and Computer Engineering, Okayama University
of Science (OUS), 1-1 Ridaicho, Kita-ku, Okayama-shi 700-0005, Japan
`{t19j061nm,t20j091yy}@ous.jp, oda@ous.ac.jp`
[3] Department of Information and Communication Engineering, Fukuoka Institute
of Technology, 3-30-1 Wajiro-Higashi-ku, Fukuoka 811-0295, Japan
`barolli@fit.ac.jp`

Abstract. The automation is a very important in manufacturing industry to improve the efficiency of production processes. Recently, there are some research works that consider the use of robot vision for inspection and early recognition of malfunctions and abnormalities in order to prevent many accidents. In this paper, we propose and implement an anomaly detection system for intelligent robot vision using LSTM. We consider an anomaly when an object contacts the robot arm. The anomaly detection is determined based on the error of acceleration values between the measured value and predicted value by LSTM. We evaluate the proposed system by an experiment. The experimental results show that proposed system can detect anomalies.

1   Introduction

In the manufacture industry, automation is very important to improve the efficiency of the production process [1–8]. There are some research work that consider the use of robot vision for inspection [9–12]. However, the operation of robot vision requires continuous maintenance such as checking of malfunctions and abnormalities. Thus, continuous maintenance takes a lot of labor work and time. In addition, early recognition of malfunctions and abnormalities can prevent many accidents. Therefore, it is necessary to detect anomalies as soon as possible.

In this paper, we propose an implement an anomaly detection system for intelligent robot vision using LSTM. We consider an anomaly when an object contacts the robot arm. The anomaly detection is determined based on the error of acceleration values between the measured value and predicted value by LSTM. We evaluate the proposed system by an experiment. The experimental results show that robot vision can detect the anomalies.

L. Barolli (Ed.): AINA 2023, LNNS 654, pp. 192–198, 2023.
https://doi.org/10.1007/978-3-031-28451-9_17

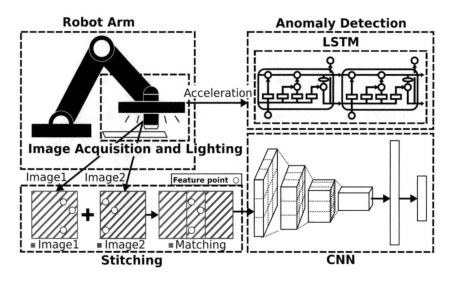

Fig. 1. Proposed system.

The structure of the paper is as follows. In Sect. 2, we present the proposed system. In Sect. 3, we present the experimental results. Finally, conclusions and future work are given in Sect. 4.

2   Proposed System

The proposed system structure is shown in Fig. 1. The implemented system has four modules: Robot Arm, Stitching, Convolutional Neural Networks (CNN) and Anomaly Detection. The robot vision system performs stitching [22–25] and object recognition [18–21]. The proposed system detects anomalies based on the measured acceleration value and predicted acceleration value by Long Short-Term Memory (LSTM) [13–17]. Also, it can visualizes the anomaly detection by Web page and and use a LED light to notify the users.

2.1   Anomaly Detection for Robot Vision

The anomaly detection is determined based on the error of accelerations values between the measured value and the predicted value by LSTM. The acceleration is represented by $a = <a_x, a_y, a_z>$ and t is the current time. The LSTM input data are shown in Eq. (1), and the predicted acceleration values are shown in Eq. (2).

$$input = [a_{t-4}, a_{t-3}, \cdots, a_t] \tag{1}$$

$$pred_{a_{t+1}} = f_{LSTM}(input) \tag{2}$$

The anomaly accelerations are determined by Eq. (3):

$$(a_t - pred_{a_t})^2 > T, \tag{3}$$

where, T is the threshold value and in the experiment is set at 0.15.

2.2   Visualization and Danger Notification for Robot Arm

In order to visualize the danger movement, the Jetson acquires acceleration values from the accelerometer attached to the hand of the robot arm. Then, it sends the data to the main server by TCP/IP protocol. The data are input to LSTM to obtain the predicted acceleration values. Then, the predicted acceleration values are stored in JSON format. The acceleration data can be displayed in real-time on Web page using the stored data. The anomalies are detected based on received and predicted acceleration data. When an abnormality is detected, a red LED lights up to notify the user of the danger situation.

3   Experimental Results

The experimental environment is shown in Fig. 2. We use as robot arm `uArm Swift Pro Standard with 4 Degrees of Freedom` and as accelerometer is used `GY-521`.

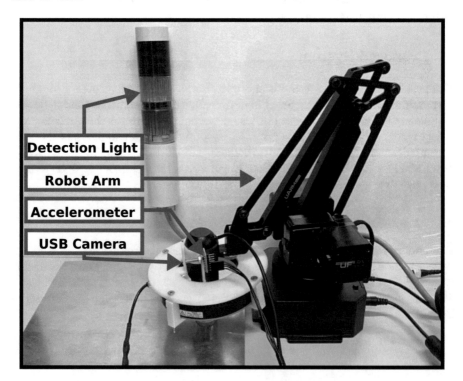

Fig. 2. Experimental environment.

The results of prediction by LSTM are shown in Fig. 3. The LSTM uses 50 movements of the robot arm as training data and learning is carried out for 5000 iterations. The learning accuracy is about 97[%]. It can be seen that there are many places where the predicted and measured accelerations values are almost the same as shown in Fig. 3.

The results of anomaly detection are shown in Fig. 4. We consider an anomaly when an object contacts the robot arm. The experimental results show that there is a contact with robot arm at about 19[sec.], and the anomaly detection was confirmed by the proposed system at that time (19[sec.]).

The visualization results are shown in Fig. 5. By our proposed systems, it is possible to visualize the predicted and measured acceleration values as a graph. Then, when the anomaly is detected, the red LED will light.

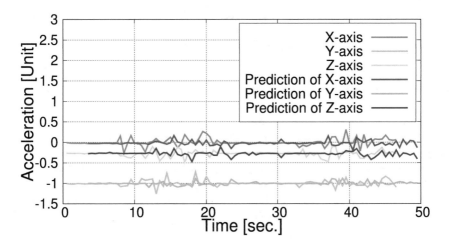

Fig. 3. Results of LSTM.

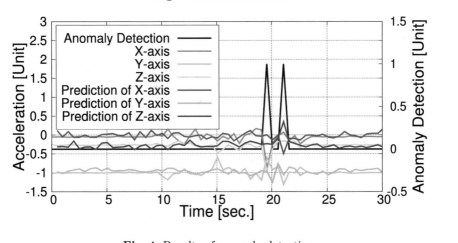

Fig. 4. Results of anomaly detection.

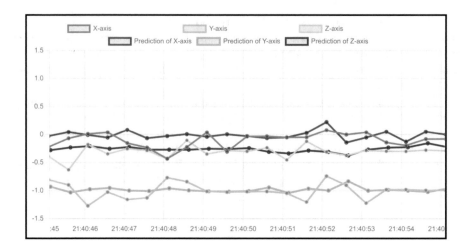

Fig. 5. Visualization results.

4   Conclusions

In this paper, we proposed an implemented an anomaly detection system for intelligent robot vision using LSTM. We considered an anomaly when an object contacted the robot arm. The anomaly detection is determined based on the error of acceleration values between the measured value and predicted value by LSTM. We evaluated the proposed system by an experiment. The experimental results show that proposed system can detect anomalies. In future work, we would like to improve the accuracy of proposed system for anomaly detection.

Acknowledgement. This work was supported by JSPS KAKENHI Grant Number 20K19793.

References

1. Dalenogare, L., et al.: The expected contribution of industry 4.0 technologies for industrial performance. Int. J. Prod. Econ. **204**, 383–394 (2018)
2. Shang, L., et al.: Detection of rail surface defects based on CNN image recognition and classification. The IEEE 20th International Conference on Advanced Communication Technology (ICACT), pp. 45–51 (2018)
3. Li, J., et al.: Real-time detection of steel strip surface defects based on improved yolo detection network. IFAC-PapersOnLine **51**(21), 76–81 (2018)
4. Oda, T., et al.: Design and implementation of a simulation system based on deep Q-network for mobile actor node control in wireless sensor and actor networks. In: Proceedings of The IEEE 31st International Conference on Advanced Information Networking and Applications Workshops, pp. 195–200 (2017)

5. Yukawa, C., et al.: Design of an intelligent robotic vision system for optimization of robot arm movement. In: Proceedings of The 17th International Conference on Broadband and Wireless Computing, Communication and Applications, pp. 353–360 (2020)

6. Yukawa, C., et al.: An intelligent robot vision system for recognizing micro-roughness on arbitrary surfaces: experimental result for different methods. In: Proceedings of The 14th International Conference on Intelligent Networking and Collaborative Systems, pp. 212–223 (2021)

7. Yukawa, C., et al.: Evaluation of a fuzzy-based robotic vision system for recognizing micro-roughness on arbitrary surfaces: a comparison study for vibration reduction of robot arm. In: Barolli, L., Miwa, H., Enokido, T. (eds.) Advances in Network-Based Information Systems. NBiS 2022. Lecture Notes in Networks and Systems, vol. 526, pp. 230–237. Springer, Cham (2021)https://doi.org/10.1007/978-3-031-14314-4_23

8. Yukawa, C., et al.: Design of a fuzzy inference based robot vision for CNN training image acquisition. In: Proceedings of The IEEE 10th Global Conference on Consumer Electronics, pp. 806–807 (2021)

9. Wang, H., et al.: Automatic illumination planning for robot vision inspection system. Neurocomputing **275**, 19–28 (2018)

10. Zuxiang, W., et al.: Design of safety capacitors quality inspection robot based on machine vision. In: 2017 First International Conference on Electronics Instrumentation & Information Systems (EIIS), pp. 1–4 (2017)

11. Li, J., et al.: Cognitive visual anomaly detection with constrained latent representations for industrial inspection robot. Appl. Soft Comput. **95**, 106539 (2020)

12. Ruiz-del-Solar, J., et al.: A survey on deep learning methods for robot vision. arXiv preprint arXiv:1803.10862 (2018)

13. Yu, Y., et al.: A review of recurrent neural networks: LSTM cells and network architectures. Neural Comput. **31**(7), 1235–1270 (2019)

14. Greff, K., et al.: LSTM: a search space odyssey. IEEE Trans. Neural Netw. Learn. Syst. **28**(10), 2222–2232 (2016)

15. Gers, F.A., et al.: Learning to forget: continual prediction with LSTM. Neural Comput. **12**(10), 2451–2471 (2000)

16. Staudemeyer, R.C., et al.: Understanding LSTM–a tutorial into long short-term memory recurrent neural networks. arXiv preprint arXiv:1909.09586 (2019)

17. Sherstinsky, A., et al.: Fundamentals of recurrent neural network (RNN) and long short-term memory (LSTM) network. Phys. D: Nonlinear Phenom. **404**, 132306 (2020)

18. Yosinski, J., et al.: How transferable are features in deep neural networks? arXiv preprint arXiv:1411.1792 (2014)

19. Dosovitskiy, A., et al.: An image is worth 16x16 words: transformers for image recognition at scale. arXiv preprint arXiv:2010.11929 (2020)

20. Sudharshan, D.P., et al.: Object recognition in images using convolutional neural network. In: 2018 2nd International Conference on Inventive Systems and Control (ICISC), pp. 718–722 (2018)

21. Radovic, M., et al.: Object recognition in aerial images using convolutional neural networks. J. Imaging **3**(2), 21 (2017)

22. Szeliski, R., et al.: Image alignment and stitching: a tutorial. Found. Trends® Comput. Graph. Vis. **2**(1), 1–104 (2007)

23. Levin, A., Zomet, A., Peleg, S., Weiss, Y.: Seamless image stitching in the gradient domain. In: Pajdla, T., Matas, J. (eds.) ECCV 2004. LNCS, vol. 3024, pp. 377–389. Springer, Heidelberg (2004). https://doi.org/10.1007/978-3-540-24673-2_31
24. Zaragoza, J., et al.: As-projective-as-possible image stitching with moving DLT. In: Proceedings of the IEEE Conference on Computer Vision and Pattern Recognition, pp. 2339–2346 (2013)
25. Li, J., et al.: Parallax-tolerant image stitching based on robust elastic warping. IEEE Trans. Multimedia **20**(7), 1672–1687 (2017)

GDLS-FS: Scaling Feature Selection for Intrusion Detection with GRASP-FS and Distributed Local Search

Estêvão F. C. Silva[1,2], Nícolas Naves[2], Silvio E. Quincozes[3(✉)],
Vagner E. Quincozes[4], Juliano F. Kazienko[5], and Omar Cheikhrouhou[6]

[1] TQI Tecnologia, Uberlândia, Brazil
`estevao.silva@tqi.com.br`
[2] FACOM/UFU, Uberlândia, Brazil
`{estevaof10,nicolasnaves}@ufu.br`
[3] FACOM/UFU, Monte Carmelo, Brazil
`sequincozes@ufu.br`
[4] UNIPAMPA, Alegrete, Brazil
`vagnerquincozes.aluno@unipampa.edu.br`
[5] CTISM/UFSM, Santa Maria, Brazil
`kazienko@redes.ufsm.br`
[6] CES Lab, University of Sfax, Sfax, Tunisia
`omar.cheikhrouhou@isetsf.rnu.tn`

Abstract. This paper presents a scalable microservice-oriented architecture, called Distributed LS (DLS), for enhancing the Local Search (LS) phase in the Greedy Randomized Adaptive Search Procedure for Feature Selection (GRASP-FS) metaheuristic. We distribute the DLS processing among multiple microservices, each with a different responsibility. These microservices are decoupled because they communicate with each other by using a message broker through the publish and subscribe paradigm. As a proof-of-concept, we implemented an instance of our architecture through the Kafka framework, two neighborhood structures, and three LS algorithms. These components look for the best solution which is published in a topic to the Intrusion Detection System (IDS). Such a process is iterated continuously to improve the solution published, providing IDS with the best feature selection solution at the end of the search process with scalability and time reduction. Our results show that using RVND may take only 19.53% of the time taken by VND. Therefore, the RVND approach is the most efficient for exploiting parallelism in distributed architectures.

1 Introduction

Nowadays, information security is a major concern for cybernetic applications across multiple domains, including billions of devices worldwide. A to the MCA Department, 98% of applications published on the web are vulnerable to cyberattacks in 2019 [25]. This problem is still more serious when the critical infrastructure (*e.g.,* gas and oil platforms, power grid, electrical substations) is

L. Barolli (Ed.): AINA 2023, LNNS 654, pp. 199–210, 2023.
https://doi.org/10.1007/978-3-031-28451-9_18

affected: in the worst cases, human lives may be at risk [3, 16, 21]. In this context, the information security area studies and applies methods to mitigate computer systems and communication network vulnerabilities, such as Intrusion Detection Systems (IDSs). However, even with the proper preventive measures, attackers have increasingly demonstrated the ability to control them and succeed in their attacks. Therefore, to properly detect the intruders' behavior it is essential to employ sophisticated IDSs (*e.g.,* using machine learning) and feed them with representative information (*e.g.,* selected features) [10].

To provide representative information for IDSs based on machine learning, employing preprocessing techniques, such as Feature Selection (FS) is necessary. Such techniques aim to select meaningful information to feed the IDSs and help it in distinguishing between legitimate and malicious traffic [9]. There are different FS approaches, each with its advantages and weakness. The filter-based FS attempts to estimate the most relevant features individually using statistical methods. Such estimation is lightweight but does not always reflect performance improvements for machine learning algorithms. On the other hand, Wrapping-based FS employs a machine learning algorithm to assess multiple combinations of features together. Wrapping-based methods are more assertive, however, they are more computationally expensive, and exploring all combinations of features may be impracticable. Therefore, using hybrid metaheuristic algorithms that combine Wrapping and Filter is an alternative to reaching suboptimal solutions in a feasible time [18].

The metaheuristic Greedy Randomized Adaptive Search Procedure for Feature Selection (GRASP-FS) generates *greedy* and *random* solutions using filter-based FS and employs a Local Search (LS) procedure using wrapping-based FS to optimize the initial solution. However, the original GRASP-FS proposal [19] needs to parameterize manually, and its execution is sequential in a single monolithic application. As the traffic pattern varies with the application context, finding the best algorithms to compose the LS phase in GRASP-FS may require extensive analysis and comparison of results. Furthermore, in case of changes in the network traffic profile, the whole process needs to be performed again. Therefore, a crucial challenge of this approach is the lack of dynamicity and scalability in the feature selection process.

In this work, we propose a microservice-oriented architecture for GRASP-FS with Distributed LS (GDLS-FS) algorithms. The proposed architecture employs the publish/subscribe paradigm to decouple the LS algorithms and address the scalability issues in metaheuristics. To evaluate our approach, we compare the VND and RVND neighborhood structures and the IWSSR, IWSS, and Bit-Flip local searches.

2   Background

In this section, we present algorithms and concepts related to the GRASP-FS metaheuristic, local search algorithms, and neighborhood structures.

The GRASP-FS [18] is an application of the GRASP metaheuristic [20] for Feature Selection (FS) purposes. The main goal of the GRASP-FS is to select a representative subset of features, discarding those that do not aggregate value to classifier models. GRASP-FS can be used in the intrusion detection context to select relevant information to represent a malicious activity. The GRASP-FS metaheuristic is divided into two main steps: (i) one for generating initial solutions and (ii) another for performing local movements and optimizing the initial solutions.

The first step of GRASP-FS is the construction phase, in which an initial reduction of the total amount of features available is performed. Such reduction is typically based on a process that employs statistical methods such as Information Gain (IG) or Gain Ratio (GR) to rank the available features according to their representativeness. After ranking, the N top-ranked features are used to compose a Restricted Candidate List (RCL). From the RCL, *greedy* and *random* solutions are generated. In this work, we call the generated feature subset as *initial solution*. These solutions are considered greedy because they consider only the statistical methods employed to filter their representativeness. They are random solutions because the selected features to compose such solutions are picked up at random from the RCL.

In the second step of GRASP-FS, the *initial solutions* are optimized through local movements performed by an LS algorithm. These LS algorithms are coordinated by a Neighborhood Structure. The feature subset optimization relies on an objective function such as F1-Score (Eq. 1) to compare the found solutions. The F1-Score metric represents the harmonic mean of Precision (Eq. 2) and Recall (Eq. 3) [18].

$$F1Score = 2 \times \frac{Precison \times Recall}{Precision + Recall} \tag{1}$$

$$Precision = \frac{TP}{TP + FP} \tag{2}$$

$$Recall = \frac{TP}{TP + FN} \tag{3}$$

The GRASP-FS metaheuristic was designed to work as a monopolistic application. The first version of GRASP-FS [19] implements a single and simple LS method, whereas its improved versions [17,18] support different neighbor structures and LS algorithms. Variable Neighborhood Descent (VND) [7] and Random Variable Neighborhood Descent (RVND) [23] are neighbor structures that were employed to handle the LS algorithms in GRASP-FS.

The VND walks on the initial solutions neighborhoods through a list of LS algorithms and moves from there to new neighborhoods after a number of iterations or maximum time criteria are reached. If improvements have been made, it moves to the first LS algorithm. Otherwise, the next LS algorithm in the list is used to explore a different neighborhood. According to [14], the resulting hybrid GRASP/VND algorithm is simple and fast, reaching solutions equal to or better than those obtained by the more complex existing procedures. The RVND neighborhood structure operates similarly to VND, but instead of going through the LS list sequentially, random movements are made every a threshold is reached.

Regardless of the neighborhood structure adopted, the LS algorithms are responsible to define the movements from one solution to one of its neighbors. The Incremental Wrapper Subset Selection (IWSS) is an algorithm in which the RCL features are pickup incrementally to compose the neighbor solutions. The added features are kept if and only if they present an improvement to the neighbor solution (*e.g.,* a better F1-Score than the previous one) [2]. The Incremental Wrapper Subset Selection with replacement (IWSSr) algorithm is a variation of IWSS which also performs sequential movements, but rather than only adding it also attempts to perform replacements between the solution and RCL features [11].

Finally, Bit-Flip (BF) is the simplest LS procedure already implemented within GRASP-FS. It operates by performing iteratively exchanges among features from the RCL and the selected to compose the initial solution [19].

3   Related Works

The main related works in the current literature can be divided into three categories: those that adopt the local search (i) within GRASP metaheuristic, (ii) without GRASP, and (iii) a distributed/parallel architecture. In this section, we discuss these works. Table 1 summarizes the works. Note that, to the best of our knowledge, this work is the only one that combines all four characteristics.

Table 1. Related Works.

Ref.	GRASP	Feature selection	Local search	Distributed local search
[5,11,19,22]	√	√	×	×
[12]	√	√	√	×
[13]	×	√	√	×
[6]	×	√	√	Parallel
[4]	×	×	√	√
[1]	×	×	√	Parallel
This work	√	√	√	√

The GRASP-based metaheuristics are found in the literature in applications such as feature selection in cardiovascular disease classification [22], feature selection in high-dimensional datasets, and feature selection to detect and prevent instructions in Cyber-Physical Systems [18] and real-time applications [5]. Subanya et al. [22] uses an artificial colony of bees for cardiovascular disease classification and reduces computational classification complexity by eliminating redundant resources. The authors implement the Support Vector Machine (SVM) classifier to perform the wrapper-based assessment in the GRASP LS. Other work [5] presents an architecture for real-time intrusion detection and prevention in eBPF switches. The authors use asynchronously optimized models through the GRASP-FS metaheuristic. Finally, an important remark is that the aforementioned works do not address the local search distribution in order to scale feature selection services.

A further work that adopts the GRASP metaheuristic is the [12]. The authors propose an approach to feature selection using GRASP with local search based on Simulated Annealing (SA) algorithm. The main goal is to reduce the classification processing time based on reducing the number of features and new parameters embedded in SA. Thus, the authors argue that their proposal is scalable. Nevertheless, the authors in fact do not use distribution and parallelism to reach scalability. Other work [11] employ GREselect a subset of features with a high-dimensional dataset. The authors apply the FICA and IWSSr algorithms in the wrapper phase to incorporate the selection of relevant features. The results indicate an improvement in terms of accuracy in the wrapper phase.

Another local search approach consists in to perform LS without GRASP. Nekkaa et al. [13] propose a local hybrid search based on harmony search (HSA) and stochastic search (SLS) for classification and feature selection. HSA-SLS algorithms are combined with a Support Vector Machine (SVM) classifier to send optimized parameters. Thus, according to the authors, it is possible to maximize the classification accuracy of the SVM. Another proposal [6] implements LS and data sorting using SVM. The goal is select and classifies features optimally. In addition, the proposal seeks to reduce the use of resources. The authors suggest diversifying the search space through multiple LS algorithms running in parallel and periodically sharing information.

Distribution and parallelism are the focus of another works in literature. Cai et al. [4] presents a distributed planning approach. The proposal allows robots to build a team plan collaboratively plan via distributed local search. The robots propose local operations based on their trajectories. Computing and communication requirements reduce using Lazy Search and Greedy Warm. Already, Araujo et al. [1] propose to use GPU for parallel processing of LS neighborhood structures instances in order to minimize the latency compared to sequential workflows. However, different from that work, a broader scaling strategy is adopted in this work based on LS execution through microservices. It allows distribution – and scaling – in different scenarios ranging from clusters to GPUs, as proposed by [1]. Particularly, our work proposes LS scaling to solve the feature selection problem for IDSs.

4   The GDLS-FS Architecture

The GRASP-FS with Distributed LS (GDLS-FS) is a novel microservice-oriented architecture that enables the distribution and parallelism of microservices to solve the scalability issues of LS algorithms in the GRASP-FS metaheuristic. GRASP-FS employs machine learning algorithms to process potentially large volumes of data in its LS phase. Therefore, these algorithms may cause overhead for processing and slowness depending on the amount of processed data: this is the main bottleneck of GRASP-FS. In this context, our approach gives scalability to the LS process, speeding up the convergence to the best solution. To reach more scalability, we distribute the LS processing among multiple microservices, each with a different responsibility. These microservices communicate with each other by using the *publish/subscribe* paradigm. Our proposed architecture is illustrated in Fig. 1.

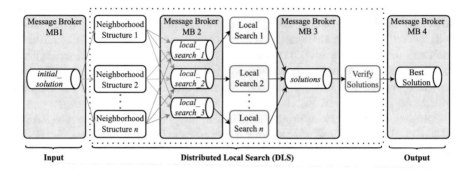

Fig. 1. The proposed micro-service oriented architecture.

Firstly, there is a Message Broker `MB1` which is responsible for initializing the proposed LS process by providing an initial solution to be optimized. In GRASP-FS the initial solutions would be provided by specific algorithms, implemented in its constructive phase (the first phase of GRASP-FS), which combines a greedy and random procedure. In this work, we focus on the LS only (the second phase of GRASP-FS). Therefore, the component responsible for providing initial solutions should be transparent to our architecture. The process starts whenever an already computed initial solution is given.

The `Neighborhood Structure` microservices are responsible for handling the LS algorithms. This includes scheduling and feeding them with the initial solution. Therefore, each `Neighborhood Structure` microservice subscribes to the *initial_solution* topic in `MB1` and publishes the received initial solutions, according to its implementation, into one or multiple *local_search* topics in `MB2`. For each initial solution consumed, the `Neighborhood Structure` microservice publishes, iteratively, multiple times until it reaches a predefined threshold (*e.g.,* number of iterations, maximum running time, etc.).

The `LS` microservices implement a set of *local movements* to derive neighbor solutions (*i.e.,* feature subsets with few features different from the initial solutions) iteratively until reaching a predefined threshold. Every solution generated is assessed by a wrapping-based approach — which employs a machine learning algorithm. Each microservice may implement a distinct LS procedure. Alternatively, they may be deployed as replicas with redundant implementation for attaining greater scalability due to the smaller overhead of each microservice. Considering replicas for this microservice is relevant since it uses machine learning algorithms to assess the quality of each generated neighbor solution, which may cause the main bottleneck of the proposed architecture. The generated solutions are published into the *solutions* topic of MB3, which is consumed by the interested microservices, such as the VerifySolutions or the Neighborhood Structure microservices that need to receive feedback to perform their next decisions.

Finally, the VerifySolutions microservice subscribes to the *solutions* topic from MB3 to process and compare all solution founds, regardless of the LS algorithm used to find such solutions. In a real and production scenario, a real-time IDS would subscribe to the *solutions* topic to receive optimized solutions whenever they are generated by our architecture. This procedure avoids the IDS being unnecessarily re-trained when no updated solutions are available and keeps the IDSs continuously updated with the current best-known feature subset.

5   Proof-of-Concept: Implementation

To evaluate the novel approach presented in this work, we implemented an instance of the proposed GDLS-FS architecture as a proof-of-concept to compare its performance with the traditional GRASP-FS metaheuristic implementation [19] and its extended version [18]. The implemented architecture components are depicted in Fig. 2 and discussed below.

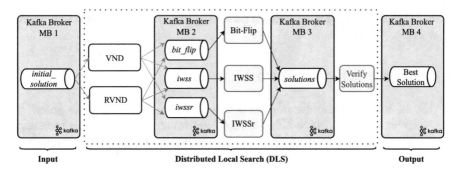

Fig. 2. Implementation of the Distribute Local Search (DLS) as a part of GDLS-FS.

Our implementation has three main categories of microservices: Neighborhood Structures (*i.e.,* VND and RVND microservices), Local Searches (*i.e.,* Bit-Flip, IWSS, and IWSSr microservices), and the Verify Solutions microservice. Whereas the two formers are aimed to handle the local movements through the solutions' neighborhood, the latter is responsible for delivering the optimized solutions to the subscribers IDSs interested in that information. As classifier algorithms for wrapping-based evaluations within the LS microservices, we used J48 algorithm with support of the Weka machine learning workbench [8]. To implement the message brokers M1, M2, M3, and M4, we adopted the Apache Kafka [24].

To assess the implemented instance of GDLS-FS in light of a critical application, we generate 2,000 samples of traffic from electrical substation networks through the ERENO Framework [15]. ERENO makes it possible to model and simulate cyberattacks to the electrical power grid infrastructure.

As the scope of this work is the DLS as a part of the GRASP-FS metaheuristic, we abstract the GRASP-FS' construction phase by using a fixed initial solution S = {1, 2, 3, 4, 5} as input to DLS process. This allows us to assess the DLS process with a unified input as a starting point. The remaining features are added to RCL = {6, 7, ..., 69} for enabling the LS algorithms to walk through the neighbor solutions.

To start the DLS process, S is published into *initial_solution* topic in MB1 and consumed by VND and RVND. From this point, messages are sent iteratively to LS microservices by using S to initialize them — which microservice is initialized depends on the internal logic of the neighborhood structure. For each message published by VND and/or RVND, IWSS and IWSSr explore all *RCL* features sequentially and *Bit-Flip* performs $N = 100$ neighbor solutions. We publish $M = 10$ messages for each neighborhood structure. Therefore, for each message published by a neighborhood structure, the explored neighborhood has:

- $M \times N$ neighbors generated by *Bit-Flip*;
- $M \times |RCL|$ neighbors generated by IWSS, where $|RCL|$ is the RCL length; and
- $M \times |RCL| + R$ neighbor solutions generated by IWSSr, where R is the number of replace movements.

6   Experiments and Results

In this section, we present and discuss the found results. Firstly, we show how the Neighborhood Structure Microservices performed when publishing messages to the LS subscribers (Sect. 6.1). Then, we analyzed one iteration from each LS algorithm to figure out how they performed (Sect. 6.2).

6.1   Neighborhood Structure Microservices

Figure 3 depicts the sequential iterations of local search microservices. We used a logarithmic timestamp in the chart to enable the visualization of all LS algorithms, as IWSSr took too much time in comparison to IWSS and Bit-Flip. The entire processing of VND and LS algorithms took 3,205,075 ms (53,4 min). Note that at the end of the execution, there are still novel solutions being reached by IWSS and Bit-Flip. This occurs because the choice of which LS will be requested by VND depends upon the results of the previous LS it requested.

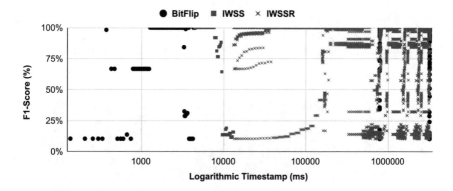

Fig. 3. VND iterations

One of the key advantages of using RVND rather than VND in a distributed architecture is the expanded exploration of parallelism. Figure 4 depicts the random and parallel iterations of local search microservices.

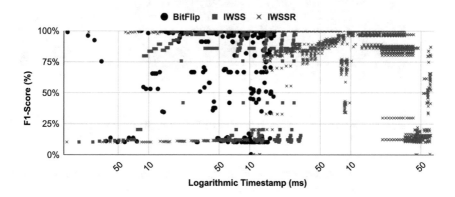

Fig. 4. RVND iterations.

The entire processing of RVND and LS algorithms took 626,096 ms (10,43 min). RVND is faster than VND because it do not need a feedback from the LS. Therefore, RVND can explore better the parallel processing among the LS microservices. In this scenario, Bit-Flip took only 17,650 ms and IWSS in 32,354 ms. The graph's logarithmic timestamp enables a better visualization of all LS algorithms.

6.2   Local Search Microservices

Figure 5 depicts the F1-Score of the solutions found within one iteration for each LS microservice. Bit-Flip reaches high and low F1-Scores more randomly because of their random movements. In contrast, IWSS has a more stable behavior which follows a sequential pattern. The IWSS performed worse at the end of the processing because those RCL features were less representative in the assessed.

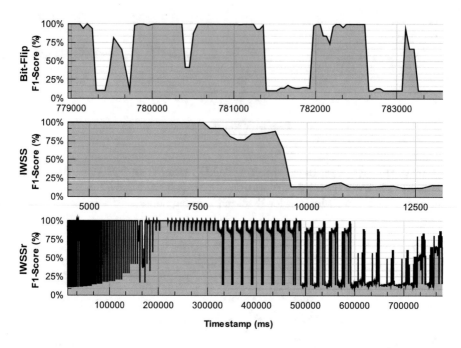

Fig. 5. Comparison of the behavior of Bit-Flip, IWSS, and IWSSr.

Finally, IWSSr presented a very unstable behavior because for each *add* movement it tested the replacement of the current solution features by another from the RCL. This may be the main reason for the high and low F1-Scores found along the processing. All methods reached a 100% F1-Score in the experimented dataset.

7   Conclusion

Adopting FS methods are fundamental to enable good detection performance for IDSs. However, a major challenge is performing an accurate feature selection in real-time applications: as the traffic pattern changes, irrelevant features may gain more value and the representative features may loose relevance for the machine learning models. Although GRASP-FS offers a trade-off between the high-quality and computational expensiveness, it is not scalable.

To address the aforementioned challenges, we presented the GDLS-FS: a novel microservices-oriented architecture for GRASP-FS with DLS algorithms.

The proposed architecture addresses the scalability issues by decoupling the LS microservices through the publish/subscribe paradigm. As a proof of concept, we implement an instance of our architecture using the Kafka framework, two neighborhood structures (i.e., VND and RVND), and three LS algorithms (i.e., IWSS, IWSSR, and Bit-Flip). Our results show that using RVND may takes only 19.53% of the time taken by VND. Therefore, the RVND approach is the most efficient for exploiting parallelism in distributed architectures.

In future work, we plan to implement the full GRASP-FS architecture, covering the construction phase. Besides, we intend to experiment in more challenging scenarios and to assess other algorithms to implement our microservices. Finally, our future experiments will use a larger dataset to assess the scalability of the proposed solution under massive data processing.

Acknowledgments. This work is financially supported by TQI Tecnologia.

References

1. Araujo, R.P., Coelho, I.M., Marzulo, L.A.J.: A multi-improvement local search using dataflow and GPU to solve the minimum latency problem. Parallel Comput. **98**, 102661 (2020)
2. Bermejo, P., Gamez, J.A., Puerta, J.M.: Incremental wrapper-based subset selection with replacement: an advantageous alternative to sequential forward selection. In: 2009 IEEE Symposium on Computational Intelligence and Data Mining, pp. 367–374 (2009)
3. Borgiani, V., Moratori, P., Kazienko, J.F., Tubino, E.R.R., Quincozes, S.E.: Toward a distributed approach for detection and mitigation of denial-of-service attacks within industrial internet of things. IEEE Internet Things J. **8**(6), 4569–4578 (2021)
4. Cai, X., Schlotfeldt, B., Khosoussi, K., Atanasov, N., Pappas, G.J., How, J. P.: Non-monotone energy-aware information gathering for heterogeneous robot teams. In: 2021 IEEE International Conference on Robotics and Automation (ICRA). IEEE, pp. 8859–8865 (2021)
5. Carvalho, D., Quincozes, V.E., Quincozes, S.E., Kazienko, J.F., Santos, C.R.P.: BG-IDPS: Detecção e prevenção de intrusões em tempo real em switches eBPF com o filtro de pacotes berkeley e a metaheurística GRASP-FS. In *XXII Simpósio Brasileiro em Segurança da Informação e de Sistemas Computacionais - SBSeg*, SBC, pp. 139–152 (2022)
6. Cura, T.: Use of support vector machines with a parallel local search algorithm for data classification and feature selection. Expert Syst. Appl. **145**, 113133 (2020)
7. Hansen, P., Mladenović, N., Brimberg, J., Pérez, J.A.M.: Variable neighborhood search. In: Gendreau, M., Potvin, J.-Y. (eds.) Handbook of Metaheuristics. ISORMS, vol. 272, pp. 57–97. Springer, Cham (2019). https://doi.org/10.1007/978-3-319-91086-4_3
8. Holmes, G., Donkin, A., Witten, I.H.: Weka: a machine learning workbench. In: Australian New Zealnd Intelligent Information Systems Conference, pp. 357–361 (1994)
9. Liao, H.-J., Lin, C.-H.R., Lin, Y.-C., Tung, K.-Y.: Intrusion detection system: a comprehensive review. J. Netw. Comput. Appl. **36**(1), 16–24 (2013)

10. Mihoub, A., Fredj, O.B., Cheikhrouhou, O., Derhab, A., Krichen, M.: Denial of service attack detection and mitigation for internet of things using looking-back-enabled machine learning techniques. Computers Electr. Eng. **98**, 107716 (2022)
11. Moradkhani, M., Amiri, A., Javaherian, M., Safari, H.: A hybrid algorithm for feature subset selection in high-dimensional datasets using FICA and IWSSR algorithm. Appl. Soft Comput. **35**, 123–135 (2015)
12. Moshki, M., Kabiri, P., Mohebalhojeh, A.: Scalable feature selection in high-dimensional data based on GRASP. Appl. Artif. Intell. **29**(3), 283–296 (2015)
13. Nekkaa, M., Boughaci, D.: Hybrid harmony search combined with stochastic local search for feature selection. Neural Process. Lett. **44**(1), 199–220 (2016)
14. Parreño, F., Alvarez-Valdés, R., Oliveira, J.F., Tamarit, J.M.: A hybrid GRASP/VND algorithm for two-and three-dimensional bin packing. Ann. Oper. Res. **179**(1), 203–220 (2010)
15. Quincozes, S.: ERENO: An Extensible Tool for Generating Realistic IEC-61850 Intrusion Detection Datasets. Ph.D thesis, Fluminense Federal University (2022)
16. Quincozes, S.E., Albuquerque, C., Passos, D., Mossé, D.: A survey on intrusion detection and prevention systems in digital substations. Comput. Netw. **184**, 107679 (2021)
17. Quincozes, S.E., Mossé, D., Passos, D., Albuquerque, C., Ochi, L.S., dos Santos, V.F.: On the performance of grasp-based feature selection for cps intrusion detection. IEEE Trans. Netw. Serv. Manage. **19**(1), 614–626 (2021)
18. Quincozes, S.E., Passos, D., Albuquerque, C., Mossé, D., Ochi, L.S.: An extended assessment of metaheuristics-based feature selection for intrusion detection in CPS perception layer. Ann. Telecommun. **77**, 1–15 (2022)
19. Quincozes, S.E., Passos, D., Albuquerque, C., Ochi, L.S., Mossé, D.: Grasp-based feature selection for intrusion detection in CPS perception layer. In: 2020 4th Conference on cloud and internet of things (CIoT)IEEE, pp. 41–48 (2020)
20. Resende, M.G., Ribeiro, C.C.: GRASP: greedy randomized adaptive search procedures. In: Burke, E., Kendall, G. (eds.) Search Methodologies, pp. 287–312. Springer, Boston (2014). https://doi.org/10.1007/978-1-4614-6940-7_11
21. Soares, A.A.Z., et al.: Enabling emulation and evaluation of IEC 61850 networks with TITAN. IEEE Access **9**, 49788–49805 (2021)
22. Subanya, B., Rajalaxmi, R.: Feature selection using artificial bee colony for cardiovascular disease classification. In: 2014 International Conference on Electronics and Communication Systems (ICECS). IEEE, pp. 1–6 (2014)
23. Subramanian, A., Drummond, L.M., Bentes, C., Ochi, L.S., Farias, R.: A parallel heuristic for the vehicle routing problem with simultaneous pickup and delivery. Comput. Oper. Res. **37**(11), 1899–1911 (2010)
24. Thein, K.M.M.: Apache kafka: next generation distributed messaging system. Int. J. Sci. Eng. Technol. Res. **3**(47), 9478–9483 (2014)
25. Vander-Pallen, M.A., Addai, P., Isteefanos, S., Mohd, T.K.: Survey on types of cyber attacks on operating system vulnerabilities since 2018 onwards. In: 2022 IEEE World AI IoT Congress (AIIoT). IEEE, pp. 01–07 (2022)

Multi-agent Deep Q-Learning Based Navigation

Amar Nath[1]([✉]), Rajdeep Niyogi[2], Tajinder Singh[1], and Virendra Kumar[3]

[1] Sant Longowal Institute of Engineering and Technology, Deemed-to-be-University, Longowal, India
{amarnath,tajindersingh}@sliet.ac.in
[2] Indian Institute of Technology Roorkee, Roorkee, India
rajdeep.niyogi@cs.iitr.ac.in
[3] Central University of Tamil Nadu, Thiruvarur, India
virendrakumar@cutn.ac.in

Abstract. Navigating an unknown indoor environment like a building is quite challenging as Global Positioning System-based solutions are hard or near impossible. However, it is a crucial problem as it has many applications, such as search and rescue. The rescue operations should start at the earliest to avoid casualties. Hence, the navigation process should be completed quickly. The navigation process can be done and accelerated with multiple agents/robots. The Q-learning-based approach can help navigation, but it is suitable only if state and action spaces are low. This paper suggests a distributed multi-agent deep Q-learning algorithm for indoor navigation in a complex environment where the number of states is large.

1 Introduction

There are many indoor navigation applications, such as rescuing human beings after a disaster [1]. Earthquakes are among the world's most devastating natural hazards. Performing rescue operations in such an unknown environment becomes a challenging and crucial problem as rescue operation needs to start at the earliest. A multi-agent system can solve the problem efficiently [2,3].

Consider a building with five rooms with four connecting doors (Fig. 1). Knowing the building's map, including the locations of its rooms and doors as well as its current state—that is, whether doors connect its rooms is necessary to perform rescue operations.

A distributed multi-agent strategy can aid in the exploration because the agents will move about the structure to determine the doors' locations and open/closed states. The problem mentioned above can be solved using a distributed multi-agent Q-learning strategy, in which each agent regularly broadcasts its observations to every other agent [4].

L. Barolli (Ed.): AINA 2023, LNNS 654, pp. 211–221, 2023.
https://doi.org/10.1007/978-3-031-28451-9_19

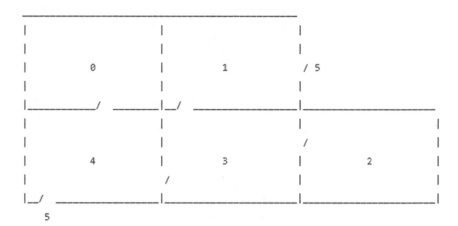

Fig. 1. Environment/Building

Q-learning is based on value iteration algorithms. When the state and action spaces are discrete, and the dimensions are not too high, a Q-table is suitable to store the Q value of each state action [17]. It is challenging to employ Q-table storage if the number of states and actions is large and continuous. Therefore, the Q-table update can be transformed into a function fitting the situation to generate Q-values corresponding to similar states through similar output actions. Deep Q-learning networks combine deep machine learning and reinforcement learning with deep neural networks to extract complicated shapes (DQN) [5]. A distributed multi-agent deep Q-learning algorithm for indoor navigation is suggested in this paper.

The remainder of the paper is structured as follows. In Sect. 2, related works are listed. The foundational ideas and notation are presented in Sect. 3. The proposed approach is described in full in Sect. 4. Results and implementation are described in Sect. 5. In Sect. 6, conclusions are drawn.

2 Related Work

Robots are now used in a variety of fields, including risky and crucial ones when direct human intervention is difficult or unfeasible, such as criminal justice, search and rescue, autonomous warehouses, space, and deep sea, thanks to advances in AI, robotics, and machine learning [6] that make it possible to work in such scenario. A search and rescue domain is critical to handle. This task or domain becomes considerably more difficult if a GPS system is not available.

Navigation is challenging, and a manual navigation system is necessary to search within a building where GPS is unavailable. Human-based navigation is risky or nearly impossible in a dangerous domain. As a result, robot-based navigation is recommended.

Work [7] covers how to generate several types of static maps of the system, such as topological and metric maps, using sequential localization and mapping procedures. It also addresses many information sources that can aid in creating an environment map. A method for centralizing Multi-Agent Deep Reinforcement Learning (MARL) training using agent communication and the model-free Deep Q-Network (DQN) as the foundational model is proposed in [8]. Work [9] introduces cooperative multi-agent exploration (CMAE), in which agents investigate a technique while working toward a common objective. They follow the traditional centralized training and decentralized execution (CTDE) paradigm, in which each agent's local observations, actions, and state are available to the learning algorithm during training. Only at test time or during execution does each agent have access to its local observation.

Recent works [5,10,11] discuss deep learning-based navigation. [5] suggest a complete navigation strategy for mobile robots using only depth image data in an uncharted environment. The method described in the article [10] blends reinforcement learning concepts with double net and recurrent neural network modules. It is founded on deep RL and RCN.

The work [11] proposes a method for deep reinforcement learning based on graphs called SG-DQN that (i) extracts a useful graph representation for the crowd-robot state using a social attention mechanism, (ii) uses a learned dueling deep Q network (DQN) to assess the coarse q-values of the raw state, and (iii) refines the coarse q-values through online planning on potential future trajectories. However, the distributed DQN-based approach to navigation is not covered in any of the studies mentioned above.

3   Preliminary Concepts

3.1   Q-Learning

An effective model-free reinforcement learning method for picking the best course of action in a given state is Q-learning [12]. A Q-table cell contains the Q-value for a specific state action. The Q-starting table's values are all zero. When an agent performs an action (a) in the state (s), it is rewarded (r), and it also observes the new state. s'. The table is then updated in accordance with Eq. (1)

$$Q(s, a) \leftarrow r + \gamma \max_{a'} Q(s', a') \qquad (1)$$

The discount factor γ ($0 \leq \gamma \leq 1$) is low and accounts for the immediate benefit, while a high discount factor accounts for the actual long-term payoff. The discount factor is used to balance out the benefits of short-term and long-term investments. Based on the Q-values, an agent uses the Q-table to guide the best course of action. An agent chooses the action a for a state s such that $Q(s, a)$ has the maximum [4] value.

However, Q-learning is a straightforward but incredibly effective algorithm for generating action for an agent. This aids the agent in determining precisely what action to take in a specific environment/building state. What if this environment/building should be shorter, though?

3.2   Deep Q-Learning

We employ a neural network to simulate the Q-value function in deep Q-learning [13]. Input states are translated into (action, Q-value) pairs via a neural network. The state is provided as the input, and the Q-value for every action that could be taken is produced as the output. The Q-learning and deep Q-learning are illustrated in Fig. 2.

Instead of using the Q-table to store the Q value, Deep Q-learning employs the neural network to anticipate the Q value and continually update the neural network to learn the best course of action. In Deep Q-learning, there are two neural networks: the main net, which has a relatively fixed parameter and is used to determine the value of Q in the main net, and the target net, which determines the value of Q for evaluation [14,15]. Equations 2 and 3 serve as the mathematical representation of the loss function of Deep Q-learning, and θ' is the parameter in the target network.

$$L(\theta) = E\{(Q_{target} - Q_{main}(s_t, a_t, \theta))^2\} \tag{2}$$

$$Q_{target} = r_{t+1} + \gamma.maxQ_{target}((s_{t+1}, a_{t+1}, \theta')) \tag{3}$$

The same value is used to pick and measure action in both Deep Q-learning and conventional Q-max learning's operation.

4   Distributed Multi-agent Deep Q-learning Algorithm

The proposed deep Q-learning algorithm aims to know the current map of the building by observing it with multiple agents. The proposed method has the benefit of being able to accelerate the process of knowing the map of the building. The proposed approach is given in Algorithm 1.

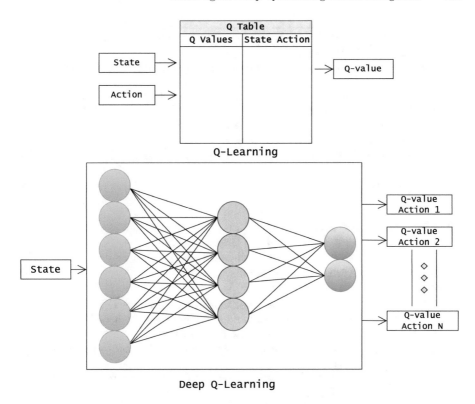

Fig. 2. Q-learning and deep Q-Learning architectures

4.1   Description of the Algorithm

The proposed algorithm starts with initializing the Q-table and weights for each agent's network. The environmental states are taken as input. Finally, we get the Q-table from which paths can be identified for navigation. Using the **epsilon**-greedy exploration technique, the agent chooses a random action and benefits from it with probability **epsilon**.

The input states are translated into a pair (action, Q-value) by the main and target neural networks. In this instance, the Q-value of each output node (indicating an action) is stored as a floating point number. As a result, the output nodes will not add up to 1 because they do not represent a probability distribution. In the illustration (shown in Fig. 3), one of the actions has a Q-value 12 while the other has a Q-value 6.

Algorithm 1. Distributed Multi-Agent Deep Q-learning Algorithm

---

Data: Input: All the states of the environment
Result: Q-value(table) of all possible actions is generated as the output

1 **for** *(Agent i = 1, ..., N)* **do**
2 | Initialize the neural networks' main and target nodes.
      | Initialize relay memory D.
      | Initialize weights of the network randomly.
      | Initialize the starting state s_i randomly.
      | An action is selected by ϵ-greedy exploration strategy.
3 **end**
    /*Sender_Function$_i$:   */
4 **for** *each episode* **do**
5 | **for** *each time step* **do**
6 | **while** *($s_i \neq Goal$)* **do**
7 | **if** *(Agent i detects a door)* **then**
8 | Agent *i* receives immediate reward r_i
               | Agent *i* observes the new state s'_i
               | $s_i \leftarrow s'_i$
               | Agent *i* store the experience in relay memory D_i.
               | Agent *i* sample random batch from relay memory D_i.
               | Agent *i* prepossess state from the batch.
               | Agent *i* passes a batch of pre-processed states to the policy network.
               | Calculate loss between output Q-values and target Q-values as per Eq. 2
               | To reduce loss, gradient descent modifies the policy network's weights.
9 | **end**
10 | **end**
11 | Agent *i* broadcast its updated memory D_i value to other agents
12 | **end**
13 **end**
    /*Receiver_Function$_j$:   */
14 **if** *(Agent j receives the D_i from Agent i)* **then**
15 | Agent *j* updates its memory D_j.
16 **end**

The best action from the network is selected as follows. The output actions of the main model and target models' output actions are mapped to the input states. These output actions represent the model's projected Q-value. In this instance, the most well-known action in that state is the one with the most significant anticipated Q-value. The best action from the network is selected as follows: The output actions of the main model and target models' output actions are mapped to the input states. These output actions represent the model's projected Q-value. In this instance, the most well-known action in that state is the one with the most significant anticipated Q-value.

Input states

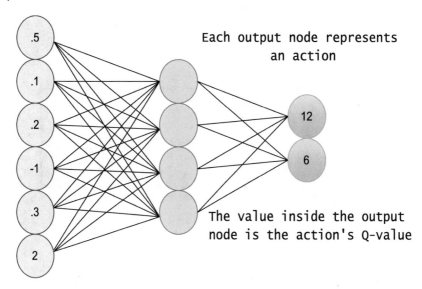

Each output node represents
an action

The value inside the output
node is the action's Q-value

Fig. 3. A neural network that converts a state's input into the corresponding pair of (action, Q-value) values

5   Implementation and Results

The implementation of the suggested approach is illustrated with a maze scenario in which the mouse is put at random into one of the it's locations and must figure out where the cheese is (reward). The example scenario is shown in Fig. 4.

First, the Deep Q-learning architecture is set up to learn the map of a building scenario given in Sect. 1. The Q-table of Q learning is replaced with a neural network. This model uses six output neurons (Actions) and one input neuron (State). Consider one hidden layer that has two extra neurons than the output neurons. The action to be taken from the state is represented by each output neuron. The input layer has a single neuron, indicating the status of our mouse at the moment. We scale the input using Q-format by dividing the current state by the highest potential state (i.e., 5, 6 states in total numbered 0–5).

For each action taken from the given state, the neural network is trained to predict the Q-value. A random selection of the most recent training samples will be injected into network N to train it after each move. We will only use a select handful of the most recent training examples because we anticipate that the skills will improve over time. Old samples, which are undoubtedly subpar, will be forgotten and removed from memory.

More specifically, we will create an episode after every motion and store it in a short-term memory sequence. An episode is a tuple of the five components we require for a single training. An environment state is "*envstate.*" It refers to an accurate representation of the maze cells in our maze (the state of each

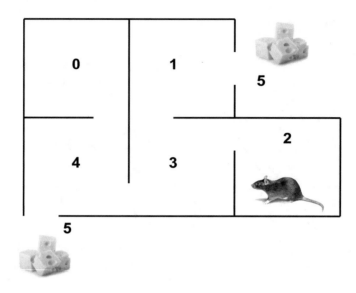

Fig. 4. Maze scenario with mouse and cheese location

cell including rat and cheese location). We reduce the maze to a 1-dimensional vector that fits the input of our neural network in order to simplify things for it. The Boolean value (True/False) *game_over* denotes whether or not the game is over. If the rat reaches the cheese cell (wins) or reaches a negative reward limit, the game is done (lose).

We create this five-element episode after each move in the game and add it to our memory sequence. If the size of our memory sequence exceeds a predetermined limit, we remove elements from its tail to keep it under this limit. Network N's weights are first set to random values. N will therefore initially yield poor results. The Bellman Equation should be solved if the parameters of our model are properly specified, hence results from subsequent experiments are anticipated to be positive. There is certainly room for improvement, however right now it seems to be quite difficult to construct a model that converges quickly.

The action corresponding to the maximum Q-value is then selected after training the neural network. The epsilon exploration factor determines the frequency level of how much exploration to do. It is set at 0.1 which translates to the agent performing an entirely arbitrary action once per 10 moves (Fig. 5).

Our neural network model uses the SReLU (the S-shaped relu) activation function. It learns both convex and non-convex functions by modeling the many function forms provided by the two fundamental laws of psychophysics and neurosciences, namely the Webner-Fechner law and the Stevens law [16]. As an optimizer, we use Root Mean Squared Propagation (RMSProp). The MSE (Mean Squared Error) is used as the loss function. The experiment is run with 100 epochs on `googlecolab`.

```
building = np.array([
                    [1, 0, 0, 0, 1, 0],
                    [0, 1, 0, 1, 0, 1],
                    [0, 0, 1, 1, 0, 0],
                    [0, 1, 0, 1, 1, 0],
                    [1, 0, 0, 0, 1, 1],
                    [0, 1, 0, 0, 1, 1]
                    )]
```

Fig. 5. Maze's representation as an input matrix to be navigated: a neural network that connects an input state to the (action, q-value) pair that corresponds

Result: The output of the program is given below. If we see the building given in Fig. 1, all the possible paths from every node to the goal state, without learning the environment, from each room to exit, i.e., 5 are as follows

- $0 \rightarrow 4 \rightarrow 5$; *or* $0 \rightarrow 4 \rightarrow 3 \rightarrow 1 \rightarrow 5$.
- $1 \rightarrow 5$; *or* $1 \rightarrow 3 \rightarrow 4 \rightarrow 5$.
- $2 \rightarrow 3 \rightarrow 1 \rightarrow 5$; *or* $2 \rightarrow 3 \rightarrow 4 \rightarrow 5$.
- $3 \rightarrow 1 \rightarrow 5$ *or* $3 \rightarrow 4 \rightarrow 5$.
- $4 \rightarrow 5$; *or* $4 \rightarrow 3 \rightarrow 1 \rightarrow 5$.

However, paths after learning the environment are as follows

- $0 \rightarrow 4 \rightarrow 5$.
- $1 \rightarrow 5$.
- $2 \rightarrow 3 \rightarrow 1 \rightarrow 5$; *or* $2 \rightarrow 3 \rightarrow 4 \rightarrow 5$.
- $3 \rightarrow 1 \rightarrow 5$ *or* $3 \rightarrow 4 \rightarrow 5$.
- $4 \rightarrow 5$.

The time taken to complete the learning with DQN is 436 s while with Q-leaning it is 83 s. More code and data are used in the DQN implementation than in the Q-learning. To exploit the benefits of DQN, we would need to use a more complicated state space.

6 Conclusion

This paper proposes a distributed algorithm for multi-agent Deep Q-learning-based navigation. Multiple agents learn the map on their own in an unknown environment. However, every agent periodically communicates with other agents to share their acquired knowledge. Exploration overlap is prevented by preventing multiple agents from navigating the same area of the environment. As a result, learning the map of the environment is done quickly.

We have assumed that there is no message loss when the agents communicate with each other. However, in a real-world environment, messages may be lost. In the future, we would like to address this issue.

Acknowledgment. The authors are grateful to the anonymous referees for their insightful criticism, which helped to make the paper better. The second author was in part supported by a research grant from Google.

References

1. Comerio, M.C.: Disaster Hits Home: New Policy for Urban Housing Recovery. University of California Press, Berkeley (1998)
2. Van der Hoek, W., Wooldridge, M.: Multi-agent systems. Found. Artif. Intell. **3**, 887–928 (2008)
3. Dorri, A., Kanhere, S.S., Jurdak, R.: Multi-agent systems: a survey. IEEE Access **6**, 28573–28593 (2018)
4. Nath, A., Niyogi, R., Singh, T., Kumar, V.: Multi-agent Q-learning based navigation in an unknown environment. In: Barolli, L., Hussain, F., Enokido, T. (eds.) AINA 2022. LNNS, vol. 449, pp. 330–340. Springer, Cham (2022). https://doi.org/10.1007/978-3-030-99584-3_29
5. Ruan, X., Ren, D., Zhu, X., Huang, J.: Mobile robot navigation based on deep reinforcement learning. In: 2019 Chinese Control and Decision Conference (CCDC), pp. 6174-6178. IEEE (2019)
6. Nath, A., Arun, A.R., Niyogi, R.: A distributed approach for road clearance with multi-robot in urban search and rescue environment. Int. J. Intell. Rob. Appl. **3**(4), 392–406 (2019). https://doi.org/10.1007/s41315-019-00111-5
7. Meyer, J.A., Filliat, D.: Map-based navigation in mobile robots:: Ii. a review of map-learning and path-planning strategies. Cogn. Syst. Res. **4**(4), 283-317 (2003)
8. Bhalla, S., Ganapathi Subramanian, S., Crowley, M.: Deep multi-agent reinforcement learning for autonomous driving. In: 33rd Canadian Conference on Artificial Intelligence (CCAI-2020), pp. 67–78. Ontario, Ottawa (2020)
9. Liu, I.J., Jain, U., Yeh, R.A., Schwing, A.: Cooperative exploration for multi-agent deep reinforcement learning. In: 38th International Conference on Machine Learning (ICML-2021). Virtual mode, pp. 6826–6836 (2021)
10. Quan, H., Li, Y., Zhang, Y.: A novel mobile robot navigation method based on deep reinforcement learning. Int. J. Adv. Rob. Syst. **17**(3), 1-11. SAGE (2020)
11. Zhou, Z., Zhu, P., Zeng, Z., Xiao, J., Lu, H., Zhou, Z.: Robot navigation in a crowd by integrating deep reinforcement learning and online planning. Appl. Intell. **52**, 1–17 (2022)
12. Watkins, C.J., Dayan, P.: Q-learning. Mach. Learn. **8**(3), 279–292 (1992)

13. Van Hasselt, H., Guez, A., Silver, D.: Deep reinforcement learning with double q-learning. In: Proceedings of the AAAI Conference on Artificial Intelligence, vol. 30, pp. 2094–2100 (2016)
14. Fan, J., Wang, Z., Xie, Y., Yang, Z.: A theoretical analysis of deep Q-learning. In: Learning for Dynamics and Control, pp. 486-489. PMLR (2020)
15. Achiam, J., Knight, E., Abbeel, P.: Towards characterizing divergence in deep q-learning. arXiv preprint arXiv:1903.08894 (2019)
16. Jin, X., Xu, C., Feng, J., Wei, Y., Xiong, J., Yan, S.: Deep learning with s-shaped rectified linear activation units. In: Proceedings of the AAAI Conference on Artificial Intelligence, vol. 30, no. 1 (2016)
17. Neves, M., Vieira, M., Neto, P.: A study on a Q-learning algorithm application to a manufacturing assembly problem. J. Manuf. Syst. **59**, 426–440 (2021)

Identifying Network Congestion on SDN-Based Data Centers with Supervised Classification

Filipe da Silva de Oliveira, Maurício Aronne Pillon, Charles Christian Miers, and Guilherme Piêgas Koslovski[✉]

Graduate Program in Applied Computing, Santa Catarina State University, Joinville, Santa Catarina, Brazil
filipedasilvadeoliveira@gmail.com,
{mauricio.pillon,charles.miers,guilherme.koslovski}@udesc.br

Abstract. Data centers have networks that provide, for a large number of users, various services such as video streaming, financial services, file storage, among others. Like any network-based system, a Data Center (DC) is subject to congestion, a fact that can cause slowness and instability, affecting the performance of applications for end users. Among the technologies for data center management, Software Defined Networking (SDN) separate the control from the data plane, allowing applications to access statistical information and other details about current state of the network through a controller. In this sense, SDN brings a new opportunity to detect congestion on DC networks in a centralized manner. However, identifying the occurrence of congestion in networks is an arduous task due to the high number of related data. With these data, this work seeks to apply supervised classification to identify congestion events. The experimental campaign demonstrated that the SDN-based supervised classification mechanisms find more congestion events then the native congestion control algorithms from Transmission Control Protocol (TCP). Moreover, among diverse supervised algorithms, we demonstrate which one has better accuracy and efficient performance when analyzing Cubic and DCTCP variants.

1 Introduction

In a utopian scenario, a computer network could transfer any desired amount of data instantly, however in reality, it has processing, queuing, and transmission delays that can even generate the loss of data. These problems depend on the degree of network congestion, in which too many sources try to send data with a resulting throughput greater than the forwarding capacity of the switching equipment, resulting in little or almost no useful workload. The Transmission Control Protocol (TCP) provides a reliable data transfer, that is, everything that was sent will reach the destination respecting the order in which it was submitted due to the mechanisms implemented to guarantee the service. In fact, TCP uses the packet conservation principle, which consists of the idea that for a connection to be in balance, new packets can only be added to the network as

L. Barolli (Ed.): AINA 2023, LNNS 654, pp. 222–234, 2023.
https://doi.org/10.1007/978-3-031-28451-9_20

old ones are leaving. The obedience of this principle determines that the collapse of congestion is an exception. Specifically, the protocol is based on information gathered at network end-points, eventually receiving support from intermediate switches, and controls a Congestion Window (CWND) for deciding the window size of segments allowed to be transferred [4].

DC networks based on Software Defined Networking (SDN) introduced a new opportunity to improve the overall network congestion and avoidance scenario [17]. The logical centralized control offered by SDN is a new opportunity to gather on-going and historical data from all network flows, organized by sources, ports, and switches. In fact, by analyzing this immense volume of data it is possible to indicate exactly the congested points of the network. However, the analysis require fast computing algorithms to deal with on-the-fly variations. Moreover, the congestion detector must support different types of traffic, analyzing different communication patterns and trends. That is exactly the point where supervised classification algorithms come in, as they will have been trained beforehand and can process large quantities of data almost in real-time [3,14]. The present work investigated the use of multiple supervised classification algorithms (Random Forest, Extremely Randomized Trees, K-nearest neighbor, Linear Discriminant Analysis, Bayesian Networks, and Support Vector Machines) to determine if the DC network is suffering from congestion. Later, the classifiers decisions are compared with two TCP scenarios based on Cubic [12] and DCTCP [2] variants.

The experimental analysis demonstrated that our prototype can correctly identify congestion events. In fact, the prototype identified more congestion events than the counterparts protocols. The results demonstrate that a centralized congestion detection mechanism is a promising approach to improve the overall network performance, once the decisions are common for all end-points. The remaining of this work is organized as follow. Section 2 presents the motivation and related work. The design and implementation details are given in Sect. 3, while the experimental analysis is discussed in Sect. 4. Finally, Sect. 5 concludes the work.

2 Motivation and Problem Definition

2.1 Congestion Control and Avoidance in SDN Data Centers

DCs provide a structure for a large number of users for the most diverse services such as video applications, file storage, social networks, cloud computing, financial services and others. Its topology is created with focus on low-latency and high data throughput. Generally, the structure of a DC is formed by a network up to thousands of servers connected to it [23]. In DCs, disputes over communication resources are common, causing problems when multiple flows tend to use the same switch port. Moreover, unfortunately the use of traditional TCP congestion control over data center networks has proven to be inefficient for specific loads [18,24].

DCs rely on optimized TCP variants, receiving network support through Explicit Congestion Notification (ECN) and Random Early Detection (RED) techniques [19]. Specifically, the DCTCP [2] variant relies on intermediate switches for marking packets, through monitoring their maximum capacities, which must be previously configured. Although this technique improved the overall performance of a DC network, it requires complex thresholds configurations. Furthermore, DC networks have applications with different purposes, employing their own congestion controls, with different flow sizes, and different versions of operating systems, which affects the use of ECN [5,22].

Contemporaneous DCs are using SDN-enabled switches to enrich the administrative options. SDN emerges as a way of separating the control from the data plane from the devices. This new organizational model creates a new opportunity to detect or even avoid congestion on DC networks. Specifically, the OpenFlow protocol [8] allows the collection of statistical information about network flows. When requested, the forwarding device passes its information about the flow traffic to the logically centralized SDN controller, which composes a dataset based on flows, ports, and switches metrics. Consequently, strategies that seek to eradicate the network congestion can work with information from all network switches and are not tied to edge devices. In summary, Fig. 1 presents a DC topology with an SDN controller logically connected to all system entities, from the rack servers to the core switches. The controller can collect statistical information from each individual and provide this information to applications. This scenario is used along the text to exemplify the design, implementation, and experimental analysis.

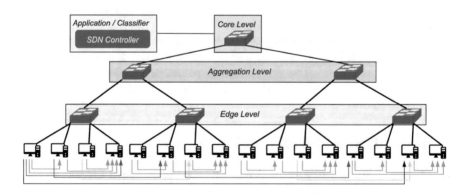

Fig. 1. A DC managed by a centralized controller using SDN.

2.2   Related Work

The use of machine learning to decrease or even to avoid network congestion is a challenging topic discussed by the specialized literature. Initially, [21] highlighted that an end-to-end tool capable of differentiating packet losses from network congestion can improve the congestion control in hybrid wired and wireless network. The authors adopt the concept of Round-Trip Time (RTT) loss pair, formed by two consecutive packets, in which one packet is lost and the other

received, and the second packet is capable of transporting the state of the network path. Losses related to the wireless network and due to congestion have
a different RTT distribution. Later, an unsupervised classifier learns these distributions and makes inferences about the cause of packet loss. The obtained
accuracy was between 44–98%, and the variation happened due to the network
configurations. In turn, [7] used the pair of RTT losses, however this time to
identify congestion and packet reordering in wired networks. Furthermore, the
RTT loss pair is used to infer the state of the network and not of a single TCP
connection. Classification is done using a supervised Bayesian algorithm, which
obtained an accuracy of 90%. Finally, in the context of optical networks, [13]
proposed two unsupervised classifiers to identify when a packet loss event was
caused by network congestion or lack of switch buffers.

Some authors used machine learning to improve the network congestion control in wireless scenarios [11, 15, 16]. In general, the authors applied several supervised learning algorithms, highlighting the use of variation of decision trees, in
which the trees are generated sequentially, and their weights are adjusted according to the interactions. For the construction of the model, offline data from synthetic random topologies composed of wired and non-wired connections were
used. The technique demonstrated greater classification accuracy compared to
solutions that do not use machine learning for the problem, such as TCP Veno [9]
and TCP Westwood [10].

The work [1] proposed a collaborative approach to address network congestion on the Internet. The authors combined Deep Reinforcement Learning (DRL)
techniques with classic congestion control algorithms, then introduced a hybrid
approach called Orca. In summary, the congestion control was divided into two
levels, in the first one the classic congestion control techniques work normally
and meanwhile, in the second one, over a time interval, the statistical data coming from the communication are monitored and at the end of each interval the
DRL calculates a new size for the CWND window and the classic techniques
start to use this value as a basis for their operations. Orca shows a greater
ability to take advantage of the communication link without generating congestion compared to other techniques. Finally, [6] proposes the use of actor-critic
reinforcement learning to identify TCP parameters improving the overall performance of TCP congestion control algorithms running on SDN DCs. In addition,
the authors used a random forest data classification technique to identify the
on-going network status.

The specialized literature demonstrated a growing interest for automating
the network management. In this sense, machine learning techniques compose a
promising approach to achieve zero-touch network management [3]. Specifically
regarding the network congestion control and avoidance, the related work used
supervised and unsupervised techniques to classify the network traffic in users'
defined categories. In the present work, we investigate the use of supervised
classification algorithms to train a neural network to identify the occurrence
of network congestion. Our methodology avoids predefined labels, and use traditional synthetic TCP loads to compose the labels set. Moreover, we offer to

the classification algorithms a powerful dataset composed of SDN-based data on flows, ports, and switches. Later, we demonstrate the proximity of algorithms predictions to real TCP variants.

3   Design and Implementation

The execution scenario is summarized in Fig. 1. A centralized SDN controller acquires data from network switches composing a dataset for training the classification algorithms. Once trained, each algorithm can be used with real workload. We opted for composing the experimental prototype atop Mininet [20] and relied on Ryu controller for managing the network resources and monitored data. Specifically, we used the OpenFlow protocol 1.3 [8]. The details on data monitoring, adjustments, and models training are discussed in the following.

3.1   Data Monitoring and Adjustments

We developed a Ryu module to retrieve and organize data from the network topology. The module monitors the network following predefined intervals, gathering data for switches ports and TCP flows. For each switch port, the module collects the number of sent and received packets, bytes, dropped packets, errors and collisions, while for each flow it collects the number of bytes and packets sent, as well as source and destination IPs and ports. It is worthwhile to mention that bytes and packets are cumulative fields, for example, the amount of bytes that traveled through a port represents the total volume so far. To identify the increase and decrease in data transfer throughput, while the switches report their statistical data asynchronously, the module saves a copy of the last analyzed data to calculate the difference. In this sense, the module accounts the current throughput alongside the cumulative values.

The module uses internal identifiers (based on switches datapaths, IPs, and ports) to correlate ports and flows data. Later, both switch ports and flows metrics are unified in a single representation to serve as input for the classification mechanisms. Specifically, we combined the source port statistics, data flow stats, and destination port data. In addition, the amount of bytes and packets between the switches was inferred, using the equation $(a - b)/a$, where a represents the transmitted amount and b represents the received amount. This value is obtained through an end-to-end path mapping where the information travels in the topology. For example, according to Fig. 1, an edge-based communication crosses only one switch, while an aggregation- and core-based requires additional hops. The logical connection (input and output data) is realized per switch's port. However, as data monitoring is asynchronously performed, there is no guarantee that the data used were captured in the same time interval, and to avoid inconsistencies such as receiving more data than was transmitted, the following logic was adopted. First, a numerical time discretizer t is created. Every time the module requests data, t is incremented by 1, and as the data is received, a copy of the last analyzed value is saved. The volume of lost bytes and packets is therefore only computed when the value of the time discretizer of the last copy is equal to $t-1$ both for information from the switch that is sending and for the one that is receiving the communication. The formula is then modified to $\frac{(a_t-a_{t-1})-(b_t-b_{t-1})}{a_t-a_{t-1}}$. Finally, the bytes, packets, and lost packets are accounted following this rationale and appended to the dataset.

To organize and prepare the data, any line with non valied values was removed, and the IP information was replaced by an equivalent and unique numerical index for each value. The identifiers columns were removed from the dataset as they are irrelevant information for machine learning. Finally, the values obtained for the bytes and packets losses calculations are normalized in a binary domain, receiving 1 when any value was identified, and zero otherwise.

3.2   Generating Classification Labels and Training the Models

Supervised machine learning methods use previously labeled data for training. Differently from the previous work (Sect. 2.2), we adopted only two labels for our formulation: 1 for congested and 0 for non-congested. We opted for composing a training protocol offering to the classification mechanisms distinct congestion scenarios. In essence, we executed the flows depicted in Fig. 1 separately, and later in a combined and competitive manner. All flows are generated with the *iperf* tool, and the configurations are detailed in Sect. 4. Intuitively, a TCP flow transferring data alone in a network path is not experience congestion, however, it is only limited by the network capacity (the related work discussion brought some classification approaches to distinct both scenarios). On the other hand, two or more flows disputing a network path tend to create a congestion scenario (as discussed in Sect. 2).

In this sense, we categorized the TCP flows according to the intermediate switches: Edge, Aggregation, and Core. Initially all Edge flows from Fig. 1 are started and marked as 0. After approximately half of the total execution scenario, the Core and Aggregation Traffic are started and labeled as congested data. Finally, to enrich the dataset, some flows are randomly started including distinct network disputes, and labeled according to their network position. For training, the dataset was partitioned between test base (40%) and training data (60%). Then, the classification method trains on the training base and with the model ready, the algorithm tries to classify the test base sample without the label. The classification algorithms adopted (covering the previously selected by specialized literature, Sect. 2.2) for training are Random Forest (RF), Extremely Randomized Trees (ERT), K-Nearest Neighbor (KNN), Linear Discriminant Analysis (LDA), and Bayesian Networks (BN), all belonging to the supervised learning paradigm and which respond very well to classification problems.

4   Experimental Analysis

The proof-of-concept prototype was analyzed running two experimental scenarios. In essence, each scenario has a different TCP congestion control algorithm,

Cubic or DCTCP. While Cubic represents the default configuration for Linux-based systems [12], DCTCP is a DC-optimized variant [2]. For training, a 2-hours labeled dataset with approximately 100 MB of information was used for each scenario. The data monitoring was performed every 5 s. The network topology depicted in Fig. 1 was implemented with Mininet [20], and all links have 1Gbps. In the following sections we present the metrics, scenarios, control traffic parameters, and discuss the results.

4.1   Metrics and Scenarios

We selected four metrics for assessing the performance of our proof-of-concept prototype and classifiers: models' accuracy, training time, CWND evolution, and total throughput. The accuracy of each model denotes the ability to predict values, accounted with a cross-validation method, in which the dataset is divided between 60% as a training base and 40% as a test base. The classifier (RF, ERT, KNN, LDA or BN) figures out how to classify the data in the training base. After training, the test base is used to predict values without knowledge about the original labels. The set of generated labels is compared with the original labels and the number of equivalent labels determines the accuracy of the model in predicting a new data sample that was not used in its training. Furthermore, the time that each model took to train on the same dataset is used for comparison and analysis of results.

The CWND evolution describes the internal TCP behavior. The values were collected only from a subset of communication pairs (termed control traffic in Fig. 1). As communications occur and the TCP identifies congestion, the variable has its value changed following the internal logic of the congestion control and avoidance algorithm. For instance, when CWND has the value 1, TCP is running a slow start phase (starting a new connection or recovering from a congestion event). On the other hand, additive increases and multiplicative decreases indicate the natural evolution of the protocol, guided by the selected variant (e.g., Cubic, DCTCP). We accounted both events separately and represented them as percentages of the total sample. Later, we compared the percentage of inferred congestion events with the percentage of events identified by the TCP. CWND values were gathered from end-hosts through a script based on Linux *ss* tool. It is noteworthy that consecutive duplicates of CWND were eliminated since they do not represent a window state change. Finally, the overall throughput achieved by TCP connections were accounted, and organized by switches levels. This information, combined with CWND values, give insights on the applications' performance.

4.2 Control Traffic

The control traffic was adopted to compare how the sender identifies network congestion with what each model inferred it. Thus, three pairs of servers initiate a TCP communication using *iperf.* The pairs were chosen such that the control traffic uses the Core, Aggregation, and Edge switches, as depicted in Fig. 1 (represented in red). It is worthwhile to mention that the control traffic is not part of the original training dataset. A new sample was created capturing both the original scenario and the control traffic, and then the prediction phase is executed. Other traffic such as Core, Aggregation, and so on, are executed exactly the same way as described in Sect. 3. We collected the CWND values for each control traffic pair during all running time.

4.3 Results and Discussions

The results are organized as follow. Initially, the accuracy and execution time of each supervised classifier are discussed. Later, the CWND values and the throughput are analyzed, as well as the models' predictions regarding the control traffic.

4.3.1 Accuracy and Execution Time

The results for Cubic- and DCTCP-based scenarios are summarized in Table 1. Most accuracies are better than 92%, the only exceptions are Cubic Bayesian Networks (BN) and K-Nearest Neighbor (KNN). Even though, the values reach higher than 83%. The best accuracies also have the worst processing times, *e.g.*, ERT and RF with Cubic and DCTCP. For the Cubic-based scenario, the results indicated that the classifier with the highest accuracy was the Extremely Randomized Trees (ERT) and the worst was the BN. In addition, training times were very different. For instance, K-Nearest Neighbor (KNN) used less than 1 second. One can conclude that for the Cubic-based scenario, ERT outperformed its counterparts regarding a combined accuracy-time analysis. In short, given such a complex scenario, the ERT algorithm required less than a minute to train and infer well, only getting it wrong 1.77% of the time. However, if time is also important to take a decision, the best cost-benefit may be Linear Discriminant Analysis (LDA).

In turn, the DCTCP is a specific algorithm for DCs, which receives indicators from the switches on the eventual occurrence of congestion. In this case, the difference between the best and worst accuracies is only 3.54%, nonetheless the time differences can be greater than 14s. For this dataset, ERT algorithm outperformed the other classifiers following the Cubic-based results. Although the LDA can be a alternative with a good accuracy and response time less than 1s. In addition, the training time of the algorithms decreased when compared to the Cubic-based scenario, probably due to data stability (the network throughput is

Table 1. Accuracy and execution time of classifiers for both Cubic- and DCTCP-based scenarios.

	Cubic		DCTCP	
Classifier	Accuracy (%)	Time (s)	Accuracy (%)	Time (s)
Random Forest (RF)	97.49	107	94.66	14.92
Extremely Randomized Trees (ERT)	98.23	45.57	96.31	7.35
K-Nearest Neighbor (KNN)	84.11	0.09	92.77	0.01
Linear Discriminant Analysis (LDA)	93.83	1.70	94.46	0.40
Bayesian Networks (BN)	83.75	0.23	93.24	0.05

analyzed in Fig. 2), in which the classifier with the longest time was 24.45 s and several algorithms took less than 1 s to train. Therefore, ERT was indicated for the DCTCP-based scenario regarding accuracy and execution time.

4.3.2   CWND Values and Throughput

The control traffic was prepared to assess the performance of supervised classifiers. The CWND values for Cubic and DCTCP scenarios are presented by Tables 2 and 3, respectively. For each switch level (edge, aggregation, and core, as defined in Fig. 1), we accounted the percentage of CWND events equals to 1 (termed CWND == 1) and the percentage of decreasing events (termed CWND−−). While the first indicates the initial point for a slow-start phase, the second accounts the events in which the TCP algorithm opted for reducing the segment's window.

Initially, for the Cubic scenario (Table 2), we can observe that the timer expired more than 44% for all switch-levels, so the network suffered with a high congestion. In this case, the ERT, which was indicated as the best accuracy algorithm, pointed out that the network was congested 61.37% for the edge communication, 31.86% for the aggregation, and 45.32% for traffic crossing the cores switches. It is worthwhile to mention that although organized per switches level, the results are cumulative, or in other words, they account the end-to-end path.

Table 2. CWND values and models' predictions for the Cubic-based scenario.

Flows	CWND == 1	CWND−−	RF	ERT	KNN	LDA	BN
Edge	48,26%	48,71%	97,37%	68,80%	61,37%	6,55%	52,91%
Aggregation	46,01%	47,68%	47,52%	31,86%	38,32%	29,27%	19,84%
Core	44,85%	46,65%	54,21%	45,32%	38,68%	38,11%	20,54%

Table 3. CWND values and models' predictions for the DCTCP-based scenario.

Flows	CWND == 1	CWND--	RF	ERT	KNN	LDA	BN
Edge	1,96%	14,37%	98,63%	78,64%	52,79%	2,39%	85,70%
Aggregation	0,11%	4,44%	46,40%	31,44%	16,33%	35,50%	30,44%
Core	00,09%	6,30%	43,13%	34,90%	9,55%	39,82%	15,17%

The results for the DCTCP scenario are summarized by Table 3. This time it is possible to observe how DCTCP is superior to the Cubic algorithm for the DC workload (long flows sent through high network links) since the timer expired up most 1.96%, while the Cubic-based scenario had a number greater than 44%. Furthermore, CWND decreased less than 15% of total events, for all flows. DCTCP aims to reduce the use of intermediate switches queues, and consequently avoid situations that could potentially lead to congestion. In DCTCP, switches inform hosts before congestion occurs, so unlike the Cubic-based scenario, in which only the end points (sender and receiver) are responsible for congestion control, the intermediate switches also have their participation. Regarding the CWND metric, most algorithms classified larger numbers of congestion events. This is possible as the models use a larger dataset of information to detect congestion, while DCTCP only reacts to switches and traditional feedbacks.

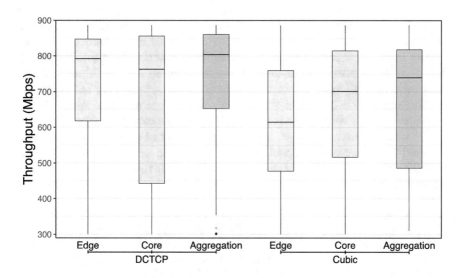

Fig. 2. Network throughput for both scenarios.

Figure 2 corroborates the performance different observed between Cubic and DCTCP scenarios in terms of total network troughput. In general, DCTCP stabilized and increased the network throughput for all switch levels. However, even improving the network usage, the congestion phenomena is still present, and was properly identified by some supervised classifiers. In fact, some classifiers identified more events than the traditional TCP congestion control and avoidance techniques.

Finally, it is notable that the classifiers constructed a different view, in most cases, from the view offered by TCP. This fact was expected, since the compared elements use different information to generate conclusions. While TCP uses information obtained only from end systems, classifier algorithms use information obtained from the network. It should be noted that the objective of the present work was to indicate the occurrence of congestion, exactly as it was carried out. Actions to eventually mitigate the source of the problems are indicated as future work.

5 Conclusion

Although data centers implement low latency and high throughput networks, the congestion phenomena is still present. Data centers host multiple applications with distinct traffic patterns, and consequently, some applications will communicate through a shared subset of links, eventually leading the network to congestion. Commonly, data centers rely on network-aided TCP versions to improve the overall performance, which receive feedback from the network to avoid congestion. However, even applying sophisticated algorithms, the decision-making regarding the congestion control and avoidance remains realized at network endpoints. This work demonstrated that the network can have a different point-of-view regarding the congestion events. The centralized management introduced by SDN brings new management opportunities. Specifically, we collected data from all switches ports and on-going flows to compose a powerful dataset. The dataset was used as input for training some classifiers, which are responsible for detecting congestion events on the network. Our experimental analysis demonstrated that the classifiers, once correctly trained, can capture the particularities of TCP variants, and moreover, found a large subset of congestion events when compared to end-points' perspective. As future work, we aim to improve classifiers with switches queues information, as well as to investigate the performance with multiple traffic patterns.

Acknowledgements. This work was funding by the National Council for Scientific and Technological Development (CNPq), the Santa Catarina State Research and Innovation Support Foundation (FAPESC), UDESC, and developed at LabP2D. This work received financial support from the Coordination for the Improvement of Higher Education Personnel - CAPES - Brazil (PROAP/AUXPE) 0093/2021.

References

1. Abbasloo, S., Yen, C.Y., Chao, H.J.: Classic meets modern: a pragmatic learning-based congestion control for the internet. In: SIGCOMM 2020, pp. 632-647. ACM, NY (2020)

2. Alizadeh, M., et al.: Data center TCP (DCTCP), vol. 40, no. 4, pp. 63-74 (2010)
3. Boutaba, R., Salahuddin, M.A., Limam, N., et al.: A comprehensive survey on machine learning for networking: evolution, applications and research opportunities. J. Internet Serv. Appl. **9**(16), 1–99 (2018). https://doi.org/10.1186/s13174-018-0087-2
4. Chiu, D.M., Jain, R.: Analysis of the increase and decrease algorithms for congestion avoidance in computer networks. Comput. Netw. ISDN Syst. **17**(1), 1–14 (1989)
5. Cronkite-Ratcliff, B., et al.: Virtualized congestion control. In: Proceedings of the ACM SIGCOMM Conference, pp. 230-243. Association for Computing Machinery, New York (2016)
6. Diel, G., Miers, C.C., Pillon, M., Koslovski, G.: Data classification and reinforcement learning to avoid congestion on SDN-based data centers. In: IEEE Global Communications Conference: Next-Generation Networking and Internet (Globecom). Rio de Janeiro, Brazil (2022)
7. Fonseca, N., Crovella, M.: Bayesian packet loss detection for TCP. In: Proceedings IEEE 24th Annual Joint Conference of the IEEE Computer and Communications Societies., vol. 3, pp. 1826–1837 (2005)
8. Foundation, O.N.: Openflow v1.3.0 (2021). https://opennetworking.org/wp-content/uploads/2014/10/openflow-spec-v1.3.0.pdf
9. Fu, C.P., Liew, S.: TCP Veno: TCP enhancement for transmission over wireless access networks. IEEE J. Sel. Areas Commun. **21**(2), 216–228 (2003)
10. Gerla, M., Sanadidi, M., Wang, R., Zanella, A., Casetti, C., Mascolo, S.: TCP westwood: congestion window control using bandwidth estimation. In: GLOBECOM'01. IEEE Global Telecommunications Conference (Cat. No.01CH37270), vol. 3, pp. 1698–1702 (2001)
11. Geurts, P., Irrthum, A., Wehenkel, L.: Supervised learning with decision tree-based methods in computational and systems biology. Mol. BioSyst. **5**, 1593–605 (2009)
12. Ha, S., Rhee, I., Xu, L.: Cubic: a new TCP-friendly high-speed TCP variant. ACM SIGOPS Oper. Syst. Rev. **42**(5), 64–74 (2008)
13. Jayaraj, A., Tamarapalli, V., Murthy, C.: Loss classification in optical burst switching networks using machine learning techniques: improving the performance of TCP. IEEE J. Sel. Areas Commun. **26**, 45–54 (2008)
14. Jiang, H., et al.: When machine learning meets congestion control: a survey and comparison. Comput. Netw. **192**, 108033 (2021)
15. Khayat, I., Geurts, P., Leduc, G.: Improving TCP in wireless networks with an adaptive machine-learnt classifier of packet loss causes, pp. 549–560 (2005)
16. Khayat, I., Geurts, P., Leduc, G.: Enhancement of TCP over wired/wireless networks with packet loss classifiers inferred by supervised learning. Wireless Netw. **16**, 273–290 (2010)
17. Kreutz, D., Ramos, F., Veríssimo, P., Esteve Rothenberg, C., Azodolmolky, S., Uhlig, S.: Software-defined networking: a comprehensive survey. ArXiv e-prints **103** (2014)
18. Kumar, G., et al.: Swift: delay is simple and effective for congestion control in the datacenter. In: Proceedings of the ACM Special Interest Group on Data Communication on the Applications, Technologies, Architectures, and Protocols for Computer Communication, pp. 514–528 (2020)
19. Kuzmanovic, A., Ramakrishnan, K., Mondal, A., Floyd, S.: RFC 5562: Adding explicit congestion notification (ECN) capability to TCP's SYN/ACK packets. IETF (2009)

20. Lantz, B., Heller, B., McKeown, N.: A network in a laptop: rapid prototyping for software-defined networks. In: Proceedings of the 9th ACM SIGCOMM Workshop on Hot Topics in Networks, pp. 1–6 (2010)
21. Liu, J., Matta, I., Crovella, M.: End-to-end inference of loss nature in a hybrid wired/wireless environment (2003)
22. Moro, V., Pillon, M.A., Miers, C.C., Koslovski, G.P.: Analysis of virtualized congestion control in applications based on Hadoop MapReduce. In: Bianchini, C., Osthoff, C., Souza, P., Ferreira, R. (eds.) WSCAD 2018. CCIS, vol. 1171, pp. 37–52. Springer, Cham (2020). https://doi.org/10.1007/978-3-030-41050-6_3
23. Noormohammadpour, M., Raghavendra, C.S.: Datacenter traffic control: understanding techniques and tradeoffs. IEEE Commun. Surv. Tutorials **20**(2), 1492–1525 (2017)
24. Rajasekaran, S., Ghobadi, M., Kumar, G., Akella, A.: Congestion control in machine learning clusters. In: Proceedings of the 21st ACM Workshop on Hot Topics in Networks, pp. 235–242 (2022)

Detection of Malicious Sites Using Graph Machine Learning

Rhayane da Silva Monteiro$^{(\boxtimes)}$ and Leonardo Sampaio Rocha

LAGIC, State University of Ceara, Av. Dr. Silas Munguba,
1700 - Campus do Itaperi, Fortaleza, CE, Brazil
`rhayane.monteiro@aluno.uece.br`, `leonardo.sampaio@uece.br`

Abstract. Information and communication technology networks are developing rapidly, and many users' daily activities, including e-banking and e-commerce, have moved to the internet. By using this anonymous framework, attackers establish new techniques such as phishing to trick victims into using fake sites to gather sensitive information. Detection of malicious sites is a challenging problem, due to its semantic-based attack framework, which exploits system and user vulnerabilities. While several approaches have been proposed to cope with phishing, new attacks constantly emerge requiring the improvement of existing techniques. This work proposes an approach to detect phishing, evaluating the structure of the graph from the connection of sites to extract features from the graph, in addition to the features commonly used in the literature such as features of URL and SSL certificate. We evaluate the impact of the use of graph features compared to existing features in the literature, from the point of view of machine learning metrics and of computational time required. The obtained results suggest that graph machine learning can be used effectively in phishing detection.

1 Introduction

With the advances in information and communication technologies, many daily activities of users, including electronic banking, social networking and electronic commerce, have been transferred to the internet. This anonymous, uncontrolled and open infrastructure provides a platform for cyber attacks [15]. Attackers use various methods to exploit vulnerabilities in a system, which helps them gain unauthorized access to personal data and engage in phishing scams.

The theft of sensitive information such as username, passwords, credit card information, social security number, etc., using a fake web page that mimics a trusted website is called phishing [16]. This type of attack can be carried out in several ways, such as email, site and malware. This paper focused on the detection of phishing from sites, which is achieved by making the user visit suspicious Uniform Resource Locator (URLs). Users fall into phishing due to five main reasons: i) they don't have detailed knowledge about URLs; ii) they don't know which pages can be trusted; iii) they don't see the full address of site,

L. Barolli (Ed.): AINA 2023, LNNS 654, pp. 235–246, 2023.
https://doi.org/10.1007/978-3-031-28451-9_21

due to redirection or hidden URLs; iv) they don't have much time to query the URL and v) they can't distinguish phishing pages from legitimate pages [17].

According to the report by the Anti-Phishing Working Group (APWG)[1], for the third quarter of 2021, the number of phishing has more than doubled since the start of 2020, when the APWG detected between 68,000 and 94,000 attacks per month. In July 2021, in turn, the organization counted 260,642 attacks, which was the highest monthly attack count recorded in the APWG reporting history. Furthermore, the software as a service and webmail sector was hit the hardest with 29.1% of all attacks. Regarding Brazil, the same report identified a total of 7,741 attacks in the third half of 2021, against 4,275 in the second quarter and 6,209 in the first quarter. Considering the sectors most affected by this type of attack in the country, there are e-commerce (40%) and the financial and banking sector (41%).

Phishing attacks can be performed manually, but overcoming them and respond effectively requires a lot of time, intelligence and manpower. This can take days or even weeks. Furthermore, these manual investigations are vulnerable due to human errors [18]. The big challenge when detecting these attacks is discovering the techniques used. Phishers continually improve their strategies and can create sites capable of protecting themselves against many forms of detection. Thus, developing robust, effective and up-to-date phishing detection methods is much needed to counter the adaptive techniques employed by attackers [25].

Several solutions to attacks on phishing sites have been proposed, such as heuristics, black and white lists and machine learning based techniques [20]. Machine Learning (ML) algorithms are popularly used to solve problems involving automatic classification of data, such as electronic fraud detection [4]. They can analyze the content of the data and extract hidden patterns from it. The problem of phishing attacks can be addressed by shifting it to classification. Parallel to this, data modeled as graphs are present in daily life. You can use graphs to model the complex relationships and dependencies between entities. Therefore, machine learning on graphs has been an important research direction for both academics and industry [24].

Based on the above considerations, the objective of this work is to propose, implement, compare and validate the impact of using graph features and features commonly used by academia, such as features of URL and Secure Socket Layer (SSL) certificate, in detecting malicious websites. The main contributions of this paper include:

- Build a dataset with legitimate and phishing URLs;
- Extract, select and explore features from the page graph, in addition to using URL and SSL certificate features;
- Analyse the impact of different sets of features, with and without graph features, on the performance.

---

[1] Available at https://apwg.org/trendsreports/.

This document is organized as follows: Sect. 2 will present the important concepts for understanding the work; previous works related to this paper will be presented in Sect. 3; the objective of this research will be explained in Sect. 4; results and discussions will be shown in Sect. 5; and, finally, the conclusion and future work will be exhibited at Sect. 6.

2 Preliminaries

This section is dedicated to presenting the concepts necessary for understanding this work.

Machine Learning. Machine learning is a type of artificial intelligence technique that can automatically discover useful information from large datasets [12]. It plays an important role in a wide scope of applications such as image recognition, data mining, qualified systems and image recognition [10].

Decision Tree (DT). It is a popular ML algorithm and the model logic is a tree structure. The classifier splits the dataset into smaller subsets and the related decision tree will be improved simultaneously. When trees grow too large, this is likely to lead to overfitting training datasets [23].

Random Forest (RF). Corresponds to a set of decision trees used for classification and regression tasks. This type of algorithm reduces the problem of overfitting by sorting or averaging the output of individual trees in the training processing [5].

Support Vector Machine (SVM). It is a supervised learning algorithm that classifies data points into two sections and predicts new data points belonging to each section. It is suitable for linear binary classification, which has two classes labeled [23].

Extreme Gradient Boosted Tree (XGBoost). It is an optimized implementation of gradient boosted trees, first introduced by Chen and Guestrin [7]. XGBoost implements a process known as boosting to improve the performance of gradient-driven trees. XGBoost has many strengths when compared to traditional gradient boost implementations [1].

Machine Learning on Graphs. A graph consists of a finite set of vertices and a set of edges that represent relations between pairs of vertices [14]. These two basic elements can describe multiple phenomena, such as connections between web pages, virtual routing network, network of biological relationships and many other relationships [8]. Recently, the academia has been interested in applying ML to data structured as graphs, with the objective of automatically learning representations to support predictions, as well as discovering new patterns, in order to improve traditional ML methods [21]. The objective of this area is to extract the desired characteristics of a graph and develop suitable techniques for graph-based datasets. Generally, methods extract relevant features from graphs by taking advantage of ML algorithms. Therefore, graph-based machine learning has gained popularity in recent years, transcending many traditional techniques [13]. Some ML tasks can be handled when working with graphs like classifying graphs, node classification, edge prediction, among others.

3 Related Works

In this section, previous works with different techniques for detecting phishing will be presented. Although all approaches achieved good performance, so far there is no effective technique that combines different types of data characteristics to detect phishing attacks. Table 1 presents a comparison between previous research and this work.

Table 1. Comparison between works presented and this research.

Proposal	URL	SSL certificate	Graph	Feature selection	Available dataset	Train and test time analysis
[9]	√	X	X	√	√	X
[3]	X	√	X	√	X	X
[22]	√	X	√	√	X	X
[6]	X	X	√	X	√	√
This paper	√	√	√	√	√	√

The work of Gupta *et al.* resulted in an anti-phishing solution that can detect malicious URLs in a real-time environment without requiring information from third parties and with a low response time [9]. The approach focused on achieving high accuracy with a limited number of URL resources. Therefore, the authors evaluated the most important features in the literature and created nine features based on the lexical structure of the URL to develop an accurate phishing detection solution.

The main point of Akanchha's paper was to explore a robust detection system in which it would be possible to detect whether a domain is legitimate or phishing, using information from the SSL certificate [3]. For that, the research investigated and evaluated the detection system by training it with different sets of resources and testing with different assumptions. The highlight of this work is that instead of using Web pages or URLs, the SSL certificate of the domains was studied to detect phishing attacks. As this type of certificate has mandatory fields that are necessary for issuing the said certificate, this information was analyzed and created as an attack identification resource.

Tan *et al.* used graph features as a more complete representation of the hyperlink of the site and network structure to increase the detection rate of phishing [22]. According to the authors, it is the first work that fully uses features of graphs to detect malicious sites. The paper specifically uses hyperlinks as a starting point for deriving new graph features, which is different from most previous work that only used URL-based features. The results show that the graph features are able to achieve a significant hit rate while allowing a significant reduction in the dimensionality of the feature.

PhishDet is a new way to detect phishing websites through Long Term Recurrent Convolutional Network and Graph Convolutional Network using URL and HyperText Markup Language (HTML) features and was the approach proposed by Ariyadasa *et al.* [6]. PhishDet's resource selection is automatic and occurs within the system as the model gradually learns URLs and HTML content resources to deal with ever-changing phishing attacks. However, PhishDet requires periodic recycling to maintain its performance over time.

4   Our Proposal

The objective of this paper consists of a comparative study of existing machine learning algorithms based on the creation of a dataset that explores the characteristics of the links graph, in addition to URL features and SSL certificate. The phases comprising this paper include: i) the collection of legitimate and phishing web pages; ii) the extraction of features; iii) the feature selection in each defined scenario; iv) the implementation of the ML models and v) the analysis of the results. Each step will be detailed below.

In the URL collection step, a dataset was built consisting of 1000 URLs and associated features, 500 of which were legitimate URLs and 500 of phishing. It is worth noting that the dataset was developed in this work, with the objective of maintaining a current dataset for application in real scenarios. The dataset and code implementation are available at GitHub[2]. Web pages were collected between October and November 2022. To obtain legitimate URLs, site The Moz Top 500 Websites[3] was used, which lists the 500 most popular sites in the world based on Domain Authority, a metric based on link that models how Google ranks sites. Regarding the URLs of malicious sites, these were obtained from site OpenPhish[4], which is an automated platform for phishing intelligence. The URLs were extracted through the API sites and the data was generated in Comma-Separated-Values (CSV) format.

The next step of the paper comprised the extraction of features from the collected URLs. This extraction took place in November 2022. The dataset has a total of 30 features, divided into the following groups: i) features extracted from the URL structure; ii) features extracted from the SSL certificate and iii) features obtained from modeling the problem in graphs. All features will be explained below.

URL features were obtained from its structure. A total of 17 features were extracted and chosen as they are commonly used by academia for phishing detection. Chosen features were based on works by [2,9,19]. Next, we list the features

---

[2] https://github.com/RhayMonteiro/phishing-detection-with-graph.git.

[3] https://moz.com/top500.

[4] https://openphish.com/.

used: *getDomain, havingIP, haveAtSign, getLength, getDepth, redirection, http-Domain, tinyURL, prefixSuffix, dns, web_traffic, domainAge, domainEnd, iframe, mouseOver, rightClick* and *forwarding*.

The SSL certificate features, which total 8, were taken from its fields. This type of certificate was chosen because it is a security strategy used especially in legitimate sites. Features in this category are i) *versionSSL*, which corresponds to the version of the SSL certificate; ii) *serialNumberSSL*, if the serial number is empty, the value assigned to this resource is 1 (phishing), otherwise 0 (legitimate); iii) *expirationDateSSL*, expiration year; iv) *startDateSSL*, year the certificate was granted; v) *validCert*, days left for the SSL certificate to expire; if the number of days is less than 0, the value is assigned 1 (phishing), as the certificate has already expired; if not, 0 (legitimate); vi) *subjectAltNameSSL*, number of domains covered by the SSL certificate; vii) *subjectNameSSL*, if the field Organization that receives the certificate is empty, the assigned value is 1 (phishing), otherwise, 0 (legitimate); and viii) *issuerNameSSL*, checks if the organization granting the certificate is one of the listed authorities; if not, the assigned value is 1 (phishing), if yes, 0 (legitimate).

$$score = \frac{2 * domain + path}{3} \tag{1}$$

Regarding the graph features, these were extracted from the modeling of the problem as a graph. Graphs are composed of a set of vertices and a set of edges between those vertices. In this paper, the vertices represent the URLs and the connection between them is calculated by the similarity between two sites. For each occurrence of the dataset, we obtained a graph with its respective features. To obtain the vertices of the graph, in turn, we obtained the internal links, which are part of the same domain as the analyzed URL, and the external links, which represent links from different domains of the URL. The calculation of similarity between two sites is given by the weighted average displayed in Equation 1. In the formula, we have the URL domain, which is identified by the target being accessed; the path is the location to find the resource, file or object within the server of that domain. It was considered that the domain has greater influence (weight 3) when analyzing two URLs, since real URLs are associated with the same real domain. As for the path, weight 1 was used because it can also influence the definition of phishing URLs. Therefore, if the similarity score between two sites is greater than 75%, an edge is established between them.

To exemplify the generation of graphs, Fig. 1 displays two examples. Figure 1a represents the graph from a legitimate URL[5], with 83 vertices and 3057 edges. It is noticed that the vertices have several edges incident on them, which points to a well-connected graph. Figure 1b represents a phishing URL[6], which tries to mimic the Apple homepage. The graph is composed of 70 nodes and 936 edges. It is noticed that the graph generated two subgraphs, based on the similarity

---

[5] https://www.apple.com/br/.
[6] http://ptxx.cc/client/apple.com/.

rule explained above. In addition, it presents a number of vertices close to the legitimate URL. However, the number of edges is less than a third of the number of edges in the legitimate URL.

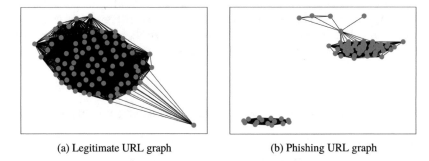

(a) Legitimate URL graph   (b) Phishing URL graph

Fig. 1. Generated graphs examples

In the group of features graph, we chose the features listed below, which total 5. They were chosen because they are numerical features and because they are relevant for the detection of phishing pages. We use the well-known graph features: i) *meanDegree*, which is the average degree, that is, the average number of edges per vertex in the graph; ii) *density*, which represents the fraction of edges present in the graph and all possible edges; iii) *average_clustering*, or average clustering coefficient, which is the average fraction of all pairs of neighbors of a node that have an edge between them; iv) *transitivity*, which is the general probability that the network has adjacent interconnected nodes, therefore revealing strongly connected communities, such as clusters; and v) *network_efficiency*, or network efficiency, which is a measure of how efficient the network's information exchanges are.

The next step was to define the approaches. To evaluate the impact of feature groups, four types of scenarios were defined. The first scenario corresponds to all features; the second scenario only considered URL and SSL certificate features, commonly used by academia; in scenario 3, only the graph features were evaluated, in order to assess the independence of these features in phishing detection; finally, in scenario 4, the feature selection technique Chi-Squared test was used, which is a statistical hypothesis test that assumes (the null hypothesis) that the observed frequencies for a categorical variable correspond to the expected frequencies for the categorical variable. This technique was chosen considering the input and output data type of the dataset to be categorical. The features selected in this model were *getLength* and *getDepth*, from URL features; *expirationDateSSL*, *startDateSSL*, *validCert*, *subjectAltNameSSL* and *subjectNameSSL*, from SSL certificate features; and *mean_degree*, *average_clustering* and *transitivity*, from graph features.

The next step comprised the implementation, training and testing of the ML algorithms. The models used were Decision Tree, Random Forest, Support

Vector Machine and XGBoost. Such algorithms were chosen because they are commonly used by academia to detect phishing. The training and testing of the models took place in November 2022.

The last step of the paper was dedicated to analyzing the results obtained from testing the ML algorithms. For this work, the following metrics were used accuracy, precision, recall, F-Score and ROC curve (Receiver Operating Characteristic Curve). Such parameters were chosen because they are common measures when analyzing the performance of ML models.

5   Results and Discussions

In this section, the results obtained from the dataset built for this paper will be presented.

The dataset was split using the traditional train-test split method, which consists of randomly splitting the dataset into two parts: training set and test set. In this work, the percentage of 70% of data for training and 30% for testing was defined. This percentage was chosen because it is a parameter accepted by the academic community, in addition to bringing more reality to the research. As this is a comparative study between the performances of ML algorithms in different scenarios, after dividing the data into training and testing, they were submitted to the ML algorithms. Metrics and time measures are explained below and obtained from the performance of the models on the test dataset.

The first results refer to metrics commonly used when analyzing the performance of ML algorithms, shown in Table 2. Considering scenario 1, in Table 2a in which all features were considered, we have that all algorithms presented similar rates. In terms of precision, XGBoost obtained the highest rate, with 90.68% and the lowest rate was for DT, with 89.98%. For recall metrics and F-Score, the lowest rates were for the RF algorithm, with 89.47% and 89.58%, respectively. The highest rates were for SVM, with 90.70% and 90.66%, respectively. In terms of accuracy, the lowest rate was for RF (89.67%) and XGBoost and SVM scored the highest hit rates, with 90.67%.

When analyzing scenario 2, shown in Table 2b, in which only the URL and SSL certificate features were considered, we observed that the SVM algorithm obtained the lowest rates for the precision metrics, recall, F-Score and accuracy (90.67%). The highest rates for these metrics were registered by the RF and XGBoost algorithms, which obtained 92.34% for precision and 92.33% for recall, F-Score and accuracy.

In scenario 3, shown in Table 2c, in which only graph features were considered, we have results below 90%, in addition to greater performance variation between models. As in this scenario only 5 features were considered, the lower performance may be due to this fact: the algorithms did not obtain enough features for a good classification. Despite this, the results are still relevant, as shown below. The lowest rates were recorded by the SVM algorithm with 84.31% precision, 82.93% recall, 82.82% F-Score and 83% accuracy. Regarding the highest values, these were obtained from the RF algorithm with 87.33% for precision, 85.94% for recall, 85.86% for F-Score and, finally, 86% for accuracy.

Table 2. Metrics analyzed

(a) Scenario 1

Algorithm	Precision	Recall	F-Score	Accuracy
Decision Tree	89,98%	90,01%	89,99%	90,00%
Random Forest	90,29%	89,47%	89,58%	89,67%
XGBoost	90,68%	90,63%	90,65%	90,67%
SVM	90,66%	90,70%	90,66%	90,67%

(b) Scenario 2

Algorithm	Precision	Recall	F-Score	Accuracy
Decision Tree	91,33%	91,33%	91,33%	91,33%
Random Forest	92,34%	92,33%	92,33%	92,33%
XGBoost	92,34%	92,33%	92,33%	92,33%
SVM	90,67%	90,67%	90,67%	90,67%

(c) Scenario 3

Algorithm	Precision	Recall	F-Score	Accuracy
Decision Tree	85,69%	84,27%	84,16%	84,33%
Random Forest	87,33%	85,94%	85,86%	86,00%
XGBoost	87,05%	85,26%	85,14%	85,33%
SVM	84,31%	82,93%	82,82%	83,00%

(d) Scenario 4

Algorithm	Precision	Recall	F-Score	Accuracy
Decision Tree	85,11%	85,04%	85,00%	85,00%
Random Forest	89,28%	88,58%	88,61%	88,67%
XGBoost	93,48%	93,48%	92,97%	93,00%
SVM	87,16%	86,74%	86,64%	86,67%

Scenario 4, shown in Table 2d, considered only the features selected by the Chi-squared test technique and observed a varied behavior between the ML models. Despite this, it was the scenario in which we obtained the highest rates of the analyzed metrics for the XGBoost algorithm. This fact was expected, since the feature selection technique is applied in order to reduce the complexity of the model and improve its performance. In this scenario, the DT model obtained the lowest accuracy rate (85.11%), recall (85.04%), F-Score and accuracy (85%), respectively. The highest rates, in turn, were obtained from the XGBoost model with 93.48% of precision and recall, respectively, 92.97% of F-Score and 93% of accuracy.

Table 3. Training and testing time analysis in seconds

(a) Scenario 1

Algorithm	Train	Test
Decision Tree	0.004	0.004
Random Forest	0.154	0.045
XGBoost	0.239	0.014
SVM	0.217	0.013

(b) Scenario 2

Algorithm	Train	Test
Decision Tree	0.003	0.006
Random Forest	0.156	0.049
XGBoost	0.251	0.014
SVM	0.101	0.012

(c) Scenario 3

Algorithm	Train	Test
Decision Tree	0.003	0.009
Random Forest	0.169	0.042
XGBoost	0.162	0.007
SVM	1.132	0.018

(d) Scenario 4

Algorithm	Train	Test
Decision Tree	0.003	0.006
Random Forest	0.367	0.086
XGBoost	0.171	0.056
SVM	0.036	0.016

Fig. 2. ROC Curve for the XGBoost algorithm

Other important results when analyzing the performance of ML algorithms are the training and testing times of the models, since they are crucial to validate their relevance. Times are shown by scenario and are computed in seconds, as shown in Table 3. In scenario 1, shown in Table 3a, DT obtained the lowest training and test time, with 0.004s. For scenario 2, shown in Table 3b, DT again obtained a shorter training and test time, with 0.003s and 0.006s, respectively. Table 3c corresponds to scenario 3, in which the shortest time was for DT (0.003s). Regarding the test, the shortest time was for XGBoost (0.007s). Finally, in scenario 4, shown in Table 3d, DT presented the shortest training and testing times, being 0.003s and 0.006s, respectively.

The next results obtained were the graphs of the ROC Curve, which is an important metric when dealing with binary problems, such as phishing detection. Figure 2 shows the ROC curves of the XGBoost model in each scenario. It should be noted that this model had the highest AUC rates, as shown in the images Fig. 2a, with 91% AUC in scenario 1, Fig. 2b, which had a 92% rate in scenario 2, Fig. 2c, with 87% in scenario 3 and, finally, Fig. 2d, with the highest rate among the scenarios, 93%, in scenario 4.

6   Conclusion and Future Works

The objective of this paper was to propose, implement and validate an approach for detecting malicious websites by analyzing the impact of using graph features, in addition to URL and SSL certificate features, from the modeling of the problem using ML in graphs. After analyzing the results, it was evident that the scenario with the best performance was scenario 4, in which the resource selection technique was applied, in which we obtained URL, SSL certificate and graphic features that proved to be effective in detecting phishing. We also believe that the scenario's superior performance was due to the use of graph features. Regarding the best algorithm, the XGBoost model stands out, which, regardless of the scenario, presented relevant rates of the analyzed metrics, in addition to the AUC values of the ROC curve and relevant training and test times. In the best scenario, we obtained an accuracy of 93% for XGBoost algorithm, which demonstrates the relevance of the work for the detection of phishing.

As future works, we will intend to evaluate the impact of other techniques based on graphs and graph neural networks, such as the Graph Convolutional Network [11]. In addition, we will use other features of graphs, in order to assess their independence in detecting malicious sites. With these adjustments, we will hope to improve the performance of our paper, which already stands out as a relevant phishing detection technique.

References

1. Abdulrahman, A.A.A., Yahaya, A., Maigari, A.: Detection of phishing websites using random forest and XGBoost algorithms. Int. J. Pure Appl. Sci. **2**(3), 1–14 (2019)
2. Adil, M., Alzubier, A.: A review on phishing website detection. EasyChair. **15**, 2020. Accessed Apr 2019
3. Akanchha, A.: Exploring a robust machine learning classifier for detecting phishing domains using SSL certificates (2020)
4. Akinyelu, A.A.: Machine learning and nature inspired based phishing detection: a literature survey. Int. J. Artif. Intell. Tools **28**(05), 1930002 (2019)

5. Almseidin, M., Zuraiq, A.A., Al-Kasassbeh, M., Alnidami, N.: Phishing detection based on machine learning and feature selection methods (2019)
6. Ariyadasa, S., Fernando, S., Fernando, S.: Combining long-term recurrent convolutional and graph convolutional networks to detect phishing sites using URL and HTML. IEEE Access **10**, 82355–82375 (2022)
7. Chen, T., Guestrin, C.: XGBoost: a scalable tree boosting system. In: Proceedings of the 22nd ACM SIGKDD International Conference on Knowledge Discovery and Data Mining, pp. 785–794 (2016)
8. Goldenberg, D.: Social network analysis: from graph theory to applications with python. arXiv preprint arXiv:2102.10014 (2021)
9. Gupta, B.B., Yadav, K., Razzak, I., Psannis, K., Castiglione, A., Chang, X.: A novel approach for phishing URLs detection using lexical based machine learning in a real-time environment. Comput. Commun. **175**, 47–57 (2021)
10. Harinahalli Lokesh, G., BoreGowda, G.: Phishing website detection based on effective machine learning approach. J. Cyber Secur. Technol. **5**(1), 1–14 (2021)
11. Kipf, T.N., Welling, M.: Semi-supervised classification with graph convolutional networks. arXiv preprint arXiv:1609.02907 (2016)
12. Liu, H., Lang, B.: Machine learning and deep learning methods for intrusion detection systems: a survey. Appl. Sci. **9**(20), 4396 (2019)
13. Negro, A.: Graph-Powered Machine Learning. Simon and Schuster, Manhattan (2021)
14. Nelson, C.J., Bonner, S.: Neuronal graphs: a graph theory primer for microscopic, functional networks of neurons recorded by calcium imaging. Front. Neural Circuits **15**, 662882 (2021)
15. Odeh, A., Keshta, I., Abdelfattah, E.: Machine learning techniques for detection of website phishing: a review for promises and challenges. In: 2021 IEEE 11th Annual Computing and Communication Workshop and Conference (CCWC), pp. 0813–0818. IEEE (2021)
16. Rao, R.S., Pais, A.R.: Jail-phish: an improved search engine based phishing detection system. Comput. Secur. **83**, 246–267 (2019)
17. Sahingoz, O.K., Buber, E., Demir, O., Diri, B.: Machine learning based phishing detection from URLs. Expert Syst. Appl. **117**, 345–357 (2019)
18. Shankar, A., Shetty, R., Nath, B.: A review on phishing attacks. Int. J. Appl. Eng. Res. **14**(9), 2171–2175 (2019)
19. Somesha, M., Pais, A.R., Rao, R.S., Rathour, V.S.: Efficient deep learning techniques for the detection of phishing websites. Sādhanā **45**(1), 1–18 (2020). https://doi.org/10.1007/s12046-020-01392-4
20. Sonowal, G., Kuppusamy, K.: PhiDMA-a phishing detection model with multi-filter approach. J. King Saud Univ. Comput. Inf. Sci. **32**(1), 99–112 (2020)
21. Stamile, C., Marzullo, A., Deusebio, E.: Graph Machine Learning. Packt Publishing, Birmingham (2021)
22. Tan, C.L., Chiew, K.L., Yong, K.S., Abdullah, J., Sebastian, Y., et al.: A graph-theoretic approach for the detection of phishing webpages. Comput. Secur. **95**, 101793 (2020)
23. Tang, L., Mahmoud, Q.H.: A survey of machine learning-based solutions for phishing website detection. Mach. Learn. Knowl. Extr. **3**(3), 672–694 (2021)
24. Zhang, Z., Wang, X., Zhu, W.: Automated machine learning on graphs: a survey. arXiv preprint arXiv:2103.00742 (2021)
25. Zuraiq, A.A., Alkasassbeh, M.: Phishing detection approaches. In: 2019 2nd International Conference on new Trends in Computing Sciences (ICTCS), pp. 1–6. IEEE (2019)

Bonet Detection Mechanism Using Graph Neural Network

Aleksander Maksimoski[1], Isaac Woungang[1(✉)], Issa Traore[2],
and Sanjay K. Dhurandher[3]

[1] Department of Computer Science, Ryerson University, Toronto, ON, Canada
{amaksimoski,iwoungan}@ryerson.ca
[2] Department of Electrical and Computer Eng., University of Victoria, Victoria, BC, Canada
itraore@ece.uvic.ca
[3] Department of Information Technology, NSUT, New Delhi, India

Abstract. A botnet is a group of computers that are infected by a malware that can be utilized to wreak havoc on other computers. Botnets have been around for quite some time and various techniques have been proposed in the literature to detect their presence in networks and systems. In today world, Intrusion Detection Systems (IDSs) and Intrusion Prevention Systems (IPSs) are capable of defending against botnets that create volumetric and fast paced traffic. But, these systems are not well suited to address prevalent real-time, long-term and stealth attacks. This paper proposes a Graph Neural Network (GNN)-based method for detecting botnets activity based on supervised learning. This work is the first ever application of AEN to build a GNN model for botnet detection purpose. The proposed model is evaluated using five different labelled datasets, namely, the CTU-MALWARE-CAPTURE-BOTNET-42, CTU-MALWARE- CAPTUREBOTNET-43, CTU-MALWARE-CAPTURE-BOTNET-48, ISCX Botnet Training and ISCX Botnet Test datasets, yielding preliminary promising results in terms of botnet prediction, using precision, recall, F1-score, and accuracy, as performance metrics.

1 Introduction

A botnet (also called robot net) is defined as a group of computers that are infected by a malware and are utilized to wreak havoc on other computers [1]. This malware is controlled by an entity known as bot-herder [1], which has access to all the computers in the chain (called zombies) and can use them simultaneously to instigate attacks such as Denial of Service (DOS) attack [1]. Various corporations utilize intrusion detection systems (IDSs) and intrusion prevention systems (IPSs) to protect their networks from being victims of botnet attacks. These systems were designed to ensure protection against attacks that are typically volumetric in nature and fast paced [3]. Nowadays, attackers often work under government institutions and they have enough time and resources to conduct long term and stealth attacks, which IDS and IPS systems) cannot detect. Therefore, there is a clear demand for new models to detect such attacks.

This paper proposes the design of a novel graph neural network (GNN)-based model for detecting botnets in networks and systems. Our approach consists of studying the

L. Barolli (Ed.): AINA 2023, LNNS 654, pp. 247–257, 2023.
https://doi.org/10.1007/978-3-031-28451-9_22

Activity and Event Network (AEN) model [3], a novel security knowledge graph model that can be utilized to model the diversity of the objects and actors of a computer network as well as the dynamic and uncertainty nature of those elements. It relies on a probability model that is meant to assign a probability of correctness value to each element in the graph. Then, the structural foundational knowledge acquired from the AEN graph is used to design the aforementioned GNN model.

The rest of the paper is organized as follows. Section 2 discusses some background and related work on botnet detection. Section 3 describes the proposed AEN-based GNN model for botnet detection. Section 4 presents the performance evaluation of the proposed GNN model using the considered datasets and few metrics. Section 5 concludes the paper.

2   Background and Related Work

2.1   Activity and Event Network Model

The Activity and Event Network (AEN) graph is built from heterogeneous data obtained from internal security perimeter such as network traffic logs, flow data, system logs, IDS alerts, anti-virus and email security logs, to name a few [3]; external data sources originated from third-trusted party services such as Domain Name Server queries, WHOIS, to name a few, and data sources that report suspicious activities (such as SIEMs, security events, named detectors). The features extracted from the data sources are of two categories: (1) First order features, collected directly from the sources - these are features obtained when analyzing the network traffic via IDS system and features obtained from data collected from the log analysis of applications, (2) Second-order features, obtained from available data by using some tools, and (3) Third-party features - which are features obtained by performing scans for collecting data about certain entities such as IP address, location, name server, email, operating system, etc.). The AEN multi-graph model can also exhibit the uncertainty and dynamicity of the activities in the network, hence it provides a solution to detect known attacks and new attacks, including stealth, hidden, and long term attacks, which are difficult to detect using traditional IDS and IPS systems.

An example of a partial view of a AEN graph obtained by using the data from the ISOC CID dataset [4] as shown in Fig. 1.

In Fig. 1, every node element has an ID and a label representing its type. It also has some associated properties as shown in Fig. 2.

Similarly, every edge in the AEN graph has an ID, label, and the associated properties' fields as illustrated in Figs. 2 and 3 for the ISOT CID dataset [1], where the data for two edges are shown, with SESSION label or type assigned to these edges and the associated properties.

Of particular interest is the use AEN graph as input to the proposed GNN model for botnets detection in networks. In this context, the AEN model consists of two main elements: the nodes (represented by their IP addresses) and the edges (called sessions). A node can be active, meaning that it is a source of botnet attack such as the botmaster or an intermediary infected device. It can be passive, meaning that it is a normal host

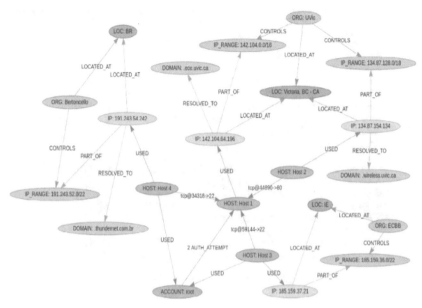

Fig. 1. Sample AEN graph based on a subset of ISOC CID dataset [1]

```
[{'id': 3505522295267017419,
  'label': 'ALERT',
  'properties': {'destIP': '172.16.1.24',
  'protocol': 'tcp',
  'sourcePort': 22,
  'destPort': 37136,
  'sourceIP': '142.104.64.196',
  'service': '',
  'classification': '',
  'priority': 0,
  'timestamp': '2016-12-09T17:49:59.205766Z'}},
{'id': -7903716220219263361,
  'label': 'ALERT',
  'properties': {'destIP': '172.16.1.24',
  'protocol': 'tcp',
  'sourcePort': 16783,
  'destPort': 23,
  'sourceIP': '14.162.253.53',
  'service': '',
  'classification': '',
```

Fig. 2. Snippet of node elements [1].

that is not a threat to botnet attack [2]. The edges represent the relationships that exist between the source and destination IPs [2].

The AEN model relies on a probabilistic model that assigns a probability of correctness to any element (edge/node) of the graph as well as a feature confidence [3]. At inception, a node or edge of the graph is allocated a correctness probability, which is based on the type of data associated to it. For instance, a graph representing two potential locations for an IP address node with different probabilities calculated using the probability model is illustrated in Fig. 4.

```
[{'id': 5712465223099226337,
  'label': 'SESSION',
  'properties': {'__maliciousLabel': False,
   'destSize': 52,
   'protocol': 'tcp',
   'sourcePort': 38040,
   'destPort': 22,
   'packetCount': 11,
   'fragmentedPacketCount': 0,
   'deltaTime': 0,
   'tcpState': 6,
   'startTime': '2016-12-16T17:18:31.940Z',
   'stopTime': '2016-12-16T17:18:31.962Z',
   'sourceSize': 22768},
  'source': -4539149150930014966,
  'destination': -5938715740280398657},
 {'id': 2267145363307886547,
  'label': 'SESSION',
  'properties': {'__maliciousLabel': False,
   'destSize': 0.
```

Fig. 3. Snippet of edge elements [1].

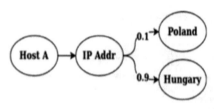

Fig. 4. Graph representing two possible locations for an IP address node with different probabilities [1]

The AEN framework implementation [3] is depicted in Fig. 5, where the AEN engine represents the main component responsible for the AEN graph maintenance and update.

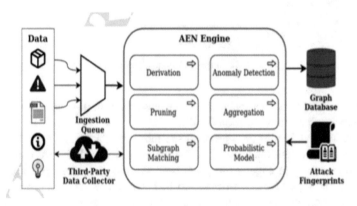

Fig. 5. AEN framework architecture [1]

The main functionalities of the AEN engine such matching the attack fingerprints, performing data processing and aggregation, checking for botnet detection, and building the AEN graph probabilistic model. In this engine, the AEN graph is stored in a customized fashion, so that it is possible to add, remove, update or search for a graph element, and frequent storage of graph snapshots can be saved on the Graph Database, then reloaded for future use when a previous graph state happened to have failed. Therefore, these snapshots can be shared with other systems/tools, providing other auxiliary functionalities such as scalability using read replicas, graph visualization, etc. The AEN graph database and engine have been implemented in Java, and can be executed in a single process, enabling a direct memory access of graph elements.

2.2 Related Work

Representative botnet detection schemes [4, 5, 7–9], are described as follows:

In [4], Masoudi-Sobhanzadeh and Emami-Moghaddam proposed a two-step machine learning-based optimization algorithm (referred to as world competitive contests algorithm), along with a support vector machine classifier to detect botnets in real-time IoT environments within a shorter period of time. In the proposed scheme, a wrapper method to derive the most effective subset of features required for training the machine learning model is implemented, based on which the considered SVM method creates the predicting model. Through experiments, the efficiency of the proposed botnet detection model is validated and compared against that of ten benchmark approaches, showing its superiority in terms of stability, convergence, and accuracy, chosen as performance metrics. However, in the design of the proposed scheme, no clue is given as to how the optimal number of selected features to yields the best possible performance has been setup.

In [5], Ostap and Antkiewicz proposed a method to detect botnets in networks and systems by identifying their synchronous actions. In this scheme, a clustering technique is used to form the groups of moments when the packets originated from the source IPs are sent to the analyzed destination IP. This helps in determining a similarity factor based on which it is determined if any of the analyzed group of moments contains a bot net or not. Experiments are conducted to validate the proposed scheme and compare it against a benchmark method (so-called BOTGAD [6], showing its efficiency in terms of botnet detection and detection of synchronous activity. However, for the proposed scheme, the possibility of relying on an online analysis that do not required the collection of network traffic is yet to be proven, and this constitute a limitation.

In [7], Alani proposed a packet-based IoT botnet detection scheme using a machine learning algorithm. Basically, the proposed approach relies on the use of a machine learning classifier method to minimize the type and number of the most effective features extracted from the network packets (instead of network flows). I doing so, the feature importance is considered as criterion. This has led to the construction of a dataset that has only seven features to train the machine learning model. Using the proposed scheme, whenever a botnet has been detected, its source node is immediately blocked. Using experiments involving multiple steps of testing and validation, it has been proven that the proposed machine learning classifier can achieve a F1-score of about 99.68%, and an accuracy of 99.76%, which are promising results. However, the authors reported that a distributed version of the proposed model that could help reducing the computational load on IoT devices is yet to be developed.

In [8], Alharbi and Alsubhi proposed a graph-based machine-learning botnet detection scheme that takes advantage of graph features to identify the botnets in the network. In this scheme. In doing so, various feature sets are explored and a filter-based evaluation mechanism is implemented to select the suitable ones to train the chosen machine learning models. Through experiments using some datasets, various types of botnets along with various types of behavioral characteristics have been detected, with accuracy scores in the range [99\%-100\%], showing the efficiency of the proposed scheme. As reveal by the authors, extending the proposed model by taking into account additional node attribute features would be desirable in practical settings.

In [9], Duan et al. proposed a botnet detection method based that makes use of a conjunction of gradient boosting-based decision trees and autoencoder neural network. In this scheme, to select the features that are appropriate for the considered machine-learning model training, a statistical analysis and a deep flow inspection techniques are implemented, in which a packet capture of the headers is utilized to convert the original traffic data into a vector data that is appropriate for use by the machine learning algorithms. On the other hand, the autoencoder scheme is meant to strengthen the machine learning model construct and reduce the features dimension. It also helps optimizing the traffic data characteristics, so as to enable the detection of botnet activities. By experiments using the ISCX-botnet traffic dataset, it is shown that the proposed scheme can achieve promising results in terms of accuracy, F1-score, recall, and area under the curve. However, as reported by the authors, from a practical perspective, the assessment of the proposed botnet detection scheme using a real-time online version invoking deep learning models is yet to be achieved.

Unlike the above-mentioned botnet detection schemes, the one proposed in this Thesis is the first ever application of the newly proposed AEN framework model [4] to design a GNN model for botnet detection purpose.

3   Proposed Graph Network Model

The proposed GNN model is depicted in Fig. 6, consists of a graph convolutional network (GCN) whose components are described as follows:

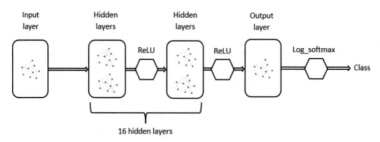

Fig. 6. Proposed GNN Model.

- An input layer: This layer takes in the AEN graph instance and propagates it to the hidden layers.
- Hidden layers: these layers are located between the input and output layers. The number of hidden layers can range from 0 to any value. In our case, we choose 16.
- An output layer: this layer receives the outputs from the preceding hidden layers and generates the final results of our binary classification, where 1 represents a botnet and 0 represents a normal session.
- ReLU: this is an activation function used to normalize the input and generate an output that is propagated to the next layer. The output of the ReLU function is 0 when the input is negative, else it is 1. We could have use other activation functions such as sigmoid or hyperbolic tangent functions. In our case, ReLU has been used in order to accelerate the model's training speed.
- Log_softmax: This is a probability function used to calculate the logarithm of the softmax function, which takes a vector of real numbers from the output layer and computes a probability as outcome.

The proposed GNN model is trained for a number of epochs based on the following parameters:

- Optimizer: In our proposed GNN model, the optimization process is meant to adjust the weights and learning rate values in such a way that the model performs and yields a better accuracy and minimal loss. In our case, the Adaptive Moment Estimation optimizer (denoted ADAM) is considered [10].
- Loss function [10]: this function is used to measure the error between the predicted values and the ground truth. At each epoch of the model training, its output represents the performance of the model according to the considered training data. In our model, *nll_loss* is used as loss function.
- Learning rate: This parameter is utilized at each epoch to reduce the loss metric value. Its value is initialized to $1e-1$.

The flowchart of the proposed GNN algorithm is shown in Fig. 7.

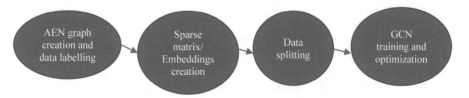

Fig. 7. Flowchart of the proposed GNN algorithm

The Algorithm 1 (AEN graph creation and data labelling) steps are as follows:

- Read the csv file and store it as a dataframe (line 1)
- Create an instance of AEN graph from dataframe (line 2)
- Extract the nodes and edges from dataframe (lines 3 to 6)
- Specify the botnets IP addresses (line 7)
- Label the sessions (lines 8 to 13)

Algorithm 1 (AEN) Graph Creation and Labelling Data

1: $df \leftarrow (CSVFILE)$
2: $G \leftarrow Digraph$
3: **for** each row in data **do**
4:      $nodes \leftarrow row[2]$
5:      $nodes \leftarrow row[3]$
6:      $edges \leftarrow row[2] + row[3]$
7:      $BotnetIP \leftarrow IPADDRESS$
8:      $labelledarray \leftarrow NULL$ ▷ labelled array represents the session labels
9:      **for** each row in rows **do**
10:         **if** SourceIP OR DestIP EQUALS BotnetIP **then**
11:            $labelledarray \leftarrow 1$
12:         **else**
13:            $labelledarray \leftarrow 0$

The Algorithm 2 (Spare matrix and embedding creation) steps are as follows:

- Create the sparse matrix (line 1)
- Create the edge index (lines 2 to 4)
- Create and normalize the embeddings (lines 5 and 6)

Algorithm 2 Spare Matrix and creation of Embedding

1: $adj \leftarrow sparsematrix(G)$
2: $row \leftarrow adj.row$
3: $col \leftarrow adj.col$
4: $edgeindex \leftarrow (row, col)$
5: $embeddings \leftarrow df['LengthColumn']$
6: $embeddings \leftarrow Normalize(embeddings) = 0$

The Algorithm 3 (Data splitting) is steps are as follows:

- Get the number of nodes of the AEN graph (lines 1 and 2)
- Get the input and expected output values of GNN model (lines 3 and 4)
- Define the number of classes (line 5)
- Split the data (lines 6 and 7)

Algorithm 3 Splitting Data into different sets

1: *Data ← Data(edgeindex ← edgeindex)*
2: *Data.num_nodes ← G.number_of_nodes*
3: *Data.x ← torch(embedding)*
4: *Data.y ← torch(labels)*
5: *Data.num_of_classes ← 2*
6: *Test_data ← 0.2(Data)*
7: *Validation_data ← 0.3(Data)*

The Algorithm 4 (GCN training and optimization) steps are as follows:

- Apply ReLU function to the output of the hidden layers (line 1)
- Apply log_softmax function to xxxx (line 2)
- Set the number of epochs (line 3)
- Run the GNN model and compute the negative log loss (lines 4 to 8)

Algorithm 4 (GCN) procedure, training and optimization

1: relu(convolution(data.x,edgeindex))
2: log_softmax(data.x)
3: Optimized value   ▷ Optimized value essentially indicates that after that epoch value the loss does not go any further down or goes down a insignificant amount it is found through the process of trial and error
4: **for** epochs>=0 AND epochs<=optimized value **do**
5:     *optimizer ← ADAM*
6:     *model ← GCN()*
7:     *Negative_Log_Loss(model,data.y)*
8:     *optimizer.step*   ▷ parameter update based on the current gradient descent

---

4   Experiments and Results

4.1   Datasets

We have considered five datasets: the CTU-MALWARE-CAPTURE-BOTNET-42 (denoted CTU42), CTU-MALWARE-CAPTURE- BOTNET-43 (denoted CTU43), and CTU-MALWARE-CAPTURE-BOTNET-48 (denoted CTU48), the ISCX Botnet Training (denoted ISCX-Train) and the ISCX Botnet Test (denoted ISCX-Test) datasets [11, 12]. The CTU datasets contain data traffic generated that consists of malware operating on various protocols, including IRC, SPAM, CF, PS, DDOS, FF, P2P, and HTTP. All datasets are pcap files, which are converted into csv files before they are used.

4.2   Performance Metrics

The following performance metrics are considered for evaluating the proposed model:

- *Precision (P):* this measures how many samples are tagged as normal and abnormal traffic, and it is obtained as $P = TP/(TP + FP)$
- *Recall (R):* this metric represents the ratio of True Positive over the sum of the True Positive and False Negative, i.e. $R = TP/(TP + FN)$
- *F1-score*: this metric provides the harmonic mean of Precision and Recall, calculated as $F1\ score = 2/((1/Recall) + (1/Precision))$
- *Accuracy*: this metric represents the number of correctly predicted data over the total number of data, which is calculated as $Accuracy = (TP + TN)/(TP + TN + FP +$

4.3   Considered Scenarios and Results

For our binary classification task, we have trained each of the five datasets using the following break of data: 20% for testing, 30% for validation, and 50% for training. In these experiments, the proposed GNN model is trained using each of the above-mentioned dataset, with the focus to investigate the capability of the considered classifier to identify the sessions as containing a botnet or not. To avoid over-fitting, we have continuously retrained the model until the accuracy on the validation set gets improved. The results obtained are given in Table 1.

Table 1. Results of experiments

Metrics	CTU42	CTU43	CTU48	ISCX-Train	ISCX-Test
Precision (%)	100	98	100	100	98
Recall (%)	75	66	77	76	78
F1-score (%)	86	79	87	86	87
Accuracy (%)	75	66	78	77	78

From Table 1, it is observed that (1) the best result over all the studied performance metrics is obtained using the Precision metric (99.2%), followed by the Recall metric (90%), the F1 score metric (85%), and the Accuracy metric (74.8%), (2) CTU42, CTU48, and ISCX-Train datasets generate the best result in terms of the Precision compared to the CTU43 and ISCX-Test datasets, (3) in terms of Recall metric, the ISCX-Test dataset yields the best result, (4) in terms of F1 score metric, the CTU48 and ISCX-Test generate the best result. Similar observation prevails in the case of the Accuracy metric, (5) for the considered datasets, CTU48 generates the best possible results in average when considering the aggregate performance over all the studied metrics (85.5%), followed by the ISCX-Test dataset (85.25%), the ISCX-Train dataset (84.75%), the CTU42 dataset (84%), and the CTU43 (77.25%).

5   Conclusion

This paper has proposed a novel AEN-based graph neural network-based scheme for botnet detection based on supervised learning. The performance evaluation of the proposed model is evaluated using five labelled datasets and four performance metrics, showing that: (1) the best possible result over all the studied performance metrics is obtained using the Precision metric, (2) CTU42, CTU48, and ISCX-Train datasets generate the best result in terms of the Precision compared to the other datasets, (3) the ISCX-Test dataset yields the best result in terms of Recall metric, (4) the CTU48 and ISCX-Test generate the best result in terms of F1 score and Accuracy metrics, and (5) for the considered datasets, CTU48 generates the best possible results in average when considering the aggregate performance over all the studied metrics. In future, we plan to compare the performance of the proposed detection model against that of other benchmark models, and test it using other known datasets.

References

1. Nie, C., Quinan, P.G., Traore, I., Woungang, I.: Intrusion detection using a graphical fingerprint model. In: 22nd IEEE International Symposium on Cluster, Cloud and Internet Computing, pp. 806–813 (2022)
2. Quinan, P.G., Traore, I., Gondhi, U.R., Woungang, I.: Unsupervised anomaly detection using a new knowledge graph model for network activity and events. In: Renault, É., Boumerdassi, S., Mühlethaler, P. (eds.) MLN 2021. LNCS, vol. 13175, pp. 117–130. Springer, Cham (2022). https://doi.org/10.1007/978-3-030-98978-1_8
3. Quinan, P.G., Traore, I., Woungang, I.: Improved threat detection through the activity and event network graph belief propagation model. In: Submitted 20 October 2022 to IEEE Transactions on Information Forensics and Security, Manuscript ID: T-IFS-15257–2022
4. Masoudi-Sobhanzadeh, Y., Emami-Moghaddam, S.: A real-time IoT-based botnet detection method using a novel two-step feature selection technique and the support vector machine classifier. Comput. Netw. **217**, 109365 (2022). ISSN 1389-1286https://doi.org/10.1016/j.comnet.2022.109365
5. Ostap, H.: Antkiewicz: R.: BotTROP: detection of a botnet-based threat using novel data mining algorithm. In: Communications of the IBIMA, vol. 2022, p. 20, (2022). Article ID 156851, ISSN: 1943–7765. https://ibimapublishing.com/articles/CIBIMA/2022/156851. https://doi.org/10.5171/2022.156851
6. Choi, H., Lee, H.: Identifying botnets by capturing group activities in DNS traffic. Comput. Netw. **56**(01), 20–33 (2012)
7. Alani, M.M.: BotStop : packet-based efficient and explainable IoT botnet detection using machine learning. Comput. Commun. **193**, 53–62 (2022). ISSN 0140–3664. https://doi.org/10.1016/j.comcom.2022.06.039
8. Alharbi, A., Alsubhi, K.: Botnet detection approach using graph-based machine learning. IEEE Access **9**, 99166–99180 (2021). https://doi.org/10.1109/ACCESS.2021.3094183
9. Duan, L., Zhou, J., Wu, Y., Xu, W.: A novel and highly efficient botnet detection algorithm based on network traffic analysis of smart systems. Int. J. Distrib. Sens. Netw. **18**(3), 15501477211049910 (2022). https://doi.org/10.1177/15501477211049910. Accessed 16 Dec 2022
10. Madurkar, A.: Graph machine learning with python part 1: Basics, metrics, and algorithms. Towards Data Science (2021). https://medium.com/towards-data-science/graph-machinelearning-with-python-pt-1-basics-metrics-and-algorithmscc40972de113. Accessed 17 Dec 2022
11. The CTU Datasets: https://github.com/davedittrich/lim-cli/blob/master/docs/ctu.rst. Accessed 18 Dec 2022
12. ISCX Datasets: Canadian Institute for Cybersecurity, University of New Brunswick. https://www.unb.ca/cic/about/contact.html. Accessed 18 Dec 2022

Object Placement Algorithm with Information Flow Control in Fog Computing Model

Shigenari Nakamura[1]([✉]), Tomoya Enokido[2], and Makoto Takizawa[3]

[1] Tokyo Metropolitan Industrial Technology Research Institute, Tokyo, Japan
`nakamura.shigenari@iri-tokyo.jp`
[2] Rissho University, Tokyo, Japan
`eno@ris.ac.jp`
[3] Hosei University, Tokyo, Japan
`makoto.takizawa@computer.org`

Abstract. In the IoT (Internet of Things), data are exchanged among objects in devices and subjects through manipulating objects. In order to reduce the network traffic and satisfy the time constraints, an FC (Fog Computing) model where a fog layer is introduced between devices and subjects is considered. Here, the fog layer is composed of fog nodes. Data from objects are processed at fog nodes and processed data are sent to subjects. Even if subjects manipulate objects in accordance with the CBAC (Capability-Based Access Control) model, the subjects can get data which are not allowed to be gotten by the subjects, i.e. illegal information flow and late information flow occur. Hence, the FCOI (FC-based Operation Interruption) and FCTBOI (FC and Time-Based OI) protocols where operations occurring illegal and late types of information flows are interrupted are implemented. In the FC model, fog nodes execute tasks to process data from objects. If the tasks are sent to fog nodes with a load balancing algorithm, data may be stored at multiple fog nodes. However, multi locations of data may cause significant security issues. Hence, an SOP (Source objects-based Object Placement) algorithm is proposed in this paper. Here, fog nodes to execute tasks are selected to reduce the number of fog nodes holding objects of common device owners.

Keywords: IoT (Internet of Things) · Device security · CBAC (Capability-Based Access Control) model · Information flow control · FC (Fog Computing) model · Object placement algorithm

1 Introduction

It is widely recognized that access control models [3] are useful to make information systems secure. For the IoT (Internet of Things) [22], the CBAC (Capability-Based Access Control) model [4,6] is considered where capability tokens which

L. Barolli (Ed.): AINA 2023, LNNS 654, pp. 258–267, 2023.
https://doi.org/10.1007/978-3-031-28451-9_23

are collections of access rights are issued to subjects. Here, subjects manipulate objects in devices in accordance with capability tokens issued themselves. Since data are got and put in objects by subjects, the data are exchanged among objects and subjects. Illegal accesses by subjects to objects in devices are prevented from occurring by the CBAC model. However, subjects might get data via other subjects and objects even if the subjects are not allowed to get the data, i.e. illegal information flow might occur [11–14]. In addition, a subject might get data which are older than the subject expects to get, i.e. the data come to the subject *late* [15].

First, the illegal and late information flows in the IoT are prevented from occurring in the OI (Operation Interruption) [14] and TBOI (Time-Based OI) [15] protocols. These protocols are implemented in a Raspberry Pi3 Model B+ [1] with Raspbian [2]. Here, subjects and devices communicate in the CoAP (Constrained Application Protocol) [24,25]. Next, an MRCTSD (Minimum Required Capability Token Selection for Devices) algorithm [17] is implemented to shorten the request processing time in the OI and TBOI protocols. An MRCTSS (MRCTS for Subjects) algorithm [16] is also implemented to reduce the communication traffic. Finally, the electric energy consumptions of these approaches are made clear [19] based on the models [8,18]. It is shown that the electric energy consumption is reduced by using the capability token selection algorithms.

An FC (Fog Computing) [5] model is a computing paradigm for the IoT. Here, a fog layer which consists of fog nodes is nearer IoT devices than cloud. Each fog node has storage, computing, and communication capabilities to provide edge services. For example, summarized data are generated at fog nodes by processing row data from objects before the raw data arrive at a subject. In this case, since only the summarized data arrive at the subject, the amount of data flow to the subject becomes smaller. If the amount of data exchanged among objects and subjects decreases, the number of illegal and late information flows also decreases. The numbers of operations interrupted in the FCOI (FC-based OI) and FCTBOI (FC-based TBOI) protocols are smaller than the OI and TBOI protocols, respectively [18].

It is significant the load balancing [26] to avoid concentrating tasks to process data at a part of fog nodes in the FC model. The load balancing approaches based on various types of parameters such as usage of computation resources in computers [8] and features of data manipulated [9] have been proposed and discussed so far. In order to balance tasks to fog nodes, an SLB (Source objects-based Load Balancing) algorithm is proposed [21]. Here, tasks to process data are sent to fog nodes based on not only computation loads but also data locality. The computation loads are estimated based on the *source* objects. A set of source objects indicates which objects the data flow from to its entity.

In the cloud computing model, multi locations of data may cause significant security issues [10]. The cloud computing model is maintained by a cloud provider which is one of the third party. If data of a client are stored in multiple computers by the cloud provider, attackers can get the data more easily. In addition, if data of a pair of clients c_1 and c_2 are maintained collectively, an attack to the data of the client c_1 may affect the data of the other client c_2.

These security issues are concerned in not only the cloud computing model but also the FC model. In the FC model, each object o_{oi}^{di} in a device d_{di} belongs to a device owner which owns the device d_{di}. On the other hand, fog nodes are maintained by a service provider which is one of the third party as well as the cloud computing model. If tasks with objects are sent to fog nodes in the SLB algorithm, an object may be stored at multiple fog nodes. Hence, an SOP (Source objects-based Object Placement) algorithm is proposed in this paper. Here, fog nodes to execute tasks are selected to reduce the number of fog nodes holding objects of common device owners.

In Sect. 2, the system model and types of information flow relations are discussed. In Sect. 3, the SOP algorithm is proposed to reduce the number of fog nodes holding objects of common device owners. In Sect. 4, the FCOI and FCT-BOI protocols are discussed to prevent both types of information flows.

2   System Model

2.1   CBAC (Capability-Based Access Control) Model

A huge number and various types of devices are interconnected in addition to computers in the IoT [5,22]. The number sn of subjects $s_1, \ldots, s_{sn}$ $(sn \geq 1)$ manipulate the number dn of devices $d_1, \ldots, d_{dn}$ $(dn \geq 1)$ in the IoT. An object o_{oi}^{di} indicates a component in the device d_{di}. Let on^{di} be the number of objects held by each device d_{di}. In this paper, three types of devices are considered. A sensor type device d_{di} stores data collected by sensing events occurring in physical environment in its object o_{oi}^{di}. An actuator type device d_{di} stores the data received from subjects in its object o_{oi}^{di}. Actions are performed based on the data on the physical environment. A hybrid type device d_{di} is equipped with both the sensors and actuators.

In order to prevent illegal accesses by subjects to objects from occurring, a CBAC model [4,6] is considered. Each subject s_{si} is issued a set CAP^{si} which consists of the number cn^{si} of capability tokens $cap_1^{si}, \ldots, cap_{cn^{si}}^{si}$ $(cn^{si} \geq 1)$. A capability token cap_{ci}^{si} is designed as shown in the papers [14,15]. An access right field in each capability token cap_{ci}^{si} indicates how the subject s_{si} can manipulate which objects. A pair of public keys of the subject s_{si} and the issuer of the capability token cap_{ci}^{si} and a signature of the issuer are generated in the ECDSA (Elliptic Curve Digital Signature Algorithm) [7]. The validity period of the capability token cap_{ci}^{si} is also indicated by its fields.

A subject s_{si} sends an access request to manipulate the object o_{oi}^{di} in an operation op to a device d_{di} with a capability token cap_{ci}^{si} which indicates the access is allowed. If the device d_{di} confirms that the access by the subject s_{si} is allowed in accordance with the attached capability token cap_{ci}^{si}, the operation op is performed on the object o_{oi}^{di}. Here, it is easier to adopt the CBAC model to the IoT than the RBAC (Role-Based Access Control) [23] and ABAC (Attribute-Based

Access Control) [27] models because the device d_{di} just checks the capability token cap_{ci}^{si} to authorize the subject s_{si},

Let a pair $\langle o, op \rangle$ be an access right. Subjects issued a capability token including an access right $\langle o, op \rangle$ is allowed to manipulate data of an object o in an operation op. A set of objects whose data a subject s_{si} is allowed to get is $IN(s_{si})$ i.e. $IN(s_{si}) = \{o_{oi}^{di} \mid \langle o_{oi}^{di}, get \rangle \in cap_{ci}^{si} \wedge cap_{ci}^{si} \in CAP^{si}\}$. Since data are got and put in objects by subjects, the data are exchanged among objects and subjects. Objects whose data flow into entities are referred to as source objects for these entities. Let $o_{oi}^{di}.sO$ and $s_{si}.sO$ are sets of source objects of an object o_{oi}^{di} and a subject s_{si}, respectively, which are initially ϕ.

Let a pair of times $gt^{si}.st(o_{oi}^{di})$ and $gt^{si}.et(o_{oi}^{di})$ be the start and end time when a subject s_{si} is allowed to get data from the object o_{oi}^{di}. These times are decided based on the validity periods of capability tokens issued to the subject s_{si}. The time when data of an object o_{oi}^{di} are generated is referred to a generation time. Let $minOT_{oi}^{di}(o_{oj}^{dj})$ and $minSBT^{si}(o_{oj}^{dj})$ be the earliest generation times of data of an object o_{oj}^{dj} which flow to an object o_{oi}^{di} and a subject s_{si}, respectively.

2.2   FC (Fog Computing) Model

An FC (Fog Computing) [5] model is a computing paradigm for the IoT. Here, a fog layer F which consists of the number fn of fog nodes $f_1, \ldots, f_{fn}$ ($fn \geq 1$) is nearer IoT devices than cloud as shown in Fig. 1. Each fog node has storage, computing, and communication capabilities to provide edge services. Let $f_{fi}.O$ be a set of objects held by a fog node f_{fi}. An object o_{oi}^{di} sent to a fog node f_{fi} from devices are included in the set $f_{fi}.O$ of the fog node f_{fi}. The fog node f_{fi} processes raw data of the object o_{oi}^{di} in accordance with the requirement of a subject s_{si}. After that, the summarized data generated by the fog node f_{fi} are forwarded to the subject s_{si}. In this case, since only the summarized data arrive at the subject s_{si}, the amount of data flow to the subject s_{si} becomes smaller. Since it is a critical issue that a large volume of sensor data are transmitted from sensors to subjects in networks in the IoT, the FC model is useful to satisfy the time constraints and reduce the network traffic.

Suppose a subject s_{si} issues a *get* operation on an object o_{oi}^{di} to a device d_{di}. First, the object o_{oi}^{di} is forwarded to a fog node f_{fi} by an edge node. Next, the data of the o_{oi}^{di} are processed and summarized data are generated at the fog node f_{fi} for the subject s_{si}. Finally, the summarized data are sent to the subject s_{si}. Let $f_{fi}(o_{oi}^{di}.sO)$ be a set of source objects whose data are included in the summarized data. The set $f_{fi}(o_{oi}^{di}.sO)$ is decided in accordance with the data processing of the fog node f_{fi}. By data processing at fog nodes, only the data of objects in $f_{fi}(o_{oi}^{di}.sO)$ flow to the subject s_{si}.

2.3   Information Flow Relations

As noted above, if a subject s_{si} issues a get operation to get data from an object o_{oi}^{di}, information flow occurs. Based on the CBAC model, types of information flow relations are defined as follows:

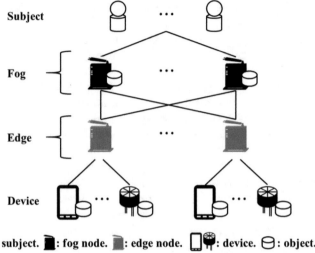

Fig. 1. FC model.

Definition 1. $o_{oi}^{di} \rightarrow^f s_{si}$ iff $f_{fi}(o_{oi}^{di}.sO) \neq \phi$ and $o_{oi}^{di} \in IN(s_{si})$.

Definition 2. $o_{oi}^{di} \Rightarrow^f s_{si}$ iff $o_{oi}^{di} \rightarrow^f s_{si}$ and $f_{fi}(o_{oi}^{di}.sO) \subseteq IN(s_{si})$.

Definition 3. $o_{oi}^{di} \mapsto^f s_{si}$ iff $o_{oi}^{di} \rightarrow^f s_{si}$ and $f_{fi}(o_{oi}^{di}.sO) \not\subseteq IN(s_{si})$.

Definition 4. $o_{oi}^{di} \Rightarrow_t^f s_{si}$ iff $o_{oi}^{di} \Rightarrow^f s_{si}$ and $\forall o_{oj}^{dj} \in f_{fi}(o_{oi}^{di}.sO)$ $(gt^{si}.st(o_{oj}^{dj}) \leq minOT_{oi}^{di}(o_{oj}^{dj}) \leq gt^{si}.et(o_{oj}^{dj}))$.

Definition 5. $o_{oi}^{di} \mapsto_l^f s_{si}$ iff $o_{oi}^{di} \Rightarrow^f s_{si}$ and $\exists o_{oj}^{dj} \in f_{fi}(o_{oi}^{di}.sO)$ $\neg(gt^{si}.st(o_{oj}^{dj}) \leq minOT_{oi}^{di}(o_{oj}^{dj}) \leq gt^{si}.et(o_{oj}^{dj}))$.

3   Object Placement Algorithm

In the FC model, subjects issue *get* operation on objects to get data in the objects. The objects are sent from devices to edge nodes in the edge layer. The edge nodes generate tasks to process data of the objects and forward the tasks to fog nodes. Here, the load balancing [26] is significant to avoid concentrating loads at a part of fog nodes. In the SLB algorithm [21], a fog node to execute a task is selected based on computation loads and data locality. Here, tasks with objects except for the objects with large data are sent to the idlest fog node. Although the objects with large data are sent to fog nodes which already hold the same objects with large portion of the data, objects held by each fog node are from various devices.

In the cloud computing model, multi locations of data may cause significant security issues [10]. A cloud model is maintained by a cloud provider. Here, data of clients are maintained by the service provider which is generally one of the third party. Therefore, the service provider may store data of a client in multiple computers. Here, although an attacker cannot get the data from a computer, the attacker may get the data from another computer. In addition, the service provider may store data of different clients in the common computer. If the data are not separately maintained by the service provider, an attack to the data of a client may affect the data of another client.

The above security issues are concerned in not only the cloud computing model but also the FC model. In the FC model, each object o_{oi}^{di} in a device d_{di} belongs to a device owner which owns the device d_{di}. On the other hand, fog nodes are maintained by a service provider which is one of the third party as well as the cloud computing model. Therefore, if an object o_{oi}^{di} is held by multiple fog nodes, attackers can get the object o_{oi}^{di} more easily. If data of an object o_{oi}^{di} and another object o_{oj}^{dj} are held by the common fog node, an attack to the object o_{oi}^{di} may affect the object o_{oj}^{dj}. Since objects belonging to various device owners are held by fog nodes and the performance of fog nodes are not high in general, fog nodes are vulnerable to attacks. Hence, it is significant to reduce the number of fog nodes holding objects of common device owners.

For these reasons, an SOP (Source objects-based Object Placement) algorithm is proposed in this paper. Let t_{ti}^{fi} be a task which will be executed at a fog node f_{fi}. Each fog node f_{fi} has a task queue $f_{fi}.Q$ and a task t_{ti}^{fi} is in the queue $f_{fi}.Q$. In the FC model, objects whose data will be processed are forwarded to fog nodes. Hence, each task t_{ti}^{fi} has a set $t_{ti}^{fi}.O$ of objects. Let $f_{fi}.D$ and $t_{ti}^{fi}.D$ be sets of devices of source objects in f_{fi} and t_{ti}^{fi}, respectively. In this paper, the meta-data of each object o_{oi}^{di} is considered. Let $o_{oi}^{di}.d$ and $o_{oi}^{di}.o$ indicate the device holding the object o_{oi}^{di} and itself, i.e. $o_{oi}^{di}.d = di$ and $o_{oi}^{di}.o = oi$. $o_{oi}^{di}.ty$ means the type of device d_{di}, i.e. $o_{oi}^{di}.ty = sensor$, $actuator$, or $hybrid\ device$. $o_{oi}^{di}.sz$ is the size of data the object o_{oi}^{di} holds. $o_{oi}^{di}.sO$ is the set of source objects of the object o_{oi}^{di}. Each task t_{ti}^{fi} has a set $t_{ti}^{fi}.sO$ of source objects which is defined to be $\bigcup_{oi=1}^{|t_{ti}^{fi}.O|} o_{oi}.sO$.

In this paper, it is assumed that a task is generated at an edge node according to an operation issued by a subject s_{si}. After that, the edge node selects a fog node to execute the task. The SOP algorithm performs as follows:

SOP(a task t, meta-data of objects held by fog nodes in F) {
    $f = f_1$;
    for each task t_{ti}^1 ($\in f_1.Q$) {
        $f_1.sO = f_1.sO \cup t_{ti}^1.sO$;
    }
    for each object o_{oi}^{di} ($\in f_1.sO$) {
        $f_1.D = f_1.D \cup o_{oi}^{di}.d$;
    }
    for each object o_{oi}^{di} ($\in t.sO$) {
        $t.D = t.D \cup o_{oi}^{di}.d$;
    }

```
add_dev_num = |t.D − f₁.D|;
dev_num = |t.D ∪ f₁.D|;
for each fog node f_fi (∈ F) { /* fi = 2, ..., fn. */
    for each task t_ti^fi (∈ f_fi.Q) {
        f_fi.sO = f_fi.sO ∪ t_ti^fi.sO;
    }
    for each object o_oi^di (∈ f_fi.sO) {
        f_fi.D = f_fi.D ∪ o_oi^di.d;
    } if |t.D − f_fi.D| < add_dev_num {
        add_dev_num = |t.D − f_fi.D|;
        f = f_fi;
    } else if |t.D − f_fi.D| = add_dev_num {
        if |t.D ∪ f_fi.D| < dev_num {
            dev_num = |t.D ∪ f_fi.D|;
            f = f_fi;
        }
    }
}
return f; /* a fog node f is selected to execute the task t. */
}
```

4   Information Flow Control in the FC Model

If subjects try to get data from objects, objects including the data arrive at the fog layer composed of fog nodes before the data are sent to the subjects. The data are processed and summarized data are generated. Finally, subjects gets the summarized data. As noted above, data are exchanged among subjects and objects in the FC model. Hence, subjects may get data which are not allowed to be gotten by the subjects even if the subjects manipulate objects according to capability tokens issued them. For example, a subject s_{si} can get data of an object o_{oi}^{di} flowing to another object o_{oj}^{dj} by accessing the object o_{oj}^{dj} even if the subject s_{si} is not allowed to get the data from the object o_{oi}^{di}, i.e. illegal information flow occurs. In addition, a subject s_{si} can get data from an object o_{oi}^{di} generated out of validity period of a capability token cap_{ci}^{si} to get the data. Here, the data are older than the subject s_{si} expects to get, i.e. information comes to the subject s_{si} *late*. In order to prevent illegal information flow and both illegal and late types of information flows, the FCOI and FCTBOI protocols are evaluated [20].

In order to prevent both illegal and late types of information flows, sets of source objects are manipulated in the FCOI and FCTBOI protocols. For example, since data flow from an object o_{oi}^{di} to a subject s_{si} via a fog node f_{fi} in a *get* operation, the set $f_{fi}(o_{oi}^{di}.sO)$ of source objects are added to the set $s_{si}.sO$ of the subject s_{si}. On the other hand, since data flow from a subject s_{si} to an object o_{oi}^{di} in a *put* operation, the set $s_{si}.sO$ are added to the set $o_{oi}^{di}.sO$. In the

FCTBOI protocol, the earliest generation time of data of every source object is also updated. The FCOI and FCTBOI protocols perform as follows:

[**FCOI protocol**] A *get* operation on an object o_{oi}^{di} issued by a subject s_{si} is interrupted if $\exists o_{oj}^{dj} \in t.O \ \neg(o_{oj}^{dj} \Rightarrow^f s_{si})$ holds.

[**FCTBOI protocol**] A *get* operation on an object o_{oi}^{di} issued by a subject s_{si} is interrupted if $\exists o_{oj}^{dj} \in t.O \ \neg(o_{oj}^{dj} \Rightarrow_t^f s_{si})$ holds.

In the FCOI and FCTBOI protocols, tasks to process data from objects are sent to fog nodes in the SOP algorithm. $t.O$ is a set of objects whose data are needed to be processed. In this paper, calculable data are assumed to be exchanged among entities. The set $f_{fi}(o_{oi}^{di}.sO)$ of source objects are decided in accordance with data calculation. For example, if a specific value such as maximum value is required by a subject s_{si}, only the data including the value are extracted from an object o_{oi}^{di} and sent to the subject s_{si}. Here, the set $f_{fi}(o_{oi}^{di}.sO)$ is composed of only one object o_{oj}^{dj}.

5 Concluding Remarks

For the IoT, the CBAC model is proposed where capability tokens which are collection of access rights are issued to subjects. The data required by subjects are processed at fog nodes in the FC model. Since data are exchanged among entities through manipulating objects, two types of illegal and late information flows occur. In order to prevent both types of illegal and late information flows from occurring, the FCOI and FCTBOI protocols were proposed. In the FC model, objects are sent to fog nodes from devices to process data of the objects before the data are sent to subjects. Here, fog nodes execute tasks including objects to process data of the objects. If the tasks are sent to fog nodes with a load balancing algorithm, an object may be stored at multiple fog nodes. However, multiple locations of data may cause significant security issues. Hence, an SOP (Source objects-based Object Placement) algorithm is proposed in this paper. Here, fog nodes to execute tasks are selected to reduce the number of fog nodes holding objects of common device owners.

Acknowledgements. This work was supported by Japan Society for the Promotion of Science (JSPS) KAKENHI Grant Number JP22K12018.

References

1. Raspberry pi 3 model b+. https://www.raspberrypi.org/products/raspberry-pi-3-model-b-plus/
2. Raspbian, version 10.3, 13 Feb 2020. https://www.raspbian.org/
3. Denning, D.E.R.: Cryptography and Data Security. Addison Wesley, Boston (1982)
4. Gusmeroli, S., Piccione, S., Rotondi, D.: A capability-based security approach to manage access control in the internet of things. Math. Comput. Model. **58**(5–6), 1189–1205 (2013)

5. Hanes, D., Salgueiro, G., Grossetete, P., Barton, R., Henry, J.: IoT Fundamentals: Networking Technologies, Protocols, and Use Cases for the Internet of Things. Cisco Press, Indianapolis, IN, USA (2018)
6. Hernández-Ramos, J.L., Jara, A.J., Marín, L., Skarmeta, A.F.: Distributed capability-based access control for the internet of things. J. Internet Serv. Inf. Secur. **3**(3/4), 1–16 (2013)
7. Johnson, D., Menezes, A., Vanstone, S.: The elliptic curve digital signature algorithm (ECDSA). Int. J. Inf. Secur. **1**(1), 36–63 (2001). https://doi.org/10.1007/s102070100002
8. Kataoka, H., Nakamura, S., Duolikun, D., Enokido, T., Takizawa, M.: Multi-level power consumption model and energy-aware server selection algorithm. Int. J. Grid Util. Comput. **8**(3), 201–210 (2017)
9. Ke, W., Xraobing, Z., Tonglin, L., Dongfang, Z., Michael, L., Ioan, R.: Optimizing load balancing and data-locality with data-aware scheduling. In: 2014 IEEE International Conference on Big Data (Big Data), pp. 119–128 (2014)
10. Minqi, Z., Rong, Z., Wei, X., Weining, Q., Aoying, Z.: Security and privacy in cloud computing: a survey. In: 2010 Sixth International Conference on Semantics, Knowledge and Grids, pp. 105–112 (2010)
11. Nakamura, S., Duolikun, D., Enokido, T., Takizawa, M.: Influential abortion probability in a flexible read-write abortion protocol. In: Proceedings of IEEE the 30th International Conference on Advanced Information Networking and Applications, pp. 1–8 (2016)
12. Nakamura, S., Duolikun, D., Enokido, T., Takizawa, M.: A read-write abortion protocol to prevent illegal information flow in role-based access control systems. Int. J. Space-Based Situated Comput. **6**(1), 43–53 (2016)
13. Nakamura, S., Enokido, T., Takizawa, M.: Information flow control in object-based peer-to-peer publish/subscribe systems. Concurr. Comput. Pract. Exp. **32**(8), e5118 (2020)
14. Nakamura, S., Enokido, T., Takizawa, M.: Implementation and evaluation of the information flow control for the internet of things. Concurr. Comput. Pract. Exp. **33**(19), e6311 (2021)
15. Nakamura, S., Enokido, T., Takizawa, M.: Information flow control based on capability token validity for secure IOT: implementation and evaluation. Internet Things **15**, 100423 (2021)
16. Nakamura, S., Enokido, T., Takizawa, M.: Traffic reduction for information flow control in the IoT. In: Barolli, L. (ed.) BWCCA 2021. LNNS, vol. 346, pp. 67–77. Springer, Cham (2022). https://doi.org/10.1007/978-3-030-90072-4_7
17. Nakamura, S., Enokido, T., Takizawa, M.: Capability token selection algorithms to implement lightweight protocols. Internet of Things **19**, 100542 (2022)
18. Nakamura, S., Enokido, T., Takizawa, M.: Energy consumption model of a device supporting information flow control in the IOT. In: Barolli, L., Kulla, E., Ikeda, M. (eds.) Advances in Internet, Data & Web Technologies. EIDWT 2022. Lecture Notes on Data Engineering and Communications Technologies, vol. 118, pp. 142–152. Springer, Cham (2022). https://doi.org/10.1007/978-3-030-95903-6_16
19. Nakamura, S., Enokido, T., Takizawa, M.: Energy consumption of the information flow control in the IoT: simulation evaluation. In: Barolli, L., Hussain, F., Enokido, T. (eds.) AINA 2022. LNNS, vol. 449, pp. 285–296. Springer, Cham (2022). https://doi.org/10.1007/978-3-030-99584-3_25

20. Nakamura, S., Enokido, T., Takizawa, M.: Evaluation of the information flow control in the fog computing model. In: Barolli, L. (eds.) Advances on Broad-Band Wireless Computing, Communication and Applications. BWCCA 2022. Lecture Notes in Networks and Systems, vol. 570, pp. 78–90. Springer, Cham (2022). https://doi.org/10.1007/978-3-031-20029-8_8

21. Nakamura, S., Enokido, T., Takizawa, M.: Load balancing algorithm for information flow control in fog computing model. In: Barolli, L. (eds.) Advances in Internet, Data & Web Technologies. EIDWT 2023. Lecture Notes on Data Engineering and Communications Technologies, vol. 161. Springer, Cham. (2023). https://doi.org/10.1007/978-3-031-26281-4_28

22. Oma, R., Nakamura, S., Duolikun, D., Enokido, T., Takizawa, M.: An energy-efficient model for fog computing in the internet of things (IOT). Internet Things 1–2, 14–26 (2018)

23. Sandhu, R.S., Coyne, E.J., Feinstein, H.L., Youman, C.E.: Role-based access control models. IEEE Comput. 29(2), 38–47 (1996)

24. Shelby, Z., Hartke, K., Bormann, C.: Constrained application protocol (CoAP). IFTF Internet-draft (2013). http://tools.ietf.org/html/draft-ietf-core-coap-18

25. Tanganelli, G., Vallati, C., Mingozzi, E.: CoAPthon: Easy development of CoAP-based IOT applications with python. In: IEEE 2nd World Forum on Internet of Things (WF-IoT 2015), pp. 63–68 (2015)

26. Willebeek-LeMair, M.H., Reeves, A.P.: Strategies for dynamic load balancing on highly parallel computers. IEEE Trans. Parallel Distrib. Syst. 4(9), 979–993 (1993)

27. Yuan, E., Tong, J.: Attributed based access control (ABAC) for web services. In: Proceedings of the IEEE International Conference on Web Services (ICWS 2005), p. 569 (2005)

An Energy-Aware Algorithm for Changing Tree Structure and Process Migration in the Flexible Tree-Based Fog Computing Model

Dilawaer Duolikun[1(✉)], Tomoya Enokido[2], and Makoto Takizawa[1]

[1] RCCMS, Hosei University, Tokyo, Japan
dilewerdolkun@gmail.com, makoto.takizawa@computer.org
[2] Faculty of Business Administration, Rissho University, Tokyo, Japan
eno@ris.ac.jp

Abstract. Since the IoT consumes a huge amount of electrical energy due to the scalability, it is critical to reduce the energy consumed by nodes. In the FTBFC (Flexible Tree-Based Fog Computing) model proposed in our previous studies, fog nodes are structured in a tree. Here, operations to change the tree structure of fog nodes and make application processes migrate among parent and child nodes are proposed to reduce the total energy consumption. In this paper, we newly propose an FTBFC algorithm to flexibly change the tree structure and processes supported by nodes to reduce the total energy consumption of the nodes. In the evaluation, we show the total energy consumption can be reduced by changing the tree in the FTBFC model.

Keywords: Green computing · FTBFC (Flexible Tree-Based Fog Computing) model · IoT · Fog computing (FC) model · Energy consumption

1 Introduction

The IoT (Internet of Things) [3] is now widely used to realize various applications in our societies. Since a scalable system like the IoT consumes a huge amount of electric energy [1], it is significant to reduce the total electric energy consumed by the IoT to decrease carbon footprint.

In the fog computing (FC) model [3,18] of the IoT, sensor data is first processed by fog nodes and the processed data is delivered to servers. Since a smaller amount of data is processed by servers and transmitted in networks than the CC (Cloud Computing) model [2]l, servers consume smaller energy. On the other hand, fog nodes additionally consume energy to process sensor data. The TBFC (Tree-Based FC) model [15,17–21] is composed of fog nodes structured in a tree. Here, the root node indicates a cloud of servers and each leaf node is a device

© The Author(s), under exclusive license to Springer Nature Switzerland AG 2023
L. Barolli (Ed.): AINA 2023, LNNS 654, pp. 268–278, 2023.
https://doi.org/10.1007/978-3-031-28451-9_24

node. Each node processes input data sent by the child nodes and send the processed data to the parent node. In order to make the TBFC model tolerant of node faults, the FTTBFC (Fault-Tolerant TBFC) model [21–23] is proposed. Here, every child node of a faulty node f is reconnected to another node which supports the same processes as the node f. In order to adapt the IoT to the change of traffic, the DTBFC (Dynamic TBFC) model [17,23] is proposed. Here, processes of a node f migrate to the parent and child nodes if the node f consumes larger energy to process input data. The NFC (network FC) model is also discussed [24]. Here, each node f does the negotiation with another fog node nf on whether or not the node nf can process the output data of the node f. In the FTBFC (Flexible TBFC) model [16], we proposed operations to change not only the tree structure but also the processes supported by nodes in the tree.

In this paper, we newly propose an FTBFC algorithm by which the tree structure and processes supported by nodes can be flexibly changed so that the total energy consumption is reduced in the FTBFC model. In the evaluation, we show the total energy consumption and delivery time of the IoT can be reduced by changing the structure and processes in the FTBFC model.

In Sect. 2, we present the TBFC model. In Sect. 3, we discuss the FTBFC model and how to change the FTBFC model. In Sect. 4, we evaluate the FTBFC model.

2   The TBFC Model

2.1   Tree Structure of Nodes

The FC (Fog Computing) model [3] of the IoT is composed of fog nodes in addition to servers and devices. Here, fog nodes support subprocesses of an application process. Fog nodes receive and process sensor data sent by devices and deliver the processed data to servers through fog-to-fog (F2F) communication.

The TBFC (Tree-Based FC) model [15,17–21] is composed of nodes structured in a tree, where a root node is a server cloud, each leaf node shows a set of devices, and the other non-root and non-leaf nodes are fog nodes. A root node f communicates with *child* nodes $f_1, \ldots, f_b$ $(b \geq 0)$ in networks. Each node f_i communicates with child nodes $f_{i1}, \ldots, f_{i,b_i}$ $(b_i \geq 0)$. An index I of a node f_I is a sequence $\langle i_1 i_2 \ldots i_{l-1} \rangle$ $(l \geq 1)$ of numbers, which means a path $\langle f, f_{i_1}, f_{i_1 i_2}, \ldots, f_{i_1 i_2 \ldots i_{l-1}} (= f_I) \rangle$ from the root node f to f_I. A node f_I is at level $l (= |I| + 1)$. Thus, each node f_{Ii} only communicates with a parent node f_I and child nodes $f_{Ii1}, \ldots, f_{Ii,b_{Ii}}$ $(b_{Ii} \geq 0)$. An edge node f_I communicates with a leaf node d_I named device node. Here, a device node denotes a collection of devices which support sensors and actuators and each device is included in one device node.

In this paper, an application process P to process sensor data is assumed to be a sequence $\langle p_1, \ldots, p_m \rangle$ $(m \geq 1)$ of processes. Here, p_1 and p_m are *top* and *tail* processes, respectively. A pair of processes exchange a data unit (DU). The tail process p_m first receives a DU dt_{m+1} of sensor data from a device node, obtains an output DU dt_m by processing the DU dt_{m+1}, and sends dt_m to the process p_{m-1}. Thus, each process p_i receives an input DU dt_{i+1} from the process

p_{i+1} and sends an output DU dt_i to the process p_{i-1}. The top process p_1 finally obtains an output DU dt_1 which shows actions to be performed by actuators.

Each node f_I supports a subsequence P_I of the application process P, i.e. $\langle p_{t_I}, p_{t_I+1}, \ldots, p_{l_I} \rangle$ where $1 \leq t_I \leq l_I \leq m$ and p_{t_I} and p_{l_I} are the top and tail processes. Here, P_0 denotes a process subsequence of the root node f. In a path $\langle f, f_{i_1}, f_{i_1 i_2}, \ldots, f_{i_1 i_2 \ldots i_l} \rangle$ from the root node f to each edge node $f_{i_1 i_2 \ldots i_l}$, a concatenation of the subsequences P_0, P_{i_1}, $P_{i_1 i_2}$, ..., $P_{i_1 i_2 \ldots i_l}$ is the process sequence P. The node f_I receives an input DU id_{Ii} from each child node f_{Ii} ($i = 1, \ldots, b_I$). Let ID_I be a collection $\{id_{I1}, \ldots, id_{I,b_I}\}$ of the input DUs. The input DUs ID_I are processed by the subsequence P_I and an output DU d_I is obtained and sent to the parent node.

2.2   Power Consumption and Computation Models

Each fog node f_I consumes electric energy [J] to receive input DUs ID_I from the child nodes, calculate the output DU od_I, and send od_I to the parent node. As power consumption models, a pair of the SPC (Simple Power Consumption) [5–7] and MLPCM (Multi-Level Power Consumption) [10–12] models are proposed to show the electric power [W] to be consumed by a whole node to perform application processes at macro level. In this paper, the SPC model is taken because a small computer like Raspberry Pi3 [25] follows the SPC model. The power consumption NE_I [W] of a fog node f_I is maximum xE if some process is performed, otherwise minimum mE in the SPC model. For example, xE_I and mE_I of a Raspberry Pi3 node f_I [25] are 3.7 and 2.1 [W], respectively. A node f_I also consumes power RE_I and SE_I [W] to receive and send data, respectively. In this paper, $RE_I = re_I \cdot xE_I$ and $SE_I = se_I \cdot xE_I$ where $re_I (\leq 1)$ and $se_I (\leq 1)$ are constants. For a Raspberry Pi3 node f_I, $se_I = 0.68$ and $re_I = 0.73$ [20].

Next, we discuss the execution time $PT_h(x)$ [sec] of each process p_h ($t_I \leq h \leq l_I$) in a subsequence P_I to process an input DU of size x. In this paper, $PT_h(x)$ is assumed to be $cc_{Ih} \cdot x$, i.e. $O1$ type or $cc_{Ih} \cdot x^2$, i.e. $O2$ type where cc_{Ih} [sec/bit] is a constant. In this paper, cc_{Ih} and cc_{Ik} are assumed to be the same cc_I for every pair of processes p_h and p_k on each node f_I. The *computation ratio* cr_I of a node f_I is the ratio of computation speed of f_I to the root node f. If the same process is performed for the same input DU on f and f_I, the computation ratio cr_I is cc/cc_I (≤ 1). The computation ratio cr_I of the Raspberry Pi3 fog node f_I [25] is 0.18 for a HP DL360 server node f [4] as discussed [18]. The *computation rate* CR_I [bps] shows how many bits a node f_I processes for one second. CR_I is $CR \cdot cr_I$ (≤ 1) where CR is the computation rate of the root node f. As discussed in papers [9, 10, 12–14], the initial computation residue R_h [bit] of a process p_h is x and x^2 if p_h is $O1$ and $O2$ types, respectively, and the initial execution time T_h [sec] is zero. For each second, R_h is decremented by the computation rate CR_I and T_h is incremented by one. If $R_h \leq 0$, the process p_h terminates. Thus, T_h gives the execution time $PT_h(x)$ to calculate on the input DU of size x. The size $|od_h|$ of the output DU od_h of a process p_h is $pr_h \cdot x$ where pr_h is a *reduction ratio*. The total execution time $ET_I(x)$ of the node f_I is $PT_{l_I}(x) + PT_{l_I-1}(pr_{l_I} \cdot x) + \ldots + PT_{t_I}(pr_{l_I} \cdot \ldots \cdot pr_{t_I+1} \cdot x)$. The size of the output

DU od_I of the node f_I is $fr_I \cdot x$ where fr_I is the reduction ratio $pr_{l_I} \cdot \ldots \cdot pr_{t_I}$ of the node f_I.

In this paper, it takes time $RT_I(x)$ and $ST_I(x)$ [sec] to receive and send a DU of size x, respectively, i.e. $RT_I(x) = rc_I \cdot x$ and $ST_I(x) = sc_I \cdot x$ where rc_I and sc_I are constants which depend on the transmission rate of the network. For example, $sc_I/rc_I = 0.22$ for a Raspberry Pi3 node f_I with a 100 Mbps network.

The total execution time from f_I receives ID_I of size x until f_I sends od_I is $TT_I(x) = RT_I(x) + ET_I(x) + ST_I(fr_I \cdot x)$ [sec]. $TE_I(x) = RT_I(x) \cdot RE_I + ET_I(x) \cdot NE_t + ST_I(fr_I \cdot x) \cdot SE_I$ [J] is totally consumed by the node f_I.

3 The FTBFC Model

3.1 Operations on the FTBFC Model

Suppose a node f_{Ii} is heavily loaded, i.e. receives more number of input DUs than the computation rate and the length of the receipt queue RQ_{Ii} is getting longer. If a new node nf is connected as a child node of the node f_I and some of the input DUs are sent to the node nf, the node f_{Ii} consumes smaller energy because f_{Ii} calculates on fewer input DUs.

Let P be an process sequence $\langle p_1, \ldots, p_m \rangle$ to handle sensor data in the IoT. Each node f_{Ii} supports a subsequence $P_{Ii} = \langle p_{t_{Ii}}, p_{t_{Ii}+1}, \ldots, p_{l_{Ii}} \rangle$ of the process sequence P. In the FTBFC model, the following operations are considered to change the tree structure and do the process migration. Here, suppose a node f_{Ii} and a parent node f_I support subsequences $\langle p_{t_I}, p_{t_I+1}, \ldots, p_{l_I} \rangle$ and $\langle p_{t_{Ii}}, p_{t_{Ii}+1}, \ldots, p_{l_{Ii}} \rangle$, respectively, where $p_{l_I} = p_{t_{Ii}-1}$. Here, it is noted the top processes $p_{t_{Ii}}$ and $p_{t_{Ij}}$ are the same for every pair of child nodes p_{Ii} and p_{Ij}.

[Operations in the FTBFC model]

1. $MU(f_{Ii})$.
2. $MD(f_I)$ and $MD^*(f_I)$.
3. $nf = SPLT(f_{Ii})$ and $nf = SPLT^*(f_{Ii})$.
4. $nf = EXPD(f_I)$ and $\langle nf_1, nf_2 \rangle = GRW(f_I)$.

By performing the operation $MU(f_{Ii})$ on a node f_{Ii}, the top process $p_{t_{Ii}}$ of the node f_{Ii} migrates to the parent node f_I and the top process $p_{t_{Ij}}$ of every other child node f_{Ij} is removed if every child node f_{Ij} supports more than one process, i.e. $l_{Ij} - t_{Ij} > 1$. Otherwise, no process on p_{Ii} migrates to p_I. After performing $MU(f_{Ii})$, the parent node f_I and every child node f_{Ij} support subsequences $\langle p_{t_I}, \ldots, p_{l_I}, p_{t_{Ij}} \rangle$ and $\langle p_{t_{Ij}+1}, \ldots, p_{l_{Ij}} \rangle$, respectively, where $t_{Ij} = l_I + 1$. Processes on a node can migrate to the parent and child nodes in a live migration way of virtual machines [11–15].

By performing the operation $MD(f_I)$ on a node f_I, the tail process p_{l_I} of the node f_I migrates to every child node f_{Ii} if the parent node f_I supports more than one process, i.e. $l_I - t_I > 1$. Then, the nodes f_I and f_{Ii} support subsequences $\langle p_{t_I}, \ldots, p_{l_I-1} \rangle$ and $\langle p_{l_I}, p_{t_{Ii}}, \ldots, p_{l_{Ii}} \rangle$, respectively, where $l_I = t_{Ii} - 1$. Here, an ancestor node af of f_I is a *least ancestor* node supporting multiple processes, where af is a parent of f_I or no node in a path from a child node of af to f_I supports multiple processes. In the $MD^*(f_I)$ operation, if a parent node pf of f_I supports multiple processes, $MD(pf)$ is performed. Otherwise, $MD^*(pf)$ is performed on the parent node pf. Thus, the operation MD^* is recursively performed until a least ancestor node is found. An ancestor node cf of f_I in the child nodes makes the tail process migrate to the child nodes. Thus, a process migrates from the parent node to the node f_I.

By performing the operation $SPLT(f_{Ii})$ on a child node f_{Ii}, a new node nf is created as a child node of the node f_I. Then, the half of the child nodes of the node f_{Ii} are reconnected to the new node nf if the node f_{Ii} is not an edge node. Hence, the node f_{Ii} is required to have multiple child nodes, i.e. $b_{Ii} > 1$. If the node f_{Ii} is an edge node, one new device node is created as a child node of the node nf. In $SPLT^*(f_{Ii})$, if the node f_{Ii} has multiple child nodes, $SPLT(f_{Ii})$ is performed. If f_{Ii} has only one child node f_{Iij}, $SPLT^*(f_{Iij})$ is performed. Thus, descendant nodes of f_{Ii} are recursively split.

By performing the operation $EXPD(f_I)$ on a node f_I, a new node nf is created as a child node of the node f_I and every child node f_{Ii} of the node f_I is reconnected to the new node nf as a child node. Here, the node f_I has one child node nf which has the child nodes $f_{I1}, \ldots, f_{Ib_I}$ and supports no process. By using the $MD^*(nf)$ operations, a process migrates from the parent node f_I. In the operation $GRW(f_I)$, $nf = EXPD(f_I)$ and then $MD^*(nf)$ are performed and then $nnf = SPLT^*(nf)$ is performed. Here, the node f_I has two new child nodes nf and nnf and supports one process fewer processes.

3.2   Flexible Algorithm

Each node f_I is equipped with a receipt queue RQ_{Ii} for each child node f_{Ii} $(i = 1, \ldots . b_I)$. Let ql_I be the total queue length of the receipt queues of the node f_I. A node f_I whose total queue length ql_I is the longest is referred to as *most congested* in the FTBFC tree T. First, a most congested node f_I is found in the tree T. If the node f_I has multiple child nodes, the node f_I can be split to f_I and a new node nf by the operation $nf = SPLT(f_I)$. If the node f_I supports multiple processes, a new node nf can be created and a tail process of the node f_I can migrate to the node nf by the $MD(nf)$ operation.

First, suppose a root node f is most congested. Since the root node f cannot be split, the energy consumption of the node f can be reduced by making a process migrate to the child nodes. If the root node f has only a device node, a new child node nf is created as a child node of f in the operation $nf = EXPD(f)$ and the tail process of f migrates to the node nf by $MD(nf)$. In the operation $\langle nf, nnf \rangle = GRW(f)$, a pair of child nodes nf and nnf are created and the tail process migrates to the new child nodes nf and nnf.

Next, suppose an edge node f_I is most congested where the device node d_I of the node f_I supports nd_I devices. In one way, the edge node f_I is split to the node f_I and a new edge node nf, i.e. $nf = SPLT(f_I)$, each of which has one device node supporting $nd_I/2$ devices. In another way, a new child node nf can be created as a child node of the node f_I by $nf = EXPD(f_I)$. If the node f_I supports only one process, the node f_I obtains a process from an ancestor node by the $MD^*(f_I)$ operation. Then, the tail process of the node f_I migrates down to the new node nf.

Then, we consider a node f_I which is neither a root node nor an edge node in the tree T. There are a pair of ways, split and expand. If the node f_I has multiple child nodes, the node f_I can be split to a pair of the node f_I and a new node nf by $nf = SPLT(f_I)$. Here, the half of the child nodes are reconnected to the new node nf as presented in the preceding section. If the node f_I has only one child node f_{Ii}, the node f_I is split after the node f_{Ii} is recursively split by $SPLT^*(f_{Ii})$. In another way, a new node nf is created as a child node of the node f_I by $nf = EXPD(f_I)$ and then the tail process of the node f_I migrates to the node nf by $MD(f_I)$. In the operation $\langle nf, nnf \rangle = GRW(f_I)$, a node f_I can have pair of new chikd nodes nf and nnf.

A node f_I is changed in the following function $ChangeNode(f_I)$. Here, a variable $root$ denotes a root node of the tree T. The function $ChangeNode(f_I)$ returns a procedure $[proc]$ to change the tree T, by which the total energy consumption can be most reduced. $Exec(f)$ gives the energy consumption of a tree whose root node is f. In the $RollBack(proc)$, a tree rolls back to one before the procedure $proc$ is performed.

[ChangeNode (f_I)]

1. The node f_I is a root node:
   (a) The tail process of the node f_I migrates to every child node if the node f_I is not an edge node,
      [P1a] $MD(f_I)$;
      $E1 = Exec(root)$; $RollBack(P1a)$;
   (b) A new node nf is created as a child node of the node f_I and the tail process migrates to nf.
      [P1b] $nf = Expand(f_I)$; $MD(f_I)$;
      $E2 = Exec(root)$; $RollBack(P1b)$;
   (c) A pair of new nodes nf_1 and nf_2 are created as child nodes of f_I are created and the tail process of f_I migrates to nf_1 and nf_2.
      [P1c] $\langle nf_1, nf_2 \rangle = GRW(f_I)$;
      $E3 = Exec(root)$; $RollBack(P1c)$;
   (d) If $E1$ is smaller than $E2$ and $E3$, return($E1$, P1a);
      If $E2$ is smaller than $E1$ and $E3$, return($E2$, P1b);
      If $E3$ is smaller than $E1$ and $E2$, return($E3$, P1c);
2. The node f_I is an edge node which has a device node d_I:
   (a) The node f_I is split to f_I and a new node nf.
      [P2a] $nf = Split(f_I)$;
      $E1 = Exec(root)$; $RollBack(P2a)$;
   (b) A new node nf is created as a child node of f_I and the tail process of f_i migrates to nf.
      [P2b] If f_I supports one process, $MD^*(f_I)$; $nf = EXPD(f_I)$; $MD(nf)$;
      $E2 = Exec(root)$; $RollBack(P2b)$;

(c) A pair of new child nodes nf_1 and nf_2 are created as a child node of f_I and the tail process of f_I migrates to nf_1 and nf_2.

[P2c] $\langle nf_1, nf_2 \rangle = GRW(f_I)$;

$E3 = Exec(root);$ $RollBack(P2c);$

(d) If $E1$ is smaller than $E2$ and $E3$, return($E1$, P2a);

If $E2$ is smaller than $E1$ and $E3$, return($E2$, P2b);

If $E3$ is smaller than $E1$ and $E1$, return($E3$, P2c);

3. The node f_I is a non-root, non-edge node.

(a) [P3a] $nf = SPLT^*(f_I)$;

$E2 = Exec(root);$ $RollBack(P3a);$

(b) [P3b] $MD^*(f_I);$ $nf = EXPD(f_I);$ $MD(f_I);$

$E2 = Exec(root);$ $RollBack(P3b);$

(c) [P3c] $MD^*(f_I);$ $nf = GRW(f_I);$

$E3 = Exec(root);$ $RollBack(P3c);$

(d) If $E1$ is smaller than $E2$ and $E3$, return($E1$, P3a);

If $E2$ is smaller than $E1$ and $E3$, return($E2$, P3b);

If $E3$ is smaller than $E1$ and $E1$, return($E3$, P3c);

We implement an FTBFC simulator $SimFTBFC$ to estimate the energy consumption [J] of an FTBFC tree T. A variable c_time denote current time which is incremented by one for each time unit [sec].. If a DU du is sent by a device node at time $time$, the field $du.stm$ of du is $time$. A node f_I starts calculating an output DU od_I from the input DUs $ID_I = \{id_{I1}, \ldots, id_{Ib_I}\}$ only if f_I receives an input DU id_{Ii} from every child node f_{Ii} where time $id_{Ii}.stm$ is the same $time$. Here, $ID_I.stm = time$ and $od_I.stm = ID_I.stm$. The transmission time tm of a DU du is $|du|/tr$ where tr is the transmission rate [bps]. In the evaluation, tr is assumed to be 200 [Kbps]. If a child node f_{Ii} sends an output DU od_{Ii} at time $time$, the parent node f_I can start calculating an output DU on the input DUs at time $time + tm$ where $tm = |od_{Ii}|/tr$. Let $od_{Ii}.cstm$ show time $time$ when the node f_{Ii} sends od_{Ii}. Suppose an input DU id_{Ii} arrives at a node f_I. If each node f_I is active, i.e. f_I is calculating preceding input DUs or does not yet receive an input DU from some child node, the input DU id_{Ii} is stored in the receipt queue RQ_{Ii} of the node f_I. In the receipt queue RQ_{Ii}, the input DU id_{Ii} is sorted in the time $id_{Ii}.stm$. If a top input DU id_{Ii} in every receipt queue RQ_{Ii} has the same time $id_{Ii}.stm$ and $id_{Ii}.cstm \geq c\_time$, id_{Ii} is dequeued from the receipt queue RQ_{Ii} and then f_I starts calculation an output DU od_I on the input DUs $ID_I(= \{id_{I1}, \ldots, id_{Ib_I}\})$.

The simulator $SimFTBFC$ is implemented in the following procedures. Here, $f.xE$, $f.mE$, $f.CR$, $f.R$, and $f.fr$ denote xE, mE, computation rate CR, computation residue R, and reduction ratio fr of a node f, respectively. c_time shows current time. In the procedure $Exex(f)$, every node in a tree whose root node is f is executed for one time unit [sec] and the energy to be consumed by the tree is obtained.

$Exec\ (f)\ \{$

$E = PerformNode\ (f);$

if f is not a device node,

for each child node cf of f, $\quad E = E + PerformNode(cf)$;
return (E); };

For each node f in the tree T, the procedure $PerformNode(f)$ is executed and the energy to be consumed by the node f for one time unit [sec] is obtained. A device node sends a DU to the parent edge node every ist time units.

$PerformNode(f)$ {
$E = mE$;
if $c\_time \% its = 0$ and f is a device node, **send** a DU to a parent node;
else
if f is not a device node, /* root or fog node */
if f is idle, {
if the receipt queue is not empty,
take input DUs ID of size x from the receipt queue; $f.R = x$; }
else { /* f is active */
$f.R = f.R - f.CR$; /* residue is decremented */
$E = f.xE$; $\qquad$ * energy consumption */
if $f.R \le 0$, { /* input data is processed */
send an output DU od whose size is $x \cdot f.fr$ to the parent node;
f gets $idle$; };
}; /* if end */
return (E); };

Every node in the tree T is periodically checked each time every device node sends ten DUs. If the total energy consumption of the tree T can be reduced by changing a most congested node f_I in a procedure $Proc$, f_I is changed in $Proc$. $SimFTBFC$ returns the total energy consumption TE.

$SimFTBFC$(T) {
$c\_time = 1$; /* initial time */ $TE = 0$;
while(some node has input DUs in the receipt queue) {
if $c\_time \% 10 \cdot ist = 0$, { /* every st transmission of DU */
f_I is a most congested node in the tree T;
$TE = TE + Exec(root)$;
$\langle NE, Proc \rangle = ChangeNode(f_I)$;
if $NE < TE$, the tree is changed by the procedue $Proc$;
$c\_time = c\_time + 1$;
}; /* if end */
}; /* while end */
reurn(TE); };

4 Evaluation

We consider an application process P of five $pO1$ rocesses $p_1, \ldots, p_5$ as shown in Table 1. Here, each element $(ptype_i, pr_i)$ denotes a process type $ptype_i (\in \{O1, O2\})$ and reduction ratio pr_i of each process p_i. I

Table 1. Processes

Process	p_1	p_2	p_3	p_4	p_5
P	$(O1, 1.0)$	$(O1, 0.8)$	$(O1, 0.6)$	$(O1, 0.4)$	$(O1, 0.2)$

The computation rate CR of the root node f is 800 [Kbps] in the evaluation. The computation rate CR_I of a Raspberry pi3 fog node f_I is $0.185 \cdot CR$, which is obtained through our experiment [18,20]. The power xE and mE of the root node f are 301.3 and 126.1 [W], respectively. The power xE and mE of the fog node f_I are 3.7 and 2.1 [W], respectively.

Each device node d_I sends a DU ID_I to the edge node f_I every ist [sec]. In the evaluation, the inter-sensing time ist is ten [sec]. Even if the number of device nodes increases by changing the tree, the total size tsd of DUs sent by the device nodes is not changed while the size of a DU sent by each device node is changed. In the evaluation, tsd is 8,000 [Kbit]. In this paper, each node is assumed not to consume energy to send and receive DUs.

In the simulation, we evaluate the FTBFC tree in terms of the total energy consumption TE [W sec] of the nodes. First, there are a root node f_1 and a device node d_2. The device node d_2 supports all the nd devices in the system. Here, the device node d_2 sends a DU of size tsd to the root node f_1 every ist [sec]. If the queue length ql_1 of the node f_1 is larger than zero after the device node d_1 sends ten DUs, i.e. $10 \cdot ist$ [sec], the root node f_1 is changed by the function $ChangeNode(f)$. By $ChangeNode(f_1)$, a pair of child nodes f_3 and f_4 are created for the root node f_1 by the operation $GRW(f_1)$, where the device node d_2 is connected to the node f_3 and a new device node d_5 is connected to the node f_9. The number of devices supported by each device node is $nd/2$. The devices d_2 and d_5 send DUs of size $tsd/2$ to the nodes f_3 and f_4, respectively. Then, the tree T is checked every $10 \cdot ist$ [sec] and a most congested node f_I is found. The node f is changed by $ChangeNode(f)$ if the total energy consumption TE can be reduced. This procedure is iterated until the total energy consumption cannot be reduced.

Table 2. Total energy consumption

steps	0	1	2	3	4	5	6	7	8	9	10
TE [KWsec]	29,828	109	105	89	83	79	74	73	57	53	50

Table 2 shows the total energy consumption TE of the FTBFC tree T. One step means the tree T is changed by $ChangeNode(f)$ for a most congested node f. Step 0 shows a CC mode where a tree T is composed of one root node f_1 and a device node d_2. At step 1, the root node f_1 has a pair of child edge nodes f_3 and f_4 in the tree T as presented. As shown in Table 2, the energy consumption of the tree can be reduced by changing the tree T.

5   Concluding Remarks

We have to reduce the energy consumption of the IoT. In order to adapt the IoT to the traffic change, we newly proposed the FTBFC algorithm to change the tree structure of nodes and make processes migrate to nodes in the FTBFC model to reduce the energy consumption of nodes. In the evaluation, we showed the total energy consumption of the nodes can be reduced by changing the tree in the FTBFC model than the CC model.

Acknowledgment. This work is supported by Japan Society for the Promotion of Science (JSPS) KAKENHI Grant Number 22K12018.

References

1. Dayarathna, M., Wen, Y., Fan, R.: Data center energy consumption modeling: a survey. IEEE Commun. Surv. Tutorials **18**(1), 732–787 (2016)
2. Qian, L., Luo, Z., Du, Y., Guo, L.: Cloud computing: an overview. In: Proceedings of the 1st International Conference on Cloud Computing, pp. 626-631, (2009)
3. Rahmani, A.M., Liljeberg, P., Preden, J.-S., Jantsch, A.: Fog Computing in the Internet of Things, 1st edn., p. 172. Springer, Cham (2018)
4. HPE: HP server DL360 Gen 9. https://www.techbuyer.com/cto/servers/hpe-proliant-dl360-gen9-server
5. Enokido, T., Aikebaier, A., Takizawa, M.: Process allocation algorithms for saving power consumption in peer-to-peer systems. IEEE Trans. Ind. Electron. **58**(6), 2097–2105 (2011)
6. Enokido, T., Aikebaier, A., Takizawa, M.: A model for reducing power consumption in peer-to-peer systems. IEEE Syst. J. **4**(2), 221–229 (2010)
7. Enokido, T., Aikebaier, A., Takizawa, M.: An extended simple power consumption model for selecting a server to perform computation type processes in digital ecosystems. IEEE Trans. Ind. Inform. **10**(2), 1627–1636 (2014)
8. Enokido, T., Takizawa, M.: Integrated power consumption model for distributed systems. IEEE Trans. Ind. Electron. **60**(2), 824–836 (2013)
9. Kataoka, H., Duolikun, D., Sawada, A., Enokido, T., Takizawa, M.: Energy-aware server selection algorithms in a scalable cluster. In: Proceedings of the 30th International Conference on Advanced Information Networking and Applications, pp. 565-572 (2016)
10. Kataoka, H., Nakamura, S., Duolikun, D., Enokido, T., Takizawa, M.: Multi-level power consumption model and energy-aware server selection algorithm. Int. J. Grid Util. Comput. **8**(3), 201–210 (2017)
11. Duolikun, D., Enokido, T., Takizawa, M.: Energy-efficient dynamic clusters of servers. In: Proceedings of the 8th International Conference on Broadband and Wireless Computing, Communication and Applications, pp.253-260 (2013)
12. Duolikun, D., Enokido, T., Takizawa, M.: Static and dynamic group migration algorithms of virtual machines to reduce energy consumption of a server cluster. In: Nguyen, N.T., Kowalczyk, R., Xhafa, F. (eds.) Transactions on Computational Collective Intelligence XXXIII. LNCS, vol. 11610, pp. 144–166. Springer, Heidelberg (2019). https://doi.org/10.1007/978-3-662-59540-4_8

13. Duolikun, D., Enokido, T., Takizawa, M.: Simple algorithms for selecting an energy-efficient server in a cluster of servers. Int. J. Commun. Netw. Distrib. Syst. **21**(1), 1–25 (2018)

14. Duolikun, D., Enokido, T., Barolli, L., Takizawa, M.: A monotonically increasing (MI) algorithm to estimate energy consumption and execution time of processes on a server. In: Barolli, L., Chen, H.-C., Enokido, T. (eds.) NBiS 2021. LNNS, vol. 313, pp. 1–12. Springer, Cham (2022). https://doi.org/10.1007/978-3-030-84913-9_1

15. Duolikun, D., Nakamura, S., Enokido, T., Takizawa, M. : Energy-Consumption Evaluation of the Tree-Based Fog Computing (TBFC) Model. In: Barolli, L. (ed.) Advances on Broad-Band Wireless Computing, Communication and Applications. BWCCA 2022. Lecture Notes in Networks and Systems, vol. 570, pp. 66-77 . Springer, Cham (2022). https://doi.org/10.1007/978-3-031-20029-8_7

16. Duolikun, D., Enokido, T., Barolli, L., Takizawa, M.: A Flexible Fog Computing (FTBFC) Model to Reduce Energy Consumption of the IoT. In: Proceedings of the 10th International Conference on Emerging Internet, Data and Web Technologies (2022)

17. Mukae, K., Saito, T., Nakamura, S., Enokido, T., Takizawa, M.: Design and implementing of the dynamic tree-based fog computing (DTBFC) model to realize the energy-efficient IoT. In: Barolli, L., Natwichai, J., Enokido, T. (eds.) EIDWT 2021. LNDECT, vol. 65, pp. 71–81. Springer, Cham (2021). https://doi.org/10.1007/978-3-030-70639-5_7

18. Oma, R., Nakamura, S., Duolikun, D., Enokido, T., Takizawa, M.: An energy-efficient model for fog computing in the internet of things (IoT). Internet Tings **1–2**, 14–26 (2018)

19. Oma, R., Nakamura, S., Enokido, T., Takizawa, M.: A tree-based model of energy-efficient fog computing systems in IoT. In: Barolli, L., Javaid, N., Ikeda, M., Takizawa, M. (eds.) CISIS 2018. AISC, vol. 772, pp. 991–1001. Springer, Cham (2019). https://doi.org/10.1007/978-3-319-93659-8_92

20. Oma, R., Nakamura, S., Duolikun, D., Enokido, T., Takizawa, M.: Evaluation of an energy-efficient tree-based model of fog computing. In: Barolli, L., Kryvinska, N., Enokido, T., Takizawa, M. (eds.) NBiS 2018. LNDECT, vol. 22, pp. 99–109. Springer, Cham (2019). https://doi.org/10.1007/978-3-319-98530-5_9

21. Oma, R., Nakamura, S., Duolikun, D., Enokido, T., Takizawa, M.: A fault-tolerant tree-based fog computing model. Int. J. Web Grid Serv. **15**(3), 219–239 (2019)

22. Oma, R., Nakamura, S., Duolikun, D., Enokido, T., Takizawa, M.: Energy-efficient recovery algorithm in the fault-tolerant tree-based fog computing (FTBFC) model. In: Barolli, L., Takizawa, M., Xhafa, F., Enokido, T. (eds.) AINA 2019. AISC, vol. 926, pp. 132–143. Springer, Cham (2020). https://doi.org/10.1007/978-3-030-15032-7_11

23. Oma, R., Nakamura, S., Enokido, T., Takizawa, M.: A dynamic tree-based fog computing (DTBFC) model for the energy-efficient IoT. In: Barolli, L., Okada, Y., Amato, F. (eds.) EIDWT 2020. LNDECT, vol. 47, pp. 24–34. Springer, Cham (2020). https://doi.org/10.1007/978-3-030-39746-3_4

24. Guo, Y., Saito, T., Oma, R., Nakamura, S., Enokido, T., Takizawa, M.: Distributed approach to fog computing with auction method. In: Barolli, L., Amato, F., Moscato, F., Enokido, T., Takizawa, M. (eds.) AINA 2020. AISC, vol. 1151, pp. 268–275. Springer, Cham (2020). https://doi.org/10.1007/978-3-030-44041-1_25

25. Raspberry pi 3 model b. https://www.raspberrypi.org/products/raspberry-pi-3-model-b (2016)

SANKMO: An Approach for Ingestion, Processing, Storing, and Sharing IoT Data in Near Real-Time

Agmar A. Torres[(⊠)] and Flávio de Oliveira Silva

Faculty of Computing, Federal University of Uberlândia, Uberlândia 38400-902, Brazil
{agmar.torres,flavio}@ufu.br

Abstract. The evolution and use of the Internet of Things created a context where billions of devices are constantly producing data daily. This high data throughput offers different challenges for consuming, processing, and using the data on time. In this sense, this work presents the SANKMO approach, capable of handling a high volume of data to collect, process, and store data in near real-time from applications or IoT devices. Using the SANKMO approach, we did an experimental evaluation using its implementation view. We conducted experiments focusing on metrics such as near real-time, fault tolerance, latency, throughput, and data output. Our qualitative analysis of SANKMO with the related work shows that SAKMO satisfies these selected metrics compared to the literature.

1 Introduction

In recent years, a significant spread of the Internet of Things (IoT) devices has been detected [1]. A Statista study reveals that the number will become 75.44 billion worldwide by 2025 [2]. In addition, by 2030, more than 125 billion smart devices will be interconnected, creating a massive network of intelligent devices, cars, gadgets, and tools [2].

According to [1], in the current context, technology plays a fundamental role in developing solutions that involve manual and repetitive work. There are many initiatives and applications aimed at large domains such as smart cities, industry 4.0, health, and agro that rely on computer systems to manage, finance, and obtain sensor data to help increase the efficiency of processes, in which in many cases, demand the functioning of platforms and infrastructure to manage, manipulate and organize large volumes and flows of data [2].

With increasingly advanced technology and initiatives to make the world more and more highly connected, the number of connections between devices has increased exponentially [3]. New technologies and platforms have been developed to support data from different domains, approaches, and needs, increasing data transmitted over the network. In addition, new applications need to be designed to support high amounts of data and process them on time [3].

L. Barolli (Ed.): AINA 2023, LNNS 654, pp. 279–291, 2023.
https://doi.org/10.1007/978-3-031-28451-9_25

In this sense, storing all these data streams is very difficult in terms of storage and processing capacity, requiring a high degree of ingestion, transformation, and availability. Another point presented as a challenge is how to improve storage logic so that information can be used in the best possible way [1]. According to [4], as the speed and amount of data increase, there is a limitation in processing, storing, and filtering on a single server, causing problems such as delay and, in the worst cases, loss of information.

Based on the environment and the capture of data, aiming at increasing the monitoring, management, and efficiency of the process, data has been the primary raw material for companies and institutions [5]. With a wide variety of sensors and data collected, companies have taken the initiative to use them to assist in the decision-making process so that, increasingly, operations and techniques have been changed or reinforced based on insights. However, this new way of using data brings numerous challenges, such as reusing the interconnection of sensors, technologies, storage, and ways of consuming real-time data in a specific domain.

In this sense, the main objective of this work is to present an approach that allows the integration of different applications and data producers in a single repository. In this work, we developed an approach called SANKMO using the three-layer model to allow the reuse and consumption of data close to real-time, capturing, processing, storing, and making them available.

One of the main contributions of this approach is that one can centralize all the IoT resources of a domain [6,7]. Once data is centralized, data sent to other environments is prevented from being redundant or dispersed, simplifying and optimizing computational resources. The domain where we applied the tests was in a fog computing environment and using other types of input data.

This work is structured as follows: Sect. 2 presents the theoretical foundation of the work. Section 3 presents the SANKMO architecture for data ingestion, processing, and consumption and describes an implementation view of SANKMO. Section 4 describes the experimental evaluation and a discussion about SANKMO when compared to the literature. Finally, we give some concluding remarks in Sect. 5.

2   Related Work

In this section, we will present a discussion about the literature related to this work. We conducted this study from three perspectives: (i) data consumption and collection, analyzing approaches regarding data acquisition (ii) data processing, from the perspective of the features associated about how data is processed, and (iii) availability and storage of data, considering also how these frameworks output the data for further use from other systems. We did a qualitative analysis considering a set of metrics relevant for this area.

Real-time indicates the processing speed matters, but a processing time before having the final data is acceptable [8,9]. In this work, we will refer also as near real-time. Real-time, or near real-time, indicates that data is available respecting the time requirements of the consumer. Fault tolerance is a property that allows systems to continue to perform operations after failures or errors [1]. Latency is the time to process data considering its entrance into the system and the instant where it is ready for output [10]. Output considers the available interfaces for further consumption from others systems [11].

The paper of [10], created a application focus services in three categories: 1) data processing; 2) data storage; and 3) publication of data. In this sense, this paper contributes by providing the interfaces for consumption, processing and transformation of static data, not working with frameworks capable of handling failures and also not being necessary to measure latency during data integration to this platform.

In another way, the paper of [12] proposed a framework called SoFA, focusing real-time data processing and data extraction or availability was resolved through the development of a Fog Computing architecture using and leveraging Spark features to provide greater system utilization, energy efficiency and scalability within the IoT paradigm. The SoFA framework layers provide a distributed processing paradigm with the help of Spark to utilize the full processing power of all available high-end devices, leading to increased energy efficiency and system utilization.

The [13], said that new paradigms and technological evolutions require different approaches and functionalities to deal with the problems that innovations bring. Thus, in they paper was proprosed a application called SIGHTED aimed at processing heterogeneous IoT data. In SIGHTED work, the use and integration of consumption, processing, storage and data availability layers is a necessity, in which each layer has roles and responsibilities. It presents in the work of [13] different layers, such as: acquisition layer, semantic annotation and ontologies layer, storage layer, data publication and data consumption. In the acquisition and transformation layers, the work did not propose to use external interfaces so as not to deal with latency [14].

The study carried out by [15], presents an architectural design using Apache Spark as the main framework to process health data on a large scale DSDS (Design Distributed System based on Spark). In this sense, it is able to work with data near real-time in addition to offering scalability. However, the data output is called open since the consumption is done by an external application, all data is created and maintained within the system, being necessary to create everything manually. The article outlined key considerations throughout the system design process, including comparing different components, finding the optimal data flow mechanics for tasks with high processing and storage demands.

Table 1. Features found in articles regarding SANKMO

Reference	Atributes				
	Near Real-Time	Fault Tolerance	Latency	Throughput	Output
[16]	v	–	v	v	v
[7]	–	–	v	v	–
[6]	v	v	v	v	–
[10]	–	–	–	–	v
[12]	v	–	–	v	v
[13]	v	–	–	v	v
[14]	–	v	–	v	–
[15]	v	v	–	v	–
[11]	v	–	–	v	–
[9]	–	v	v	–	–
[3]	–	v	v	–	–
SANKMO	v	v	v	v	v

In the work of [11], a Distributed Uniform Stream (DUST) Framework was proposed, which implements an elastic streaming platform. It optimizes resource allocation in a modern IoT environment. The main component and core of the framework is called DUST Core, it allows different application components to transmit events to each other. This results in a system of application components that can flexibly adapt to changes in event streaming rates. This mechanism aims to increase scalability between components that produce data and also among data consumers, thus addressing both the issue of near real-time, latency and fault tolerance, but leaving as a gap or improvements new connections with presentation components or availability data.

3   SANKMO

This section presents the specification of the conceptual, architectural and implementation proposal to deal with data from different scenarios applied to efficient resource management. For this, details of the components, mechanisms, resources and their functionalities are discussed. In the current scenario, there are different types of approaches to ingest, receive and process data. In addition, different applications use approaches to consume data from specific contexts and, however, not all data are processed in a way that simplifies its subsequent reuse as cited in papers by [11,13].

3.1 Architectural Vision

In this section, we present the conceptual model of SANKMO's architecture of the SANKMO. In this sense, Fig. 1 briefly illustrates its three main layers.

The *Production Layer* concerns data ingestion from different sources and provides connectors to receive data from files, other applications, or frameworks. The *Transformation Layer* is responsible for the lifecycle of the information captured in the data production layer and for applying different operations over the data. The *Consumption* layer controls and manages resources and provides the APIs to share the data with other applications.

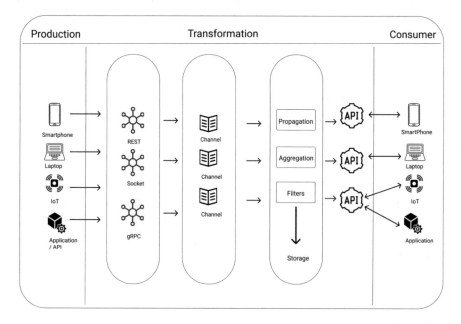

Fig. 1. Architectural vision about layers

The production layer in this scenario offers a perspective of receiving data from a given domain in a flexible way [11]. However, the data input abstraction and data format are flexible. In addition, this layer can be expanded or present different components to scale the operation and not allow the framework not to crash if the data supply is more than consumed. Another scenario is that this layer can be used more tightly where devices can be registered before use. Another point that this layer has as a possible solution, and that instead of receiving the event, it can directly consume data and values from other resources to complement the information if necessary.

In this sense, the IoT data was previously registered on the platform to know the data source and what data will be transmitted. Once the data mapping exists, Apache Nifi provides input data through processors and interface adapters. Other applications, such as SIGHTED [13], use Apache Storm to integrate and offer data entry services. Nevertheless, based on the tests carried out in our local environment, Apache Nifi was used because it offers different interfaces, modules, and services, such as real-time data traffic. In this sense, we choose Apache Nifi in our SANKMO implementation.

The transformation layer is a layer dedicated to data processing. It is a layer capable of performing complex operations according to the needs and business rules. This layer communicates directly through a predefined contract. It has features aimed at scalability, rigidity, and speed to process information to be stored or made available. This way, this layer is decoupled from the others. It offers the possibility of integration with other auxiliary layers, if necessary, to compose and reproduce the data from arrival to output. One of the primary goals of this layer is to offer information integrity, ensuring that the incoming and outgoing data has been changed correctly and that no information is lost.

The consumption layer is related to data availability. In this sense, the main objective of this layer is to make all the information available so that other applications can consume it and offer integrations to other applications. The layer supports numerous strategies to make the data available such as events, files, and APIs. Thus, any system can easily consume the data processed by the transformation layer.

3.2   Implementation View

This section presents a SANKMO implementation. The goal was to implement SANKMO architectural vision using freely available technologies to create the three layers' functionality. Figure 2 visually presents all the used technologies.

3.2.1   Data Ingestion

When several applications and IoT devices produce data, it is possible to use different approaches to capture the information, such as JSON, XML, and Sockets data types. SANKMO has a driver to search for data at every moment in a

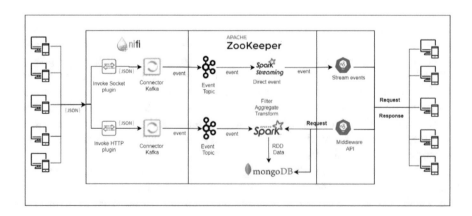

Fig. 2. Layer implementation diagram

given time. In this work, as shown in Fig. 1, we used Apache Nifi to consume the information through the HTTP GET method.

The Apache Nifi can be used to consume streaming of data in online transactions. According to [15], it can be considered as continuous streaming of data. The large amount of streaming transactions flow at high speed over the network and is processed quickly when data transactions arrive. The streaming processing based on a big data framework can handle high-speed real-time data. In another way, the streaming analysis is found in various industries, such as, intrusion detection and data generated by Internet of Things (IoT) sensors.

Apache Nifi offers different resources to consume through connection and communication interfaces. It is possible to structure different *endpoints* simultaneously and through channels created with the Kafka framework, which provides redundancy functions to synchronize and structure all data through topics. Each topic can consume a type of information, and this separation helps to organize the formation of data used in Spark in the data transformation stage.

According to [1], Kafka can store streaming events durably and reliably for a predefined time. In this sense, it can provide functionalities and availability of information in a distributed, highly scalable, elastic, fault-tolerant, and secure manner where each instance can have more than one topic.

3.2.2   Data Processing and Transformation

The Spark context keeps looking at the Kafka topic data: it is necessary to type the data input and define contexts and the amount of cluster needed to capture the Kafka information. Each mini-application corresponds to a driver to be processed, including the functions needed to make the transformations and receive multiple topics.

At this stage, Spark receives, processes, and structures the data using several operations. It is possible to create several different data transformation functions using Java. In this sense, functionalities are developed and configured to capture the input data and the distribution environment. In this layer, it is possible to use the output data from one stage as the input to the next one, thus creating a chain of functionalities according to the needs of each application.

This layer filters the data to discard the irrelevant part from the point of view of one specific processing purpose. The aggregation decreases the number of tuples handled in the analytic layer. In general, this component requires stateful stream processing. The data join that it is possible to connect different data entries or topics with static content or through data frames loaded from files. Creating the storage layer is concerned with persisting the data and processing the results in the data store. The filtering, aggregation, and analysis steps involve data from the current layer. Moreover, the storage layer is concerned with persisting the data and processing the results in the data store. Furthermore, the data from the current layer is involved in the filtering, aggregation, and analysis steps.

3.2.3   Data Consumption

After the connection, ingestion, and processing of the data, it is crucial to provide a layer of presentation or availability to the data. In this sense, there are different types of approaches used in the literature, such as the use of sockets, application programming interface (API) through Representational State Transfer (REST) using the HTTP protocol, and frameworks that encapsulate all these processes as Content Management System (CMS). In this work, we used a CMS in Python using REST to meet the need for data consumption to manage and support the management of all information. In developing, services and URLs for consumption, filters, parameters, and specific routes can be applied to seek only the necessary information. These routes can be consumed by any device capable of connecting to the internet.

4   Results and Discussion

This section presents an experimental evaluation of SANKMO and compares it with related work using the metrics presented in Sect. 2.

4.1   Experimental Evaluation

To verify the SANKMO approach presented in Sect. 3.1, we implemented, deployed, and evaluated SANKMO experimentally according to the view detailed in Sect. 3.2. We used the technologies presented in Table 2 in this experimental evaluation.

Table 2. Technologies used to the Experimental Evaluation

Name	Version	Description
Apache Nifi	1.1	Used as an adaptation to receive external events
Kafka	2.11	Used for topic ingestion and management
Zookeeper	3.4.8	Used as server for Spark services
Apache Spark	3.0.3	Used for event processing
Java	11	Used to develop Spark functions
MongoDB	3.6.3	Used for storage
Flask	3	Used to create the data consumption API

4.2   Production Layer

To test the number of records collected in an external API, we used the API https://randomuser.me/api/0.8 for the experiment. Consumption lasted 10 min, and 13 GBs were obtained, focused on the user and product data sets. Based on

Table 3. Results obtained after consuming external API data

API	Input	Throughput	Time	Payload
/users	82.234	13 MB/s	600 s	6 GB
/products	45.903	9 MB/s	600 s	4 GB
/meet	30.041	5 MB/s	600 s	1.5 GB

Table 4. Results obtained after consuming internal API data

API	Input	Throughput	Time	Payload
/users	147.021	40 MB/s	600 s	12 GB
/products	103.687	25 MB/s	480 s	6 GB

this data, it is possible to calculate the throughput when simulated in an API outside the domain and consuming data in a real or production environment available on the internet. All information gathered is listed in Table 3.

In another scenario to consume data, we developed a local API to ignore the network's latency and timeout and thus test data consumption in a single node to estimate the amount of processing supported by the Invoke service HTTP. The consumption lasted 10 min, and 12 GB of data was obtained in routes related to user domains and 6 GB related to the product's domain. Based on this data, it is possible to calculate the throughput. All information gathered is listed in Table 4 based on data consumption in the local environment.

There is a big difference between data sent over the network or internet and data that does not have to wait or use external services. The local API had a better performance in the speed of consumption since it managed to capture 147,021 records compared to the production environment, which managed to capture 82,234 records in 10 min. Another interesting point to be highlighted in this analysis is that, with SocketIO configuring the environment that receives public data in external environments, it is difficult to test accurately since it depends on real-time data in production.

One of the main mechanisms applied in this work was that development is dynamic. It aims to integrate and interact with different platforms with different methods and approaches, which is why it was separated into layers. However, each layer presents its own technological difficulties, complexity, and possibilities for improvement.

4.3 Data Transformation

It can be seen from this experiment that there is a computational gain when using more nodes, as is the case with six nodes in which 500,000 records were processed in less than 1 min. In this sense, a configuration with only two nodes or monolith would take more than 7 min to process this information. One point that has yet to be thoroughly tested is the memory cost for this type of environment or tool, as the processed data is not stored in RDDs, and in some cases, it is necessary to store it in local files to free up space in RAM. These settings can be programmed and passed as parameters to SANKMO, thus increasing performance for large data or file sizes. The performance gain of 6 nodes compared to 2 nodes with 50k input size and 500k records is a time difference of 40 s and can increase with the amount of input data.

Table 5. Processing based on the quantity

Data	2 node	4 node	6 node
50k	18 s	12 s	10 s
100k	30 s	23 s	20 s
200k	63 s	30 s	26 s
500k	100 s	60 s	56 s

Table 6. Results obtained after consuming internal API data

	1 batch	5 batch	10 batch
50k	40 s	8 s	5 s
100k	82 s	18 s	16 s
250k	226 s	40 s	34 s
500k	460 s	80 s	64 s

These experiments obtained elapsed time and the number of partitions between batch intervals. We followed some standards presented in [8]'s work to carry out these tests since the amount of data and storage can be changed according to the payload size and processing time. In this example, data batches or batches were used to store the data as a kind of buffer [16]. One of the main challenges was that the storage itself takes little time, but opening the connection based on every transfer and amount of logging can be time-consuming and thus performance slow, as illustrated in Table 6.

The processing time decreases slightly when the number of partitions is increased in three different batches. According to the results, the 50-s batch interval performs better than the 10-s batch interval. The experiment's 10-s, 30-s, and 50-s batch intervals work with less scheduling delay. The programming delay is 1 or 2 ms for each batch. However, the larger batch interval can cause a longer scheduling delay.

All works analyzed in this work have similarities with SANKMO or bring information regarding the proposed solution, with alternatives for modeling and construction of similar architectures. Table 1 contrasts the constructed solution with its peers. It contains efforts noted in the literature to approach applications capable of supporting the architecture of consumption, processing, and data availability.

4.4   Discussion

One of the main mechanisms applied in this work was that development is dynamic. It aims to integrate and interact with different platforms with different methods and approaches, which is why it was separated into layers. However, each layer presents its own technical difficulties, complexity, and possibilities for improvement. For example, the data production layer, which is the main reference for collection, with the evolution of frameworks and drivers capable of supporting high demands, this solution can become better with the development of new technologies, thus improving speed, ease and mitigating errors for the ingestion of large masses of data.

The development of new platforms with the SANKMO approach can be expanded according to the need, proposing improvements by layers instead of a system as a whole. New related technologies can be used, implemented, and tested in this sense. All the metrics used to test SANKMO are similar to some of the works presented in Table 1. We will make a parallel of this information, exploring the results obtained.

The work of [16] performed data processing by comparing three different processing frameworks with the number of nodes or threads. The values obtained in the tests he carried out were similar to those presented in Table 6 after converting time and the amount of data processed. In the processing layer, back to storage, the throughput obtained during the experiments by Spark proved to be better than Storm and Flink. However, this does not make it impossible to use these means of processing in SANKMO, where it would be enough to change the adaptations of the data production and consumption layer to process such information.

Like the work by [16], the work by [15] uses Spark to process information, but the approach of queuing messages and events is through Kafka, which makes them more resilient data. However, the proposed application needed to bring performance-oriented values and tests. However, it brought a flow and operation to treat data obtained through sensors and IoT devices in a distributed way.

Using other already validated frameworks and technologies mitigates the limitations of each layer and thus reduces the risk of bugs, errors, and security in the framework. Another characteristic that this work provides is the development cycles to be integrated by the community since the layers are independent but integrated for the common purpose of transiting and making available data from different sources to imply access and integration with other APIs and integrations.

5   Concluding Remarks

This work proposed and evaluated SANKMO. An approach capable of working with data in near real-time in IoT scenarios. For this, we proposed an architecture with three decoupled and independent layers. In addition, we integrated technologies and applications to create a platform capable of consuming, processing, storing, and making data available to other applications.

Our evaluation shows that it becomes efficient in the context of data streaming and fills some gaps when compared to other available solutions in the literature. The *Data Consumption Layer* is responsible for capturing or receiving data in different formats and with high demand, including data processing close to real-time. The *Data Processing Layer* is responsible for aggregating, filtering, and creating new types of information based on data captured at the consumer's home was implemented. Once the API is connected and receiving data, historical data storage mechanisms are created to store sensitive data that can be used as a source of new information to other consumers.

We plan to perform additional measurements in future work, using other technologies and improving the transformation layer by creating more features.

In addition, we could exploit this layer by adding AI to handle data more accurately and improve application performance.

Acknowledgements. The authors thanks the Brazilian Agency for Industrial Development (ABDI) for the partial support for this work inside the project TIED4.0.

References

1. Kim, Y., Son, S., Moon, Y.-S.: SPMgr: dynamic workflow manager for sampling and filtering data streams over apache storm. Int. J. Distrib. Sens. Netw. **15**(7), 1550147719862206 (2019)
2. Morfino, V., Rampone, S.: Towards near-real-time intrusion detection for IoT devices using supervised learning and apache spark. Electronics **9**(3), 444 (2020)
3. Wang, S., Zhong, Y., Wang, E.: An integrated GIS platform architecture for spatiotemporal big data. Futur. Gener. Comput. Syst. **94**, 160–172 (2019)
4. Greco, L., Ritrovato, P., Xhafa, F.: An edge-stream computing infrastructure for real-time analysis of wearable sensors data. Futur. Gener. Comput. Syst. **93**, 515–528 (2019)
5. Curry, E., Derguech, W., Hasan, S., Kouroupetroglou, C., Ul Hassan, U.: A real-time linked dataspace for the internet of things: enabling "pay-as-you-go" data management in smart environments. Futur. Gener. Comput. Syst. **90**, 405–422 (2019)
6. Lnenicka, M., Komarkova, J.: Developing a government enterprise architecture framework to support the requirements of big and open linked data with the use of cloud computing. Int. J. Inf. Manage. **46**, 124–141 (2019)
7. Kolajo, T., Daramola, O., Adebiyi, A.: Big data stream analysis: a systematic literature review. J. Big Data **6**(1), 47 (2019)
8. Tun, M.T., Nyaung, D.E., Phyu, M.P.: Performance evaluation of intrusion detection streaming transactions using apache Kafka and spark streaming. In: 2019 International Conference on Advanced Information Technologies (ICAIT), pp. 25–30. IEEE (2019)
9. Velasco, L., et al.: Monitoring and data analytics for optical networking: benefits, architectures, and use cases. IEEE Netw. **33**(6), 100–108 (2019)
10. Macêdo, J.J.d.: OpenEasier: a CKAN extension to enhance open data publication and management. Master's thesis, Brasil (2018)
11. Vanneste, S., et al.: Distributed uniform streaming framework: an elastic fog computing platform for event stream processing and platform transparency. Futur. Internet **11**(7), 158 (2019)
12. Maleki, N., Loni, M., Daneshtalab, M., Conti, M., Fotouhi, H.: Sofa: a spark-oriented fog architecture. In: IECON 2019-45th Annual Conference of the IEEE Industrial Electronics Society, vol. 1, pp. 2792–2799. IEEE (2019)
13. Nagib, A.M., Hamza, H.S.: SIGHTED: a framework for semantic integration of heterogeneous sensor data on the internet of things. In: ANT/SEIT, pp. 529–536 (2016)
14. Scolati, R., Fronza, I., El Ioini, N., Elgazazz, A.S.A., Pahl, C.: A containerized big data streaming architecture for edge cloud computing on clustered single-board devices. In: Closer, pp. 68–80 (2019)

15. Tu, Y., Lu, Y., Chen, G., Zhao, J., Yi, F.: Architecture design of distributed medical big data platform based on spark. In: 2019 IEEE 8th Joint International Information Technology and Artificial Intelligence Conference (ITAIC), pp. 682–685. IEEE (2019)
16. Karimov, J., Rabl, T., Katsifodimos, A., Samarev, R., Heiskanen, H., Markl, V.: Benchmarking distributed stream data processing systems. In: 2018 IEEE 34th International Conference on Data Engineering (ICDE), pp. 1507–1518. IEEE (2018)

A Polystore Proposed Environment Supported by an Edge-Fog Infrastructure

Ludmila Ribeiro Bôscaro Yung[✉], Victor Ströele,
and Mario Antônio Ribeiro Dantas

Federal Universidade of Juiz de Fora (UFJF), Juiz de Fora, Brazil
yung.ludmila@estudante.ufjf.br, {victor.stroele,mario.dantas}@ice.ufjf.br

Abstract. The grand computational challenge observed nowadays can be considered in how to conceive an environment to deal with billions of IoT devices. The major objective is how to get and process data from these devices, which have challenges such as access, manipulation, and consultation of this digital data in a more transparent fashion to final users. Therefore, the necessary utilization of a contemporaneous architecture should be conceived to support the relation between the edge, fog and cloud layers. Targeting to tackle this issue, we have a proposal which targets an initial integration of data. The proposal considers that applications utilize a Polystore environment which provides facilities to reduce time and processing costs, in comparison to ordinary cloud configurations.

1 Introduction

Some well-known steps on the Internet of Things (IoT) approach, such as data integration, pre-processing, and storage, are being considered on several systems as their major challenges.

According to [5] IoT applications has been increasingly used in various types of applications: healthcare, agriculture [6], city, smart homes and autonomous vehicles, for example.

Many of these challenges involve performance, energy, data received in real-time, latency, and computational cost [9], for example. In IoT environments, we have layers of action for each part of the system/architecture, such as cloud, fog, and edge, where each plays a different role.

There are several scenarios for using heterogeneous and distributed Databases (DB), which are growing increasingly, specially in blockchain [7], for example. With this, the need to manipulate data, perform queries, possibilities of inference of results, and research, becomes essential.

Some points are essential: access to DBs, search queries, manipulation, and operations on different types of DBs. For this, the Polystore architecture comes in as a differential that promotes gathering the DBs and enabling access at runtime, thus allowing the construction of several integrated and heterogeneous storage mechanisms [3].

L. Barolli (Ed.): AINA 2023, LNNS 654, pp. 292–302, 2023.
https://doi.org/10.1007/978-3-031-28451-9_26

Our general objective with this work is to promote the use of the Polystore architecture within the computational fog so that we can increasingly take this processing to the edges of the network.

There is the difficulty of processing the data, organization, integration, and storage in the cloud as a matter of time and cost. Therefore, it is interesting to consider using an architecture in which we consider the cloud, edge, and fog layers. To this end, we use an idea/paradigm of these layers. We will use a Polystore solution to facilitate these integrations within these layers without impacting this heavy use in the cloud.

Our specific objectives are: to validate the architecture, promote assistance to end users with good results, perform data integration and pre-processing in the fog layer, and reduce the Polystore processing cost; This leads us to some research questions:

- What impact can the proposed architecture be expected to have on a large scale?
- What are the steps to carry out the validation of the proposed architecture?
- How will the steps be done within each of the layers?

This research work is divided as follows: the second section deals with a brief literature review; the third section shows the proposed environment, and the last section includes the conclusions and future work.

2 Related Work

The interest of the scientific, civil, and academic community in the Internet of Things devices, Computational Layers (Cloud, Edge, and Fog Computing), and Databases, and their challenges, is growing [2]. After all, it is increasingly part of the reality of many users, whether they are people who want to use these devices or professionals in the health, engineering, and computing.

Therefore, it is clear that IoT devices and concepts related to them are through Smart Cities (and other associated classifications), Scaling, and Low Latency, among others, are recurring aspects that revolve around applications that involve Cloud, Edge, and Fog [4].

Furthermore, we also tend to work increasingly with heterogeneous Databases that play an essential role in storage, processing, delivery of queries, data manipulation, and pertinent analysis.

As a result, there is a need to gather in one place the possibility of working with heterogeneous databases, which can vary according to the type of each DB; in addition to the need to store the data and facilitate the use on these bases without harming performance so that access is facilitated. Operations are carried out more quickly at the edges of the network.

Polystore is not a Data Base Management System (DBMS) but instead has a distinct storage mechanism that promotes access to data through its query means, that is, through a query [1].

The main thing about [8] is to work in an environment so that the sources of lead time, delay, and cost are executed to meet user requirements that minimize lead times. Because it has a time-sensitive focus, as applications cannot be processed in just one cloud, it is necessary to prioritize the use of Fog.

According to [8], IoT-based systems require Fog and Cloud Computing. The advanced sensors and tools that enhanced the efficiency and advantages of the system were observed in developed and developing countries. For example, [8] report many studies that work with the Fog and Cloud-based computing system and that in these studies was found "convenient by the patients due to its cost-effectiveness, reliability, and safety."

There are issues working with heterogeneous data from polysources that have been researched in the context of multi-database systems, Polystore systems, and data integration systems [9]. Some datasets have an enormous amount of data from various data sources, so there can be multiple languages to obtain access to these different datasets.

On the other hand, [10] brings comparisons of the main architectures used in Big Data Systems and one of these architectures is precisely the Polystore. In addition to explaining how Polystore works, [10] demonstrates how it works in a basic scenario of a Big Data System, including volume, speed, veracity, variety and variability.

In [11] there is a proposal to join an S-Store with the BigDawg Polystore. The authors proposed the use of the S-Store in order to satisfy some requirements of this system, such as: providing low latency, push-based processing seamlessly integrated with ACID data management. In addition to ensuring that all states are in stream state, windows, or relational tables, may only be accessed within the context of a transaction.

However, many works related to the context of this work discard the use of Polystore, not only because it is something new but also because of the difficulty of implementing a proposal capable of carrying out actions in an IoT environment working in Cloud and Fog efficiently and with end-user friendly search environment.

3   Proposed Environment

Given a hypothetical and generalized scenario of the use of IoT in the matter of Smart Cities, as pointed out, for example, in [4], we can verify that the devices that send data through the sensors to the other layers, these devices are located in the Edge, at the outermost Edge of the network. In Fog, there is the processing of what was sent by the devices.

In [12], the experiments were based on a model that allows you to capture the location of students and their body signals, so that the collected data can be analyzed and recommended activities within the university campus, as well as giving tips on the that students can do while they are at university. The model used is based on emotional contexts using IoT and Machine Learning.

Figure 1 represents an overview of our proposal in which the objective is to facilitate the use of DBs for both experienced and non-experienced users. This stage will be done through the data integration technology that will be performed in the Fog layer. To reach the Core layer, the data undergoes processing within the architecture.

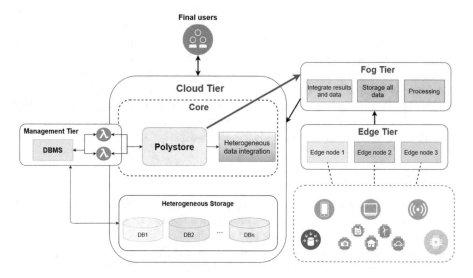

Fig. 1. The proposal three essential tiers: data storage, and access, processing, and edge nodes.

In the cloud, we have a great deal of processing that can last even days, as the load received is enormous, and it is the most external layer to the network. Therefore, accessing and working with the cloud is more expensive, which lacks classic IoT system operations and needs such as mobility, geographic distribution, low latency, and fast decision-making response. Therefore, it becomes unfeasible to work with Polystore within the cloud.

Data is sent through the most diverse types of devices, be they servers, computers, smart devices, whether bracelets, cell phones, or electronics that have technology aimed at using the Internet of Things. This data is allocated in the layer we call Edge Tier, where each node corresponds to a device that generated the collected data.

The proposed architecture allows end users to consult the databases by accessing the Core; the Cloud will require using Polystore and the data integration results of heterogeneous DBs.

In Fig. 1, the highlighted red arrow represents the use of the Polystore inside the fog. In theory, as seen in Sect. 2 of this work, the Cloud would run the Polystore. However, our proposal brings the collaboration of fog as the layer of execution, integration, and pre-processing of data.

The data are transferred to the Fog Tier, where the processes of integration of the results of the data collected and obtained from the Edge Tier take place. In this layer, there is the refinement where the critical stage of merging the Databases is carried out, thus allowing the integration of the bases.

Finally, the Cloud Tier is the one that will deliver the results of surveys carried out by end users. It contains the Polystore abstraction that will scan the results refined in the previous layers and deliver the search result to the user using a unique language. This process brings more transparency to the application, in addition to not forcing and overloading the Cloud that, here, has the role of delivering the results-leaving the other layers closer to the edge of the network to carry out the most expensive processes.

3.1   Development

The proposal implementation process began with the separation of each step so that the organization of the DBs was correct for the use of Polystore. The steps that were taken so far were: classes and code level.

In classes, we separate what is fundamental for the basic functioning of our proposal, that is, the implementation of Polystore and n DBs to bring the heterogeneity of data sources to the application.

As seen throughout the 3 section, at the code level the steps are made in specific steps for each essential point highlighted in the classes. For better visualization, the Fig. 2 represents what each of these steps within the application will perform.

For example purposes, we have separated in Fig. 2 the types of databases that we initially thought of working with. Note that, as we are dealing with databases from heterogeneous sources, this does not imply that we will work with different types of DBs.

During the process of elaborating our proposal, we noticed that some points are essential to be implemented. In general, what was presented in the Fig. 2 represents the basic scheme of communication between Polystore and DBs, this part is in the cloud within our proposal.

In addition to this information, we would like to highlight the following features shown in the Fig. 2:

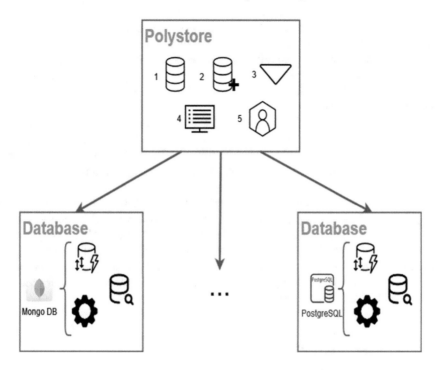

Fig. 2. The basic scheme of implementation of the Polystore communication with DBs

- Polystore:
  - In 1: Database Verification that checks among the registered DBs which are available;
  - In 2: Adding access to DBs that returns which DBs and their settings are available;
  - In 3: Merge data;
  - Em 4: Exhibition;
  - Em 5: Permission to access queries;
- Databases:
  - Initial configurations;
  - Access to the DB;
  - Search on DB;
  - Output formatting for Polystore;

It is worth mentioning that the process of formatting the data output is essential for the use of Polystore, as it allows a unification of the search queries within the application.

The data used in this initial experiment of the proposal were generated randomly and their pre-processing was made so that they could be used later by the application. This pre-processing step is important because we can have there a simulation of what would normally occur in a real environment. Thus, the data is located in the Edge layer as they are the closest to the edge of the network and are captured through the data collection made by the Edge node applications.

Our proposal aims to take advantage of working with Polystore in the cloud, since this access to the cloud, as seen previously in 1, becomes expensive if working entirely in the cloud, from data access to the final return. of searches and inferences from DBs.

Therefore, our objective is to use the implementation of our proposal to provide execution comparisons ranging from access to Polystore, DBs, data processing and the use of fog that would bring a gain in time and costs involved, whether running costs on the machine, as well as energy and money saving, for example.

The initial idea is to execute tests from the perspective of the end users of our proposal. That is, users who do not have access to the implemented layers, aiming only to return searches and queries to DBs. To this end, these comparisons mentioned above would be carried out without the addition of data integration, which would be one more step than Polystore would do, this integration would be due to the fog layer that would perform what Polystore already does, but faster , delivering faster results.

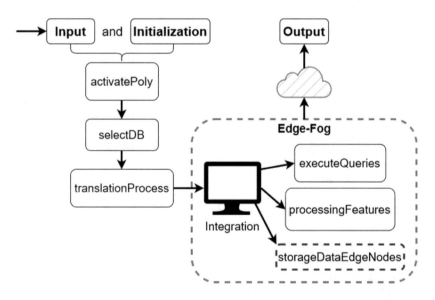

Fig. 3. An illustration of the algorithm which runs Polystore on Edge-Fog infrastructure

In the Fig. 3 we have the beginning of the execution in Input and Initialization that, respectively, represent the Polystore queries and a variable q that will be a future query that will be executed by a user. The application goes ahead and activates the Polystore in *activatePoly* which executes all the already pre-established configuration of the Polystore. Later, in *selectDB*, the selection of the DB that you want to carry out the procedure is carried out, which can be more than one. In *translationProcess* the translation of the query to the unified model is carried out.

After this initial execution process, the algorithm will perform the requested queries on the machine, as well as process the configurations defined in the previous steps, in addition to processing the data that arrived from the edge-nodes. Furthermore, it will store the data coming from the external sources of IoT devices. The end of processing will return to the cloud what was executed at the request of the user who will receive his application query.

3.2   Experimental Environment

In our initial experiment, we chose to configure the DBs, the Polystore environment and configure the initial data randomly, in order to validate these primary configurations. In Fig. 4 below there are the primary configuration that we choosed to our experiments. The primarily testbed, using diverse DBs, were those mentioned in the Fig. 2.

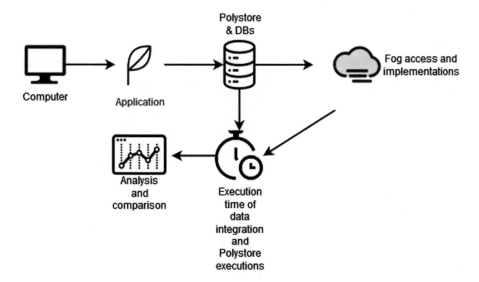

Fig. 4. Initial experiment plan for the main propose through the execution

The Fig. 4 illustrates a general idea behind the initial executing proposals. All experiments were performed using the same machine ordinary, configured with the following characteristics:

- AMD Ryzen Processor^{TM}5 1600, CPU @ $3.2GHz$;
- Asrock motherboard, $AB350M - HDV$;
- RAM $16GB(2*8GB)$, DDR4;

- Operational System Linux Ubuntu 20.04.1 LTS, Focal Fossa;
- Radeon™RX 570 Series (POLARIS10 DRM 3.35.0, 5.4.0–45-generic LLVM 10.0.0)
- IDE *Spyder, Python3*.

In Fig. 4 the idea of the process is to start with the user's request for the application, demonstrated by the image of a computer at the beginning of the experiment.

The application requests access to the already configured DBs, so the access user and password are already configured and requested in this access. As well as the Polystore is started and already associates the DBs to the actual context.

There are two important phases from the Polystore request and the DBs, the one that goes down to the execution time of a certain action requested by the user, for example, searching for some data; and the second that follows for the access to Fog and its respective implementations.

The first phase mentioned in the previous paragraph would be the complete execution that a user would do only through Polystore, that is, a search, for example, that will be returned to the user. The second phase would move towards Fog in which Polystore actions would be carried out closer to the edge of the network. Data integration would not be carried out by Polystore, for example, but by Fog which, in addition to performing this step, would also pre-process the data.

Therefore, it is evident that the phases that are separated from the Polystore step and the DBs seen in Fig. 4, would be a means of comparison in relation to the direct execution time and with the intervention of the Fog layer.

Finally, in Fig. 4 we see the step of analysis and comparison of the results obtained. It is important to know that, for purposes of example and validity of our proposal, we chose to run tests related to the execution time that we will be gaining by implementing the proposal of this paper. There are many other ways to analyze performance, gain, loss, etc. in any application. We chose to do time execution in order to validate and demonstrate that it is possible to have better performance at run time.

4   Conclusion and Future Work

We have seen that IoT systems have been increasingly used as research objectives, improving the population's quality of life (health, well-being, and intelligent environments) and providing the ease of capturing data through sensors.

However, we also have the challenge of bringing processing, execution speed, data capture, and integration of multiple heterogeneous databases, among others, to this IoT environment. Systems like this demand many tasks and quick responses.

With this, our proposal seeks to help users, with or without expertise in IoT/Cloud, to access an IoT scenario in general and aims to assist these users in carrying out searches, and decision making, through the easy use of Polystore being executed and powered by Fog.

The Polystore architecture aims to facilitate access for users who do not need to have expertise and notion of DBs, in addition to having a unique way for users to do research and operations within an application. The exciting thing is that in addition to delivering this uniqueness, it also promotes the integration of databases, facilitating storage and providing the use of distributed DBs.

Our proposal brings an innovation that promotes changing the way of executing Polystore so that the results of research involving IoT data reach end users faster, thus reducing the computational cost of execution and thus delivering what an IoT application needs: agility, results, and decision-making in a faster way with lower cost. Validation will be done through execution comparisons using our proposal that intends to execute many of the functions that a Polystore has into the Edge-Fog infrastructure and comparing with the raw execution of the Polystore inside the cloud.

For future work, we intend to implement and increase the amount of complexity in our proposal, for that, we intend to study one or more ways to validate our proposal. The validation process will take place by executing the implemented part and using it in architectures, frameworks, etc. that already exists in the literature, as seen in 2, for example. It is worth emphasizing that the execution will be based on some tests that justify the change in the choice of a large part of the execution of Polystore outside the cloud. In our ongoing research we are executing several diverse types of experiment which are providing results from different applications which are indicating how to deal with large-scale systems.

References

1. Vijay, G., et al.: The BigDAWG polystore system and architecture. In: 2016 IEEE High Performance Extreme Computing Conference (HPEC). IEEE (2016)
2. Amadeo, M., Molinaro, A., Paratore, S.Y., Altomare, A., Giordano, A., Mastroianni, C.: A cloud of things framework for smart home services based on information centric networking. In: 2017 IEEE 14th International Conference on Networking, Sensing and Control (ICNSC), pp. 245–250 (2017)
3. Jennie, D., et al.: The BigDAWG polystore system. ACM Sigmod Rec. **44**(2), 11–16 (2015)
4. Gomes, E., Costa, F., De Rolt, C., Plentz, P., Dantas, M.: A survey from real-time to near real-time applications in fog computing environments. Telecom **2**, 489–517 (2021)
5. Vu Khanh, Q., et al.: IoT-enabled smart agriculture: architecture, applications, and challenges. Appl. Sci. **12**(7), 3396 (2022)
6. Farahani, B., Firouzi, F., Luecking, M.: The convergence of IoT and Distributed Ledger Technologies (DLT): opportunities, challenges, and solutions. J. Netw. Comput. Appl. **177**, 102936 (2021)
7. Shafaq Naheed, K., et al.: Blockchain smart contracts: applications, challenges, and future trends. Peer-to-peer Netw. Appl. **14**(5), 2901–2925 (2021). https://doi.org/10.1007/s12083-021-01127-0
8. Naha, R.K., et al.: Deadline-based dynamic resource allocation and provisioning algorithms in fog-cloud environment. Futur. Gen. Comput. Syst. **104**, 131–141 (2020)

9. Manoj, P., et al.: Processing analytical queries over polystore system for a large astronomy data repository. Appl. Sci. **12**(5), 2663 (2022)
10. Davoudian, A., Liu, M.: Big data systems: a software engineering perspective. ACM Comput. Surv. **53**(5), 1–39 (2020)
11. Zdonik, M., et al.: Integrating real-time and batch processing in a polystore. In: IEEE High Performance Extreme Computing Conference, pp. 1–7 (2016)
12. Di iorio Silva, G., Sergio, W.L., Ströele, V., Dantas, M.A.R.: ASAP - academic support aid proposal for student recommendations. In: Barolli, L., Woungang, I., Enokido, T. (eds.) AINA 2021. LNNS, vol. 226, pp. 40–53. Springer, Cham (2021). https://doi.org/10.1007/978-3-030-75075-6_4

OASL: SPARQL Query Language
for OpenAPI Ontologies

Nikolaos Lagogiannis, Nikolaos Mainas, Chrisa Tsinaraki,
and Euripides G. M. Petrakis[✉]

School of Electrical and Computer Engineering,
Technical University of Crete (TUC), Chania, Greece
{nlagogiannis,ctsinaraki}@tuc.gr, {nmainas,petrakis}@intelligence.tuc.gr

Abstract. OpenAPI descriptions detail the actions exposed by REST APIs in YAML or JSON syntax. The semantics of valid OpenAPI descriptions can be further exploited if they can be mapped to an OpenAPI ontology, as we have done in our previous work. However, queries become complex and require the user to be familiar with the peculiarities of the ontology. We introduce the OpenAPI SPARQL Language (OASL), an RDF query language for OpenAPI semantic descriptions. To formulate an OASL query, a user needs only a basic understanding of SPARQL and no knowledge of the OpenAPI ontology. OASL builds on top of SPARQL and simplifies query complexity, so even highly complex SPARQL queries can be expressed using only a few OASL statements. The run-time performance of OASL has been assessed experimentally on a Virtuoso database with OpenAPI ontologies of real-world services. A critical analysis of its performance is also presented along with several query examples.

1 Introduction

Software vendors and cloud providers publish web services in service registries on the Web so that they can be easily discovered and used. The OpenAPI Specification [4] is a powerful framework for the description of REST APIs. An OpenAPI service description is a machine-understandable technical document that provides guidance on the effective use and integration of APIs in applications. A service is described by a JSON (or YAML) document specifying requests, responses, and security information such as authentication and authorization rules for an action.

OpenAPI service descriptions can become particularly complex, also complicating the expression of queries on OpenAPI catalogs. To address this problem, OpenAPI-to-GraphQL [7] generates a GraphQL interface for a given OpenAPI schema that allows clients to write queries on API collections and retrieve descriptions of service resources. OpenAPI QL [6] introduces an abstract layer to N1QL[1] (i.e. the JSON query language of Couchbase) that allows users to express queries in OpenAPI catalogs with minimal or no knowledge of OpenAPI.

---

[1] https://www.couchbase.com/products/n1ql

L. Barolli (Ed.): AINA 2023, LNNS 654, pp. 303–317, 2023.
https://doi.org/10.1007/978-3-031-28451-9_27

The semantics of valid OpenAPI RESTful service descriptions can be further exploited if they can be mapped to ontologies. This is facilitated by a reference OpenAPI ontology that we have developed [3]. OpenAPI description translation has been incorporated into a Web Application[2]. However, queries on top of these descriptions cannot be specified with the languages introduced in the previous paragraph, but with SPARQL [1] only, resulting in lengthy and complex queries. The use of SPARQL would require the user to be familiar with the OpenAPI syntax and especially with the axioms (including rules) that have been defined in the OpenAPI ontology. SPARQL queries expressed using ontology axioms may become particularly complicated, involving many statements that an ordinary SPARQL user (with no understanding of the ontology) is almost impossible to express.

We introduce OpenAPI SPARQL (OASL), a SPARQL-like query language for OpenAPI descriptions without the syntax complexity of SPARQL queries. OASL has been evaluated in a relatively large database with 100 ontologies using a set of example queries of different query complexity levels (e.g. simple, complex, very complex queries). To the best of our knowledge, a query language for OpenAPI ontologies has not been proposed elsewhere in the literature.

Related work is discussed in Sect. 2. The proposed query model and the OASL syntax are discussed, respectively, in Sect. 3 and Sect. 4. Section 5 outlines the translation from OASL to SPARQL, followed by experimental results and issues for future research in Sect. 6 and Sect. 7 respectively.

2   Related Work and Background

An OpenAPI service description comprises many objects, with every object having a list of properties with values that can also be objects. Objects and properties defined under the `Components` unit of an OpenAPI document can be reused by other objects, and they can be linked to each other (e.g. using the keyword `$ref`). Figure 1 outlines the structure of an OpenAPI service description.

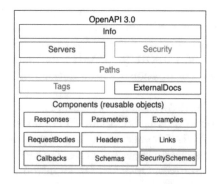

Fig. 1. OpenAPI document structure.

---

[2] https://www.intelligence.tuc.gr/semantic-open-api/.

The `Servers` object provides information on the API server locations. Servers can be defined for different operations (locally declared servers override global servers). The service description contains an `Info` object with some non-functional information for the service, an `External Documentation` object, and possibly `Tag` objects (i.e. tags that refer to resources described by a Schema object and are used to group operations either by resource or by any other qualifier). Tags are optional in OpenAPI and are commonly used to group services by endpoints.

The `Path` object holds all the available service paths (i.e. endpoints) and their operations. It provides information about expressing HTTP requests to the service and about the service responses. It describes the supported HTTP methods (e.g. get, put, post, etc.) and defines the relative paths of the service endpoints (which are appended to a server URL in order to construct the full URL of an operation).

The `Response` object describes the responses of an operation, its message content, and the HTTP headers that a response may contain. The `Parameters` object describes parameters of operations (i.e. path, query, header, and cookie parameters). OpenAPI responses can include custom headers to provide additional information about the result of an API call. Every operation must have at least one response. A response is described using its HTTP status code and the data returned in the response body or headers.

Service operations accept and return data in different formats (e.g. JSON, XML, text, or images). These formats are defined by `Media Type` objects within a request or response. A `Media Type` object is identified using the content keyword. Request bodies in operations (e.g. put, post, get) specify the message that will be sent in a request. It is defined using the `request- Body` keyword and its contents are media types or `Schema` objects (using the `Schema` keyword). The response body can be a `Schema` object which is defined under the `Components` object. The `Schema` object describes the request and response messages based on JSON Schema[3]. A `Schema` object can be a primitive (string, integer), an array or a model, or an XML data type and may have properties on its own (i.e. `externalDocs`). Schema properties do not have semantic meaning, and their meaning can be vague [3].

The `Components` object lists reusable objects and includes (among others) definitions of schemas, responses, headers, parameters, and security schemes. The `Security` object lists the security schemes of the service (declared using the `security scheme` keyword). The specification supports HTTP authentication, API keys, OAuth2 common flows or grants (i.e. ways of retrieving an access token), and OpenID Connect. If an operation authentication scheme is oauth2, the value of a flow is a string with values one of `authorizationCode`, `implicit`, `password` and `clientCredentials`. The value of a scope is a string defining an action like `read` or `write`. OpenID Connect security declares a sign-in flow that enables a client application to obtain user information via authentication. If the security scheme is of type `openIdConnect` then the value is a list of scope

---

[3] https://json-schema.org.

names required for the execution. An API key is actually a token that a client provides when making API calls. In OpenAPI 3.0, API Keys are described as a combination of name and location, specified in the `name` and `in` properties, respectively. Finally, if the security scheme is HTTP, OpenAPI 3.0 includes numeric cases of security definitions built in the HTTP protocol (e.g. bearer authentication). Each HTTP scheme defines the property `type` with value `http` and `scheme` with values like `bearer`, `basic`, or another arbitrary string.

3   OpenAPI Ontology

An overview of the OpenAPI ontology is presented in Fig. 2. The `Document` class represents the documentation and the entry point of the service described, provides general information (`Info` class) about it, and also specifies the service paths and the entities that it supports. The `Path` class represents (relative)

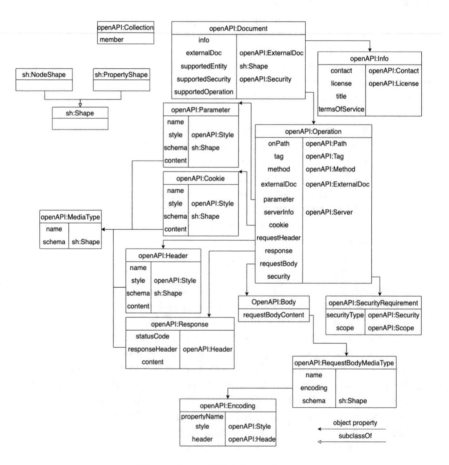

Fig. 2. OpenAPI 3.0 ontology.

service paths (in the `pathName` property). The `Operation` class provides information for sending HTTP requests. Request bodies are represented by the `Body` class, while, responses are declared in `Response` class specifying the status code and the data returned. The `MediaType` class describes the format of request and/or response body data. Class `Operation` refers to security schemes specified according to the `SecurityRequirement` class. OpenAPI parameters are separate classes for every parameter type. The `Header` class provides all the definitions of header parameters. The `Cookie` class defines the cookies that are sent with HTTP requests. The `Parameter` class describes the parameters of the operations and has the `PathParameter` and `Query` sub-classes that refer to the corresponding path and query parameters. Request and response bodies, are defined as classes and so is defined their media type. Class `Encoding` defines keywords denoting serialization rules for media types with primitive properties (e.g. `contentType` for nested arrays or JSON). Figure 3 shows the security schemes supported by OpenAPI. Class `Security` has security schemes as sub-classes. Class `OAuth2` has different flows (grants) as sub-classes. If the security scheme is of type `OAuth2` or `OpenID Connect` [5], then scope names are defined as properties.

OpenAPI documents include information spread among objects and properties in a deeply nested (i.e. tree) structure and their RDF serializations, that have been mapped to the OpenAPI ontology, require many triples to describe some OpenAPI objects and/or properties. When it comes to querying, the challenge is to query an RDF/graph database using SPARQL a path of property triples leading to a node expressing the requested object. The reason is that SPARQL queries may become confusing and complicated, since they may require multiple triples to describe some OpenAPI objects and properties, or use the same paths to reach a group of these. Moreover, the user should have good knowledge of the OpenAPI specification and the triple paths s/he should follow.

OASL applies flattening (i.e. the reduction of long path expressions into simple statements with the properties of interest) to resolve this issue.

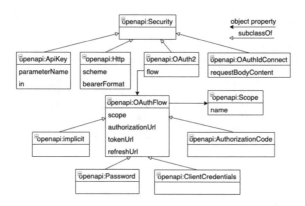

Fig. 3. OpenAPI 3.0 security class.

This simplifies the query model a lot, allowing all OpenAPI objects/properties to be "reached" with no more than two triples (in most cases, one is enough). Table 1 lists the triples of the OASL query model.

4   The OASL Query Language

The OASL syntax is very similar to that of SPARQL and it supports the SPARQL clauses: (a) the (optional) PREFIX clause, which specifies the namespaces corresponding to specific URIs; (b) The SELECT clause, which contains the variables with values that will be included in the query results; (c) The WHERE clause, which contains all conditions and rules in triple form; and (d) The (optional) rearrange clause, which may include the different clauses that rearrange the query result, i.e. LIMIT, ORDER BY and OFFSET. In addition to the standard SPARQL features, OASL also supports range queries (by introducing the BETWEEN clause) and comparison operators integrated in the WHERE clause. The Backus Naur form (BNF) of the high-level OASL syntax is presented in Listing 1. The full BNF of the OASL syntax is available at [2].

As an example, Listing 2 describes an OASL query where the user wants to retrieve for every query result (i.e. OpenAPI document) the service title, the operation name and the response status code. The resulting OASL query is concise, with three triples only.

Listing 1. BNF form of the high-level OASL syntax.

```
1    <query> ::=<prefixes> <select query> <where query> <
         rearrange clause> | <select query> <where query> <
         rearrange clause> | <prefixes> <select query> <where
         query> | <select query> <where query>
2    <prefixes> ::= PREFIX preflx_name: URI | PREFIX
         prefix_name: URI <prefixes>
3    <select query> ::= SELECT <select clause> | SELECT
         DISTINCT <select clause>
4    <where query> ::= WHERE{ <where Body> }
5    <where Body> ::= <triples> {<triples>} OR{<triples >} |
         <triples> <where Body> {<triples>} OR {<triples>} <
         where Body>
6    <rearrange clause> ::= (<order clause> | <limit clause>
         | <offset clause>) ( <order clause> (<limit clause>
         | <offset clause>)) ( <limit clause> (<order clause>
         I <offset clause>)) ( <offset clause> (<limit clause
         > | <order clause>)) ( (<order clause> <limit clause
         > ) | (<limit clause> <order clause> ) <offset
         clause>) ((<order clause> <offset clause> ) | (<
         offset clause> <order clause> ) <limit clause>) ( (<
         offset clause> <limit clause> ) (<limit clause> <
         offset clause> ) <order clause>)
```

Table 1. Triples of OpenAPI objects (subjects), properties (predicates) and values (objects).

Subject	Predicate	Object
Service	title	string
	description	string
	email	string
	externalDocUrl	string
	externalDocDescription	string
	contactUrl	string
	contactName	string
	version	string
	licenseName	string
	licenseDescription	string
Request	method	'get', 'put', 'post', 'delete'
	path	string
	operationName	string
	schema	string
	contentType	string
	header	string
	extDocUrl	string
	extDocDescription	string
	summary	string
	bodyDescription	string
	tag	string
Server	description	string
	url	string
SecurityScope	description	string
	name	string
Parameter	name	string
	required	boolean
	schema	string
	description	string
	style	'simple', 'form', 'label', 'matrix', 'pipeDelimited', 'spaceDelimited'
	allowEmptyValue	boolean
	allowReserved	boolean
	deprecated	boolean
	location	'path', 'header', 'query','cookie'
Tag	name	string
	description	string
	extDocUrl	string
Security	apiKeyIn	string
	apiKeyName	string
	httpBearrerFormat	string
	httpBearrerScheme	string
	OpenIdConnectUrl	string
	OAuth2AuthUrl	string

(*continued*)

Table 1. (*continued*)

Subject	Predicate	Object
	OAuth2RefreshUrl	string
	OAuth2TokenUrl	string
	OAuth2flowType	'Implicit', 'Client Credentials', 'Password', 'AuthCode'
	securityType	'Oauth2', 'Api Key', 'OpenIdConnect', 'Http'
Response	schema	string
	header	string
	statusCode	integer
	statusCodeRange	'1xx', '2xx', '3xx', '4xx', '5xx'
	description	string
ResponseHeader	name	string
RequestHeader	required	boolean
Header	schema	string
	description	string
	style	'simple', 'form', 'label', 'matrix', 'pipeDelimited', 'spaceDelimited'
	allowEmptyValue	boolean
Schema <variable>	dataType	'integer', 'number', 'string', 'boolean', 'array'
	type	string
	property	string
	style	'simple', 'form', 'label', 'matrix', 'pipeDelimited', 'spaceDelimited'
	title	string
	minProperties	number
	maxProperties	number
	minLength	number
	maxLength	number
	minCount	number
	maxCount	number
	minimum	number
	maximum	number
	description	string
Property <variable>	dataType	'integer', 'number', 'string', 'boolean', 'array'
	name	string
	type	string
	description	string
	minLength	number
	maxLength	number
	minimum	number
	maximum	number
	minCount	number
	maxCount	number

Listing 2. OASL query that retrieves for every query result the service title the operation name and the response status code.

```
1    SELECT ?serTitle, ?statCode, ?opName
2    WHERE {
3      Service title ?serTitle.
4      Response statusCode ?statCode.
5      Request operationName ?opName.
6    }
```

A more complex example is provided in Listing 3, where in the query Listing 2 the user wants the status code of the query results to be in the range [100, 400] and the query results to be ordered by ascending service title values.

Listing 3. OASL query that retrieves the service title the operation name and the response status code of the OpenAPI descriptions with response status code in the range 100–400. The query results to be ordered by ascending service title values.

```
1    SELECT ?serTitle, ?statCode, ?opName
2    WHERE {
3      Service title ?serTitle.
4      Response statusCode ?statCode.
5      Response statusCode BETWEEN(100,400).
6      Request operationName ?opName.
7    }
8    ORDER BY asc ?serTitle
```

5   Query Translation

Figure 4 outlines the OASL query translation system architecture. The input to the system may be an OASL query or a semantic OpenAPI service description (in ttl form) to be stored. The queries are sent to the OASL-to-SPARQL query translator and the resulting SPARQL query is sent by the SPARQL Execution Handler to the underlying Virtuoso server. The Virtuoso server sends the query results to the SPARQL Execution Handler to format and send them to the user.

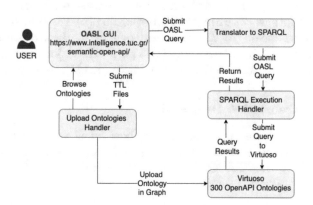

Fig. 4. OASL Architectural overview.

The most important part of our work lies with the OASL-to-SPARQL query translator, after the lexicographic analysis. The translation algorithm is outlined in Algorithm 1. To illustrate the algorithm operation, Listing 4 presents the result of translating Listing 2.

Listing 4. Result of the translation of the OASL query of Listing 2 in SPARQL.

```
1    PREFIX rdf: <http://www.w3.org/1999/02/22-rdf-syntax-ns
         #>
2    PREFIX owl: <http://www.w3.org/2002/07/owl#>
3    PREFIX xsd: <http://www.w3.org/2001/XMLSchema#>
4    PREFIX rdfs: <http://www.w3.org/2000/01/rdf-schema#>
5    PREFIX myOnt: <https://www.example.com/service/
         googleBooks_API#>
6    PREFIX openapi: <http://www.intelligence.tuc.gr/ns/open-
         api#>
7    PREFIX sh: <http://www.w3.org/ns/shacl#>
8    SELECT ?serTtile ?statCode ?opName
9    WHERE {
10     ?service a openapi:Document.
11     ?service openapi:info ?info.
12     ?info openapi:serviceTitle ?serTtile.
13     ?service openapi:supportedOperation ?operation.
14     ?operation openapi:response ?response.
15     ?response openapi:statusCode ?statCode.
16     ?service openapi:supportedOperation ?operation.
17     ?operation a openapi:Operation.
18     ?operation openapi:name ?opName.
19   }
```

6   Evaluation

We have shown in the previous sections that the OASL queries are simpler and shorter than the equivalent SPARQL ones and that OASL does not require the user to be familiar with the OpenAPI ontology, which is a requirement for querying semantic OpenAPI descriptions with SPARQL. In this section, we evaluate the performance of OASL on top of a Virtuoso server with 100 semantic OpenAPI descriptions (27.6 MB storage) regarding (a) The translation overhead that OASL introduces; (b) The overhead due to the lack of optimization in the resulting SPARQL queries; and (c) The overhead introduced due to the duplication of schema information in semantic OpenAPI descriptions.

6.1   Query Translation Overhead

The first issue that we have examined is how much we have to pay for the ease-of-use that OASL introduces. To measure this, we use as metrics the *Query Translation Time*, the *SPARQL Execution Time* and the *Total Execution Time* for each of the evaluation queries.

Algorithm 1. Query translation algorithm.

---

addInitialPrefixes()
if PREFIX clause is detected **then**
    while new prefix exists **do**
        ParsePrefixStatement()
Ensure: Prefix is unique
        PrefixList ← addnewprefixes()
    generatePrefixSPARQLCode()
Ensure: next clause = SELECT
  ParseVariables()
  generateSelectSPARQLCode(identified variables)
Ensure: next clause = WHERE
    while current clause = WHERE **do**
        structure ← RecogniseTripleStructuring
        ParseTriples()
        if structure = OPTIONAL **then**
Ensure: Optional syntax rules
        else if structure = NOT **then**
Ensure: NOT syntax rules
        else if structure = UNION **then**
Ensure: UNION syntax rules
        generateWHERESPARQLCode(structure)
    while next clause = rearrange **do**   ▷ LIMIT , OFFSET or ORDER BY
        if rearrange = LIMIT **then**
            Parse LIMIT clause
        if rearrange = OFFSET **then**
            Parse OFFSET clause
        if rearrange = ORDER BY **then**
            Parse ORDER BY clause
            correctPositioning ← checkIfClausespositionAreInOrder()
    if correctPositioning is false **then**
        return Syntax Error
    else
        generateRearrangeSPARQLCode()

We have run queries of different complexity in terms of numbers of triples and SPARQL clauses. We distinguish three levels of complexity: (a) *Simple queries*, like the query of Listing 2, containing no more than three triples and only the SELECT and WHERE clauses; (b) *Complex queries*, like the query of Listing 3, with many triple endpoints and any type of clause; and (c) *Extended and complex queries*, which are useful mostly for testing, that contain all the OASL clauses and a high number of triples. The difference between the complex and the extended complex queries is that the former could still be average user queries, while the later are more useful for testing, since their complexity aims to stress the system.

Listing 5. Complex extended query example.

```
1   PREFIX myOntb1:<https://www.example.com/service/
        petstorellb
2   PREFIX myOntb2:<https://www.example.com/service/
        petstore2#>
3   PREFIX myOntb3:<https://www.example.com/service/
        petstore2#>
4   PREFIX myOnth4:<https://www.example.com/service/
        petstore4#>
5   PREFIX myOnth5:<https://www.example.com/service/
        petstore5#>
6   PREFIX myOntb6:<https://www.example.com/service/
        petstore6#>
7   PREFIX myOntb7:<https://www.example.com/service/
        petstore7#>
8   PREFIX myOntb8:<https://www.example.com/service/
        petstore8#>
9   PREFIX myOntb9:<https://www.example.com/service/
        petstore9#>
10  PREFIX myOntb10:<https://www.example.com/service/
        petstore10#>
11  PREFIX myOntb11:<https://www.example.com/service/
        petstorell#>
12  PREFIX myOntb12:<https://www.example.com/service/
        petstore12#>
13  PREFIX myOntb:<https://www.example.com/service/petstore
        #>
14  SELECT DISTINCT ?serviceName ?serDesc ?opName ?tgName ?
        serverUrl ?parName ?headName ?headDesc
15  WHERE {
16    Service title ?serviceName.
17    Service description ?serDesc.
18    Parameter name ?parName.
19    OPTIONAL {Parameter required TRUE.}
20    Response statusCode BETWEEN(100,450).
21    {Response schema myOnt:Pet.}
22    OR
23    (Response schema myOntb:Pets.}
24    Request operationName ?opName.
25    OPTIONAL {Tag name ?tgName.}
26    OPTIONAL {Server url ?serverUrl.}
27    ResponseHeader name ?headName.
28    OPTIONAL {ResponseHeader description ?headlesc.}
29    ResponseHeader style simple.
30    ORDER BY asc ?serviceName
31    LIMIT 60
32    OFFSET 5
```

The evaluation results are presented in Table 2. Columns 2–5 contain the measurements for the different query categories. The fourth and the fifth column refer to the extended complex queries, but the former presents the results of queries with no results and the later of queries with results.

It is clear in the second row of Table 2 that the average translation time is in the range of 35–41 ms for the different query types and does not exceed 2.45% (with the minimum being 0.19%) of the total query execution time for any query type. As expected, the average translation time is slightly less for simple queries (35 ms) and higher for complex (39 ms) and very complex (41 ms) queries, since the query complexity influences the query translation time.

Table 2. Overhead introduced by the OASL-to-SPARQL query translation.

Query type	Simple	Complex	Complex & Extended (no results)	Complex & Extended (results)
Translation time	35 ms	39 ms	41 ms	41 ms
Execution time	248 ms	1643 ms	1451 ms	16215 ms
Total time	2193 ms	8341 ms	1547 ms	21003 ms
Translation/Total time %	1.59%	0.46%	2.45%	0.19%
Execution/Total time %	11.3%	19.7%	93.8%	77.2%

Since the query translation time is much less than the query execution time for all query types, it is expected that the highest ratio of query translation time over total query execution time happens for the query types that require less time in total, as is the case for complex and extended queries with no results (with total query execution time 1547 ms and ratio 2.45%) and simple queries (with total query execution time 2193 ms and ratio 1.59%). These observations show that the ease-of-use that OASL facilitates comes with an insignificant and sometimes negligible overhead.

6.2 Lack of Resulting Query Optimization

In the current version of the OASL-to-SPARQL query translator no optimization takes place at the query triple analysis level, resulting in triples that may repeat in the resulting SPARQL queries. This happens for properties that may belong to different objects, as is the case of the `openapi:supportedOperation` property that belongs to the **Request**, **Response**, and **Tag** objects.

To examine the overhead that the lack of query optimization at the triple level introduces, we run on our Virtuoso server an extreme (and rather unrealistic) SPARQL query with a triple pattern that repeats 300 times and compared its execution time with that of the optimized at the triple level query, that contains 3 triples only. The overhead introduced in that case was of 1 ms only, which is negligible compared to the average SPARQL query execution times for all query types (evidence can be found in the third row of Table 2) and indicates that this type of optimization should not be a priority.

6.3 Schema Information Duplication Overhead

An issue worth exploring was the overhead that schema information duplication introduces in storage space since larger files would cause delays in query execution. To get a first feeling of the extent of this issue, we analyzed YAML OpenAPI files together with the semantic OpenAPI files (from the Web Application[4]) that convey the same information. The files that we have analyzed are:

- File1: Cisco521_SaaS-Connect-Provision
- File2: EliteFintech_list_management

---

[4] https://www.intelligence.tuc.gr/semantic-open-api/.

- File3: youtube_API
- File4: petV3_Anotado
- File5: googleBlogger_API

The analysis results are presented at Table 3. In the fourth row of the table the number of ontology schema lines is almost constant (in the range 2107–2287), while the lines describing individual triples vary a lot (in the range 10–1128). Moreover, we notice that the number of individual triple lines increase together with the YAML file lines. The ratio of individual triple lines over schema lines in the semantic OpenAPI files ranges between 0.47% and 33% and indicates that the query execution time would be reduced if the schema information of the OpenAPI ontology is not duplicated.

Table 3. Analysis of YAML ontology OpenAPI files that convey the same information.

Files selected	File 1	File 2	File 3	File 4	File 5
Yaml file lines	20	493	170	844	552
Ontology individual lines	10	580	357	1031	1128
Ontology schema lines	2107	2173	2180	2192	2287
Individual triples %	0.47%	26.69%	14.07%	32%	33%
Yaml file size (KB)	0.505	16.8	6.5	23.5	19.8
TTL file size (KB)	90.3	127.1	116	174.5	179

7   Conclusions and Future Work

We have introduced the OpenAPI SPARQL Language (OASL), an RDF query language for OpenAPI semantic descriptions. OASL is SPARQL-like, while the OASL queries are less complex and much shorter than their SPARQL equivalents and it doesn't require in-depth knowledge of the OpenAPI ontology. The experimental evaluation we performed has shown that the OASL-to-SPARQL query translation is fast and takes very little time compared to the execution of the resulting SPARQL query. Our future research plans include the use of indexing in order to speed up query execution, the support of OpenAPI Links, Callbacks and Webhooks as well as the support of the SPARQL FROM clause and GRAPH function.

References

1. Harris, S., Seaborne, A., Prud'hommeaux, E.: SPARQL 1.1 query language (2013). https://www.w3.org/TR/sparql11-query/. W3C Recommendation
2. Lagogiannis, N.: OpenAPI SPARQL: querying OpenAPIOntologies in OWL. Diploma thesis, School of Electrical and Computer Engineering, Technical University of Crete (TUC), Chania, Crete (2022). https://dias.library.tuc.gr/view/93679

3. Mainas, N., Bouraimis, F., Karavisileiou, A., Petrakis, E.G.: Annotated OpenAPI descriptions and ontology for REST services (2022). https://doi.org/10.1142/S0218213023500173

4. Miller, D., Whitlock, J., Gardiner, M., Ralphson, M., Ratovsky, R., Sarid, U.: OpenAPI specification v3.1.0 (2021). https://github.com/OAI/OpenAPI-Specification/. (Latest editor's draft)

5. Octa: What is OAuth 2.0? (2022). https://auth0.com/intro-to-iam/what-is-oauth-2/

6. Stergiou, I.-M., Mainas, N., Petrakis, E.G.M.: OpenAPI QL: searching in OpenAPI service catalogs. In: Barolli, L., Hussain, F., Enokido, T. (eds.) AINA 2022. LNNS, vol. 450, pp. 373–385. Springer, Cham (2022). https://doi.org/10.1007/978-3-030-99587-4_32

7. Wittern, E., Cha, A., Laredo, J.A.: Generating GraphQL-wrappers for REST(-like) APIs. In: Mikkonen, T., Klamma, R., Hernández, J. (eds.) ICWE 2018. LNCS, vol. 10845, pp. 65–83. Springer, Cham (2018). https://doi.org/10.1007/978-3-319-91662-0_5

Orchestrating Fog Computing Resources Based on the Multi-dimensional Multiple Knapsacks Problem

Daniel E. Macedo[1]([✉]), Marcus M. Bezerra[1], Danilo F. S. Santos[2], and Angelo Perkusich[2]

[1] Electrical Engineering Graduate Program, Federal University of Campina Grande, Campina Grande, PB, Brazil
{daniel.macedo,marcus.bezerra}@ee.ufcg.edu.br
[2] VIRTUS RDI Center, Federal University of Campina Grande, Campina Grande, PB, Brazil
{danilo.santos,perkusic}@virtus.ufcg.edu.br

Abstract. Cloud computing offers computational resources remotely. However, some applications require low latency response, making them unfeasible to allocate resources physically far away from users. As an option, the edge computing paradigm proposes moving computational resources from the cloud to the edge of the network, physically closer to the user, allowing low latency. However, resource orchestration presents new challenges in edge computing, as computational resources at the edge are usually limited compared to the cloud. Another alternative is fog computing, which connects these two environments by distributing computational resources both in the cloud and at the edge. In fog computing, resource allocation can consider user requirements as latency and availability of computational resources on servers. In this context, we show an approach to resource orchestration using the multi-dimensional knapsack problem with constraints applied to fog computing scenarios. We allocate resources through booking, ensuring that resources are allocated only for the necessary time, respecting the maximum waiting time and latency for each request. We analyzed simulation results, comparing the performance of our proposed approach with a traditional one. Furthermore, results demonstrated that our proposed orchestration method decreases the overall cost and guarantees user requirements.

1 Introduction

Cloud computing offers computational resources remotely for users. Resources are physically separated from the local network, providing high availability of resources, and controlling factors such as physical facilities, temperature, and energy, among others. However, offering resources to applications that demand low latency is still challenging because communication delays can happen between the cloud servers and users due to distance. Different from cloud computing, edge computing allocates resource servers near the user to decrease the latency time. Another difference between cloud and edge computing is the availability of resources. Edge computing usually provides fewer resources than the cloud [14].

It is possible to integrate these two environments by maintaining servers distributed between the cloud and the edge. This model is called fog computing, which can assess user requirements allocating requests with low latency at the edge. Thus, it is essential to choose when and where to use the resources efficiently, considering the involved costs and requests [9,14].

Orchestrating resources in fog computing is essential to match the requirements, providing the computational resources quickly and efficiently [15,18,24]. A strategy to guarantee resource provision and allocation is anticipating the demands, and booking the required resources. However, this reservation must be efficient to avoid wasting resources and increasing the costs related to their allocation.

We propose an orchestration resources method in fog computing that matches user requests for computational resources, evaluating requirements requests and the cost of allocating them. In this context, studies analyzed the orchestration problem as an extension of the knapsack problem [3,10,22,23].

The main contributions of our proposed method are summarized as follows:

- introduction of a mathematical model with constraint equations considering the time variable;
- analysis of resource availability subject to different latency times and based on different requirements;
- implementation of cost optimization for guaranteed resource availability.

The remaining of this work is organized as follows: in Sect. 2, we present previous works related orchestration. In Sect. 3, we introduce the Multi-dimensional Multiple Knapsack problem. In Sect. 4, we describe our mathematical approach to orchestrating resources in fog computing scenarios. In Sect. 5, we evaluated the orchestration method with simulation data. Lastly, we conclude this work in Sect. 6.

2   Review Work

Different researchers are suggesting applications that take advantage of computational resources at the edge. These works permeate the area of online games [20], autonomous vehicles [17], industrial processes [4], different Internet of Things (IoT) applications [16,19,21], and others. Recognizing the importance of computational resource orchestration in environments with available computational resources near the edge, several studies in the literature have already evaluated the use of paradigms such as edge computing in their applications and techniques for resource orchestration management.

Guangshun et al. (2018) [12] proposed an edge computing framework, which, through a penalty factor and the Gray matrix, improves the accuracy of similarity matching and the selection of resources that match users' needs. The proposed methodology was based on two stages. In the first, a similar matching algorithm is used to establish a QoS grade matrix of resources between users and resources. In the second stage, a Markov regression prediction method analyzes the change in the resource load, and among the resources that satisfy the needs defined in the first stage, the ideal resource is selected.

To effectively support the largest number of demands, Guangshun et al. (2019) [11] also proposed a method of scheduling resources for edge computing. In this method, a discrete objective function is defined between a set of demands and a set of servers whose binary decision variable validates the specific relationship between a demand and a resource server. In the second stage, fuzzy clustering methods with particle swarm optimization are used to divide the resources, reducing the resource's scale to be researched.

Zhou et al. (2019) [25] also used predictive mathematical models to improve the distribution of computational resources. Applied directly to the mobile edge computing (MEC) context, it deploys MEC servers straightly to base stations using generic computing platforms. Unlike other solutions in which a mobile device (MD) is associated with just one MEC server, the author proposes a methodology in which an MD can choose to download its tasks to several nearby MEC servers with computational resources in addition to only a MEC server. The work focuses on system optimization in scenarios where a single MD can scale its CPU frequency and allocate computing tasks to multiple MEC servers, exploring the diversity in terms of task assignment and CPU frequency to make optimal control decisions to minimize the trade-off between task execution time and mobile power consumption. In addition, a problem was formulated by combinatorial optimization, an NP-hard problem that can be optimized through efficient approximation with increased computational performance [25].

Fog computing also plays a vital role in cloud computing with the IoT. In this context, the fog assumes the function of managing resources and performing data filtering, pre-processing, and security measures. Mohammad et al. (2015) proposed an effective resource management model, covering the problems of resource forecasting, resource estimation, and reservation based on the type of customer; booking and prices for new and existing IoT customers, based on their features [1].

Aazam et al. (2016) [2] extended the model to propose a Probabilistic Resource Estimation model considering different customer characteristics for edge computing, but it also focused on IoT environments. This weighting made the model more flexible and scalable. In addition, for most mobile devices, such as *smartphones*, *laptops*, *tablets*, where most of the time the video content is played or the storage services are used, efficient resource management is required.

Other authors proposed solutions based on variations of the classic knapsack optimization problem [3, 10, 22, 23]. Amarante et al. (2013) [3] aimed to develop an orchestration proposal to match the users' requirements that guarantee minimum energy consumption. For this, the work improved the modeling of the multiple knapsack problem, addressing energy-saving restrictions. With an emphasis on orchestrating resources under 5G mobile networks, Ketykó et al. (2016) [10] provided a general model of the system considering the point-to-point computational latency of MEC applications. For this purpose, a reducible multiuser MEC download model was proposed for the Multiple Knapsack Problem (*MKP*) [10].

Wang et al. [23] evaluated the data unloading on the edge servers to improve the service quality and future network efficiency. A multilayer computing data download architecture was proposed in this work, consisting of the user layer, mobile fog layer, fixed fog layer, and cloud layer. The unloading problem is formulated as a generalized multidimensional *MKP*, in which each layer is considered a large backpack, and computation tasks are treated as items. This work did not evaluate temporal aspects, with the influence of the time needed for unloading in the servers' occupation.

Based on our review, we did not find solutions that simultaneously apply restrictions to the mathematical model's orchestration proposals that assess restrictions on the resources' time allocation or the maximum waiting time for this request to be met. Thus, we believe that adding the time variable in the mathematical model and evaluating the scenarios with multi-heterogeneous servers (with different latency times and availability of computational resources) contributes to orchestration solutions that are based on the backpack optimization problem.

3   Multi-dimensional Multiple Knapsack Problem

The Multi-dimensional Multiple Knapsack Problem (*MMKP*) is an integer programming problem and it is a generalization of the classic knapsack optimization problem. *MMKP* corresponds to a resource allocation model, whose objective is to select a disjoint set of n items that produce the most significant profit, considering capacity constraints r of multiple containers m [6,7].

Each item n has a profit g_n and consumes a different amount of resources $d_{n,i}(i = 1, ..., r)$, where r is any resource. Resources are available in different knapsacks with a total capacity to provide $p_{m,i}(i = 1, ..., r)$ resources. The objective is to maximize the total profile of the selected objects in all knapsacks. Formally the *MMKP* is formulated by:

$$max \sum_{i=0}^{m} \sum_{j=0}^{n} x_{ij} * g_j, \tag{1}$$

$$with\ x_{ij} \in \{0, 1\}$$

The variable x_{ij} indicates whether item j is allocated in knapsack i. $x_{ij} = 1$ indicates that the item j was allocated in the knapsack i and $x_{ij} = 0$ indicates the opposite.

The Eq. 1 solution must respect the constraints of Eq. 2 and 3. In Eq. 2, the sum of each resource r of all items allocated in a knapsack should not exceed the maximum capacity of the knapsacks.

$$\forall m \forall r, \sum_{j=0}^{n} x_{mj} * d_{j,r} \leq p_{mr} \tag{2}$$

In Eq. 3, any item can only be allocated in a knapsack.

$$\forall n \sum_{i=0}^{m} x_{in} \leq 1 \tag{3}$$

The resource orchestration model can be compared with an *MMKP* because the servers (knapsacks) allocate resources based on users' requests (items), evaluating the requirements to improve the users' service quality and seeking the lowest cost to make resources available. Therefore, extending the mathematical model to the resource orchestration context in a network with applications that require resources is possible [23].

4   Proposed Method

The proposed method explores the interconnection between the user and fog servers. Resources are orchestrated by the fog server, which manages the communication between users and the cloud. Figure 1 shows this scenario.

Figure 1 exposes small networks as different applications, such as traffic light networks for smart cities and industries networks, connecting different equipment. The computational resources are available on servers in the fog and cloud, and users can request resources to match computational requests for solution algorithms, data storage, and other requirements.

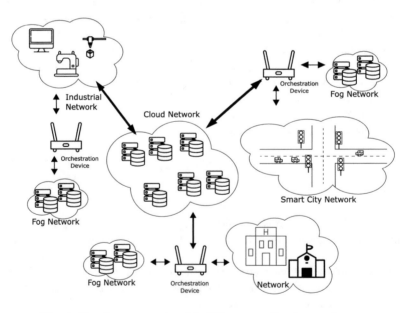

Fig. 1. Topology example with different applications and users

4.1   Problem Formulation

For the orchestration device to analyze resource requests, the fog servers must inform the ability to make resources available. Applications and users must also inform the orchestration device of the resources needed to supply the task requested.

Without loss of generality, we define a set of resources R that users can request or make available by servers. The request or availability of any resource r $(r \in R)$ can be quantified by an integer.

We define any request n as a tuple $(t_n^c, t_n^e, t_n^a, t_n^l, [d_i(i = 1, ..., r)])$, where t_n^c is the creation time, t_n^e is the maximum waiting time, t_n^a is the time needed for the allocation of resources, t_n^l is the maximum latency time that n is amenable to being assigned to a server, and $[d_i(i = 1, ..., r)]$ is a vector with r requested resources, such as $d_r \in \Re$.

The total availability of network resources at time t can be represented by $P = [p_{m,r}]_{|M|x|R|}$, where $|M|$ is the number of servers. $p_{m,r}$ is quantification of availability of resource r by server m, such as $p_{m,r} \in \Re$, and each row of the matrix 4 represents the availability of a server's resources at time t, and each column represents a resource r.

$$P = \begin{bmatrix} p_{1,1} & p_{1,2} & \cdots & p_{1,|R|} \\ p_{2,1} & p_{2,2} & \cdots & p_{2,|R|} \\ \cdots & & & \\ p_{|M|,1} & p_{|M|,2} & \cdots & p_{|M|,|R|} \end{bmatrix} \tag{4}$$

Similarly, the set of tasks to be orchestrated is represented by $D = [d_{n,r}]_{|N|x|R|}$, where $|N|$ is the number of requisitions to be processed in the orchestration. $d_{n,r}$ is quantification of availability of resource r by server n, such as $d_{n,r} \in \Re$. In matrix D, each line represents any task d and each column a resource r.

$$D = \begin{bmatrix} d_{1,1} & d_{1,2} & \cdots & d_{1,|R|} \\ d_{2,1} & d_{2,2} & \cdots & d_{2,|R|} \\ \cdots & & & \\ d_{|N|,1} & d_{|N|,2} & \cdots & d_{|N|,|R|} \end{bmatrix} \tag{5}$$

The objective function defines the allocated computational resources cost by Eq. 6, where x_{mnt} is a binary decision variable, and the value 0 disregards request assignment n to a resource server m. Similarly, the value 1 assigns the execution of this task to the server.

$$min(\sum_{t=0}^{T} \sum_{i=0}^{M} \sum_{j=0}^{N} x_{ijt} * (\sum_{r=0}^{|R|} p_{i,r} - d_{j,r}), \tag{6}$$

$$with, \ x_{ijt} \in \{0, 1\}$$

In Eq. 6, the variable T represents the discretization of time and defines as the time window that the orchestration is analyzed, and it is directly related to the precision in the optimization and the computational time of problem-solving. Increasing the value of T implies a more accurate one since more requests are evaluated.

Equation 6 expresses the total cost sum, weighting the allocated resources with their respective individual costs per server. Therefore, in order to optimize the allocation of resources, reducing costs and ensuring latency, the Eq. 6 should be minimized, respecting the constraints defined in 7, 8, 9 and 10.

Constraint 7 guarantees that every resource request n has the necessary resources allocated between its emergence t_n^c and the maximum instant of its completion $t_n^c + t_n^e + t_n^a$, considering the susceptible latency time.

$$\forall n, \sum_{i=0}^{M} \sum_{t=t_n^c}^{t_n^c+t_n^e+t_n^a} x_{int} = t_n^a \tag{7}$$

Constraint 8 guarantees that every request n is assigned to a single server m. Therefore, the problem's mathematical modeling prevents a request from being partially partitioned and served by different servers. Thus, if necessary, the users are responsible for partitioning large requests into smaller subrequests.

$$\forall n \forall m, (\sum_{t=0}^{T} x_{mnt} - t_n^a) * \sum_{t=0}^{T} x_{mnt} = 0 \tag{8}$$

Constraint 9 defines that no server takes on too many requests, preventing it from being able to match all requirements simultaneously. In its turn, Constraint 10 specifies that requests can only be assigned to servers that can match all resource requirements, regardless of the time t

$$\forall t \forall m \forall r, p_{i,r} - \sum_{j=0}^{N} x_{mjt} d_{j,r} \geq 0 \tag{9}$$

$$\forall t \forall m \forall n \forall r, x_{mnt}(p_{m,r} - d_{n,r}) \geq 0 \tag{10}$$

Finally, Constraint 11 evaluates the maximum latency time required for each request, enabling critical latency time systems. In this equation, t_{mn}^l is the value of communication latency between a resource server m and the the origin of the request n.

$$\forall t \forall m \forall n, x_{mnt}(t_{mn}^l - t_n^l) \leq 0 \tag{11}$$

5   Simulation and Evaluation

In this section, we show results obtained from simulations. We simulated a network with users and servers, generating random events, evaluating the proposed method's performance, and comparing it with other methods.

The methodology's objective is to guarantee the resources' availability for applications, optimizing costs. Thus, to evaluate the proposal's effectiveness, two metrics are analyzed, assessing the necessary cost to match the requests and the number of requests that match.

The first metric is the average cost invested in allocating resources to match the requirements. The metric is defined by c_d/cm_d, where c_d is the cost applied in executing a requirement d and cm_d represents the executing cost in the cloud. Therefore, a greater c_d/cm_d implies a higher cost in running the demands.

The second metric analyzes the application load in the fog. Considering that the strategy of architecture with resources close to the edge is to supply the need for low latencies, this metric evaluates the percentage of requests that were not allocated on servers to ensure the maximum time required by them. The metric is defined by $|\overline{d_l}|/|D|$, where $\overline{d_l}$ is the number of unmatch requests $|D|$ is the number of total requests generated on the network. Therefore, this metric has limits between $[0, 1]$.

The proposed method was compared with traditional variations of the knapsack optimization problem. The first variation is a greedy proposal in which requests are received by the orchestrator $(T = 1)$. The second method considers analysis windows larger than $T > 1$ and does not consider the time variable. Therefore, it performs the resource reservation during the entire T interval.

5.1 Background

The solution to the problem permeates areas of linear programming and mixed-integer programming. The system implementation for solving the problem was developed in Python, Pyomo [8], and GLPK (GNU Linear Programming Kit [13]) packages. With the computational complexity of solving the problem and the possibility of unresolvable instances occurring during the simulation, we set a time limit for each problem interaction in $10 minutes$ on the penalty of interruption and return of the most appropriate solution until the interruption.

The proposed method ensures the computational resources requested, except for maximum latency, because the cloud has plenty of resources, accepting new computational demands whenever requested. Therefore, we evaluate only latency constraints.

We considered a scenario with 4 servers and 10 users for simulation. Users create requests by uniform distribution with rate $\mu = 0.5$ request/h. The simulation interval starts at 0 h and ends at 1000 h.

Besides, simulation, costs, latencies, and resource availability were randomly generated, considering the constraints:

- the cost of a resource in the cloud is 10^2 times higher than in the fog network;
- the latency between a user and the cloud is 10^4 times higher than with the fog network;
- the cloud has enough computing resources to match all requests created by users.

5.2   Results

Figure 2 shows the behavior of requirements that did not match the respective latency requirements. It is possible to verify that the proposed methodology's curve is always below the other curves. It is also possible to notice that the proposed methodology does not reach 0, implying that not all latency requirements were matched.

The occurrence of unmatch requirements in the simulation can occur for three causes:

- the impossibility of providing the necessary total resources regardless of the arrangement for prioritizing care. In this case, the problem has no optimal solution;
- the orchestrator's inability to find the optimal solution promptly. In this case, the search for the optimal solution is interrupted, and the best local solution found is chosen as the answer;
- the evaluation time window's length is not suitable for the context, preventing the resolution of the problem to get enough data to find the optimal solution.

Although the method did not achieve the perfect result in this simulation, it was superior to the other methods in that we can adequately attend less than 10% of requests. On the other hand, the proposed method fulfills the requirements of approximately 78% of users' requests.

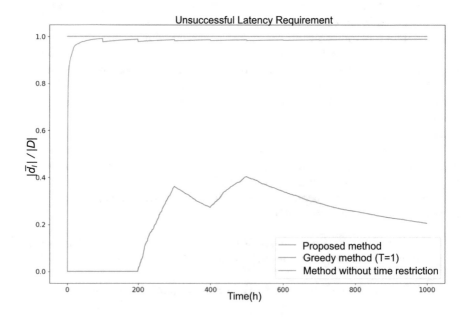

Fig. 2. Analysis of successful requirements

It is possible to identify through Fig. 3 that the curve of the proposed methodology remained below the other curves and obtained a lower cost than the others. The proposed mathematical modeling had a cost reduction of $20, 6\%$ compared to the method without time restrictions and $19, 8\%$ less than the greedy method. It is also important to emphasize that despite the decrease in costs, the proposed method obtained more satisfactory results in match requirements than the others.

The proposed method's cost is lower because, unlike other methods, the objective function aims to reduce costs, respect the maximum waiting time restrictions for a request, and organize the pre-order to match requests.

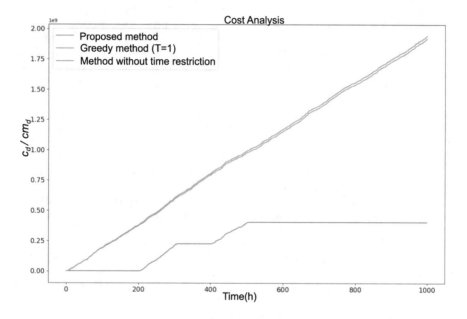

Fig. 3. Cost analysis

6   Conclusions

Our results show that minimizing resolution of the objective function 6 has exponential complexity [23], which makes the proposed method computationally unfeasible when increasing servers and user requests. This obstacle can be overcome by adopting approximate solutions, such as Genetic Algorithms, Ant Colonies [3], and Particle Swarm [5].

Therefore, the implementation of orchestration through an *MMKP* guarantees the costs' optimization, considering the restrictions if the servers have, minimally, the capacity to make resources available. However, the proposed method must be accompanied by approximation methods.

Acknowledgements. The authors would like to thank the Virtus Research, Development and Innovation Center and the Electrical Engineering Graduate Program (COPELE) from the Federal University of Campina Grande (UFCG) for supporting this research.

References

1. Aazam, M., Huh, E.N.: Fog computing micro datacenter based dynamic resource estimation and pricing model for IoT. In: 2015 IEEE 29th International Conference on Advanced Information Networking and Applications, pp. 687–694. IEEE (2015)
2. Aazam, M., StHilaire, M., Lung, C.H., Lambadaris, I.: Pre-fog: IoT trace based probabilistic resource estimation at fog. In: 2016 13th IEEE Annual Consumer Communications & Networking Conference (CCNC), pp. 12–17. IEEE (2016)
3. Amarante, S.R.M., Roberto, F.M., Cardoso, A.R., Celestino, J.: Using the multiple knapsack problem to model the problem of virtual machine allocation in cloud computing. In: 2013 IEEE 16th International Conference on Computational Science and Engineering, pp. 476–483. IEEE (2013)
4. Barzegaran, M., Cervin, A., Pop, P.: Towards quality-of-control-aware scheduling of industrial applications on fog computing platforms. In: Proceedings of the Workshop on Fog Computing and the IoT, pp. 1–5 (2019)
5. Chaturvedi, K.T., Pandit, M., Srivastava, L.: Particle swarm optimization with time varying acceleration coefficients for non-convex economic power dispatch. Int. J. Electr. Power Energy Syst. **31**(6), 249–257 (2009)
6. Dell'Amico, M., Delorme, M., Iori, M., Martello, S.: Mathematical models and decomposition methods for the multiple knapsack problem. Eur. J. Oper. Res. **274**(3), 886–899 (2019)
7. García, J., Crawford, B., Soto, R., Castro, C., Paredes, F.: A k-means binarization framework applied to multidimensional knapsack problem. Appl. Intell. **48**(2), 357–380 (2018)
8. Hart, W., et al.: Pyomo-Optimization Modeling in Python, vol. 67. Springer, Heidelberg (2017)
9. Hong, C.H., Varghese, B.: Resource management in fog/edge computing: a survey on architectures, infrastructure, and algorithms. ACM Comput. Surv. (CSUR) **52**(5), 1–37 (2019)
10. Ketykó, I., Kecskés, L., Nemes, C., Farkas, L.: Multi-user computation offloading as multiple knapsack problem for 5G mobile edge computing. In: European Conference on Networks and Communications, pp. 225–229 (2016)
11. Li, G., Liu, Y., Wu, J., Lin, D., Zhao, S.: Methods of resource scheduling based on optimized fuzzy clustering in fog computing. Sensors **19**, 2122 (2019)
12. Li, G., Song, J., Wu, J., Wang, J.: Method of resource estimation based on QoS in edge computing. Wirel. Commun. Mob. Comput. **2018**, 1–9 (2018)
13. Makhorin, A.: GLPK (GNU linear programming kit) (2008). http://www.gnu.org/s/glpk/glpk.html
14. Mao, Y., You, C., Zhang, J., Huang, K., Letaief, K.B.: Mobile edge computing: survey and research outlook. arXiv preprint arXiv:1701.01090 (2017)
15. Markakis, E.K., et al.: Efficient next generation emergency communications over multi-access edge computing. IEEE Commun. Mag. **55**(11), 92–97 (2017)
16. Ngo, M.V., Chaouchi, H., Luo, T., Quek, T.Q.: Adaptive anomaly detection for IoT data in hierarchical edge computing. arXiv preprint arXiv:2001.03314 (2020)

17. Ning, Z., Huang, J., Wang, X.: Vehicular fog computing: enabling real-time traffic management for smart cities. IEEE Wirel. Commun. **26**(1), 87–93 (2019)
18. Orsino, A., et al.: Effects of heterogeneous mobility on D2D-and drone-assisted mission-critical MTC in 5G. IEEE Commun. Mag. **55**(2), 79–87 (2017)
19. Pham, Q., et al.: A survey of multi-access edge computing in 5G and beyond: fundamentals, technology integration, and state-of-the-art. IEEE Access **8**, 116974–117017 (2020)
20. Prodan, R., Nae, V.: Prediction-based real-time resource provisioning for massively multiplayer online games. Futur. Gener. Comput. Syst. **25**(7), 785–793 (2009)
21. Redondi, A.E.C., Arcia-Moret, A., Manzoni, P.: Towards a scaled IoT pub/sub architecture for 5G networks: the case of multiaccess edge computing. In: 2019 IEEE 5th World Forum on Internet of Things (WF-IoT), pp. 436–441 (2019)
22. Tchendji, V.K., Yankam, Y.F.: Dynamic resource allocations in virtual networks through a knapsack problem's dynamic programming solution. Rev. Africaine Recherche Inform. Math. Appl. **31** (2020)
23. Wang, J., Liu, T., Liu, K., Kim, B., Xie, J., Han, Z.: Computation offloading over fog and cloud using multi-dimensional multiple knapsack problem, pp. 1–7 (2018)
24. Zhang, Q., Fitzek, F.H.P.: Mission critical IoT communication in 5G. In: Atanasovski, V., Leon-Garcia, A. (eds.) FABULOUS 2015. LNICST, vol. 159, pp. 35–41. Springer, Cham (2015). https://doi.org/10.1007/978-3-319-27072-2_5
25. Zhou, W., Fang, W., Li, Y., Yuan, B., Li, Y., Wang, T.: Markov approximation for task offloading and computation scaling in mobile edge computing. Mob. Inf. Syst. **2019** (2019)

Profit Maximization for Resource Providers Using Dynamic Programming in Edge Computing

Rajendra Prajapat[1] and Ram Narayan Yadav[2(✉)]

[1] Indian Institute of Information Technology Dharwad, Dharwad, Karnataka, India
[2] Institute of Infrastructure Technology, Research and Management,
Ahmedabad 380026, Gujarat, India
narayanram.1988@gmail.com

Abstract. Edge Computing provides mobile and Internet of Things (IoT) application users with a new distributed computing paradigm that allows them to offload their tasks at edge servers. This way, the application users can reduce latency and energy consumption. To realize edge computing services in practice, it is necessary to design a mechanism that motivates edge service providers by promising higher utility in return (higher payment) for providing edge computing facilities. So, in this paper, we formulate an optimization problem called application assignment for servers' profit maximization (SPM-AA) that aims to maximize servers' profit in the system. First, we show that it is NP-hard to find the optimal solution for SPM-AA. Then using a dynamic programming approach, we propose a distributed algorithm using dynamic programming in a synchronized edge computing environment for maximizing servers' profit called *SPM-AAA*. Finally, we conduct simulations and compared existing schemes to demonstrate the performance improvement of our algorithms in terms of servers' profit, the number of applications assigned, server utilization, and resource request service ratio.

Keywords: Edge computing · Application assignment · Profit maximization · Dynamic programming

1 Introduction

In recent years, mobile and Internet-of-Things (IoT) devices have become popular. The growth in these mobile and IoT devices has fueled a variety of mobile and IoT applications such as face recognition, natural language processing, interactive online gaming, and various ITS applications in autonomous driving [1,2]. Many of these applications are computation-intensive and require high resources. Due to limited resources present in mobile and IoT devices, it has become difficult and impractical for mobile and IoT devices to execute such users' applications on-site.

To address this problem, IoT devices can offload their computation tasks. Edge computing allows real-time computation tasks to be offloaded from resource

L. Barolli (Ed.): AINA 2023, LNNS 654, pp. 330–342, 2023.
https://doi.org/10.1007/978-3-031-28451-9_29

constraints IoT devices to edge servers [3]. These edge servers may be deployed at base stations that are geographically distributed and close to users. Users can offload computation tasks to nearby edge servers. Many research works are reported in the past on computation offloading in the edge computing environment focusing on latency and energy [4,5]. However, limited work is reported on the profit maximization of edge service providers [6,7]. This work focuses on how to maximize the profit of edge service providers so that they can get motivated to participate in edge computing infrastructure.

To realize edge computing services in practice, it is necessary to design a mechanism that motivates edge service providers for participating in providing their services by promising higher utility in return. So, in this paper, we formulate an optimization problem called the application-assignment problem that aims to maximize servers' profit in the system. A few works on service provider's profit maximization (by minimizing the energy consumption of the edge computing system) have been reported so far [8,9] while they have not considered dynamic scenarios where users' applications can randomly arrive. So, we aim to maximize the profit of servers that may be maintained by different service providers by ensuring resource constraints of edge servers.

The remainder of the paper is organized as follows. Section 2 presents the related work. Section 3 presents the system model. Section 3.1 formulates the profit maximization problem. Section 4 presents the distributed algorithm for finding the optimal set of users' applications. Section 5 presents the numerical results. Finally, Sect. 6 concludes this paper and discusses future work.

2   Related Work

Computational resource allocation in an edge computing environment is nontrivial. Ketykó, István, et al. have discussed a general model considering latency, load balancing, and fairness. They showed the problem of multi-user mobile edge computing (MEC) offloading is NP-hard by reducing the Multiple Knapsack Problem (MKP) to it [10]. In [8], Wang, Quyuan, et al. proposed a mechanism for profit maximization to edge cloud service providers. They used market based profit maximization pricing model for establishing the relationship between resource providers such as edge clouds and the price charged to users. They formulated an optimization problem to maximize the profit of resource providers and also guarantee the quality of experience (QoE) of users/mobile devices. In [11] Moro, Eugenio et al. have addressed the problem of efficient and fair joint management of radio and computing resources in a mobile edge computing environment. They have proposed a market model for resource allocation and fairness. They also proposed a resource allocation mechanism that uses convex programming to find the unique market equilibrium point that maximizes fairness while ensuring that all buyers receive their preferred resource bundle. In [12], Mahmud, Redowan, et al. have proposed a profit-aware application placement policy for Fog-Cloud environments. They formulated the problem using Integer Linear Programming considering profit and QoS simultaneously for application placement on computing instances.

3   System Model

We consider different edge service providers in the system and assume that there are total m edge servers represented as set $\mathcal{S} = \{1, 2, \ldots m\}$, and n users and represented as $\mathcal{A} = \{u_1, u_2, \ldots u_n\}$. Each edge server covers a particular geographical region. Application users that are outside the coverage area of an edge server cannot be allocated to it due to the proximity constraint.

The available computing capacities of an edge server and app users' required capacity are represented by vectors <cpu, memory, bandwidth>[1]. An app user cannot be allocated to an edge server if its requirements are more than the available resources at that edge server. This is called a capacity constraint. The available capacity of an edge server j is denoted by $c_j = \{c_{jk}\}$ and the requirements of an application user u_i is denoted by $r_{u_i} = \{r_{u_i}^k\}$, where $k \in \{$cpu, memory, bandwidth$\}$[2]. Further, we assume that c_{jk}^z denotes the zth instance of the resource type k at the server j. For example, let for server with ID 1 its capacity $c_1 = \{c_{11}, c_{12}, c_{13}\} = \{4, 5, 2\}$ then it has 4 instances of CPU ($z \in \{1, 2, 3, 4\}$) and 5 instances of RAM ($z \in \{1, 2, 3, 4, 5\}$) and 2 instance of bandwidth ($z \in \{1, 2\}$). Here c_{11}^2 represents the second instance of CPU at the server with ID 1.

We assume the application users remain unchanged (their resource requirements and locations) during the allocation.

Coverage Constraint. An app user u_i can be allocated to an edge server j only if it is under the proximity of j.

Further, it is assumed that the system operates in a time-slotted fashion. In each time slot, each edge server announces its presence. Then, each application user broadcasts its computation task offload requirements/requests to the edge servers in its coverage, and then where each user selects its targeted edge server in each time slot while satisfying the computing and coverage constraints. Now, we formulate the problem in the following section.

3.1   Problem Formulation

We assume that each edge server offers three types of resources for application users i.e. cpu or core ($k = 1$), memory ($k = 2$), and bandwidth ($k = 3$). We also assume that request r_{u_i} of an application u_i is in the form of <cpu, memory, bandwidth>. Therefore, the request of an user's application u_i can be denoted as $r_{u_i} = <r_{u_i}^1, r_{u_i}^2, r_{u_i}^3>$. From our system model, an user's application u_i with the request of size r_{u_i} needs to pay w_{ij1}, w_{ij2} and w_{ij3} per unit price for cpu, memory, and bandwidth respectively, to the service provider of the edge server j, if its request is assigned to the edge server j.

Now, we formulate the problem of servers' profit maximization by considering their operating cost (energy consumption) and QoS requirements (in terms of bandwidth requirements) of application users. The profit is mainly determined

---

[1] We consider three types of resources in this paper.
[2] CPU or core ($k = 1$), RAM or memory ($k = 2$), and bandwidth ($k = 3$).

by two factors. First is the operating cost of edge nodes and second is how the resources are allocated to different applications. The operating cost of a powered-on edge server j is mainly determined by the energy consumption of each resource type $k \in \{1, 2, 3\}$ and given by

$$E_j = \begin{cases} \sum_{k=1}^{3} E_{j,k}^{0}, & \text{when edge server is idle} \\ \sum_{k=1}^{3} (E_{j,k}^{0} + E_{j,k}^{f} \cdot u_{j,k}), & \text{otherwise} \end{cases} \quad (1)$$

where $E_{j,k}^{0}$ is the energy consumption at edge node j when the resource type k at the server is idle, $E_{j,k}^{f}$ is the energy consumption at edge server j when the resource type k is fully utilized at the server, and $u_{j,k}$ the utilization rate of resource type k of edge server j.

The utilization rate of resource type k of the edge server j is given as the ratio of the amount of resource type k assigned to applications and the total capacity of resource type k at edge server j. Therefore, the utilization rate is defined as follows,

$$u_{j,k} = \frac{\sum_{i=1}^{|A|} x_{ijk} \cdot r_{u_i}^{k}}{c_{jk}}, \quad \text{for each } k \in 1, 2, \ldots 3 \quad (2)$$

The variable x_{ijk} is a binary variable which indicates that the resource type k of the edge server j is assigned to the application i. The value $r_{u_i}^{k}$ represents the demands for resource type k, by the user application u_i.

So, the main question that remains to be addressed is how to allocate users' application requests to the edge server to maximize the profit of edge service providers. We call this problem an application assignment for servers' profit maximization (SPM-AA).

$$\max_{x} \sum_{u_i \in A} \sum_{j \in S} \sum_{k \in \{1,2,3\}} w_{ijk} \cdot r_{u_i}^{k} \cdot x_{ijk}^{z} - \sum_{j=1}^{m} E_j \cdot y_j \quad (3)$$

subject to:

$$\sum_{u_i \in A} \sum_{k \in \{1,2,3\}} r_{u_i}^{k} \cdot x_{ijk}^{z} \leq c_{jk}, \quad \forall z \in \{1, 2 \ldots c_{jk}\}, \forall j \in S \quad (4)$$

$$\sum_{j \in S} \sum_{k \in \{1,2,3\}} x_{ijk}^{z} \leq 1 \quad \forall u_i \in A \quad (5)$$

$$x_{ijk}^{z} \in \{0, 1\} \quad \forall u_i \in A, j \in S, k \in \{1, 2, 3\}, z \in \{1, 2 \ldots c_{jk}\} \quad (6)$$

where y_j is a binary variable that represents the status of the edge server j. The value of variable y_j is 1 if the server j is powered-on, and 0 otherwise.

x_{ijk}^{z} is a binary variable and is equal to 1, if the zth instance of the resource type k of the edge server j is assigned to the application u_i, 0 otherwise.

The Eq. (4) (also called capacity constraint) satisfies the constraint that for each edge server j that is selected as a service provider, the total amount of resources assigned of each type k cannot exceed its available capacity c_{jk}. The Eq. (5) ensures that an instance z of resource type k of an edge server can be assigned to at most one user application.

Now, we prove that SPM-AA problem is NP-hard.

Theorem 1. *SPM-AA problem is NP-hard.*

We omit the proof due to space constraints.

Now, we discuss the algorithm called the application assignment algorithm for servers' profit maximization (SPM-AAA) using a dynamic programming approach.

4   Offloading Algorithm

As we have mentioned in the system model that system operates in a time-slotted fashion. Each time slot is divided into five non-overlapping periods corresponding to five operation modes, i.e., edge server broadcasting (ESB), users' application requests broadcasting (URB), edge server decision making (ESDM) at the edge server for selecting application users and informing them, user application decision (UD), and user request execution and returning the result (URE).

- Edge server broadcasting (ESB): Each edge server announces its presence, and price rate per unit of its resource types.
- User's application requests broadcasting (URB): After getting the information of edge servers, each application user broadcasts its offloading requests to the edge servers as per coverage constraints.
- Edge server decision making (ESDM): After receiving the offloading requests from users, edge servers select the desired set of users aiming at maximizing their profit under the constraint of resource capacity and give offers to the selected set of users.
- Users' application decision (UD): Based on the offers received by edge servers (if a user is in the coverage of more than one edge server, it may receive more than one offer), users select the edge server based on price and QoS requirements (average response time, maximum response time, failure rate) to offload their tasks and finally transmit their data to them for computation.
- Users' requests execution and returning the result (URE): After receiving the applications' data, servers assign their resources and execute the task of the application and return the results to users.

The length of the one-time slot is a design parameter that is application-dependent [13]. Each time slot consists of the time for decision-making and the time for computation processing. Now, we discuss the algorithm that edge servers execute to select the set of users' applications that maximizes their profit under capacity constraints. Now, we discuss an algorithm for application assignments based on dynamic programming.

4.1 Application Assignment Algorithm for Servers' Profit Maximization (SPM-AAA) Using Dynamic Programming Approach

In this section, we present an algorithm used by servers to maximize their profit (in terms of payment received by clients) while satisfying the bandwidth requirements of applications to satisfy the delay constraints. The definition for the profit $P_j(|A_j|, R, c_j)$ of a server j is defined in the Eq. (7).

$$
P_j = \begin{cases}
0, & \text{if } |A_i| = 0 \\
P_j(|A_j| - 1, R, c_j) & \\
\quad \text{if } (r^1_{u_{|A_j|}} > c_{j1}) \text{ or } (r^2_{u_{|A_j|}} > c_{j2}) \text{ or } (r^3_{u_{|A_j|}} > c_{j3}) & (7) \\
\max\{(p[u_{|A_j|}] + P_j(|A_j| - 1, R - r_{u_{|A_j|}}, & \\
\quad c_j - r_{u_{|A_j|}}), P_j(|A_j| - 1, R, c_j)\}, & \text{otherwise}
\end{cases}
$$

A_j denotes the current set of users' apps requested to the server j. For each user app u_i, its request is denoted as r_{u_i}. R is the request vector for A_j (in the form of $\{r_{u_1}, r_{u_2}, \dots r_{u_{|A_i|}}\}$), and c_j is resource capacity of the server j (in the form of $\{c_{j1}, c_{j2}, c_{j3}\}$. $p[u_{|A_j|}]$ is the payment from the user $u_{|A_j|}$.

Now, we discuss our procedure based on dynamic programming that each server executes and it returns a set of users that maximizes their profit. The procedure is presented in Algorithm 1.

Let T be the length of one time-slot for the algorithm SPM-AAA. T consists of five non-overlapping periods corresponding to five operation modes, i.e., an edge server broadcasting (ESB), users' application requests broadcasting (URB), edge server decision-making (ESDM) at the edge server for selecting users and informing them, users application decision (UD), and user request execution and returning results (URE) as follows:

$$
T = ESB + URB + ESDM + UD + URE \tag{8}
$$

SIT is a small time interval between any two-time periods.

Now we discuss the steps at the server side and user side during each time slot. The steps at the server side is presented in the Algorithm 2. These steps are repeated by servers and users applications after every time slot T. Initially, when a new user application arrives it sets MATCHED variable as FALSE. All the applications for which matched variable is FALSE will execute the Algorithm 3.

5 Results

We conduct a set of experiments to evaluate the performance of our proposed algorithms. We compare the performance of SPM-AAA with the JTOPD [14] and Anchor [15] approaches in terms of server's profit, the number of users' application/clients matched, server resources (memory, cpu, and bandwidth) utilization,

---

Algorithm 1: Max_Profit(A_j, R, C)

Input: Set A_j of applications requested to the server j, R is the set of request size r_{u_i}, $\forall i \in A_j$, c_j is available resource at edge server j

Output: set V which holds selected users, variable P maximum profit of server j

begin

    initialize $P \leftarrow 0$ and $V \leftarrow \emptyset$ /* variable P returns max profit and V returns the set of selected users apps   */

    for $i = 1$ to $|A_j|$ **do**

        | $P[i, 0, 0, 0] = 0$

    end

    for $l = 1$ to c_{j1} **do**

        | $P[0, l, 0, 0] = 0$

    end

    for $m = 1$ to c_{j2} **do**

        | $P[0, 0, m, 0] = 0$

    end

    for $n = 1$ to c_{j3} **do**

        | $P[0, 0, 0, n] = 0$

    end

    for $i = 1$ to $|A_j|$ **do**

        for $l = 1$ to c_{j1} **do**

            for $m = 1$ to c_{j2} **do**

                for $n = 1$ to c_{j3} **do**

                      if $r^1_{u_i} \le l$ *and* $r^2_{u_i} \le m$ *and* $r^3_{u_i} \le n$ **then**

                        | $P[i, l, m, n] \leftarrow$

                          $\max\{P[i-1, l, m, n], p[u_i] + P[i-1, l-r^1_{u_i}, m-r^2_{u_i}, n-r^3_{u_i}]\}$

                        /* $p[u_i]$ is payment value that server j receives if user app i is offloaded to server j */

                    end

                  else

                    | $P[i, l, m, n] \leftarrow P[i-1, l, m, n]$

                  end

                end

            end

        end

    end

    /* Finding the selected users' app   */

    set $i \leftarrow |A_i|$, $l \leftarrow c_{j1}$, $m \leftarrow c_{j2}$, $n \leftarrow c_{j3}$

    while $i, l, m, n > 0$ **do**

        if $P[i, l, m, n] \ne P[i-1, l, m, n]$ **then**

            | $V \leftarrow V \cup i$ /* mark user app i as selected   */

            | $i \leftarrow i-1$, $l \leftarrow l-r^1_{u_i}$, $m \leftarrow m-r^2_{u_i}$, $n \leftarrow n-r^3_{u_i}$

        end

        else

            | $i \leftarrow i-1$

        end

    end

end

---

and resource request service ratio. JTOPD focuses on joint task offloading and payment determination in mobile edge computing. This paper aims to optimize social welfare. The limitation of JTOPD is that it takes more number of rounds

---

Algorithm 2: Steps at server j side

---

1. During edge server broadcasting (ESB)
   (a) Server j broadcasts its presence and available resources in its coverage range
2. During user application requests broadcasting (URB) Server j waits and receives the request messages from the users in its range.
3. During edge server decision making (ESDM) After URB, each server will acknowledge the received requests to corresponding users with one of these flags as follows:
   - *NOT_ENOUGH_RESOURCES* when a server can not fulfill the requirements of an application (so that application does not wait for this server)
   - *CONSIDERING* when the server has enough resources and will consider the request while decision making
   (a) During ESDM time, each server will run the algorithm `Max_Profit()` to maximize the profit and for a selected application does the following:
   (b) Reserves the resources for the application
   (c) Sends an `OFFER` message to the application with resources cost and wait for offered applications acceptance message
   (d) Sends `REJECT` message to other applications which are not selected
4. During users decision (UD) Server j wait for offered users' applications to respond to their final decision
   (a) If server a receives `APPROVE` message from an application, it allocates blocked resources to it and sends an `ALLOCATED` message to the application
   (b) If a server receives a `DECLINE` message from an application, it will unblock resources, that it blocked for that application, and add to the free resources set.
5. User request execution and returning results (URE) Server j executes the tasks of the assigned applications and returns the results to them.
6. Server will enter ESB and start from step 1.

Algorithm 3: Steps at application i side

---

1. During ESB Application u_i waits for getting information about the available servers in its range
2. During URB
   (a) Application u_i broadcasts its requirements to all servers in the range
3. During ESDM
   - Application u_i listens for *NOT_ENOUGH_RESOURCES* or *CONSIDERING* message from servers
4. During UD
   (a) If the application u_i has received an `OFFER` message from a server with resources cost during *ESDM* time interval, then it decides the least cost offer among all the received offers, the application selects the least cost offer and sends `APPROVE` message to the respective server. The application will also send a `DECLINE` message to the remaining offers.
       i. Set $MATCHED \Longleftarrow TRUE$
   (b) Else application u_i will enter ESB and start from step 1
5. During URE User u_i waits for results from the matched server.

---

to generate an offloading match and may not be optimal for servers in terms of their profit. Our algorithms aim to achieve a matching on the server side by maximizing their benefits in offloading services. In our experiments, we consider heterogeneous systems, where edge servers have different resource capabilities and applications have heterogeneous resource requirements. Our proposed algorithm is one-to-many matching, where an application can match to one edge server whereas an edge server can serve requests of many applications. The algorithms are implemented in Python and experiments are conducted on a machine with Core i5 with 12 GB RAM. Edge servers are allocated randomly using a uniform distribution.

To evaluate our proposed scheme, we performed several experiments under various scenarios by changing the following parameters: 1) the number of users, and 2) the number of edge servers. The available capacities on each edge server are randomly generated following normal distributions. We randomly generate the resource requirement $r_{u_i}^1$ within $[1, 4]$, $r_{u_i}^2$ within $[4, 16]$ and $r_{u_i}^3$ within $[5, 10]$ while resource availability c_{j1}, c_{j2} and c_{j3} is generated within $[4$–$16]$, $[64$–$256]$, and $[30$–$50]$ respectively, similar to work reported in [14]. Further, we assume that the Running/maintenance cost per CPU core at a server (idle, fully utilized) is $[2, 4]$, per GB of RAM (idle, fully utilized) is $[1, 3]$ and per Mbps of bandwidth (idle, fully utilized) is $[2, 5]$.

We consider the following two scenarios for simulation. *Scenario (a): Under random arrivals of users' applications with a fixed number of edge servers:* In this scenario, we allow users' applications to generate requests randomly. The number of requests varies with time. We fixed the maximum number of users apps to 70. We fix 25 edge servers and then allow the random arrival of users' apps within $[10$–$70]$ with random request sizes for $r_{u_i}^1$ within $[1$–$4]$, $r_{u_i}^2$ within $[4$–$16]$ and for $r_{u_i}^3$ within $[5$–$10]$ in the system. We present the results for the total assigned number of applications.

Scenario (b): Under the varying number of servers with the fixed maximum number of users' applications: This scenario simply simulated scenario (a) for different numbers of servers. We fix the number of users' applications to 70. Then, we performed matching by generating 5 to 35 edge servers with increments of 5 servers. We present the results for $[10, 15, 20, 25, 30, 35]$ to see the variations.

5.1 Performance Comparison on the Profit of Edge Servers

To investigate the advantages of task offloading of users' applications to edge servers, we compute the profit of servers in the system. The overall profit of the servers in the system is defined as follows (sum of profit of all edge servers):

$$\sum_{j=1}^{m}\{P_j - E_j \cdot y_j\} \tag{9}$$

where P_j is the total revenue received by the server j and E_j is energy consumption and $y_j = 1$ if it is ON. m denotes the number of servers in the system. To analyze the performance of considered schemes in terms of the total profit of edge servers, we simulate two scenarios as discussed above. To see the effect of the different numbers of participating applications on servers' profit, we consider the random arrival of users' apps within $[10$–$70]$ under scenario (a). From Fig. 1a, it can be observed that SPM-AAA gives the maximum profit than other considered schemes. It can also be observed from Fig. 1a that total profit increases as the number of participating users' applications increases in the system under all the schemes since the system will get more opportunities to serve more applications.

For scenario (b), we fix the number of users' applications to 70. Then, we performed matching by generating 5 to 35 servers with increments of 5 servers in each experiment. We present the results for [10, 15, 20, 25, 30, 35] to see the variations. It can be observed that SPM-AAA generates the maximum servers' profit among considered schemes, see Fig. 1b. Total server profit is also increasing on increasing the number of servers in the system under all the schemes.

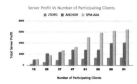

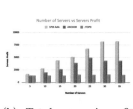

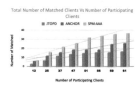

(a) Total servers' profit under scenario (a) (fixed number of edge servers and random arrival of users app request with different requirements)

(b) Total servers' profit under scenario (b)

(c) Total number of assigned users' applications vs number of participating users' applications under scenario (a)

Fig. 1. Servers' profit and total number of assigned users

5.2 Performance Comparison on the Number of Users' Applications Assigned

To see the efficiency of algorithms in terms of the number of users' applications matched to edge servers for offloading their services, we count the number of applications matched to edge servers for task offloading under both the above-discussed scenarios. For scenario (a), it can be observed from Fig. 1c that SPM-AAA is able to assign more applications to edge servers as compared to other approaches because the dynamic programming approach leads to optimized use of resources. It can also be noted that when the number of participating users' applications increases SPM-AAA was able to assign more applications to servers. For scenario (b), SPM-AAA generates better performance as compared to other schemes as shown Fig. 3a. Because of these results SPM-AAA generates higher profits as compared to other schemes.

5.3 Performance Comparison on Average Edge Servers' Utilization

To maximize the profits of servers, it is required that their resources should be utilized maximally. So, in this subsection, we conduct experiments to measure the utilization rate of edge servers under both scenarios (a) and (b) (to see the effectiveness of schemes against resource utilization of edge servers). The average edge servers' utilization rate is computed as the ratio of total resources assigned and total resources available in the system. We present the utilization data for both resources. Figures 2a and 2b show the CPU and RAM utilization ratio respectively under scenario (a). It can be observed that as the number of participating users' applications increases utilization of resources also increases. For scenario (b), when the number of servers is less their resources are fully utilized as shown in Figs. 3b and 3c.

It can be noticed that most of the resources of edge nodes are being used when there are fewer edge servers available or more applications requesting offloading (see Figs. 3b, and 3c). Note that among the considered schemes *SPM-AAA* achieves maximum server utilization under both scenarios. We observed similar results for bandwidth resources also. Due to space constraints, we omit the bandwidth resource utilization results.

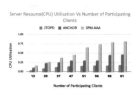  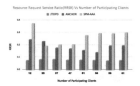

(a) Servers' resource(CPU) utilisation vs number of participating users under scenario (a)

(b) Servers' resource(RAM) utilisation vs number of participating users under scenario (a)

(c) Resource request service ratio(RRSR) vs number of participating users under scenario (a)

Fig. 2. Servers' resource utilisation under scenario (a)

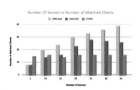 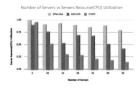 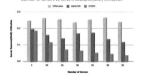

(a) Total number of assigned users' applications vs number of participating users' applications under scenario (b)

(b) Servers' resource(CPU) utilisation vs number of participating users under scenario (b)

(c) Servers' resource(RAM) utilisation vs number of participating users under scenario (b)

Fig. 3. Servers' resource utilisation under scenario (b)

5.4 Performance Comparison on Service Ratio of Edge Computing System

In this subsection, we analyze the overall task-offloading service ratio of edge servers. For this, we compute the ratio of the total resource request served by servers and the total resource request received from applications in the system. This measures the overall offloading requests served ratio. The term resource request service ratio is defined as follows:

Resource Request Service Ratio (RRSR) = Total resource request served/ total resource request received.

We measure the RRSR under the above-mentioned scenarios. For scenario (a), we can observe from Fig. 2c that *SPM-AAA* outperforms all the considered schemes. For scenario (b), we noticed that when the number of servers is less in the system the RRSR is very low because the served requests are low and received requests are higher. With SPM-AAA, as the number of servers increases, the RRSR value also increases. This implies that SPM-AAA tries to serve more requests. We omit the figure due to space constraints.

6 Conclusion

In this paper, we proposed a distributed algorithm for edge user allocation from the edge servers' perspective in the edge computing environment. We show that the optimal edge user allocation problem is NP-hard. Then we presented a distributed algorithm using dynamic programming in a synchronized time-slotted edge computing environment. We conduct experiments under two scenarios to evaluate the performance of the proposed scheme. The simulation results confirm that the proposed algorithm performs well as compared to the considered schemes in terms of servers' profit, the number of applications assigned, server utilization, and resource request service ratio. In the future, this work can be extended by considering dynamic pricing mechanisms to further enhance the understanding in case of varying edge servers' resource demands. The question that how service providers determine the prices of their resources in dynamic participation (including the arrival of new app users and the departures of existing app users) of IoT users can also be addressed. Further, users' applications experiences and the quality of services such as response time, and failure rate can also be considered in edge user allocation.

Acknowledgments. This work was supported by the SERB-DST, Government of India under Grant SRG/2020/000575.

References

1. Satyanarayanan, M.: The emergence of edge computing. Computer **50**(1), 30–39 (2017)
2. Liu, S., Liu, L., Tang, J., Yu, B., Wang, Y., Shi, W.: Edge computing for autonomous driving: opportunities and challenges. Proc. IEEE **107**(8), 1697–1716 (2019)

3. Lin, H., Zeadally, S., Chen, Z., Labiod, H., Wang, L.: A survey on computation offloading modeling for edge computing. J. Netw. Comput. Appl. **169**, 102781 (2020)
4. Badri, H., Bahreini, T., Grosu, D., Yang, K.: Energy-aware application placement in mobile edge computing: a stochastic optimization approach. IEEE Trans. Parallel Distrib. Syst. **31**(4), 909–922 (2019)
5. Feng, M., Krunz, M., Zhang, W.: Joint task partitioning and user association for latency minimization in mobile edge computing networks. IEEE Trans. Veh. Technol. **70**, 8108–8121 (2021)
6. Li, Z., Chang, V., Hu, H., Yu, D., Ge, J., Huang, B.: Profit maximization for security-aware task offloading in edge-cloud environment. J. Parallel Distrib. Comput. **157**, 43–55 (2021)
7. Teng, H., Li, Z., Cao, K., Long, S., Guo, S., Liu, A.: Game theoretical task offloading for profit maximization in mobile edge computing. IEEE Trans. Mob. Comput. (2022)
8. Wang, Q., Guo, S., Liu, J., Pan, C., Yang, L.: Profit maximization incentive mechanism for resource providers in mobile edge computing. IEEE Trans. Serv. Comput. **15**, 138–149 (2019)
9. Yuan, H., Zhou, M.: Profit-maximized collaborative computation offloading and resource allocation in distributed cloud and edge computing systems. IEEE Trans. Autom. Sci. Eng. **18**, 1277–1287 (2020)
10. Ketykó, I., Kecskés, L., Nemes, C., Farkas, L.: Multi-user computation offloading as multiple knapsack problem for 5G mobile edge computing. In: 2016 European Conference on Networks and Communications (EuCNC), pp. 225–229. IEEE (2016)
11. Moro, E., Filippini, I.: Joint management of compute and radio resources in mobile edge computing: a market equilibrium approach. IEEE Trans. Mob. Comput. (2021)
12. Mahmud, R., Srirama, S.N., Ramamohanarao, K., Buyya, R.: Profit-aware application placement for integrated fog-cloud computing environments. J. Parallel Distrib. Comput. **135**, 177–190 (2020)
13. Zhang, X., Zheng, K., Chen, J., Li, Y.: QoE-based scheduling for mobile cloud services via stochastic learning. In: 2014 IEEE 80th Vehicular Technology Conference (VTC2014-Fall), pp. 1–5. IEEE (2014)
14. Wang, X., Wang, J., Zhang, X., Chen, X., Zhou, P.: Joint task offloading and payment determination for mobile edge computing: a stable matching based approach. IEEE Trans. Veh. Technol. **69**(10), 12148–12161 (2020)
15. Xu, H., Li, B.: Anchor: a versatile and efficient framework for resource management in the cloud. IEEE Trans. Parallel Distrib. Syst. **24**(6), 1066–1076 (2012)

Stability and Availability Optimization of Distributed ERP Systems During Cloud Migration

Gerard Christopher Aloysius[1], Ashutosh Bhatia[2(✉)], and Kamlesh Tiwari[2]

[1] Schneider Electric, Bangalore, Karnataka, India
gerardchristopher.aloysius@se.com
[2] Birla Institute of Technology and Science, Pilani, Pilani, Rajasthan, India
{ashutosh.bhatia,kamlesh.tiwari}@pilani.bits-pilani.ac.in

Abstract. Enterprise resource planning is a powerful software package that enables businesses to integrate a variety of disparate functions, it is a concept whereby an organization enables effective communication and data exchange within the business units and associated third parties such as vendors and banks. SAP is the de-facto ERP software for a company that manages business operations and customer relations. Many organizations have started migrating their ERP to infrastructure-as-a-service (IAAS) providers like amazon web services and Microsoft Azure. Though there is a significant increase and positive sentiment to migrate productive workloads to the cloud, the availability and reliability of applications remain a huge concern. No comprehensive study specifically examines the stability and availability of productive ERP workload in the cloud with large databases. Cloud brings specific advantages like faster time-to-value, moving out of expensive hardware, ease of spinning up virtual machines, and accessibility of the latest technologies in one place. There is a need for a study to specifically look into moving productive ERP workload to the cloud and a roadmap to achieve optimum stability and availability while leveraging all the advantages we get from cloud computing. The paper's objective is to take in the experience of migrating several productive ERP landscapes with a database size ranging from 2 TB to 20 TB and propose a design roadmap to attain optimum Stability and 99.9% Availability. The design was implemented and measured for its effectiveness post-migration by comparing the application with the data before migration. SAP pre-go-live and post-go-live reports show a 33% reduction in dialog response time and a 40% improvement in response time during peak hours. The overall database grew from 8449 GB to 8703 GB during the analysis. When we extrapolate the analysis to 21 days post-migration, we see that the response time improves even further. We see 100% applications available during the analysis period.

1 Introduction

Many customers currently using ERP Solutions on-premise would prefer to take advantage of the benefits of the cloud, like avoiding hardware cost and

---

The original version of this chapter was revised: Acknowledgement has been added below figures 1 and 2. The correction to this chapter is available at
https://doi.org/10.1007/978-3-031-28451-9_50

L. Barolli (Ed.): AINA 2023, LNNS 654, pp. 343–354, 2023.
https://doi.org/10.1007/978-3-031-28451-9_30

maintenance, labour savings, reduction in lost sales, performance improvements, global presence, and sophisticated availability solutions [1]. These migrations are typical of a "Lift and Shift" nature where there is no change in the vendor/software though there may be changes in the target operating systems/database and hardware. The lift and Shift migration path dominates enterprise workload migration to the public cloud. In an IDC survey, respondents indicated it to be the most common migration path for almost all types of enterprise workloads [2]. Customers prefer this approach due to its simplicity in taking their workload to the cloud. However, there is a concern that there is no significant improvement in the performance or their availability and application response post moving to the cloud [3]. The fundamental problem with a "Lift and Shift" approach is that the landscape is moved "AS-IS" and retains the limitations of an on-premise deployment. On the other hand, cloud infra has some unique capabilities and advantages which should be leveraged without carrying the limitations of an on-premise deployment. In the following, we discuss some issues and constraints related to the on-premise deployment of ERP solutions, which should not be carried over to the cloud.

On-premise deployment involves High Capital expenses in purchasing the Physical hardware and much effort in building and managing the virtual systems [4]. This results in customers preferring a central installation over a distributed deployment. Cloud liberates the customers from these constraints by providing a platform where they can spin up virtual machines at low cost and with less effort, so there is no need to carry this on-premise limitation to the cloud.

Moving systems as low-availability solutions is another on-premise deployment constraint which is usually the result of a lack of flexibility and limitations of the data center. So it does not make sense to carry over these solutions to the Cloud but rather architect it again to leverage the high availability options provided by Cloud [5]. Many on-premise production deployments run in a Single Node solution without High availability but only a less efficient Disaster recovery solution that usually has a high RTO and RPO.

Due to the lack of elasticity in scaling, the on-premise deployments are can be oversized with a huge storage buffer to avoid the need to extend the hardware in the near future. This means customers typically pay for something that they do not use. The lift and shift mentality is to move the workload and size it later [6]. This is a constraint that need not be carried over to the cloud. The same applies to many production systems that are over-utilized with no room to add more CPU or memory, and it is typical of them to run on outdated hardware where there is no room to scale. So overall, the sizing mentality of on-premise should not be carried over to the cloud; hence, there is a need to approach resizing from a cloud standpoint.

In an enterprise with internal users and customers across the globe, the infrastructure needs to be closer to the user base. In an on-premise scenario, that means having to invest in multiple data centers and infrastructure components and is extremely ineffective from a cost standpoint. However, using an enterprise cloud service provider with a presence across different geography makes the process simple, and cost-effective [7]. For example, AWS has a presence across regions like North America, South America, Europe, the middle east/Africa, and the Asia Pacific, with several data centers within these regions. So, to architect an enterprise solution for a global organization is much simpler in the cloud.

Migrating an SAP landscape cannot be done in isolation. We will still need the integration between external systems, which will require much tweaking from both the migrated SAP System and the external system in question. There is a huge reliance on individuals and no systemic approach to handling this. One of the important tasks when planning an SAP migration is to ensure SAP interfaces are documented. Some of the interfaces may have hardcoded IPs that need manual correction or moved to hostnames with DNS configured as a best practice. External interfaces need to be tested, requiring coordination from external partners, so it needs to be carefully planned [8]. There have been enumerable situations where external interfaces are critical but not captured as part of the migration. Raising firewall requests and implementing them during a run is extremely cumbersome and will directly affect availability since a non-working functionality is still a downtime from a customer standpoint.

After several years of having a solution On-Premise with substantial money invested in tools and solutions, customers get tied up to a solution without questioning its efficiency vs. other available technology options. Moving to the cloud reopens this conversation; instead of carrying over their old backup and recovery and monitoring solution, it makes much sense to work with the cloud vendor and identify tools and solution that gives better performance and reliability [9]. In addition, cloud vendors usually support multiple customers with a similar setup, so they should have the performance metrics of these tools handy [10].

This paper proposes a method to ensure stability, availability, and optimum response in the migrated system for a typical "Lift and shift" scenario with or without Unicode conversion. The proposal significantly improves the stability, availability, and performance of the SAP ERP environment.

2   Proposed Methodology for AS_IS On-Premise to Cloud ERP Migrations

The paper presents an approach to ensure stability, availability, and performance in the migrated system for a typical Lift and shift scenario with or without Unicode conversion. This approach arrived after migrating several ERP productive landscapes from on-premise data centers to the cloud. The effect of the migration approach was measured after migrating the productive landscapes and significant improvements were observed in the availability and performance of the SAP ERP system.

The steps involved in the proposed approach are as follows:

1. Convert the Central installations to a Distributed SAP ERP landscape
2. Resize the system based on on-premise CPU usage and workload analysis.
3. Move SAP Landscape to regions closer to the Userbase

4. Design a High availability solution leveraging Cloud
5. Approach interface setup by Scanning ports and testing using pre-emptive tools
6. Measure the effectiveness of Load distribution, High-availability, and Application response post-migration.

We take an example scenario of migrating an SAP ERP system running with an 8 TB database and five application servers. The system is built using an SAP central installation with the central instance and database running on a single node with five application servers. We implement the proposed methodology in this box. Migration methodology is purely system specific and is irrelevant to our discussion here.

2.1   Converting the Central Installations to a Distributed SAP ERP Landscape

In a Distributed SAP system design, the following components of the SAP instance will be built on dedicated virtual machines. Since the central instance and database are installed on a single node that will be separated as part of the target system build. We will have the following components installed separately on dedicated virtual machines.

- The Primary application server (PAS) - 16 cores/128 GB RAM
- Central services (SCS) - 2 Cores 16 GB RAM
- Application servers - 16 Cores/128 GB RAM
- Database Server - 64 Cores/128 GB RAM

This ensures that they do not serve as a bottleneck to each other. The following file systems are shared across the VMs using NFS share; 1) /sapmnt - SAP Mount directory, which holds the SAP profile/ kernel, 2) /usr/sap/trans - SAP Transport directory. The components are separated and will be installed in dedicated virtual machines in a private subnet. The firewall is set at the VM level so that each instance can communicate with the other (Table 1).

Table 1. Production server inventory

SN	SAP ver.	AWS instance	System type	Cores	Memory (GB)
1	SAP ERP6	r5.large	ASCS	2	16
2	SAP ERP6	r5.large	ERS	2	16
3	SAP ERP6	r5.4xlarge	AAS1	16	128
4	SAP ERP6	r5.4xlarge	AAS2	16	128
5	SAP ERP6	r5.4xlarge	PAS	16	128
6	SAP ERP6	r5.4xlarge	AAS3	16	128
7	SAP ERP6	r5.4xlarge	AAS4	16	128
8	SAP ERP6	r5.4xlarge	AAS5	16	128
9	DB Server Primary	c5a.16xlarge	Primary DB	64	128
10	DB Server Secondary	c5a.16xlarge	Secondary DB	64	128

ASCS: Abap Central Services, PASS: Primary Application Server, ERS: Enque Replication Server, AAS: Additional Application Server

2.2   Advantages of Building a Distributed System Deployment

- Every instance of the application servers, as well as the central services and database server, will have dedicated memory and CPU and will not create a resource crunch.
- DB node will need High IO, CPU, and memory to run on a separate large instance so that it is not deprived of system resources. The diagram above shows that the database host will run on a separate host. It also shows the instance type/Sizing details of the system: All the app servers will be running on "xlarge" AWS instances, and the central services will run on a "Large" environment.

2.3   Resize the System Based on On-Premise CPU Usage and Workload Analysis

Rightly sizing the virtual machines based on current performance metrics is extremely important. This can be done in tandem with the vendor's proprietary sizing tools. One reason for inefficiency is the mindset to overprovision that many IT professionals bring with them when they build their cloud infrastructure. The traditional mindset of overprovisioning is unnecessary in the cloud. Cloud provisioning happens based on average usage rather than peak usage, so by rightly sizing, organizations can save up to 70% percent on their monthly bill [11]. By right-sizing before migration, you can significantly reduce your infrastructure costs. Improper sizing can result in higher cloud infrastructure spending for a potentially long time.

- Take a monthly average and peak CPU usage/memory usage and add the buffer to ensure there is at least 40% average idle CPU time/memory usage. For example (if an 8 CPU/64 GB RAM machine has an average CPU utilization of 80–90% and average/peak memory usage of 80/90%, it makes sense to move the target instance to a 16 CPU/128 GB).
- Take SAP workload analysis data (ST03n) and look at the Dialog and Background response times. If the response times are unsatisfactory, there could be a CPU/Mem or DB Bottleneck. So it is important to find the bottleneck and add additional resources to the target system to mitigate this. For example, if we observe an application response time of 2200 ms with a high database response time contributing to 80% of the response time. It makes sense to bump up the resources in the database server and reduce the overall response time to less than 1000 ms.
- Look at SAPS (SAP Application Performance Standard (SAPS) which is a hardware-independent unit of measurement that describes the performance of a system configuration in the SAP environment. Calculate the SAPS delivered by the On-premise system and optimize the target system to ensure it delivers the calculated SAPS with adequate CPU and memory buffers.
- Estimate the IOPS usage (baselines IOPS required) and maximum throughput on the on-premise Database server and ensure adequate idle time and sufficient throughput on the target database servers.

2.4   Move SAP Landscape to Regions Closer to the Userbase

Cloud migration can be a great opportunity to enhance the overall user experience because you can move the services closer to your user base. Based on migration done across different regions in Europe, Asia, and the Americas, we clubbed the landscape based on regions. For example, Europe was placed in "Paris" as the primary region and "Ireland" as the Disaster recovery center. Singapore served as the primary region for Asia and "Sydney" as the disaster recovery. For the landscape built-in "Paris," Since the user base was predominantly based out of Europe, this enhanced the overall user experience, and this is a unique advantage of the cloud since you are not tied to a dedicated data center in a single region. Further, the latency can be tested with utilities such as ping to ensure the server is placed in a region with acceptable latency.

2.5   Design a High Availability Solution Leveraging Cloud

The action aims to build an SAP application using a highly available architectural pattern. This will involve identifying a "Single Point of Failure" in the system and ensuring redundancy for those components. High availability usually entails having a "zero" for the Recovery point objective (RPO) and several minutes as a recovery time objective (RTO) as shown in Fig. 1. The SAP production environment was built, considering the system could handle spikes in workload. At the same time, the infrastructure is positioned to fail safely in case of failures at any layer, be it a database, application, hardware, or a complete data center failure.

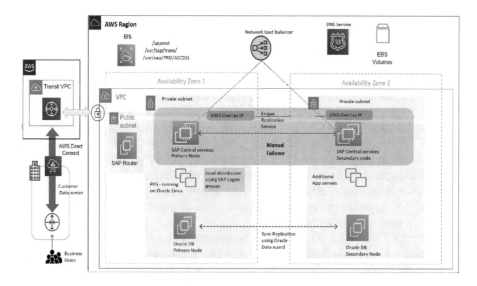

Fig. 1. Design a high availability solution using AWS single region architectural pattern: adapted from Pattern 1: A single Region with two AZs for production mentioned in [12]

Key Features of the Architecture

The Architecture will provide redundancy to the following single point of failures in the system

- SAP Central services (holds the message server and Enque)
- Application Servers
- Database Instance
- /sapmnt file system

SAP HA Components

As shown in Fig. 2, the SAP production system is built across two separate data centers (let us call them DC1 and DC2) in Paris (Region) as active/passive. The compute deployed for the productive Database, SAP central services, and application servers are of the same type on both data centers. SAP Central services (SAP ASCS) is a single point of failure, so it will be replicated by building Active central services in DC1 and a passive central service in DC2. Central services also hold the application locks through the Enqueue service; this is a single point of failure and will be addressed by building an active Enqueue replication service in DC2. In the event of a failure, the cloud network load balancer will relocate the Enqueue service to the active enqueue replication service. This will ensure the locks are not lost in the process. As noted here, we are not leveraging any OS-level cluster solutions for auto-failover but rather a replication solution used in tandem with

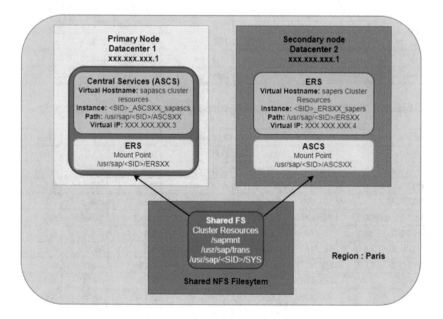

Fig. 2. SAP HA components diagram Adapted from "SAP ASCS High Availability using ERS explained" https://blogs.sap.com/2021/10/28/sap-ascs-high-availability-using-ers-explained/

the cloud network load balancer to move between nodes manually. The solution deployed in a Single region - Multiple data center architecture with redundancy built within the region. If we need an auto-failover to datacenter2, we can add a cluster solution into the mix. There are specific services like FTP, SFTP, and specific ports which can be set up for redirection in the cloud network load balancer. This can be used as an additional application service load balancing tool along with sap logon groups. The load balancing between the application servers is managed using a standard SAP logon group solution (SMLG). Redundant application servers will be available in both data centers and will serve as a backup to each other. Database instances will be replicated by a synchronous replication setup between the databases in both data centers. Logs from the primary database are shipped across the network and applied to the secondary database in real time. So if the primary database fails, the secondary is ready to handle the load. This requires reliable network availability between the two data centers.

Assumptions in the Architecture We require a minimum availability of 99.9%. This will require the system to be replicated to another node within the same region. Since the zones are within the Paris region, we will use a Sync

Table 2. CCMS: load distribution

Instance	St	Resp. time (ms)	Threshold	User	Thrshd	Sample	Quality	Dialog steps
1		393	60000	99	600	14:23:12	578	100
2		774	–	143	–	14:23:17	577	124
3		741	–	150	–	14:20:23	576	115
4		0	–	0	–	00:00:00	0	0
Summary		–	–	392	–	–	–	339

Table 3. Workload overview: average time per step in ms - 21 days of data

Task type name	# steps	Time	Avg. proc. time	CPU time	DB time	Wait time
ALE	210,570	59,6	32,0	9,6	14,9	0,3
AUTOBAP	22,746	1.364,7	599,6	266,5	599,7	1,9
AUTPCCMS	113,618	2,7	2,6	0,3	0,0	0,1
AUTOTH	101,705	2,6	1,3	0,4	0,0	1,3
BACKGROUND	3204,851	6.660,6	2.291,2	931,2	4.262,9	0,0
BUFFER SYNC	56,846	11,7	3,2	1,9	8,4	0,1
DOLOG CLEANUP	56,846	14,1	2,1	0,3	12,0	0,1
DEL. THCALL	581	235,2	173,4	48,5	61,3	0,5
DIALOG	3986,076	848,2	144,9	98,2	441,6	0,8
HTTP	307,392	61,4	17,7	8,1	18,2	0,6
OTHER	145	15.206,0	157,8	140,8	2,5	0,2
RFC	16228,825	2.259,5	433,7	59,9	351,9	0,9
RPCTH	1,861	94,7	49,0	34,6	45,7	0,0
SPOOL	530,113	198,5	128,8	25,2	28,0	38,5

replication to ensure immediate data availability on the other node. We assume Business is fine with the additional cost incurred by deploying the same set of systems with the same capacity in an additional data center. This is the price that we are paying to ensure a stable system. We assume Business has approved the data transfer cost incurred by replicating the data across the data centers. In the event of a failure of a single data center, we are fine with running the productive system on the compute of the alternate data center.

2.6 Approach Interface Setup by Scanning Ports and Testing Using Pre-emptive Tools

The migrated ERP system cannot function in silos. All the connections to the external interfaces directly or through the middleware should work. The first step is to develop a diagrammatic representation of the interfaces using existing documents and inputs from the stakeholders. As a best practice, we recommend starting with the SAP RFC connection using transaction code SM59 and using the data as a starting point to develop the integration diagram.

Capture Ports On-Premise
Set up Port scanners for one month on the on-premise environment to capture incoming and outgoing ports; cloud vendors have their proprietary tools to do this. This will give a list of used ports/IP addresses of incoming and outgoing connections.

Retain On-Premise Fully Qualified Domain Name (FQDN)
Retaining the same FQDN as a source for migration ERP system can save a lot of effort because as long as the firewall is raised and the DNS is modified to reflect the new IP, there won't be any changes required in the backend systems to redirect to the new VM. Since the FQDN is retained, they should be able to talk to the migrated system without any backend reconfiguration. This can lift a lot of burden during interface testing.

Raise Firewall Requests
Based on how the network is configured, we need firewall requests raised to open the ports to allow connection unidirectionally or bi-directionally. All the requested rules need to be implemented by the network focal.

Test Firewall Rules Beforehand
Once the Firewall rules are implemented and the virtual machines are ready, we can test whether the implemented rules are working by using a simple telnet/curl test to check for the response on specific ports since the SAP system is not installed at this point. In addition, we can use scripts to listen on specific sap ports like 32XX 33XX 5XX00, etc., and test the incoming from backend systems. This will ensure that the implemented rules are working.

Perform Interface Testing
As part of the mock migration, we can get all the interfaces tested beforehand; the final testing can follow this during the cutover. In addition, performing a mock migration helps to ensure the interfaces are talking from and to SAP ERP.

2.7   Measure the Effectiveness of the Road Map

The best part of the road map is that all the attributes addressed can be effectively measured using standard application tools. Performing this productive ERP migration, we measured the attributes as shown below.

Load Distribution
The CPU Load on the app servers and DB server are below the threshold and well distributed across the landscape.

Table 2 shows SAP Workload distribution during peak business hours as we can see, the workload is optimally distributed across the application servers.

Availability Measurements
There was not a single unscheduled downtime since the start of the application on April 7, 2022–April 23, 2022 (day of measurement). HA Movement Drill was carried out, and the total time to move to Node 2 was measured as less than 5 mins. Measurement since April 7th till date: Total Uptime = 372 h, Total time = 372 h Application Availability = Total Uptime/Total Time * 100 = 100% Availability.

Table 4. Pre migration performance indicators and values

Area	Indicators	Value
System Performance	Active users (>400 steps)	455
System Performance	Avg. Availability per week	100%
System Performance	Avg. Response Time in Dialog Task	1442 ms
System Performance	Max. Dialog Steps per Hour	31114
System Performance	Avg. Response Time at Peak Dialog Hour	1298 ms
System Performance	Avg. Response Time in RFC Task	2162 ms
System Performance	Max. Number of RFCs per Hour	52712
System Performance	Avg. RFC Response Time at Peak Hour	2602 ms
Hardware Capacity	Max. CPU Utilization on DB Server	64%
Hardware Capacity	Max. CPU Utilization on Appl. Server	25%
Database Performance	Avg. DB Request Time in Dialog Task	1136 ms
Database Performance	Avg. DB Request Time for RFC	496 ms
Database Performance	Avg. DB Request Time in Update Task	92 ms
Database Space Management	DB Size	8449.04 GB
Database Space Management	DB Growth Last Month	206.44 GB

Analysis of Response Time Post Migration
Table 4 and Table 5 show the migration analysis performed by SAP AG (Pre and Post migration). The report shows a 33% reduction in Dialog response time and a 40% improvement in response time during peak hours. Furthermore, when we extrapolate the data for 21 days, we see the dialog response shows a 41% reduction as indicated in Table 3. We see this response pattern in almost all the migrations done using the same road map and architectural pattern.

Table 5. Post migration performance indicators and values

Area	Indicators	Value	Trend
System Performance	Active users (>400 steps)	466	⟶
System Performance	Avg. Availability per week	100%	⟶
System Performance	Avg. Response Time in Dialog Task	957 ms	⟶
System Performance	Max. Dialog Steps per Hour	27602	⟶
System Performance	Avg. Response Time at Peak Dialog Hour	777 ms	⟶
System Performance	Avg. Response Time in RFC Task	2412 ms	⟶
System Performance	Max. Number of RFCs per Hour	51183	⟶
System Performance	Avg. RFC Response Time at Peak Hour	1086 ms	⟶
Hardware Capacity	Max. CPU Utilization on Appl. Server	28%	⟶
Database Performance	Avg. DB Request Time in Dialog Task	472 ms	⟶
Database Performance	Avg. DB Request Time for RFC	432 ms	⟶
Database Performance	Avg. DB Request Time in Update Task	47 ms	⟶
Database Space Management	DB Size	8703.88 GB	⟶
Database Space Management	DB Growth Last Month	224.84 GB	⟶

3 Conclusion

The proposed method to migrate workloads from different regions has consistently resulted in sustained stability and availability of the ERP system, showing significant improvement in the application performance. These migrations were of a "Life and shift" nature, but the underlying principles can be applied across any migration of productive workload to the cloud. Gartner Says More Than Half of Enterprise IT Spending in Key Market Segments Will Shift to the cloud by 2025, which translates to a significant workload moving to the cloud in the coming years. The lessons learned and the proposed method can serve as a starting point to make the project, investment, and technical decisions for companies venturing into moving their productive workload to the cloud.

Acknowledgement. This dissertation was carried out as part of requirement for Mtech (Software Systems) program under Work integrated learning program (WILP) in BITS Pilani, I would like to thank my reporting manager Xavier Deprat (Head of Global SAP operations, Schneider Electric) for his tremendous support throughout the process.

References

1. A Forrester Total Economic Impact™ study commissioned by Amazon. https:// pages.awscloud.com/Amazon_Connect_Forrester_TEI_Report.html. Accessed 06 Jan 2023
2. Subramanian, S.: Whither Goest Thou, Enterprise Workloads? IDC (2020). https://blogs.idc.com/2020/03/02/whither-goest-thou-enterprise-workloads/. Accessed 04 Jan 2022

3. Upadrista, V.: Formula 4.0 for Digital Transformation: A Business-Driven Digital Transformation Framework for Industry 4.0. CRC Press (2021)
4. Khajeh-Hosseini, A., Greenwood, D., Smith, J.W., Sommerville, I.: The cloud adoption toolkit: supporting cloud adoption decisions in the enterprise. Softw. Pract. Experience **42**(4), 447–465 (2012)
5. Lin, M., Wierman, A., Andrew, L.L., Thereska, E.: Dynamic right-sizing for power-proportional data centers. IEEE/ACM Trans. Netw. **21**(5), 1378–1391 (2012)
6. Shen, D., et al.: Stochastic modeling of dynamic right-sizing for energy-efficiency in cloud data centers. Futur. Gener. Comput. Syst. **48**, 82–95 (2015)
7. Tang, L., Dong, J., Zhao, Y., Zhang, L.J.: Enterprise cloud service architecture. In: 2010 IEEE 3rd International Conference on Cloud Computing, pp. 27–34. IEEE (2010)
8. Morgan, N., Jarkowski, B.: SAP on Azure Implementation Guide: Move Your Business Data to the Cloud. Packt Publishing (2020). https://books.google.co.in/books?id=OAdYzQEACAAJ
9. Gastermann, B., Stopper, M., Kossik, A., Katalinic, B.: Secure implementation of an on-premises cloud storage service for small and medium-sized enterprises. Procedia Eng. **100**, 574–583 (2015)
10. Rashid, A., Chaturvedi, A.: Cloud computing characteristics and services: a brief review. Int. J. Comput. Sci. Eng. **7**(2), 421–426 (2019)
11. AWS Whitepaper: Right Size Before Migrating. https://docs.aws.amazon.com/whitepapers/latest/cost-optimization-right-sizing/right-size-before-migrating.html. Accessed 07 Jan 2022
12. AWS General SAP guide: Single Region architecture patterns. https://docs.aws.amazon.com/sap/latest/general/arch-guide-single-region-architecture-patterns.html. Accessed 06 Apr 2022

Prison Break: From Proprietary Data Sources to SSI Verifiable Credentials

Katja Assaf[(⊠)], Alexander Mühle, Daniel Köhler, and Christoph Meinel

Hasso Plattner Institute, Potsdam, Germany
{katja.assaf,alexander.muehle,daniel.koehler,christoph.meinel}@hpi.de

Abstract. Despite extensive efforts, smaller companies and organisations often fail to be GDPR compliant. GDPR demands that the data subject's information is available to the data subject in a simple and structured way. One option to provide the data with additional benefits is issuing verifiable credentials (VCs) following the W3C standard and, thus, introducing the data provider as an issuer into a Self-Sovereign Identity (SSI) system. We show that this can be achieved with limited overhead by introducing a middleware component, which is only loosely coupled with the existing ecosystem. To enhance user acceptance, we define our design goals as usability, security, and privacy, which we manage to achieve partially. During our work, we identified several challenges, such as revocation, verifiability of verifiers, and legal regulations, which provide options for future research in developing Self-Sovereign Identity solutions towards real-world applicability.

1 Introduction

Since May 2018, the General Data Protection Regulation (GDPR) has been binding for all European Union member states. It requires businesses with customers within the European Union to process data in a secure, transparent and accountable way. However, according to [7], roughly half of the asked small businesses are not yet GDPR compliant despite investing heavily in compliance. In contrast, big companies such as google already provide specialised tools to enable customers to create a (data) takeout, which is a downloadable package of the user's data. Two central points within GDPR are Article 15 *Right of access by the data subject* and Article 20 *Right to data portability*. The *right of access by the data subject* requires organisations to make user data available to the data subject after authentication. The *right to data portability* goes further and requires organisations to provide the data in a "structured, commonly used and machine-readable format" [5]. We suggest enhancing GDPR compliance by providing a data takeout in a verifiable format, which grants data subjects the additional benefit of being reusable within the Self-Sovereign Identity (SSI) system. Verifiable credentials (VCs) are beneficial as learner records, health data, e.g. vaccination certificates or customer data, although they require more caution from a data privacy perspective.

L. Barolli (Ed.): AINA 2023, LNNS 654, pp. 355–366, 2023.
https://doi.org/10.1007/978-3-031-28451-9_31

Self-Sovereign Identity (SSI) is a design principle that emphasises the user's control over their data. It is at the stage where a minimum user base still needs to be established [9]. In addition to establishing a standard, as currently done by the W3C[1], it is crucial to enable individuals and organisations to participate in the system with minimal friction. Besides raising the number of credential holders and verifiers, also called relying parties, the number of credential issuers within the system also needs to rise. According to Schmidt et al. [19], in 2021, only 11 out of 147 investigated projects were classified as issuers. Consequently, we will answer the question:

> **RQ:** How to enable a service provider to become an issuer in an SSI system and, thus, integrate an SSI system with an existing ecosystem with minimal friction?

Organisations already have an infrastructure for identity and customer information management, and these existing data sources are the logical starting point for a potential credential issuance process. For this purpose, we present an approach utilising existing identity and customer information management systems (as shown in Fig. 1) for issuing VCs, lowering the barrier of entry for issuers to SSI ecosystems.

Self-Sovereign Identity (SSI), as stated above, is a design goal for an identity management system giving the credential holder full control over their digital identity [13]. However, SSI often refers to concrete implementations that usually rely on blockchain and only partially fulfil the design goals at best. These self-claimed SSI systems differ from what we mean when referring to an SSI system.

An SSI system consists of three actors: *issuer, relying party* and *credential holder.* The credential holder owns credentials containing claims about them or entities related to them. The credential holder receives or retrieves credentials

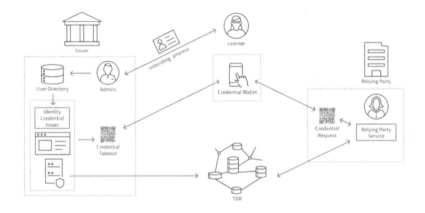

Fig. 1. System overview

---

[1] https://www.w3.org/TR/vc-data-model/.

from an issuer, who signs the claim. The holder bakes one or more credentials into a credential presentation and presents the presentation to a relying party, which verifies all signatures and looks up the current status of the entities in a trusted data registry.

Due to the primary purpose of the relying party, it is called *a verifier* in particular contexts. We will use both terms interchangeably to emphasise either the role within the system or the protocol carried out. We will use the term *customer* for individuals who have an established relationship with an organisation but have yet to become actors within the SSI system. Additionally, we will use the term *user* to refer to any participant of an SSI system regardless of its specific role.

Customer Information Management is necessary for all services, especially those available online, allowing customers to access their data via a login-protected website. This can be access to a bank account, the last eCommerce order or learning achievements at a learning platform. These service providers have in common that they store data in their database and run a web server connected to their database and an identity provider, which checks access control. Most use the same communication standards: oAuth 2.0 is the standard for federated identity management, meaning it is used for signing in with Facebook, Google or Twitter on another web page. Access to the data is provided via a RESTful API, although many are not publicly available but restricted to the company's network.

Our Contribution. To enable organisations to offer SSI credentials to their customers and enhance their GDPR compliance, we propose a new scheme to enable users and organisations to transform existing data into SSI credentials. As described above, identity management systems already have a harmonised landscape for accessing user profiles. In our proof of concept, we show the viability of using OIDC to enable organisations as credential issuers easily. We restrict assumptions about the SSI system and the organisation's infrastructure to a minimum (Sect. 3.2) to offer a solution as generic as possible in compliance with our defined design goals (Sect. 3.1). We describe our reference implementation to ensure the feasibility of our suggestion (Sect. 3.3) and assess our solution against our design goals while pointing out its limitations (Sect. 4).

2   Related Work

In general, research interest in Self-Sovereign Identity steadily increased in recent years from 5 papers in 2018 to 37 in 2021, according to Schardong et al. [18]. Additionally, Schardong et al. [18] provide a taxonomy to classify practical problems discussed in the literature. The work of Grüner et al. [9] is classified as protocol integration, identity derivation and trust policy evaluation. Grüner et al. [9]

tackle the problem of achieving a minimal user base in Self-Sovereign Identity. Thus, having a similar focus as our work but considering it from a different angle. While our solution strives to include an organisation with existing data as an issuer, Grüner et al. enable an organisation as a verifier. In terms of classical identity management, we can say that Grüner et al. developed a solution for federated identity verification, while our solution describes federated identity issuance.

Regarding the taxonomy developed by Schardong et al. [18], our solution would also fit into the category of protocol integration. However, from the eight papers in this category, only Jurado et al. [12] consider integrating an existing data provider as an issuer into the SSI ecosystem, while the other seven paper work on enabling service providers as SSI verifiers by enabling authentication with SSI. Jurado et al. [12] are firmly set in the European Health Insurance Card (EHIC) use case and its integration with eIDAS. Thus, their solution contains more details than ours and considers issuer and verifier integration simultaneously, as well as the underlying trust framework. In contrast, our solution makes fewer assumptions on the existing infrastructure and tries to provide a more flexible model focused on only issuer integration.

The survey of Schardong et al. [18] covers nearly all of the relevant literature since it was published in August 2022. The existence of two more research papers [1,11] advancing the research on SSI integration, also published last summer and thus not considered within the survey, emphasises computer scientists' current interest in the topic.

Kuperberg et al. [11] provide a state-of-the-art survey about bridging the gap between SSI and traditional IAM systems. Most of the over 40 SSI solutions were excluded since they were not concerned with integration. The remaining sources were either concerned with authentication via SSI by enabling the identity provider (Pattern A) or translating VCs into the OIDC protocol (Pattern B and C).

Bolgouras et al. integrated the FIDO protocol and the eIDAS framework into an SSI framework to allow for authentication using FIDO and verification across countries [1], extending authentication options for SSI solutions.

3   Proof of Concept

To show the feasibility of our idea, we developed a proof of concept enabling a MOOC provider to issue verifiable credentials (VCs). Since we strive to achieve flexibility and broaden user acceptance, we defined our design goals accordingly and restricted assumptions to the necessary minimum.

3.1   Design Goals

Usability. Our overall goal is to provide a solution allowing users to retrieve their existing data as a VC. Since we aim for a wider adoption of SSI and most users are reluctant to change, usability is deemed crucial. While this topic is typically viewed from the credential holder's perspective, we focus on the issuer instead. While the holder requesting the VC needs an intuitive way of performing the request, storing received credentials and managing them securely, this is a topic for a separate research project. As we aim to increase the number of active issuers, the administrators in charge of the organisation's network require a service which is easy to integrate and maintain.

DG 1: Integration with an existing network shall be frictionless.

Privacy and Security. Using SSI as an individual (credential holder) is a means to gain more control over one's data and thus protect one's privacy. Otherwise, federated identity management with single sign-on would be a better solution since it is a mature system providing good usability in general. With SSI, the responsibility for the security and privacy of the data is shifted from a trusted third party to the individual user, the credential holder. Thus, it is necessary to design a system which guides the user and minimises the options for bad choices, such as publishing private keys.

DG 2: The proof of concept architecture shall follow the security-by-design principle.
DG 3: To ensure the holder's privacy, they shall have full control over their data.

3.2   Assumptions and Design Decisions

For the proof of concept, we decided to utilise a micro-service architecture, keeping the integration efforts minimal and allowing the replacement of components with any potentially already existing systems of the implementing organisation later on [10]. Modularisation also allows us to build connectors more easily with the organisation's existing infrastructure. The key goal of the proof of concept is to develop approaches to integrate existing systems and processes. For this purpose, we took the following assumptions about the organisations:

AS 1: Customers (potential credential holders) are already known to the organisation and are managed in a user database.
AS 2: Authentication is realised via OpenIDConnect[2] (OIDC).
AS 3: Customer data is accessible through a well-defined API.

---

[2] OpenIDConnect is a widely used extension of the oAuth 2.0 protocol.

3.3   Architecture

We developed a service that enables customers to receive two types of credentials. A high level overview can be seen in Fig. 2. On the one hand, the new credential holder can receive so-called identity credentials, encapsulating their base user profile data such as name, birth date or other personally identifiable information (PII). Additionally, we integrated an example of customer data accessible through an API, in our case learning achievement data of openHPI[3].

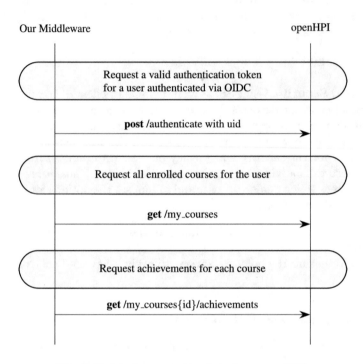

Fig. 2. Retrieving customer data from openHPI

The basic functionalities needed for issuing and verifying VCs of arbitrary content are core services in every such deployment. Therefore, the interfaces of these services are already standardised by the VC-API working group[4]. We have implemented these basic building blocks compliant with the standard. As we have utilised a micro-service architecture, we have a separate service (issuer middleware) for interacting with the credential holder and handling customer data from other services. The issuer middleware will use two data sources for our proof of concept: an identity provider and the openHPI system.

---

[3] open.hpi.de.

[4] w3c-ccg.github.io/vc-api/.

Identity Provider. Identity management protocols have converged towards a few established ones. Especially on the web, OIDC has emerged as the prevailing solution for federated identity management. However, these identity providers cannot issue VCs necessary for customers to participate in the SSI ecosystem. In our proof of concept, we utilise Keycloak[5] as an example of a popular identity provider that supports OIDC, SAML and other identity management protocols. The user database of Keycloak was filled with test data, mimicking the user management of an organisation. However, in a real-world deployment, this process would be defined by the organisation's onboarding process. Long-standing onboarding procedures might require in-person contact to fulfil specific bureaucratic needs and can be implemented upstream. In the end, user data will be put into a database and integrated with an identity provider.

openHPI API. While interacting with identity management data has been relatively homogenised through standard protocols such as OIDC, customer data is much more diverse, and the same kind of overarching protocol rarely exists. Each use case and organisation has its own APIs exposing user-relevant data. As an example of integrating this kind of data source, we have integrated openHPI, an open online learning platform. From this platform, we can retrieve information on enrolled courses, details of these courses and, most importantly, for the issuance of verifiable credentials, the learning achievements earned by the learner. The learner, in our setting, is the customer and, as such, the credential holder of the SSI system.

Initially, we utilise the user ID of the learner, which we have previously obtained through single sign-on with the identity provider, to receive a valid authentication token. The token can be used to request all enrolled courses for the learner. The course data is then enriched with further details from the general course API so that a visual choice of available courses can be presented to the learner. With a further request, the achievement for each course can be retrieved, which is then available for the learner to select.

Credential Holder Flow. The process of receiving a VC from an established data source is described in Fig. 3. When requesting an identity credential or a VC from the data sources, the credential holder visits the frontend of our middleware. Here they need to authenticate themselves using the Keycloak identity provider. Once they have successfully done so, their profile is available to the middleware. The credential holder is presented with the possible data to export using the retrieved profile. After selecting the desired attributes, the VC issuance process itself is started. Once the middleware has issued the VC using the Issuer Service, it is sent to the credential holder for storing in a wallet.

---

[5] keycloak.org.

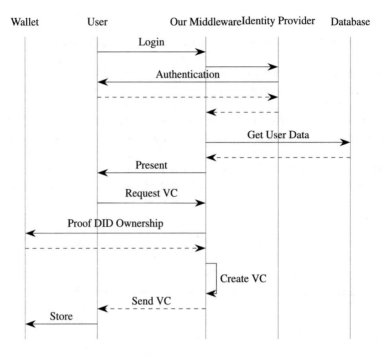

Fig. 3. Sequence diagram: data flow

4   Discussion and Improvements

As discussed below, we achieved our design goals as far as practicable in our restricted proof of concept setting. Although our work is a reasonable first step towards federated identity issuance in SSI, we identified additional technical and conceptual challenges, providing options for further research.

4.1   Implementation of Design Goals

Usability. Following the argumentation of Venters et al. [21], usability is a requirement for sustainable software. In the sense of Venters et al. [20], sustainable software is software capable of being maintained and, thus, continues to exist. From a software architecture standpoint, *separation of concerns* is required [6,21], an idea describing the modularisation of software. Software is broken into modules according to their functionality and with well-defined interfaces between the modules.

Consequently, the focus on modularisation of our solution by introducing an additional middleware improves sustainability and, thus, usability. The existing organisation's infrastructure does not need an extension but rather an additional configuration. Fewer changes to the software require less testing, documentation, and verification during the implementation and set-up phase, limiting cost and effort for a potential new issuer compared to a system extension.

Security. Our solution's security heavily depends on the organisation's network security. Optimally, the API towards the database allows exactly one necessary request, which fetches the data of the currently logged-in customer. Further, our solution uses only the standard functionality of the identity provider. Thus, the damage that can be done to the organisation's network if our solution is corrupted is limited to stealing the data of customers whose key material has been compromised.

A bigger problem is the access to the organisation's signing key. If our solution gets compromised, so is the signing key. Consequently, revocation of all certificates issued after the attack is crucial.

Privacy. As shown in Fig. 3, no additional data is added directly to the organisation. However, the organisation can collect metadata about the customer with every request sent, but the same holds true if the customer requests data via the organisation's website. Thus, we consider the customer's privacy risk towards the organisation a minor one for our use case.

Additionally, the VC holds a risk for the holder's privacy since the available data is now verifiable, making it more valuable to data collectors and thieves. If the data is not verifiable, many customers give false information to protect their privacy [16]. Again this holds true for all certifications in general, and thus SSI systems in particular. Credential holders need to be educated on the value of data and how to protect it. As a first step, our system provides a technical solution for enabling the individual toward data protection. In traditional systems, the content of a credential is entirely under the issuer's control. In contrast to the traditional system, the credential holder chooses which data to combine into a VC in our solution. Thus, they can leave out critical data they do not want to disclose from the beginning or request different VCs for different purposes.

4.2   Known Limitations

Our proposed solution is a proof of concept and will need adaptions before it is deployable for real-world applications. Some of the identified shortcomings, such as *revocation* and *credential management*, are technical and can be solved on an individual project level. Others, such as *semantics standardisation* and *legal regulations*, require a broader approach pushed forward by official institutions. *Verifiability of verifiers* can be seen as an in-between since it is possible to solve for an individual project. At the same time, an overarching approach would be more efficient and favourable in terms of interoperability.

Revocation. As seen in Sect. 4.1, revocation is a necessary feature to mitigate the threat of a compromised signing key. Revocation is still an open research field in cryptography [4,8,22] as well as in SSI [3,14], with many possible solutions known. They all have advantages and drawbacks, often regarding performance versus privacy concerns. Thus, a general recommendation cannot be given.

Credential Management. As control is shifted from the issuer to the credential holder, so is responsibility. A credential holder has to manage key material and backups for secure and privacy-preserving usage of an SSI system. Therefore wallets are employed. Despite their increased popularity in recent years, wallets are still in an early stage and need to improve, especially in usability and interoperability [17]. At the time of writing, the W3C Universal Wallet specification is still in a draft state[6]. One of the more mature solutions is the DCC's Learner Credential Wallet[7]. However, the Learner Credential Wallet focuses on the academic credential use case, and consequently, it is not sufficient for all use cases.

Verifiable Verifiers. Following the principles of SSI, it is favourable that verifiers or relying parties requesting data from a credential holder authenticate themselves to the credential holder. The relying party's authentication would empower credential holders to make informed decisions about with whom they share which data. Consequently, a complete infrastructure to manage public keys, establish trust in relying parties and supervise, which attributes a relying party can request, would be required.

Semantics Standardisation. The claims in a VC are stored in a JSON[8] or JSON-LD[9] file. The content of a JSON can be automatically validated with the help of schemas[10], while JSON-LD offers less flexibility but semantic interoperability. Different fields are working on schemas fitting their needs, such as the ELMO-xml format in the Education sector. However, most standardisation activities are still in a draft state.

Legal Regulations. The usefulness of digital certificates is foremost dependent on laws and regulations, despite the topic not being widely recognised in research. However, Brown [2] described the possible advantages of digital signatures over handwritten signatures in 1993. According to [2], digital signatures provide a higher level of assurance and can supplement or even surpass the analogue form if they get legally recognised. Pattiyanon and Aoki [15] performed a systematic review of laws, regulations and standards applicable to SSI in general, excluding domain-specific regulations. Within the 28 sources identified, 8 were only applicable on a national level, fragmenting the legal landscape further. The eIDAS regulation has tried to lay down a common legal ground for digital signatures within the EU over the past years. However, the acceptance of digital signatures is only increasing slowly. Still, the increase is an accomplishment, improving the situation from a fragmented international regulation landscape in Europe towards a more homogeneous acceptance.

---

[6] https://w3id.org/wallet.

[7] https://github.com/digitalcredentials/learner-credential-wallet.

[8] https://www.json.org/.

[9] https://json-ld.org/.

[10] https://w3c-ccg.github.io/vc-json-schemas/.

5   Conclusion

We presented a proof of concept for integrating an organisation into the SSI ecosystem as a credential issuer, which can be a building block for achieving a minimum user base for SSI systems. We tested our solution by integrating it with a MOOC platform.

However, additional real-world integration projects are necessary to solve the technical known limitations Sect. 4.2 and identify further challenges when integrating existing infrastructure with SSI. The presented proof of concept can encourage organisations to put SSI to the test with limited risk since our solution is only loosely coupled with the existing organisations' infrastructure and evaluate whether being an SSI credential issuer provides additional value for them and their customers.

Acknowledgements. This work has been funded through the Federal Ministry for Education and Research (BMBF) under grant M534800. We want to thank our partners at the TU Munich and the German Academic Exchange Service (DAAD) for the discussions on the topic.

References

1. Bolgouras, V., Angelogianni, A., Politis, I., Xenakis, C.: Trusted and secure self-sovereign identity framework. In: Proceedings of the 17th International Conference on Availability, Reliability and Security, pp. 1–6 (2022)
2. Brown, P.W.: Digital signatures: can they be accepted as legal signatures in EDI? In: Proceedings of the 1st ACM Conference on Computer and Communications Security, pp. 86–92 (1993)
3. Chotkan, R., Decouchant, J., Pouwelse, J.: Distributed attestation revocation in self-sovereign identity. In: 2022 IEEE 47th Conference on Local Computer Networks (LCN), pp. 414–421. IEEE (2022)
4. Emura, K., Takayasu, A., Watanabe, Y.: Generic constructions of revocable hierarchical identity-based encryption. Cryptology ePrint Archive (2021)
5. EU: Regulation (EU) 2016/679 of the European parliament and of the council of 27 April 2016 on the protection of natural persons with regard to the processing of personal data and on the free movement of such data, and repealing directive 95/46/EC (general data protection regulation). Technical report, European Union (2016)
6. Garlan, D.: Software architecture: a roadmap. In: Proceedings of the Conference on the Future of Software Engineering, pp. 91–101 (2000)
7. GDPR.EU: 2019 GDPR small business survey. Technical report, Proton AG (2019). https://gdpr.eu/wp-content/uploads/2019/05/2019-GDPR.EU-Small-Business-Survey.pdf
8. Ge, A., Wei, P.: Identity-based broadcast encryption with efficient revocation. In: Lin, D., Sako, K. (eds.) PKC 2019. LNCS, vol. 11442, pp. 405–435. Springer, Cham (2019). https://doi.org/10.1007/978-3-030-17253-4_14
9. Grüner, A., Mühle, A., Meinel, C.: An integration architecture to enable service providers for self-sovereign identity. In: 2019 IEEE 18th International Symposium on Network Computing and Applications (NCA), pp. 1–5. IEEE (2019)

10. Jamshidi, P., Pahl, C., Mendonça, N.C., Lewis, J., Tilkov, S.: Microservices: the journey so far and challenges ahead. IEEE Softw. **35**(3), 24–35 (2018)
11. Kuperberg, M., Klemens, R.: Integration of self-sovereign identity into conventional software using established IAM protocols: a survey. Open Identity Summit 2022 (2022)
12. Martinez Jurado, V., Vila, X., Kubach, M., Henderson Johnson Jeyakumar, I., Solana, A., Marangoni, M.: Applying assurance levels when issuing and verifying credentials using trust frameworks. Open Identity Summit 2021 (2021)
13. Mühle, A., Grüner, A., Gayvoronskaya, T., Meinel, C.: A survey on essential components of a self-sovereign identity. Comput. Sci. Rev. **30**, 80–86 (2018)
14. Mühle, A., Hoops, F., Assaf, K., Meinel, C.: Manuscript: universal statuslist: making a case for more middleware in self-sovereign identity (2023)
15. Pattiyanon, C., Aoki, T.: Compliance SSI system property set to laws, regulations, and technical standards. IEEE Access **10**, 99370–99393 (2022)
16. Polat, H., Du, W.: SVD-based collaborative filtering with privacy. In: Proceedings of the 2005 ACM Symposium on Applied Computing, pp. 791–795 (2005)
17. Sartor, S., Sedlmeir, J., Rieger, A., Roth, T.: Love at first sight? A user experience study of self-sovereign identity wallets. In: ECIS 2022 Proceedings (2022)
18. Schardong, F., Custódio, R.: Self-sovereign identity: a systematic review, mapping and taxonomy. Sensors **22**(15), 5641 (2022)
19. Schmidt, K., Mühle, A., Grüner, A., Meinel, C.: Clear the fog: towards a taxonomy of self-sovereign identity ecosystem members. In: 2021 18th International Conference on Privacy, Security and Trust (PST), pp. 1–7. IEEE (2021)
20. Venters, C., et al.: The blind men and the elephant: Towards an empirical evaluation framework for software sustainability. J. Open Res. Softw. **2**(1) (2014)
21. Venters, C.C., et al.: Software sustainability: research and practice from a software architecture viewpoint. J. Syst. Softw. **138**, 174–188 (2018)
22. Yu, T., Xie, H., Liu, S., Ma, X., Jia, X., Zhang, L.: CertRevoke: a certificate revocation framework for named data networking. In: Proceedings of the 9th ACM Conference on Information-Centric Networking, pp. 80–90 (2022)

A Post-quantum Cryptosystem with a Hybrid Quantum Random Number Generator

Maksim Iavich[1]([⊠]), Tamari Kuchukhidze[2], and Razvan Bocu[3]

[1] Department of Computer Science, Caucasus University, Tbilisi, Georgia
miavich@cu.edu.ge
[2] Georgian Technical University, Tbilisi, Georgia
[3] Transylvania University of Brasov, Brasov, Romania

Abstract. In recent years, a large amount of research has been conducted on quantum computers. If we ever get a large-scale quantum computer, they will be able to break many of the public key cryptosystems currently in use. Today, such cryptosystems are integrated into many commercial products. Alternatives are created that seems to protect us from quantum attacks, but due to safety and efficacy issues, they cannot be used in practice. We have presented an improved hash-based digital signature scheme that meets recommended standards. The scheme uses a secure quantum pseudo-random number generator as the small random initial values. It uses a random numbers that are generated using our hybrid quantum random number generator. We have discussed the method of random number generation. We have obtained a post-quantum cryptosystem, the initial random values of which are obtained using our generator. Our system is efficient and secure against quantum attacks.

Keywords: Quantum · Quantum cryptography · Post-quantum · Cryptosystem · Quantum random number generator

1 Introduction

Post-quantum cryptography, also called quantum encryption, is a cryptographic system for classical computers that can prevent attacks by quantum computers. If computers are able to exploit the unique properties of quantum mechanics, they will be able to perform complex calculations much faster than classical computers. It is clear that a quantum computer could perform certain types of complex calculations in a matter of hours. It is significant that a classical computer takes several years to perform these calculations [1].

Our goal is to create a cryptosystem that will work with both classical and post-quantum cryptography. The world's leading scientists are creating and developing quantum computers, but even the improved systems are subject to effective attacks [2].

We will reach a state where quantum computing will dominate and become more common. Quantum computers will probably destroy most if not all conventional cryptosystems that are widely used in practice. In particular, systems based on the integer

L. Barolli (Ed.): AINA 2023, LNNS 654, pp. 367–378, 2023.
https://doi.org/10.1007/978-3-031-28451-9_32

factorization problem (RSA). The RSA cryptosystem is used in many different products in many areas. Today, this cryptosystem is integrated into many commercial products, the number of which is increasing daily. Since encryption technology mainly uses the RSA algorithm, it can be considered one of the most common public key cryptosystems that evolves with the development of technology [3].

Various alternatives to RSA systems have been proposed, but they still have security or efficiency issues, so they cannot be used in practice. One of those proposed is hash-based signature schemes. Their security relies on the collision resistance of the hash function, and development requires random numbers, which are used as the initial random sequence of systems. A considerable amount of work is required to design and implement secure and efficient post-quantum cryptosystems.

As quantum computing dominates, RSA and similar asymmetric algorithms will no longer protect our private data. This is why we are trying to create post-quantum systems. And before it becomes a standard, it needs to be tested, software and hardware implementation [4, 5].

Using our novel quantum random number generator, we have created a new post-quantum cryptosystem. To construct a secure one-time signature scheme, we used one-way functions that are necessary and sufficient for secure digital signatures. We use Merkle's idea: the use of a hash tree, which reduces the validity of many unique verification keys (the leaves of the hash tree) to the validity of a single public key (the root of the hash tree). We have obtained a new post-quantum cryptosystem, which we obtain using our novel quantum random number generator.

Our post-quantum cryptosystem is based on Merkle's idea, but much more efficient and secure. We use the NIST standard Hash_drbg generator based on hashing. This generator is safe against quantum attacks, efficient and fast. We integrated the quantum random values obtained by a new hybrid quantum random generator. The scheme stores only the initial values obtained by the generator and does not require storing a large number of unique key pairs. A post-quantum cryptosystem is fast, efficient, and secure against quantum attacks.

We present an improved hash-based digital signature scheme. The scheme uses a secure pseudo-random number generator as the small random initial values of this quantum random number generator. As is well known, random numbers are frequently employed in many different domains [6]. We also have algorithmically generated numbers, which simulate random numbers but are not genuinely random. These numbers are produced by pseudo-random number generators, which are computer algorithms that generate random number sequences using mathematical formulas [7–9]. We employ true random number generators because we are unable to use pseudo random generators when actual randomness is required [10].

Not all true random number generators are cryptographically secure for our hash-based digital signature approach. We focus on a particular QRNG of the TRNG that draws its randomness from the inherent randomness of quantum processes. The majority of QRNGs in use today are based on quantum optics. Many detectors can access light particles, which are employed as a source of quantum randomness. Consequently, optical quantum random generators are swifter and more effective [11, 12]. Our scheme uses a truly random number that is generated using our novel hybrid quantum random number generator.

With the aid of a quantum random number generator, we were able to create a post-quantum cryptosystem. Our system is quick, effective, and protected from quantum attacks. We got a post-quantum cryptosystem that is superior to other systems in terms of security and efficiency.

The rest of this paper is organized according to the following structure: The second section describes the related works. The 3, 4, 5 sections describe the needed technical details. The 6 and 7 sections describe our novel approach. The 8 section offers the conclusion.

2 Literature Review

The cryptographic algorithms that are currently in use are easily cracked by quantum computers. As a result, attacks based on quantum computers are now capable of breaking traditional encryption methods. This article [1] presents digital signature schemes that are resistant to attacks from quantum computers. The paper [2] describes a McEliece public-key encryption system implementation with algorithmic choices and parameter selections, as well as the current level of cryptanalysis. One-time signature schemes and one-way functions are all covered in paper [3].

Quantum computers are being studied by the authors of article [4]. Quantum computers can break cryptosystems based on the integer factoring problem. It implies that one of the most well-known public-key cryptosystems, the RSA system, is open to attack by quantum computers. Different QRNG integration methods into Merkle are introduced in [5].

In [6], a number of pseudo-random number generators are considered, each of which makes use of a different method to ensure the randomness of the sequences and a higher level of security.

The paper [7–9] describes the various quantum random number generating technologies and the numerous applications for using this to obtain entropy with a quantum origin. Many applications make use of random numbers. The authors examine two existing quantum random number systems using their broad methodology for evaluating the quantum randomness of quantum random number generators in their study [10–12]. The authors of the articles [13–16] discuss various hash-based digital signature schemes that use a quantum number generator.

Different protocols are introduced in [17–19] to secure quantum channels to ensure confidentiality and security.

3 Hash-Based One-Time Signature Schemes

We consider signature schemes whose security is based only on the collision resistance of the cryptographic hash function. These schemes are particularly good options for the post-quantum era. Consider the Lamport–Diffie one-time signature scheme (LDOTS) [13].

It is assumed that computers have access to a stream of truly random bits when designing randomized algorithms and protocols (that is a sequence of independent and unbiased coin tosses). In actual implementations, a sample is taken from a "source of randomness" to create this sequence [14].

We assume that the security parameter n of LDOTS is a positive integer. LDOTS uses a one-way function $f : \{0, 1\}^n \rightarrow \{0, 1\}^n$ and cryptographic hashing function $g : \{0, 1\}^n \rightarrow \{0, 1\}^n$ to generate an LDOTS key pair, the LDOTS signature key X consists of a string of 2n bits of length n, chosen randomly:

$$X = (x_{n-1}[0], x_{n-1}[1], \ldots, x_1[0], x_1[1], x_0[0], x_0[1]) \epsilon R\{0, 1\}^{(n,2n)} \tag{1}$$

The LDOTS verification key is Y

$$Y = (y_{n-1}[0], y_{n-1}[1], \ldots, y_1[0], y_1[1], y_0[0], y_0[1]) \epsilon \{0, 1\}^{(n,2n)} \tag{2}$$

To calculate the key, we use the one-way function f:

$$y_i[j] = f(x_i[j]), 0 \le i \le n - 1, j = 0, 1. \tag{3}$$

So LDOTS key generation requires 2n evaluations of f. The signature and verification keys are n-length 2n-bit strings. In case of LDOTS signature generation, document $M \epsilon \{0, 1\}^*$ is signed using LDOTS with signature key X. let's $g(M) = d = (d_{n-1}, \ldots, d_0)$ is the message digest of M. The LD $-$ OTS signature is $sign = (x_{n-1}[d_{n-1}], \ldots, x_1[d_1], x_0[d_0]) \epsilon \{0, 1\}^{(n,n)}$.

This signature is a sequence of n bit strings of length n. They are selected as message digest function d. The number of cryptographic functions that a processor can perform at a given time is usually calculated in hashes per second [15]. The i-th bit string of this signature is $x_i[0]$ if the i-th bit in d is 0, $x_i[1]$ otherwise. The signature does not require the evaluation of f. The signature length is n^2.

In the case of LDOTS verification, if we want to verify the signature of M, $sign = (sign_{n-1}, \ldots, sign_0)$, verifier calculates the message digest $d = (d_{n-1}, \ldots, d_0)$. After that it is checked whether it is or not

$$(f(sign_{n-1}), \ldots, f(sign_0)) = (y_{n-1}[d_{n-1}], \ldots, y_0[d_0]). \tag{4}$$

LDOTS generates keys and signatures fairly quickly, however the signature size is quite huge. To decrease the size of signatures, the Winternitz one-time signature scheme (WOTS) is suggested. The concept is to sign multiple bits in a message digest using a single string, or a single string in a one-time signature key. WOTS utilizes a cryptographic hashing function and a one-way function, just like LDOTS.

Because each key pair can only be used once for a signature, one-time signature techniques are insufficient for the majority of real-world scenarios. An answer to this issue was put forth by Ralph Merkle. He proposes to employ a full binary hash tree. The concept is to limit the validity of an arbitrary but fixed number of one-time verification keys to a single public key, the hash tree's root, by using a full binary hash tree.

4   Merkle Tree Authentication Scheme

Merkle signature scheme (MSS) works for any cryptographic hash function and any one-time signature scheme. We assume that $g : \{0, 1\}^* \rightarrow \{0, 1\}^n$ is a cryptographic hashing function. We also assume that the one-time signature scheme is already selected.

When generating the MSS key pair, the signer chooses $H \in \mathbb{N}, H \geq 2$. Then the key pair is generated. By means of them, it will be possible to sign/verify 2^H documents. Note that this is a significant difference from signature schemes such as RSA and ECDSA, where many documents can potentially be signed/verified with just one key pair. However, in practice this number is also limited by the devices with which the signature is created or by certain policies [16].

The signer will generate 2^H unique key pairs $(X_j, Y_j), 0 \leq j < 2^H$. Here, X_j is the signature key and Y_j is the verification key. Both of them are bit strings. The leaves of the Merkle tree are $g(Y_j), 0 \leq j < 2^H$. The internal nodes of a Merkle tree are calculated according to the following rule: a parent node is the hash value of the concatenation of its left and right children. The MSS public key is the root of the Merkle tree. The MSS secret key is a sequence of 2^H one-time signature keys (Fig. 1).

This figure shows an example where the height of the Merkle tree is $H = 3$.

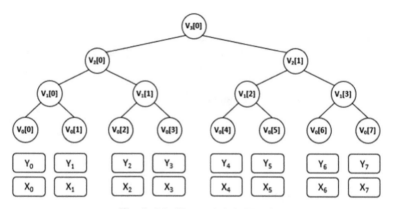

Fig. 1. Merkle tree height $H = 3$

Generating an MSS key pair requires computing 2^H unique key pairs and evaluating a $2^{H+1} - 1$ hash function.

It is not necessary to store the full hash tree to compute the root of a Merkle tree. Instead, the tree hash algorithm is used. The basic idea of this algorithm is to sequentially compute the leaves and when we can compute their parents as well (Fig. 2).

This shows the order in which Merkle tree nodes are computed by the tree hash algorithm. In this example, the maximum number of nodes stored on the stack is 3. This happens after node 11 is created and pushed onto the stack. To compute the root of a Merkle tree of height H, the tree hash algorithm requires 2^H calls, and $2^H - 1$ evaluations of the hash function.

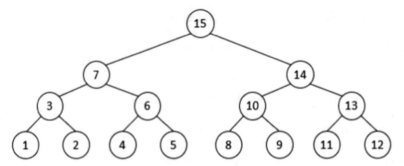

Fig. 2. Merkle tree, tree hash algorithm

MSS successfully uses one-time signing keys for signature generation. To sign a message on M, we must first compute the n-bit $d = g(M)$. The signer then generates a one-time signature $sign_{OTS}$ using the s-th one-time signature key X_s, $s \in \{0, \ldots, 2^H - 1\}$. A Merkle signature contains this one-time signature and the corresponding one-time verification key Y_s. To prove the authenticity of Y_s, the signer also appends the index s and the authentication path to the verification key Y_s. This index and the authentication path allow the verifier to construct a path from the leaf $g(Y_s)$ to the root of the Merkle tree. A node h in the authorization path is a sibling node of height h, which is the path from the leaf $g(Y_s)$ to the root of the Merkle tree.

For $h = 0, \ldots H - 1$ Fig. 3 shows an example of $s = 3$. So, the s-th merkle signature is

$$sign_s = \left(s, sign_{OTS}, Y_{s,}, \left(sign_0, \ldots, sign_{H-1}\right)\right) \tag{5}$$

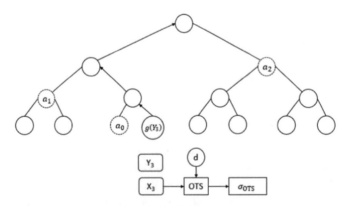

Fig. 3. Generation of Merkle's signature when $s = 3$.

The dashed nodes denote the authentication path of the leaf $g(Y_3)$. Arrows indicate the path from the leaf $g(Y_3)$ to the root.

Merkle's signature verification involves two steps. In the first step, the verifier uses the one-time verification key Y_s to verify d's signature $sign_{OTS}$ by means of the corresponding one-time signature scheme verification algorithm. At the second stage, the verifier checks the reliability of the one-time verification key Y_s.

5   Generating a Unique Key Pair Using a PRNG

MSS private key consists of 2^H one-time signature keys. Such a large amount of data cannot be stored for most practical applications. We can save space by using a deterministic pseudo-random number generator (PRNG). In this case, we need to save the PRNG seed. Each one-time signature key is then generated twice, once to generate the MSS public key and once during the signing phase [17].

Let PRNG be a cryptographically secure pseudorandom number generator whose input value is the n-bit seed S_{in}. Outputs random number RAND and updated seed S_{out}. The length of both of these is n.

$$\text{PRNG} : \{0, 1\}^n \quad \rightarrow \{0, 1\}^n \times \{0, 1\}^n$$
$$S_{in} \mapsto (\text{RAND}, S_{out}) \tag{6}$$

MSS key pair generation works using PRNG. The first step is to choose an n-bit seed S_0 uniformly random. We use the sequence of seeds $S - OTS_j$, $0 \leq j < 2^H$ to generate one-time signature keys. They are calculated iteratively, using the following:

$$\left(S - OTS_j, \ S_{j+1}\right) = \text{PRNG}(S_j), 0 \leq j < 2^H. \tag{7}$$

Here, $S - OTS_j$ is used to calculate the j-th one-time signature key. For example, in the case of WOTS, the j-th signature key is $X_j = (x_{t-1}, \ldots, x_0)$. Strings of length n, t bits in this signature key are generated using $S - OTS_j$.

$$\left(x_i, S - OTS_j\right) = \text{PRNG}(S - OTS_j), i = t - 1, \ldots, 0 \tag{8}$$

The seed $S - OTS_j$ is updated each time the PRNG is called. This shows that to calculate the signature key X_j only the knowledge of S_j is needed. And when $S - OTS_j$ is calculated, then a new seed S_{j+1} is defined for the signature key X_{j+1}. Figure 4 illustrates the generation of a one-time signature key using a PRNG.

If we use this method, the MSS secret key is initially S_0. Its length is n. It is replaced by S_{j+1} seeds defined during the generation of signature key X_j.

Unlike the original MSS signature, the unique signature key must be calculated before the signature is generated. When the signature key is calculated, the seed will be updated for the next signature.

6   Novel Quantum Random Number Generator

For our scheme we want to produce random numbers inexpensively and quickly, also significant amount of randomization is necessary at the same time. True randomness results from the breakdown of any quantum process, although the frequency of production depends on the detector output [18].

We use multiple source extractors whose model is similar to single source extractors. We gather information from two or more weak sources. They can be cultivated to generate a nearly equal sequence. Depending on the distribution of certain input values, the number of sources, and the desired properties of the output sequence, there are many ways that can be used. We get data from a number of unreliable sources, process it, and create a nearly equal sequence.

Based on the time of arrival QRNG, we present an updated quantum random number generator. Each photon that is detected only yields one random bit at most; the probability is decreased by detector inefficiency or dead time. Because different detectors have varying efficiency, if we combine them to produce more random bits, we will have a bias. We can eliminate this bias by utilizing a single detector and comparing the three successful events of detection time. Utilizing the straightforward detectors, which have only a few criteria, is pretty convenient.

We propose to employ a weak light source and an optical quantum random number generator where the probability of photon creation or non-generation is the same. So that the state of one photon must be:

$$\frac{|0>_1 + |1>_1}{\sqrt{2}} \tag{9}$$

If no detection occurs, we can assign 0, and if a click is made, we can assign 1. How many photons are used is irrelevant to us. The following can be used to express any superposition:

$$\frac{1}{\sqrt{2}}|0>_1 + \sum_{c=1}^{\infty} \alpha_c |c>_1 \tag{10}$$

where $\sum_{c=1}^{\infty} |\alpha_c|^2 = \frac{1}{2}$ is valid. No matter whether it was brought on by a single photon or many, we take it at the first click. The likelihood of detecting a photon in a coherent condition with amplitude is zero.

$$pr(n = 0) = e^{-|\alpha|^2} \tag{11}$$

likelihood of discovering one or more photons

$$pr(n \geq 1) = (1 - e^{-|\alpha|^2}) \tag{12}$$

Finding α for which $pr(n = 0) = pr(n \geq 1)$, which in this formula is $\alpha = \sqrt{ln2}$ the simplest concept. The poissonian source, where $\psi \, T = \ln2 \approx 0.693$, provides the likelihood of the desired finding.

The detector needs to have an average effective photon number of $\eta \, \psi \, T$, where η is the efficiency. Von Neumann extraction must be employed to resolve this issue so that there is no bias present during the procedure. The output result for two detections with photon numbers n_0 and n_1 is 1, if n_0 is greater than 0 and n_1, equals 0, and 0 otherwise. If two consecutive blank periods or two clicks are produced, the results must be disregarded. These values are equivalent for the poissonian source, therefore $pr(n > 0)pr(n = 0) = e^{-\eta\psi T}(1-e^{-\eta\psi T})$ is presumably the case. Although the resulting bit rate is slower, it is bias-free.

We advise employing a generator that produces many random bits after detecting a photon to increase efficiency. They are quantum random number generators with photon counting. The received findings will be split into groups with equal probabilities. In this instance, data creation is possible with just one detector. A quantum random variable can be considered to be the moment at which photons arrive. Successful photon time can be split into time blocks by a meter that operates in parallel with the detector. We receive a few bits per discovery for the specified discovery time interval. This process is a poissonian process where the events emerge independently [19].

We propose measurements in high-dimensional quantum space, such as photon temporal and spatial mode, to improve the frequency of random number generation. We obtain random bits by detecting two occurrences in the time interval t by determining the time at which a photon arrives. In the case of temporal mode, we can detect one photon and obtain many random bits. We can simultaneously assign random numbers to the detector matrix by using the spatial mode of the photon. The detector counter's speed is affected by dead time; thus, it is advisable to be aware of it when using this method [20]. A higher level of randomness can be obtained by selecting the appropriate amount of bits from the counted photons with the aid of improved rate.

Our novel quantum random number generator allows for fairly efficient generation. Our generator is based on a quantum random number generator that uses time of arrival. It is effective because it makes advantage of the less complex detectors, which have fewer needs. With the novel OQRNG, each photon detection results in the production of several random bits.

7 Generating a One-Time Key Pair Using a Novel Quantum Random Number Generator

Our idea is to insert Hash_drbg instead of pseudo random numbers. Because this generator is a NIST standard and based on hashing. We favor a hashing-based generator because the entire Merkle tree is hashing-based.

In order to provide output values of high quality and as close to equal distribution as is practical, the resulting raw bits sequence must be processed. We need random extractors for this.

But all random number generators need a random seed. In our approach, we integrate quantum seeds instead of random seeds. Our idea is to inject a quantum seed of our own creation, obtained by a new hybrid quantum random generator. We have discussed the generation method, as well as its certification and extraction method.

A PRNG can be a cryptographically secure pseudo-random number generator. We use the NIST standard, hashing-based generator Hash_drbg. In this case too, the input value is n-bit seed $QRNG_{in}$ and outputs random number RAND and updated seed $QRNG_{out}$. The length of both of these is n.

$$\text{Hash\_drbg} : \{0, 1\}^n \rightarrow \{0, 1\}^n \times \{0, 1\}^n$$

$$QRNS_{in} \mapsto (RAND, QRNG_{out}) \tag{13}$$

The process of generating an MSS key pair with Hash_drbg is similar to generating a pseudo-random number generator. First we choose an n-bit seed $QRNG_0$ equally random. To generate one-time signature keys, we use the sequence of seeds $QRNG - OTS_j$, $0 \leq j < 2^H$. They are calculated iteratively, using the following:

$$\left(QRNG - OTS_j, \; QRNG_{j+1}\right) = \text{Hash\_drbg}(S_j), 0 \leq j < 2^H. \tag{14}$$

Here, $QRNG - OTS_j$ is used to calculate the j-th one-time signature key. For example, in the case of WOTS, the j-th signature key is $X_j = (x_{t-1}, \ldots, x_0)$. Strings of length n, t bits in this signature key are generated using $QRNG - OTS_j$.

$$\left(x_i, QRNG - OTS_j\right) = \text{Hash\_drbg}(QRNG - OTS_j), i = t - 1, \ldots, 0 \tag{15}$$

The seed $QRNG - OTS_j$ is updated every time Hash_drbg is called. This shows that only $QRNG_j$ is needed to calculate the signature key X_j. And when $QRNG - OTS_j$ is calculated, then a new seed $QRNG_{j+1}$ is defined for the signature key X_{j+1}. The figure below illustrates the generation of a one-time signature key using Hash_drbg.

If we use this method, the MSS secret key is initially $QRNG_0$. Its length is n. It is replaced by the $QRNG_{j+1}$ seeds defined during the generation of the signature key X_j.

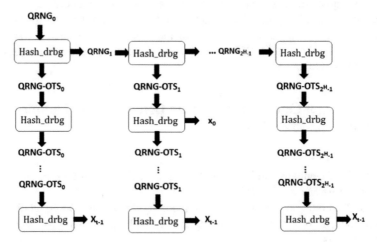

Fig. 4. Generate an one-time signing key using Hash_drbg

Our scheme is secure, all signatures issued before cancellation remain valid. MSS is secure because the actual MSS secret key can only be used to generate one-time signature keys for future signatures, but not to forge previous ones.

8   Conclusion

Our goal is a post-quantum cryptosystem that works for both classical and post-quantum cryptography. Our post-quantum cryptosystem is based on Merkle's idea, but much more efficient and secure. We use the NIST standard Hash_drbg generator based on hashing. This generator is safe against quantum attacks, efficient and fast. We do not use pseudo-random numbers to get the initial random value, because they are easy to guess during a quantum attack. We integrate quantum random values obtained by a new hybrid quantum random generator. The scheme stores only the initial values obtained by the generator and does not require storing a large number of unique key pairs. A post-quantum cryptosystem is fast, efficient, and secure against quantum attacks.

We discussed currently available random number generation tools for both classical and quantum cases. Also, as a result of evaluating various certification and random number processing methods, we have created a variant that can be used in post-quantum cryptography. We have created systems that can withstand quantum attacks.

We discussed the systems used in post-quantum cryptography. We discussed hashing-based one-way functions, their efficiency, created a new model and integrated it into Merkle. On the basis of post-quantum systems and received information, a new, more balanced, fast, safe, usable scheme was developed. The result was integrated into the software implementation.

We used the random numbers obtained during the research period as initial quantum random values obtained by means of a new hybrid quantum random generator. Basically, pseudo-random initial values are used, instead of which we integrated a quantum analogue, which makes our system resistant to quantum attacks.

Acknowledgement. This work was supported by Shota Rustaveli National Science Foundation of Georgia (SRNSF) [STEM – 22 -1076].

References

1. Chen, L., et al.: Report on post-quantum cryptography (Vol. 12). Gaithersburg, MD, USA: US Department of Commerce, National Institute of Standards and Technology (2016)
2. Biswas, B., Sendrier, N.: McEliece cryptosystem implementation: theory and practice. In: Buchmann, J., Ding, J. (eds.) PQCrypto 2008. LNCS, vol. 5299, pp. 47–62. Springer, Heidelberg (2008). https://doi.org/10.1007/978-3-540-88403-3_4
3. Buchmann, J., Dahmen, E., Szydlo, M.: Hash-based digital signature schemes. In: Bernstein, D.J., Buchmann, J., Dahmen, E. (eds.) Post-Quantum Cryptography. Springer, Berlin, Heidelberg (2009).
4. Gagnidze, A., Iavich, M., Iashvili, G.: Novel version of merkle cryptosystem. Bull. Georgian National Acad. Sci. **11**(4), 28–33 (2017)
5. Gagnidze, A., Iavich, M., Iashvili, G.: Advantages and challenges of QRNG integration into Merkle. Sci. Pract. Cyber Secur. J. (SPCSJ) **4**(1), 93–102 (2020)
6. Kabiri Chimeh, M., Heywood, P., Pennisi, M., et al.: Parallelisation strategies for agent based simulation of immune systems. BMC Bioinform. **20**, 579 (2019). https://doi.org/10.1186/s12 859-019-3181-y
7. Lewis, P.A.W., Goodman, A.S., Miller, J.M.: A pseudo-random number generator for the System/360. IBM Syst. J. **8**(2), 136–146 (1969). https://doi.org/10.1147/sj.82.0136
8. Lambić, D., Nikolić, M.: Pseudo-random number generator based on discrete-space chaotic map. Nonlinear Dyn. **90**(1), 223–232 (2017). https://doi.org/10.1007/s11071-017-3656-1
9. Mcginthy, J.M., Michaels, A.J.: Further analysis of PRNG-based key derivation functions. IEEE Access **7**, 95978–95986 (2019). https://doi.org/10.1109/ACCESS.2019.2928768

10. Wayne, M.A., Kwiat, P.G.: Low-bias high-speed quantum random number generator via shaped optical pulses. Opt. Express **18**, 9351–9357 (2010)

11. Herrero-Collantes, M., Garcia-Escartin, J.C.: Quantum random number generators. Rev. Mod. Phys. **89**, 015004 (2016). https://doi.org/10.1103/RevModPhys.89.015004

12. Okhrimenko, T., Tynymbayev, S., Iavich, M.: High-speed and secure PRNG for cryptographic applications (2020). mecs-press.org

13. Lamport, L.: Constructing digital signatures from a one way function (1979)

14. Iavich, M., Bocu, R., Arakelian, A., Iashvili, G.: Post-quantum digital signatures with attenuated pulse generator, vol 2698 (2020). ceur-ws.org

15. Iavich, M., Gagnidze, A., Iashvili, G., Okhrimenko, T., Arakelian, A., Fesenko, A.: Improvement of merkle signature scheme by means of optical quantum random number generators. In: Hu, Z., Petoukhov, S., Dychka, I., He, M. (eds.) ICCSEEA 2020. AISC, vol. 1247, pp. 440–453. Springer, Cham (2021). https://doi.org/10.1007/978-3-030-55506-1_40

16. Iavich, M., Gagnidze, A., Iashvili, G.: Hash based digital signature scheme with integrated TRNG. In: CEUR Workshop Proceedings (2018)

17. Iavich, M., Iashvili, G., Gnatyuk, S., Tolbatov, A., Mirtskhulava, L.: Efficient and secure digital signature scheme for post quantum epoch. In: Lopata, A., Gudonienė, D., Butkienė, R. (eds.) ICIST 2021. CCIS, vol. 1486, pp. 185–193. Springer, Cham (2021). https://doi.org/10.1007/978-3-030-88304-1_15

18. Gnatyuk, S., Okhrimenko, T., Iavich, M., Berdibayev, R.: Intruder control mode simulation of deterministic quantum cryptography protocol for depolarized quantum channel. In: Proceedings of 2019 IEEE International Scientific-Practical Conference: Problems of Infocommunications Science and Technology, PIC S and T 2019, Kyiv, Ukraine, 8–11 October 2019, pp. 825–828

19. S. Gnatyuk, T. Zhmurko, P. Falat, Efficiency increasing method for quantum secure direct communication protocols. In: Proceedings of the 2015 IEEE 8th International Conference on Intelligent Data Acquisition and Advanced Computing Systems: Technology and Applications (IDAACS'2015), Warsaw, Poland, 24–26 September, vol. 1, pp. 468–472 (2015)

20. Qoussini, A.E., Daradkeh, Y.I., Al Tabib, S.M., Gnatyuk, S., Okhrimenko, T., Kinzeryavyy, V.: Improved model of quantum deterministic protocol implementation in channel with noise. In: Proceedings of the 2019 10th IEEE International Conference on Intelligent Data Acquisition and Advanced Computing Systems: Technology and Applications (IDAACS 2019), pp. 572–578 (2019)

21. Iavich, M., Kuchukhidze, T., Gagnidze, A., Iashvili, G.: Advantages and challenges Of qrng integration into merklE. Sci. Pract. Cyber Secur. J. (2020)

22. Iavich, M., Gnatyuk, S., Odarchenko, R., Bocu, R., Simonov, S.: The novel system of attacks detection in 5G. In: Barolli, L., Woungang, I., Enokido, T. (eds.) AINA 2021. LNNS, vol. 226, pp. 580–591. Springer, Cham (2021). https://doi.org/10.1007/978-3-030-75075-6_47

23. Iavich, M., Kuchukhidze, T., Gnatyuk, S., Fesenko, A.: Novel certification method for quantum random number generators. Int. J. Comput. Netw. Inf. Secur. **13**(3), 28–38 (2021)

24. Iavich, M., Kuchukhidze, T., Iashvili, G., Gnatyuk, S.: Hybrid quantum random number generator for cryptographic algorithms. Radioelectronic Comput. Syst. **4**, 103–118 (2021)

Supervised Machine Learning and Detection of Unknown Attacks: An Empirical Evaluation

Miguel S. Rocha[1], Gustavo D. G. Bernardo[1], Luan Mundim[1],
Bruno B. Zarpelão[2], and Rodrigo S. Miani[1(✉)]

[1] Federal University of Uberlândia, Uberlândia, Brazil
{miguelsr,gustavo.bernardo,luanmundim,miani}@ufu.br
[2] State University of Londrina, Londrina, Brazil
brunozarpelao@uel.br

Abstract. Intrusion Detection Systems (IDS) have become one of the organizations' most important security controls due to their ability to detect cyberattacks while inspecting network traffic. During the last decade, IDS proposals have increasingly used machine learning techniques (ML-based IDS) to create attack detection models. As this trend gains traction, researchers discuss whether these IDS can detect unknown attacks. Most ML-based IDS are based on supervised learning, which means they are trained with a limited collection of attack examples. Therefore, detecting attacks that were not covered during the training phase could be challenging for these systems. This work evaluates the ability of ML-based IDS to detect unknown attacks. Our general idea is to understand what happens when a detection model trained with a particular attack A receives incoming data from an unknown attack B. Using the CIC-IDS2017 dataset, we found that supervised intrusion detection models, in most cases, cannot detect unknown attacks. The only exception occurs with DoS attacks. For example, an intrusion detection model trained with HTTP Flood DoS (GoldenEye) samples could detect a different HTTP DoS attack type (Slowloris).

1 Introduction

The increasing number of people connected to the Internet and the massive use of this network have turned this environment into a growing target for cybercriminals. For this reason, having a solid cybersecurity structure is necessary to maintain and sustain organizations' computing resources. In this scenario, tools such as Firewalls, Antivirus, and Intrusion Detection Systems (IDS) are some of the main alternatives capable of mitigating attack attempts to keep computing systems safe.

An IDS is a device that monitors computing systems and detects attack attempts [19]. Regarding its deployment, an IDS can be classified as host-based or network-based. Host-based IDS monitors host characteristics such as CPU usage, accessed system files, or applications. Network-based IDS focuses on monitoring network traffic and identifying malicious behavior in the network [1]. The focus of this work is network-based IDS.

There are two types of network-based IDS [11]. Signature-based IDS require a database with information on attack signatures, which they compare with the collected network packets. When the characteristics of a network packet match an attack signature, the IDS generates an alert, and the human operator might take the necessary actions. The second type is referred to as anomaly-based IDS. These systems aim to distinguish normal from malicious behavior. An advantage of anomaly-based IDS, compared to signature-based ones, is that it can detect new or unknown attacks due to its characteristic of analyzing the system's behavior.

Anomaly-based IDS are usually created using a labeled dataset in batch mode. This means that the machine learning algorithm is applied once to a static training dataset, and after that, the produced model is used to make predictions for incoming data. Therefore, when the supervised approach is adopted, the "unknown attack detection" assumption of anomaly-based IDS relies heavily on the quality of training data and the similarities of unknown attacks to the already known ones.

Despite the vast literature on this topic, only some studies seek to analyze the behavior of supervised machine learning algorithms in detecting unknown attacks. For example, Zhang et al. [21] proposed a specific model based on deep learning techniques to identify unknown attacks. The proposed approach was evaluated in the datasets "DARPA KDDCUP 99" [18] and CIC-IDS2017 [15]. The results were satisfactory, but the work only investigated a small sample of attacks and did not evaluate simpler algorithms commonly found in the literature, such as Random Forest, Support Vector Machine (SVM), and Decision Tree. Besides, they used a hybrid approach for the learning model (supervised/unsupervised) and also a method to update the classifier. Other studies such as [6,17], and [20] also propose hybrid and specific intrusion models for detecting unknown attacks. Our main contribution is providing a common ground to understand the impact of the most common setup for anomaly-based IDS for detecting unknown attacks: a simple supervised model with no incremental learning techniques.

Therefore, our goal is to evaluate the performance of Machine-Learning (ML) based IDS in detecting unknown attacks. First, we investigated the similarities between attacks using an unsupervised technique called t-Distributed Stochastic Neigh or Embedding (t-SNE). The idea here is to understand whether some attacks share common characteristics. If this is true, we expect the supervised model to learn some of these common patterns in the training phase, facilitating the detection of attacks with similar patterns. Next, we evaluated the performance of a classification algorithm (Random Forest) in detecting attacks that were not covered in the training dataset. These experiments were carried out on the CIC-IDS2017 dataset, and the results suggest that supervised models cannot detect unknown attacks in most cases.

The rest of the paper is organized as follows. Section 2 focuses on related work about ML-based IDS and its ability to detect unknown attacks. Section 3 presents the dataset and discusses the preprocessing methods applied. Section 4 introduces our investigation on the similarities of different attack types. Section 5 describes the performance of supervised intrusion detection models in detecting unknown attacks, and finally, Sect. 6 presents the concluding remarks and future work.

2   Related Work

Louvieris et al. [8] present a technique for detecting unknown attacks by identifying attack resources. This technique relies on k-means clustering, Naive Bayes feature selection, and C4.5 decision tree classification methods. Potnis et al. [12] focus on detecting DDoS-type attacks. The authors propose a hybrid web intrusion detection system (HWIDS) to detect this attack. Five types of DDoS that target the application layer are covered, as well as their unknown variants. The proposed system has an accuracy of 93.48% and a false negative rate of 6.52% in detecting unknown attacks. Alzubi et al. [3] introduce a new ensemble-based system called Unknown Network Attack Detector (UNAD), which proposes a training workflow composed of heterogeneous and unsupervised anomaly detection techniques. According to the study, this approach performs better when detecting unknown attacks and achieves promising results.

Shin et al. [16] point out that although anomaly detection is a good approach for detecting unknown attacks, false positives are highly probable. The work proposes a hybrid form of intrusion detection (signature-based and anomaly-based) using the Fuzzy c-means technique (FCM) alongside other ones, such as Classification and Regression Trees (CART), to avoid this problem.

Al-Zewairi et al. [2] reiterate that the problem of detecting completely unknown attacks on a system is still an open field of research as these attacks represent a complex challenge for any IDS. The work emphasizes that some definitions for unknown attacks are inconsistent and proposes a categorization into two types of attacks (Type-A and Type-B). In this case, the first type represents new attacks, and the second one represents unknown attacks but in already known categories. Experiments were carried out with IDS based on neural networks to detect Type-A and Type-B attacks as a binary classification problem. The results on two datasets (UNSW-NB15 and Bot-IoT) showed that the evaluated models (deep and shallow ANN classifiers) had poor overall generalization error measures - the classification error rate for several types of unknown attacks was around 50%.

Serinelli et al. [14] use a supervised approach to investigate unknown attack detection in the following datasets: KDD99, NSL-KDD, and CIC-IDS2018. To mimic an unknown attack, the authors performed two simple attacks (DoS using the Hping tool and PortScan using the Nmap tool) in a VirtualBox environment. Then, they recorded the packet capture files and inputted them into three intrusion detection models trained with the following algorithms: SVM, Random Forest, and XGBoost. The results show that, in most scenarios, both attacks have misclassification errors. In some cases, when the unknown attack exhibits a network traffic profile similar to the well-known attacks, the model can identify the attack type correctly. In our work, we investigated the similarity of different attack types to understand whether some attacks have similar network traffic profiles.

In [5], Ferreira and Antunes propose an evaluation of the CIC-IDS2018 dataset and compare the performance of some supervised and bioinspired algorithms, namely: CLONALG Artificial Immune System, Learning Vector Quantization (LVQ), and Back-Propagation Multi-Layer Perceptron (MLP). They also investigated how this approach can deal with some unknown attacks. The authors only work with two scenarios: a) detect different DoS attack types (training with GoldenEye and test with Slowloris), and b) train with data from one attack type and test with another type (DoS traffic × DDoS traffic). According to the authors, the proposed IDS performed better for scenario b) when it comes to identifying unknown attacks.

In comparison to the studies mentioned above, the following differences can be found in our work: i) only some specific attack types were investigated (DoS/DDoS and PortScan), ii) some works proposed specific methods (hybrid techniques - supervised/unsupervised, for instance) tailored to identify unknown attacks, iii) some works still use outdated datasets such as KDD99 and NSL-KDD and iv) absence of a benchmark for detection of unknown attacks using supervised IDS model built with traditional ML classifiers instead of Deep Learning.

3   Dataset and Preprocessing

When selecting the dataset for our experiments, several factors were considered, including the availability of labels, diversity of attack types, and recent attack data [7]. We chose the CIC-IDS2017 [10,15] dataset, as it is one of the most complete and consolidated options among datasets for IDS [13].

The dataset contains 2,830,743 records comprising over 78 network flow features with both normal traffic and attacks observed during a week (Monday to Friday). The dataset encompasses seven attack classes and 16 attack types. Table 1 shows the attack classes and types that were used in this work. The dataset also includes raw network packet data (PCAP format) and network flows labeled as normal (benign) attacks. There is only one day (Monday) with only normal traffic.

For better performance and correct functioning of the classification algorithms, we conducted the following pre-processing procedures using Python language, version 3.9.10: i) Encode categorical variables, ii) Treating infinity and null values, and iii) Normalization.

Concerning the categorical variables, we first converted benign and attack labels to 0 and 1. We then handled the "DestinationPort". This attribute represents the destination port requested by the attacker, which required proper treatment. As a result, we created a column for the following ports: 80, 53, 21, 22, 123, and 443. These ports are frequently targeted by attackers and associated with HTTP, DNS, FTP, SSH, NTP, and HTTPS protocols. We also created other columns for the following TCP/UDP ports: i) under 1024 but not the previous seven values and ii) greater than 1024. There are no attributes for the source port. We did both treatments for categorical variables using one-hot encoding, so we deleted the original column.

Network flows with infinite and null values were removed from the model since they were very few concerning the original dataset size. Finally, we applied the divide-by-maximum normalization over all features except the "Destination-Port" by dividing the attribute value by the highest value in its category.

4   Similarity Analysis of Different Attacks

4.1   General Idea and Motivation

Supervised models tend to be out of calibration with the increasing number and variety of attacks. Consequently, they require recurrent evaluations to ensure their effectiveness, which can be pretty costly due to the volume of information. However, if different attacks have similar features, we can suppose that the supervised models will continue to perform well. For example, suppose a supervised model is trained with samples from attack A, and a new attack B, similar to attack A, is fed into the classifier. In that case, we expect that the model would be able to detect attack B.

The main idea here is to look for evidence of similarity between different attacks. For this purpose, we performed t-Distributed Stochastic Neigh or Embedding (*t-SNE*) [9]. *t-SNE* is a dimension reduction technique developed to facilitate the visualization of high-dimensional datasets [9]. Our idea is to identify groups of attacks that share similar characteristics.

By applying this technique, we seek to analyze how the network flows associated with attacks are visually distributed in the search space and check if different types of attacks appear within the same groups formed by *t-SNE*, therefore, sharing similarities. If groups are composed of different attack types, then this would indicate that such attack types might have similar network traffic characteristics. In that case, we have more confidence that the rule or the logic used to detect attack types in the same group is also similar.

4.2   Results

To understand the similarities across network flows associated with different attacks, we applied a reduction to two dimensions of *t-SNE* using only data points of the following attack classes in the CIC-IDS2017 dataset: Brute force, DoS, and Web attack. We do not use the benign class, as the objective is to compare only attacks. Finally, we used the entire feature set available in the dataset. We run this experiment using R version 4.2.1 with library M3C version 1.20.0 [4]. Figure 1 shows the visual representation of CIC-IDS2017 attacks through a two-dimensional reduction using *t-SNE*.

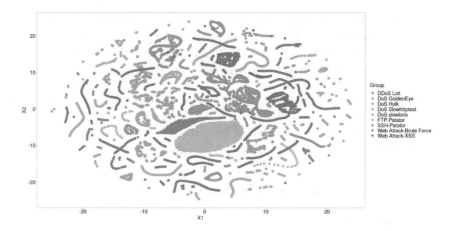

Fig. 1. *t-SNE* visualization in two dimensions

As our goal is to look for similarities between the attacks, what we should observe in the *t-SNE* plot, if it existed, would be groupings of different attacks. Figure 1 shows us that different attacks, represented by colors, are not in the same groupings; we see a clear separation. We can also see that the same type of attack has different patterns, such as DDoS Loit, represented by the orange color, has several groups distributed by the graph. Therefore, this indicates that the different attacks form isolated groups, and even within the same attack type, we observe different behaviors.

Consequently, the analysis shows no conclusive evidence of similarity between different attacks, and the data patterns may differ even within the same type. In the next section, we will conduct some supervised analyzes to see how the intrusion detection model behaves when instances that do not belong to the training set are presented to the classifier.

5   Supervised Models Performance in Detecting Unknown Attacks

The results presented in the previous section indicated that network traffic flows belonging to different types of attacks did not have clear similarities in their features. This may suggest that an intrusion detection model trained with certain attack types might not be able to detect attacks that were not previously seen. The main goal of this section is to empirically evaluate the performance of supervised intrusion detection models when presented to unseen/unknown attack types. Next, we show the experimental methodology and the results obtained using the CIC-IDS2017 dataset.

5.1   Experimental Methodology

Our experimental methodology consists of the following steps:

1. Organization of attack classes and establishment of a baseline dataset;
2. Selection of classification algorithms;
3. Creation of different training/testing sets according to attack classes and types;
4. Performance evaluation of supervised intrusion detection models in different scenarios.

Each data instance in the dataset will have a label B (benign) or A (attack). Attack instances might be grouped in classes $C_1, C_2, ..., C_i$ where i denotes the number of attack classes. Each attack class is composed of a type $t_{i,j}$ where j denotes the number of attack types. In this work, we investigated three classes of attacks: DoS, Brute Force, and Web Attacks (WA). Inside each class, we have several attack types. The DoS attack class is composed of DoS GoldenEye, DoS Hulk, DoS SlowHTTPTest, DoS SlowLoris, and DDoS Loit. The Brute Force class consists of the following attacks: FTP-Patator and SSH-Patator. Finally, the Web Attack class contains the Web Attack Brute Force and Web Attack XSS types. The Baseline dataset consists of samples from both labels B (benign) and A (attack). We selected random samples from CIC-IDS2017 to compose each class. Table 1 details the Baseline dataset composition.

Table 1. Number of attack samples organized by class and type

Type	Class	Number of samples
Benign	Benign	174421
FTP-Patator	Brute Force	794
SSH-Patator	Brute Force	590
DDoS Loit	DoS	12803
DoS GoldenEye	DoS	1043
DoS Hulk	DoS	23107
DoS Slowhttptest	DoS	550
DoS slowloris	DoS	580
Web Attack - Brute Force	Web Attack	151
Web Attack - XSS	Web Attack	65

We selected the following classification algorithms as potential candidates for developing an intrusion detection system: Decision Tree Classifier, Stochastic Gradient Descent Classifier, Multi-layer Perceptron, Gaussian Naive Bayes, K-Nearest Neighbors Classifier, Random Forest Classifier, Linear Support Vector Classifier, Logistic Regression, and Extra-trees Classifier. These algorithms represent some of the most used ones in the ML-based IDS literature. Then, we evaluated the prediction performance of each classification algorithm using the Baseline dataset. We adopted a hold-out strategy of 90/10, and the goal was to classify attack and benign samples (binary task) correctly. Our preliminary results showed that the Random Forest algorithm had a good performance in terms of Precision, Recall, AUC, and training time. For this reason, we selected Random Forest for the rest of the experiments. It is noteworthy that we used Python 3.9.10, scikit-learn, and default parameters for all algorithms in every experiment.

The next step involves creating different training/testing sets according to attack classes and types. We first created six datasets to evaluate the performance of intrusion detection models in detecting unknown attack classes. The forming rule for these sets can be summarized as follows:

- Select an attack class C_k;
- The training set will be composed of all samples from the selected attack class C_k plus benign samples;
- The testing set will be composed of a different attack class C_j and benign samples.
- Repeat the process until all attack classes are evaluated in training and testing sets.

For example, if we select the attack class $C_1 = \text{DoS}$, we will have the training set composed of all DoS samples and the testing set consisting of samples from $C_2 = \text{Web Attack}$. In both sets (train/test), we add a number of benign samples proportional to the number of attack samples. Table 1 details the number of samples for each class.

We also created 24 other datasets to observe the results from a more detailed perspective, moving our focus from attack classes to types. The forming rule for these sets is similar to the previous one and can be summarized as follows:

- Select an attack type $t_{k,j}$;
- The training set will be composed of all samples from the selected attack type $t_{k,j}$ and also of benign samples;
- The testing set will be composed of a different attack type from the same attack class $t_{k,l}$ and also of benign samples.
- Repeat the process until all attack types inside every attack class are evaluated in training and testing sets.

For example, if we select the attack type $t_{1,1} = \text{DoS GoldenEye}$, we will have the training set composed of all DoS GoldenEye samples and the testing set composed of samples from $t_{1,2} = \text{DoS Hulk}$. In both sets (train/test), we add a number of benign samples proportional to the number of attack samples. Table 1 details the number of samples for each class.

We evaluated the models using the following metrics: Precision, Recall, and AUC (Area Under the Curve). In our experiments, the positive class represents an attack, and the negative class represents a benign network flow. Recall represents the ratio of correctly classified positive samples (attacks) to the total positive examples in the dataset. Precision quantifies the number of positive class predictions that belong to the positive class. Finally, we plotted each model's receiver operating characteristic (ROC) curves. The ROC curve plots the True Positive Rate (TPR) vs. the False Positive Rate (FPR) of a given model. The area under the ROC curve is defined as the Area Under the Curve (AUC). The AUC is a value between 0 and 1. Efficient models have AUC values closer to 1.

5.2 Results and Discussion

Table 2 shows the performance of supervised models when detecting unknown attack classes. Precision, Recall, and AUC rates for all models differ significantly from the baseline. For example, an intrusion detection model trained with only DoS samples cannot detect unknown attacks related to Web Attack and Brute force. The same holds when analyzing intrusion detection models trained with Web Attack and Brute Force. The low Recall values for all experiments indicate that supervised models trained with a particular attack class cannot identify a new (unseen) attack class.

Table 2. Performance of supervised models according to attack class

Training	Test	Precision	Recall	AUC
Baseline	Baseline	0.9257	0.9454	0.9791
DoS	Web Attack	0.2303	0.0973	0.5663
DoS	Brute Force	0	0	0.4742
Web Attack	DoS	0.7442	0.0001	0.6361
Web Attack	Brute Force	0.0185	0.0001	0.4966
Brute Force	DoS	0	0	0.4836
Brute Force	Web Attack	0	0	0.8366

The subsequent analysis refers to investigating the behavior of the intrusion detection model when it is trained with specific attacks inside a class and what happens when a new variant of such attack type is presented to the classifier. The results presented in Sect. 4.1 showed no conclusive evidence of similarity between different attack types. We intend to check whether the same conclusion holds when we use supervised models. Table 3 summarizes the performance of each intrusion detection model regarding the attack type.

The scenario here is somewhat different compared to intrusion models trying to detect a new attack class. When analyzing the 20 scenarios of DoS attack types, we have mean values of 0.83, 0.41, and 0.91 for Precision, Recall, and AUC, respectively. This result indicates that a model trained with a DoS type might be able to identify a new/unseen DoS attack type. In some cases, the intrusion

Table 3. Performance of supervised models according to attack types

Training	Test	AC	Prec.	Recall	AUC
Baseline	Baseline	–	0.925	0.945	0.979
DoS GoldenEye	DoS Hulk	DoS	0.988	0.336	0.947
DoS GoldenEye	DoS Slowhttptest	DoS	0.994	0.830	0.983
DoS GoldenEye	DoS Slowloris	DoS	0.997	0.830	0.994
DoS GoldenEye	DDoS LOIT	DoS	0	0	0.778
DoS Hulk	DoS GoldenEye	DoS	0.743	0.999	0.971
DoS Hulk	DoS Slowhttptest	DoS	1	0.022	0.981
DoS Hulk	DoS Slowloris	DoS	0.996	0.587	0.985
DoS Hulk	DDoS LOIT	DoS	0.999	0.168	0.999
DoS Slowhttptest	DoS GoldenEye	DoS	0.999	0.747	0.957
DoS Slowhttptest	DoS Hulk	DoS	0.979	0.209	0.979
DoS Slowhttptest	DoS Slowloris	DoS	0.997	0.995	0.997
DoS Slowhttptest	DDoS LOIT	DoS	0.972	0.138	0.972
DoS Slowloris	DoS GoldenEye	DoS	0.994	0.742	0.992
DoS Slowloris	DoS Hulk	DoS	0.976	0.352	0.968
DoS Slowloris	DoS Slowhttptest	DoS	0.988	0.952	0.993
DoS Slowloris	DDoS LOIT	DoS	0	0	0.946
DDoS LOIT	DoS GoldenEye	DoS	0	0	0.628
DDoS LOIT	DoS Hulk	DoS	0.978	0.034	0.742
DDoS LOIT	DoS Slowhttptest	DoS	0.991	0.044	0.838
DDoS LOIT	DoS Slowloris	DoS	0.996	0.214	0.653
FTP-Patator	SSH-Patator	BF	0	0	0.637
SSH-Patator	FTP-Patator	BF	0	0	0.746
WA Brute Force	WA XSS	WA	0.934	0.998	0.983
WA XSS	WA Brute Force	WA	0.852	0.983	0.975

detection model's performance is even better than the baseline, e.g., in a scenario where we have DoS Slowhttptest in the training set and DoS Slowloris in the testing set. Both attacks are related to the Application Layer (HTTP protocol) in this case. However, in other cases, the intrusion detection model could not detect an unknown DoS attack type, for example, DoS Slowloris (training) and DoS LOIT (test).

We found similar results for Web attacks. In both cases, Web Attack Brute Force and Web Attack XSS, the supervised model could detect a new/unseen attack of such type. The only attack class where the trained models were not able to identify unknown attacks was Brute Force. Precision and Recall values were null in both cases (FTP-Patator and SSH-Patator). Since both attacks target different Application Protocols (FTP and SSH), the network traffic similarity between these two attacks is supposedly low.

To better understand the results, we conducted a new round of experiments. We created newer datasets according to the following rationale: 1) what happens if the training set has several attack classes/types and the test set has only one attack class/type that was not part of the training set? and 2) what happens if the training set has only one attack class/type and the training set has several attack classes/types that were not part of the training set? The idea here is to understand whether some unseen attacks can be detected using previous knowledge from other attacks. In case i), for example, if the training set is composed of $C_1 = \text{DoS}$ and $C_2 = \text{Web Attack}$, the test set will have only attack samples from $C_3 = \text{Brute Force}$ (i.e., the only attack class that was not part of the training set). In case ii), for example, if the training set is composed of $C_1 = \text{DoS}$, the test set will have attack samples from $C_2 = \text{Web Attack}$ and $C_3 = \text{Brute Force}$ (i.e., the attack classes that were not part of the training set). The same logic applies to the tests considering attack types. As discussed in Sect. 5.1, benign samples are also part of the training and testing sets.

The results of these new experiments showed that even when we trained the intrusion detection model with multiple attack classes/types, it was still challenging for the model to detect unseen attacks. Only two scenarios were close to the baseline. In the first one, we have a training set composed of all attack types except DoS Slowloris, and the testing set is composed of only DoS Slowloris attacks. In this scenario, Random Forest achieved values greater than 90% for Precision, Recall, and AUC. In the second one, we have a training set composed of all attack types except Web Attack - Brute Force and the testing set composed of only Web Attack - Brute Force. In this scenario, we had values higher than 90% for Precision and AUC and 83% for Recall.

Our results indicate the following implications for practice: i) the performance of supervised IDS models is directly related to the attacks presented in the training set, ii) supervised IDS models may be successful in detecting unknown/unseen variants of the same attack types, for example, DoS and some Web Attacks and iii) due to the potential similarity between some attack types, it might be interesting to develop intrusion detection models tailored for them.

6   Conclusion

We proposed a study of ML-based IDS and their ability to detect unknown/ unseen attacks. Our findings show that a supervised model trained with a specific attack type can identify unseen samples of the same attack type in some situations. For example, supervised models trained with DoS GoldenEye could identify attack samples from DoS Slowloris. We also conclude that supervised models did not efficiently detect unseen attack classes. Two different experiments indicate this: i) when a training set is composed of one attack class and a test set has one different attack class, and ii) when a training set is composed of several attack classes, and a testing set has one unseen attack class.

Future work includes conducting similar experiments on datasets such as CSE-CIC-IDS2018, UNSW-NB15, and UGR'16. We also want to apply transfer learning methods to evaluate the capability of ML-based IDS to detect unknown attacks collected from different datasets.

References

1. Molina-Coronado, B., Mori, U., Mendiburu, A., Miguel-Alonso, J.: Survey of network intrusion detection methods from the perspective of the knowledge discovery in databases process. IEEE Trans. Netw. Serv. Manage. **17**, 2451–2479 (2020)
2. Al-Zewairi, M., Almajali, S., Ayyash, M.: Unknown security attack detection using shallow and deep ANN classifiers. Electronics **9**(12), 2006 (2020)
3. Alzubi, S., Stahl, F., Gaber, M.M.: Towards intrusion detection of previously unknown network attacks. Commun. ECMS **35**(1), 35–41 (2021)
4. Christopher John [Aut, Cre]: M3c (2017). https://doi.org/10.18129/B9.BIOC. M3C, https://bioconductor.org/packages/M3C
5. Ferreira, P., Antunes, M.: Benchmarking behavior-based intrusion detection systems with bio-inspired algorithms. In: Thampi, S.M., Wang, G., Rawat, D.B., Ko, R., Fan, C.-I. (eds.) SSCC 2020. CCIS, vol. 1364, pp. 152–164. Springer, Singapore (2021). https://doi.org/10.1007/978-981-16-0422-5_11
6. Jongsuebsuk, P., Wattanapongsakorn, N., Charnsripinyo, C.: Network intrusion detection with fuzzy genetic algorithm for unknown attacks. In: The International Conference on Information Networking 2013 (ICOIN), pp. 1–5. IEEE (2013)
7. Kenyon, A., Deka, L., Elizondo, D.: Are public intrusion datasets fit for purpose characterising the state of the art in intrusion event datasets. Comput. Secur. 102022 (2020)
8. Louvieris, P., Clewley, N., Liu, X.: Effects-based feature identification for network intrusion detection. Neurocomputing **121**, 265–273 (2013)
9. van der Maaten, L., Hinton, G.: Viualizing data using t-SNE. J. Mach. Learn. Res. **9**, 2579–2605 (2008)
10. University of New Brunswick, U.o.N.B.: Intrusion detection evaluation dataset (cic-ids2017) (2017). https://www.unb.ca/cic/datasets/ids-2017.html
11. Otoum, Y., Nayak, A.: AS-IDS: anomaly and signature based ids for the internet of things. J. Netw. Syst. Manage. **29**(3), 1–26 (2021)
12. Potnis, M.S., Sathe, S.K., Tugaonkar, P.G., Kulkarni, G.L., Deshpande, S.S.: Hybrid intrusion detection system for detecting DDoS attacks on web applications using machine learning. In: Fong, S., Dey, N., Joshi, A. (eds.) ICT Analysis and Applications. LNNS, vol. 314, pp. 797–805. Springer, Singapore (2022). https://doi.org/10.1007/978-981-16-5655-2_77
13. Ring, M., Wunderlich, S., Scheuring, D., Landes, D., Hotho, A.: A survey of network-based intrusion detection data sets. Comput. Secur. **86**, 147–167 (2019)
14. Serinelli, B.M., Collen, A., Nijdam, N.A.: On the analysis of open source datasets: validating ids implementation for well-known and zero day attack detection. Proc. Comput. Sci. **191**, 192–199 (2021)
15. Sharafaldin, I., Lashkari, A.H., Ghorbani, A.A.: Toward generating a new intrusion detection dataset and intrusion traffic characterization. In: ICISSp, pp. 108–116 (2018)
16. Shin, G.Y., Kim, D.W., Kim, S.S., Han, M.M.: Unknown attack detection: combining relabeling and hybrid intrusion detection. CMC-Comput. Mater. Continua **68**(3), 3289–3303 (2021)

17. Song, J., Ohba, H., Takakura, H., Okabe, Y., Ohira, K., Kwon, Y.: A comprehensive approach to detect unknown attacks via intrusion detection alerts. In: Cervesato, I. (ed.) ASIAN 2007. LNCS, vol. 4846, pp. 247–253. Springer, Heidelberg (2007). https://doi.org/10.1007/978-3-540-76929-3_23
18. Stolfo, S.J., Fan, W., Lee, W., Prodromidis, A., Chan, P.K.: Cost-based modeling for fraud and intrusion detection: results from the jam project. In: Proceedings DARPA Information Survivability Conference and Exposition, DISCEX 2000, vol. 2, pp. 130–144. IEEE (2000)
19. Tsai, J.J., Yu, Z.: Intrusion Detection: A Machine Learning Approach, vol. 3. World Scientific (2011)
20. Xu, M.F., Li, X.H., Miao, M.X., Zhong, C., Ma, J.F.: An unknown attack detection scheme based on semi-supervised learning and information gain ratio. J. Internet Technol. **20**(2), 629–636 (2019)
21. Zhang, Z., Zhang, Y., Guo, D., Song, M.: A scalable network intrusion detection system towards detecting, discovering, and learning unknown attacks. Int. J. Mach. Learn. Cybern. **12**, 1649–1665 (2021)

Machine Learning Anomaly-Based Network Intrusion Detection: Experimental Evaluation

Ahmed Ramzi Bahlali[1,2(✉)] and Abdelmalik Bachir[2]

[1] IMATH Laboratory, University of Toulon, La Garde, France
a.ramzi.bahlali@gmail.com
[2] LESIA Laboratory, University of Biskra, Biskra, Algeria
abdelmalik.bachir@univ-biskra.dz

Abstract. The use of Machine Learning (ML) approaches to design anomaly-based network intrusion detection systems (A-NIDS) has been attracting growing interest due to, first, the ability of an A-NIDS to detect unpredictable and previously unseen network attacks, and second, the efficiency and accuracy of ML techniques to classify normal and malicious network traffic compared to other approaches. In this paper, we provide a comprehensive experimental evaluation of various ML approaches including Logistic Regression (LR), Decision Tree (DT), Random Forest (RF), and Artificial Neural Network (ANN), on a recently published benchmark dataset called UNSW-NB15 considering both binary and multi-class classification. Throughout the experiments, we show that ANN is more accurate and has fewer false alarm rates (FARs) compared to other classifiers, which makes Deep Learning (DL) approaches a good candidate compared to shallow learning for future research. Moreover, we conducted our experiments in a way to be served as a benchmark results since our used approaches are trained and tested on the configuration deliberately provided by the authors of UNSW-NB15 dataset for the purpose of direct comparison.

Keywords: Intrusion detection system · Machine learning · Network anomaly detection · Artificiel neural network · Cyber security

1 Introduction

An IDS is a proactive mechanism that monitors inbound and outbound traffic to identify malicious traffic. An IDS could be categorized according to its placement and its detection technique [5, 27]. There are mainly two types of IDSs with respect to its placement: Host-Based IDS (HIDS) and Network-Based IDS (NIDS). An HIDS runs on a local host machine and uses system activities to identify malicious behavior locally, whereas NIDS is located at the network entry to protect the internal network from external cyber threats. IDSs can also be classified according to the technique used to detect intrusions into two categories: signature-based

© The Author(s), under exclusive license to Springer Nature Switzerland AG 2023
L. Barolli (Ed.): AINA 2023, LNNS 654, pp. 392–403, 2023.
https://doi.org/10.1007/978-3-031-28451-9_34

and anomaly-based IDSs. Signature-based IDSs rely on a preexisting database of well-known attack patterns, and any packet or flow that matches one of those patterns is flagged as malicious [2,15]. This category has a major drawback as it requires continuous manual update of the database of signatures by adding up-to-date attacks and thus cannot detect unseen malicious patterns [2,15]. In contrast, Anomaly-based NIDS (A-NIDS) is more efficient than Signature-based IDSs as they can detect new unseen attacks by creating a profile or a model after observing several examples of normal behavior [2,5]. As a result, there has been a plethora of research attempting to design an A-NIDS using a variety of techniques [5,6,15]. However, A-NIDSs suffer from performance issues such as low accuracy and a high false alarm rate compared to Signature-IDSs [5,14]. Therefore, designing an A-NIDS with a low false alarm rate and high accuracy is a challenging task which attracted extensive research in the literature [5].

In area of A-NIDS, various ML techniques have been used and can be categorized into two categories: unsupervised (clustering [1,4]) and supervised (classification [13,22,28]). Supervised techniques often show superior performance compared to unsupervised approaches, since the latter are generally sensitive to the initial parameters (e.g. number of clusters, distance metric) and the nature of data at hand. The use of unsupervised learning was motivated by the lack of hand-labeled data and the high cost of doing it manually, hence, the recourse to semi-supervised learning [11,23] as it reduces the cost of manually labeling the data. However, the obtained results remained below what supervised techniques achieved.

In this work, we conduct an exprimental evaluation using the UNSW-NB15 benchmark dataset [16] leveraging both traditional ML approaches and DL such as ANN. We have shown that the DL approach appears to be significantly superior to shallow learning because of its ability to learn a useful deep representation of the data, which is demonstrated by projecting the last hidden layer of our ANN architecture using the UMAP algorithm.

2 Related Work

The development of A-NIDS based on ML requires large and comprehensive datasets. NSL-KDD, a refined version of KDDCUP'99 [24] is one of the most widely used datasets in the literature for the evaluation of these systems. However, in [16,17], it has been stated that NSL-KDD suffers from several issues, such as the lack of modern low-footprint attacks and the smaller number of attack types compared to existing real-life attacks. To overcome these limitations, two new benchmark datasets have been generated, UNSW-NB15 [16] and CSE-CIC-IDS2018 [20]. In the latter, a train and test sets are not configured and provided to the community to allow a direct comparison between proposals, as is the case with UNSW-NB15. Furthermore, simple classifiers such as DT perform well in CSE-CIC-IDS2018, which means that it is less difficult and complex compared to the UNSW-NB15 dataset.

In [10], a performance analysis of the NSL-KDD benchmark data set has been performed using ANN in binary and multiclass classification with 81.2%

and 79.9% detection accuracies, respectively. More sophisticated deep learning algorithms were explored in the literature, such as Autoencoder (AE), and Recurrent Neural Network (RNN). The authors of [28] and [22] proposed the use of RNN by considering samples from the KDD dataset as a sequence. We know from [24] that individual training samples such as bidirectionnel flows are not necessarily related (i.e., they do not come from the same sequence). Therefore, the use of a sequence model, in this case, is not justified. In [17], an evaluation of the performance of A-NIDS using several classifiers such as DT, LR, Naive Bayes (NB), ANN, and Expectation Maximization Clustering (EMC) has been carried out using UNSW-NB15 dataset. The results show that the DT classifier achieves the highest accuracy with 85.56% and the lowest false alarm rate with 15.78%.

In [3], in addition to the hand-crafted header and statistical features extracted from the raw traffic, a new set of features has been computed using the proposed supervised Adaptive Clustering (AC) that are identical to the data samples (traffic flow) belonging to the same traffic class. Despite the good performance claimed by the authors, it remains an expensive architecture because of its complexity that rises exponentially as the number of classes increases (i.e. for n classes $n \times 2$ ANN architecture is used, each of which has its own set of weights and biases). The performance of the proposed technique has been evaluated using five synthetic data sets and three public benchmarks (KDD Cup'99 [9], ISCX-IDS 2012 [21] and CSE-CIC-IDS 2018 [20]). The authors claimed a 100% accuracy and 0% FAR in both binary and multiclass classification. However, the reproducibility of the results has not been achieved using the same settings. Furthermore, the authors did not test their solution on the UNSW-NB15 dataset, as it is widely used in the literature.

In [11], the authors proposed a semi-supervised technique that requires only 10 labeled data per flow type and relies on Generative Adversarial Networks (GANs) [7] to generate more data synthetically and reduce the cost manually labeling the data. In their proposal, they used a semi-supervised GAN (SGAN) [18,19] and achieved an accuracy of 88.7%.

In [13], the authors tried to address the problem of low performance of a standalone classification model by leveraging the use of Ensemble Learning (EL) in CPS (Cyber-Physical Systems) IDS. They evaluated three known EL techniques, such as aggregation, boosting, and stacking [29]. Throughout the experience, they observed that the ensemble mòdels performed much better than the standalone models. When it comes to EL techniques, they showed that the stacking technique outperforms the others with an F1 score of 0.503 for the small imbalanced test set and a 0.996 for the large balanced test set. Although EL methods achieve good results, they have several drawbacks caused by their higher cost of creation, training, and deployment.

In [23], the authors proposed a semi-supervised approach using AEs to overcome the problems of supervised learning techniques such as the difficulty of detecting novel attack types and the need for large amount of labeled data (particularly the abnormal traffic) in the supervised training process. The learning process consists in training an AE to reconstruct only normal samples by minimizing the reconstruction error of the model's input samples. When an abnormal sample is passed to the model, its reconstructed error would be higher compared to normal samples because it has not been seen in the training phase. Although the authors obtained encouraging results, their models cannot be generalized to multi-class anomaly detection and have not been tested in recent databases.

In [8], the authors proposed an early detection approach to address the supposed issues related to the detection of pos-tmortem attacks. Typically, the existing approaches for A-NIDS using ML take the full biflow packets (i.e. until the biflow ends) to extract the handcrafted features to train and test ML models. This approach is considered by the authors as weak defense strategy since it can not detect the attack on its early stage. They have experimented with a variable number of packets that first arrive in which belong to the same biflow using different ML algorithms. In their results, they showed that the use of RF classifier with as few as the first 10 packets allows to obtain a good performance-complexity trade-off. However, it should be noted that no comparison has been done with other existing A-NIDSs.

3 Dataset Description

The design of A-NIDS based on ML is heavily dependent on the availability of datasets that represent a wide area of Internet traffic, including modern and sophisticated attacks. There have been a lot of efforts in the literature to generate datasets. One of the earliest dataset released publicly to the community is NSL-KDD [24]. However, this dataset was collected more than decades ago and does not represent current normal and traffic patterns in the real world.

Several datasets have been released since then [16,20] in order to provide a modern and more inclusive view of Internet traffic.

In this work, we consider the UNSW-NB15 [16] in which lower performance has been obtained in the community compared to other datasets (as there is room for improvement) and the availability of training and testing sets that have been defined by the authors of the dataset so our work can be compared with other according to the same benchmarks defined by the authors.

The UNSW-NB15 dataset was created using the IXIA PerfectStorm tool [25] using the testbed configuration described in [16] to generate a mixture of modern real normal as well as synthetic attack traffic.

In this dataset, the packets belonging to the same bidirectional flow collected from the original captured raw data are grouped together and used to handcraft features about the biflow with the help of tools such as *Argus, Bro-IDS, and custom scripts*. These features are categorized into five groups: *Flow, Content, Time, Basic Features, and Generated Features which contains other features*. These handcrafted features are used to form a record in the constructed dataset.

Every record in the dataset is labeled as normal, or malicious in the case of binary labaling or into several types of malicious attacks in the case of multi-class labeling as can be shown in Table 1 which lists 9 attack categories of malicious biflows (for more details about attack types, see the original paper of the dataset [16]).

The entire UNSW-NB15 dataset has a little over 2.5 million records (2,540,044 records). However, the authors deliberately configured a partition of this dataset as a training and testing set for a fair apple-to-apple future comparison, which contains 175341 and 82332 records, respectively. In Table 1, we show an illustration of the distribution of normal and malicious samples in the train and test sets.

Table 1. The number of records and the percentage of each class category in the USNW-NB15 train and test sets

(a) Classes	(b) Train set		(c) Test Set	
Category	Nbr of records	Percentage	Nbr of records	Percentage
Normal	56000	31.93%	37000	44.93%
Generic	40000	22.8%	18871	22.92%
Exploits	33393	19%	11132	13.52%
Fuzzers	18184	10.3%	6062	7.36%
DoS	12264	7%	4089	4.96%
Reconnaissance	10491	6%	3496	4.26%
Analysis	2000	1.14%	677	0.85%
Backdoor	1746	1%	583	0.70%
Shellcode	1133	0.6%	378	0.45%
Worms	130	0.07%	44	0.05%

4   Experimental Design

4.1   Dataset Pre-processing

Features Transformation: Among the 49 features of the data set, there are four categorical features such as *protocol*, *service*, *state*, and *attack-category*. Thus, a conversion of categorical features to numerical ones is required using one of the existing techniques such as the *One-Hot-Encoding*.

Features Scaling: It is a technique used to rescale the features of the data within a particular range, often between 0 and 1 or between −1 and 1. It improves the convergence speed of optimization algorithms such as gradient descent and its variants [26]. In this work, the Min-Max scaling technique has been considered to scale the data in the range of $[0, 1]$. Note that the feature scaling step is done using the train and test set separately to prevent data leakage from the test to the train set.

4.2 Evaluation Metrics

Generally, in IDS literature, two primary metrics are used to measure and evaluate the performance of the proposed solutions, such as the Accuracy and the False Alarm Rate (FAR). Accuracy is defined as the ratio of correctly classified samples to the total number of instances. The FAR is the average ratio of erroneously misclassified samples to correctly classified ones, which are calculated using the formula (1) and (2) respectively.

TP (True Positive): The number of malicious instances is correctly classified as malicious. TN (True Negative): The number of normal instances is correctly classified as normal. FP (False Positive): The number of normal instances is incorrectly classified as malicious. FN (False Negative): The number of malicious instances is incorrectly classified as normal.

$$Accuracy = \frac{TP + TN}{TP + TN + FP + FN} \tag{1}$$

$$FNR = \frac{FN}{FN + TP} \quad FPR = \frac{FP}{FP + TN} \quad FAR = \frac{FPR + FNR}{2} \tag{2}$$

In addition to the two important metrics mentioned above, there are two other metrics that help further evaluate the proposed solutions, namely Precision and Recall. Precision is the ratio of correctly classified malicious samples to all samples classified as malicious, as illustrated in the formula (3). The recall as shown in (4) is the ratio of correctly classified malicious samples to the number of all malicious samples.

$$Precision = \frac{TP}{TP + FP} \tag{3}$$

$$Recall = \frac{TP}{TP + FN} \tag{4}$$

5 Results and Discussion

This section is organized into two parts: in the first part we show the empirical results of binary classification, and in the second part we show the results of multi-class classification experiments using classical ML techniques and the ANN. All experiments were implemented using Python 3.7.0. Pandas and Numpy libraries were used for data manipulation. Sklearn version 0.24 and Keras version 2.4.3 with Tensorflow back-end version 2.4.0 for the implementation of the classical ML algorithms (LR, DT, RF) and DL (ANN), respectively.

For each part, after the preprocessing phase (see Sect. 4.1), and performing feature scaling in the interval [0,1] and feature transformation according to the one-hot technique, we obtained a new dataset with 196 normalized numerical features.

Table 2. ANN configuration in binary and multi-class classification

Hyper-parameters	Binary settings	Multi-class settings
Hidden layers	2	1
Neurons on hidden layer 1	128	64
Neurons on hidden layer 2	64	None
Neurons on output layer	1	10
Activation function in the hidden layers	Sigmoid	ReLu
Activation function in the output layer	Sigmoid	Softmax
Optimizer	RMSprop	RMSprop
Learning rate	0.01	0.001
Batch size	140	250
Epochs	40	100
Loss Function	Cross entropy	Cross entropy
Regularizer	L2 = 0.001	None
Kernel-initializer	Glorot uniform	Variance scaling

For training and evaluation, we use the official UNSW-NB15 benchmark train and test sets to allow a direct comparison with future research.

$$\mathcal{L}_{CE} = \begin{cases} -\frac{1}{M} \sum_{i=1}^{M} (y_i \log(\hat{y}_i) + (1 - y_i) \log(1 - \hat{y}_i)), & \text{for Binary} \\ \\ -\frac{1}{M} \sum_{i=1}^{M} \sum_{c=1}^{C} y_{ic} \log(\hat{y}_{ic}), & \text{for Multi-class} \end{cases} \quad (5)$$

Table 3. Results of our models in the test set

(a) Binary results						(b) Multi-class results	
classifier \ metric	Accuracy	Precision	Recall	FAR		classifier \ metric	Accuracy
LR	80.61%	74.94%	97.32%	21.21%		LR	69.14%
DT	86.13%	82.34%	95.22%	14.89%		DT	72.76%
RF	87.57%	84.74%	96.63%	12.33%		RF	70.47%
ANN	**88.77%**	**84.88%**	**96.86%**	**12.13%**		**ANN**	**73.13%**

5.1 Binary Classification

In this work, the classical ML classifiers were trained using the default hyper-parameters except for the LR which has been trained using the Limited-memory BFGS solver and a maximum iteration number of 500, and RF with a number of estimators (trees) equal to 200.

After experimenting with different hyper-parameters on the ANN classifier using the grid search technique, we end up with the binary settings shown in

Table 2. The network architecture has been selected to be 4 layers, including 2 hidden layers, each of which contains 128 and 64 artificial neurons, respectively. In the hidden and output layer, the activation function *sigmoid* is used. The optimization algorithm *RMSprob* has been chosen to minimize the Cross-Entropy loss function shown in formula (5), which differs from binary classification to multiclass.

The detailed results for binary classification indicate that ANN outperforms the other classifiers with 88.77% accuracy and 12.13% FAR as shown in Table 3a, while LR has the lowest performance in terms of accuracy and FAR with 80.61% and 21.21%, respectively.

The performance supremacy of ANN compared to traditional ML techniques is based on its ability to implicitly find other representations of data known as deep representation learning. To show how the ANN learns deep representation after the training phase, a UMAP 3-D projection has been used before training (i.e., before representation learning) with the initial set of features (194 features) and after training where the output of the second hidden layer neurons (64 features) is used as shown in Fig. 1a and 1b, respectively.

It is worth mentioning that almost all proposed ML techniques in the literature for A-NIDS have been trained and tested on randomly selected sets of UNSW-NB15 datasets despite the availability of a configuration from the authors to allow future research solutions to be compared in a fair manner (apple-to-apple comparison). We have used the architecture and the hyper-parameters of our best performing ANN model and we have trained and tested it by picking random samples from both sets. Repetition of the experiments several times has always shown an accuracy higher than 90% with a large variance (between [90%–98%] accuracy). This means that the proposed solutions can not be compared since their performance are directly affected by the chosen samples for train and test sets. For the reasons mentioned above, we are only able to invoke the experiments previously conducted on the same train and the test splits provided by the authors of the UNSW-NB15 data set and compare them with our results. As shown in Table 4, our ANN model clearly outperforms the performance of the proposed methods with an accuracy of 88.77% and a FAR of 12.13%.

Table 4. A comparison of our proposed models with other previous works in binary classification

Paper	Technique	Accuracy	FAR
R. Vinayakumar *et al.* [27]	ANN	76.5%	//
T. Kim *et al.* [12]	Encoding + 2D CNN	80%	//
N. Moustafa *et al.* [17]	ANN	81.34%	21.13%
N. Moustafa *et al.* [17]	Decision Tree	85.56%	15.78%
Ours	**ANN**	**88.77%**	**12.13%**

5.2   Multi-class Classification

ML-based techniques have been implemented using the by default hyperparameters except LR where the solver, maximum iteration, and multi-class hyperparameters have been set to Limited-memory BFGS solver, 1100 and multinomial, respectively. Moreover, the RF classifier has been implemented with 30 estimators.

The ANN classifier has been implemented with the multi-class settings shown in Table 2 after experimenting with different hyper-parameters using the grid search technique. The classifier have been trained using 1 hidden layer with 64 neurons and the *ReLu* as an activation function. In the output layer, the number of neurons was set to 10 equally to the number of existing classes and the *softmax* function as an activation function that normalizes the input into a probability distribution.

The computation of the evaluation metrics on multi-label classification is a bit different from that on binary classification. Accuracy is calculated by summing the correctly classified samples for each class divided by the sum of all elements of the confusion matrix. However, precision, recall, and false alarm rate can not be computed regarding the performance of the overall model as in binary classification. As a result, the confusion matrix (See Fig. 3) is given to assess the classifier in depth.

The experimental results shown in Table 3b indicate that the accuracy of the ANN classifier (73.13%) is better than that of the classical ML techniques. In addition, the accuracy of the classifiers on multi-class classification has declined compared to binary classification.

As the confusion matrix (see Fig. 3) indicates, our model performs much better in some classes than in others. For instance, the proportion of correctly predicted samples from the Generic class is 96% and 79% for the Normal class while 2%, 19%, and 1% for Worms, Shellcode, and Backdoor, respectively.

The primary reason behind this phenomenon is the unbalanced nature of the observations on the training set of UNSW-NB15 dataset. Normal and Generic categories make 31.93% and 22.80% of trainig set, whereas Worms, Shellcode, and Backdoor combined make only 1.67% as mentioned on the Table 1 which is a huge difference on the distribution. Therefore, the unbalanced nature of the data has led the model to learn well the patterns of more populated classes compared to less populated in which has resulted to a decrease of the performance in multi-class classification.

Finally, a 2-D projection of 1000 random samples from the training set is conducted before and after training the ANN classifier (Fig. 2a, 2b respectively) to show that the learned representation during the training process using ANN classifier are much more helpful in distinguishing between biflow classes.

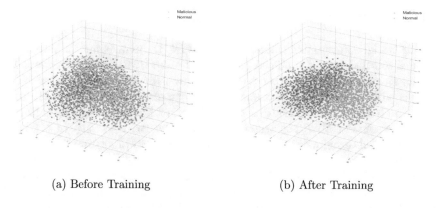

(a) Before Training   (b) After Training

Fig. 1. A 3-D projection of 10000 randomly selected samples before training (a) and after training (b) by selecting the second layer output of our ANN architecutre in binary classification using UMAP algorithm.

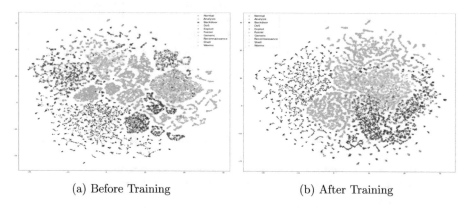

(a) Before Training   (b) After Training

Fig. 2. A 3-D projection of 10000 randomly selected samples before training (a) and after training (b) by selecting the second layer output of our ANN architecutre in multi-class classification using UMAP algorithm.

Actual Biflow Class	Analysis	Backdoor	DoS	Exploits	Fuzzers	Generic	Normal	Reconnaissance	Shellcode	Worms
Analysis	62%	0%	0%	38%	0%	0%	1%	0%	0%	0%
Backdoor	64%	1%	1%	32%	0%	0%	0%	1%	0%	0%
DoS	53%	1%	2%	41%	2%	0%	0%	1%	0%	0%
Exploits	25%	0%	2%	70%	2%	0%	1%	1%	0%	0%
Fuzzers	35%	0%	0%	10%	36%	0%	17%	2%	0%	0%
Generic	1%	0%	0%	2%	1%	96%	0%	0%	0%	0%
Normal	7%	0%	0%	1%	12%	0%	79%	1%	0%	0%
Reconnaissance	16%	0%	0%	13%	0%	0%	1%	69%	0%	0%
Shellcode	42%	0%	0%	9%	2%	0%	2%	26%	19%	0%
Worms	18%	0%	0%	75%	0%	0%	2%	2%	0%	2%

Predicted Biflow Class

Fig. 3. The Confusion Matrix of our ANN model in multi-class classification

6   Conclusions

In this work, we evaluated different ML and DL techniques for the developpement of A-NIDS using the official public benchmark UNSW-NB15 data set. The results show that the ANN model performs well on previously unseen biflow attacks with high accuracy and low FAR compared to the traditional ML techniques on both binary and multi-label classification. Furthermore, a visualization technique named UMAP has been used to show the core advantage of ANN over the traditional techniques in extracting useful representations from the data that ultimately improve the performance of the model in the detection of malicious biflows, which makes it eligible with its variants to improve the performance of A-NIDS in future works. Moreover, we have empirically shown that the imbalanced nature of the data set greatly affects the performance of the model by learning well on the populated classes compared to the less populated ones. In future research, we believe that reformulating the problem of A-NIDS is required to leverage some more sophisticated DL approaches. Moreover, we aim to apply deep generative models techniques such as GANs (Generative Adversarial Networks) and VAEs (Varational Auto Encoders) to address the issue of data imbalance by generating synthetic data samples for the classes that have insufficient samples that would theoretically improve the performance of the classifier.

References

1. Bigdeli, E., Mohammadi, M., Raahemi, B., Matwin, S.: Incremental anomaly detection using two-layer cluster-based structure. Inf. Sci. **429**, 315–331 (2018)
2. Buczak, A.L., Guven, E.: A survey of data mining and machine learning methods for cyber security intrusion detection. IEEE Commun. Surv. Tutor. **18**(2), 1153–1176 (2016)
3. Diallo, A.F., Patras, P.: Adaptive clustering-based malicious traffic classification at the network edge. In: IEEE INFOCOM, pp. 1–10. IEEE (2021)
4. Dromard, J., Roudière, G., Owezarski, P.: Online and scalable unsupervised network anomaly detection method. IEEE Trans. Netw. Serv. Manag. **14**(1), 34–47 (2017)
5. Fernandes, G., Rodrigues, J.J.P.C., Carvalho, L.F., Al-Muhtadi, J.F., Proença, M.L.: A comprehensive survey on network anomaly detection. Telecommun. Syst. **70**(3), 447–489 (2019). https://doi.org/10.1007/s11235-018-0475-8
6. García-Teodoro, P., Díaz-Verdejo, J., Maciá-Fernández, G., Vázquez, E.: Anomaly-based network intrusion detection: techniques, systems and challenges. Comput. Secur. **28**(1–2), 18–28 (2009)
7. Goodfellow, I.J., et al.: Generative adversarial networks (2014)
8. Guarino, I., Bovenzi, G., Di Monda, D., Aceto, G., Ciuonzo, D., Pescapé, A.: On the use of machine learning approaches for the early classification in network intrusion detection. In: IEEE M&N, pp. 1–6 (2022)
9. Hettich, S., Bay, S.D.: KDD Cup 1999 Data
10. Ingre, B., Yadav, A.: Performance analysis of NSL-KDD dataset using ANN. IN: SPACES, pp. 92–96 (2015)

11. Jeong, H., Yu, J., Lee, W.: Poster abstract: a semi-supervised approach for network intrusion detection using generative adversarial networks. IEEE Infocom, pp. 31–32 (2021)

12. Kim, T., Suh, S.C., Kim, H., Kim, J., Kim, J.: An encoding technique for CNN-based network anomaly detection. In: Big Data, pp. 2960–2965 (2019)

13. Li, H., Chasaki, D.: Ensemble machine learning for intrusion detection in cyber-physical systems. In: INFOCOM WKSHPS, pp. 12–13. IEEE (2021)

14. Liao, H.J., Richard Lin, C.H., Lin, Y.C., Tung, K.Y.: Intrusion detection system: a comprehensive review. J. Netw. Comput. Appl. **1**, 16–24 (2013)

15. Mishra, P., Varadharajan, V., Tupakula, U., Pilli, E.S.: A detailed investigation and analysis of using machine learning techniques for intrusion detection. IEEE Commun. Surv. Tutor. **21**(1), 686–728 (2019)

16. Moustafa, N., Slay, J.: UNSW-NB15: a comprehensive data set for network intrusion detection systems (UNSW-NB15 network data set). In: MilCIS, pp. 1–6. IEEE (2015)

17. Moustafa, N., Slay, J.: The evaluation of Network Anomaly Detection Systems: statistical analysis of the UNSW-NB15 data set and the comparison with the KDD99 data set. Inf. Secur. J. **25**(1–3), 18–31 (2016)

18. Odena, A.: Semi-supervised learning with generative adversarial networks. arXiv preprint arXiv:1606.01583 (2016)

19. Salimans, T., et al.: Improved techniques for training GANs, vol. 29. Curran Associates, Inc. (2016)

20. Sharafaldin, I., Lashkari, A.H., Ghorbani, A.A.: Toward generating a new intrusion detection dataset and intrusion traffic characterization. In: ICISSP 2018 (Cic), pp. 108–116 (2018)

21. Shiravi, A., Shiravi, H., Tavallaee, M., Ghorbani, A.A.: Toward developing a systematic approach to generate benchmark datasets for intrusion detection. Comput. Secur. **3**, 357–374 (2012)

22. Staudemeyer, R.C.: Applying long short-term memory recurrent neural networks to intrusion detection. S. Afr. Comput. J. **56**(56), 136–154 (2015)

23. Švihrová, R., Lettner, C.: A semi-supervised approach for network intrusion detection. In: ACM International Conference Proceeding Series (2020)

24. Tavallaee, M., Bagheri, E., Lu, W., Ghorbani, A.A.: A detailed analysis of the KDD CUP 99 data set. In: 2009 IEEE Symposium on Computational Intelligence for Security and Defense Applications, pp. 1–6 (2009)

25. Tool, I.P.: https://www.keysight.com/fr/en/products/network-test/network-test-hardware/perfectstorm.html

26. Tsakalidis, S., Doumpiotis, V., Byrne, W.: Discriminative linear transforms for feature normalization and speaker adaptation in HMM estimation. IEEE Trans. Speech Audio Process. **13**(3), 367–376 (2005)

27. Vinayakumar, R., Alazab, M., Soman, K.P., Poornachandran, P., Al-Nemrat, A., Venkatraman, S.: Deep learning approach for intelligent intrusion detection system. IEEE Access **7**, 41525–41550 (2019)

28. Yin, C., Zhu, Y., Fei, J., He, X.: A deep learning approach for intrusion detection using recurrent neural networks. IEEE Access **5**, 21954–21961 (2017)

29. Zhang, C., Ma, Y.: Ensemble Machine Learning. Springer, New York (2012). https://doi.org/10.1007/978-1-4419-9326-7

Applied Machine Learning for Securing the Internet of Medical Things in Healthcare

Wei Lu[✉]

Department of Computer Science, Keene State College, USNH, Keene, NH, USA
wlu@usnh.edu

Abstract. Integrating mobile computing technologies and human health activities using IoMT devices can accelerate biomedical discovery and improve the efficiency of healthcare research and delivery. However, the quality of the collected health data is critical for the success of these efforts. In this paper, a hybrid intrusion detection system is proposed to identify cyberattacks in real time on medical devices. The system combines a logistic regression-based detector using network traffic features with a gradient-boosted tree-based detector using medical sensor features. Evaluation of the system using a publicly available dataset shows an accuracy score of 95.4% using only 11 features, compared to the current best accuracy of 92.98% achieved by artificial neural networks using 40 features. Additionally, by combining the decisions of the two individual detection systems, the number of attacks detected is increased from 111 to 305 out of a total of 423 attack instances, improving the sensitivity score and addressing the challenge of effectively and efficiently integrating different detection technologies in a hybrid intrusion detection system. To the best of the authors' knowledge, this is the first attempt to combine multiple misuse detection models in a hybrid system to secure both IoMT devices and their networking equipment across the entire healthcare spectrum of the IoMT.

1 Introduction

There has been a significant overlap between computing, communication, and human activities, particularly in the healthcare industry, leading to the evolution of four generations of healthcare, from traditional healthcare 1.0 to the current system that incorporates the widespread use of medical devices and the incorporation of advanced biomedical technology in clinical and hospital networks [1]. A recent Deloitte report estimated that more than 500,000 medical technologies are currently available; many of them, if not all, are so-called Internet of Medical Things (IoMT) devices [2]. In a typical patient-centric personal healthcare scenario using these IoMT devices, the wearable medical sensors collect different types of data (e.g., EEG, ECG, blood pressure) [3] and then transmit them to an intermediate smartphone device over the body area network using one of several wireless communication protocols such as Zigbee [4], Bluetooth [5], or Wi-Fi [6].

IoMT has allowed for the development of innovative healthcare applications, but it has also brought about many security and privacy concerns that could hinder its growth.

L. Barolli (Ed.): AINA 2023, LNNS 654, pp. 404–416, 2023.
https://doi.org/10.1007/978-3-031-28451-9_35

A recent SANS Institute report concluded that over 94% of healthcare organizations had been victims of cyberattacks on infrastructure and medical devices [7]. There have been several studies in this field that have highlighted the need for proper security measures and have proposed solutions that focus on designing secure healthcare systems using prevention-based security technologies [8–12]. However, these standard security protection methods (such as access control) are insufficient because they are not practical for resource-constrained medical devices.

To fully realize the potential benefits and use of IoMT devices in healthcare, intrusion detection techniques have been suggested to address the pressing need for securing these devices [13, 14]. Traditional intrusion detection techniques can be broadly classified into misuse detection and anomaly detection [15]. Misuse detection involves detecting attacks by comparing actual behaviors recorded in audit trails with known patterns of suspicious behavior. This approach is effective at detecting known attacks but is not useful for detecting novel or unknown forms of attack for which there are no available signatures [16–18]. Furthermore, it can be challenging to define a signature that covers all possible variations of known attacks [19–21]. Errors in defining these signatures can also lead to an increase in false alarms and a decrease in the overall effectiveness of the detection technique [22, 23]. Anomaly detection, on the other hand, involves establishing normal activity profiles for the system [24, 25]. Any behavior that deviates from the normal profile is flagged as potentially suspicious and is further investigated [26, 27]. Anomaly detection studies start by defining what is considered normal behavior for the system or object being observed and then determine which activities should be flagged as potentially intrusive and how to make those decisions [28–30].

In this paper, we propose a comprehensive practical framework based on a novel hybrid intrusion detection system to assure the security of IoMT devices and detect malicious attacks from inside and outside of the healthcare context. The proposed hybrid system includes two individual detection components: IoMT misuse detection based on network traffic and IoMT misuse detection based on medical sensors. In IoMT misuse detection based on network traffic, we conduct packet payload inspection and use network traffic features to form the signatures of attacks on medical devices [33]. A comparative study does this with six typical machine learning algorithms, namely, multivariate logistic regression (MLR), decision trees (DT), random forests (RF), support vector machines (SVM), artificial neural networks (ANN), and extreme gradient boosted trees (XGBoost), from which we select the best-fit attack model. In IoMT misuse detection based on medical sensors, we apply the same methodology but with features selected from medical sensors, where we have another best-fit attack modeling with medical sensor data. The hybrid model then integrates outputs from the two misuse detectors, where a simple voting machine using the logical digital gate OR is developed to make real-time decisions on malicious attacks against medical devices.

The main contributions of this paper include: (1) We conduct a detailed analysis of the publicly available dataset WUSTL-EHMS-2020. The WUSTL-EHMS-2020 dataset was created using a real-time Enhanced Healthcare Monitoring System (EHMS) testbed IoMT network deployed at Washington University in St. Louis [31]. By conducting descriptive statistical analysis on this dataset, we find that some records of data instances are apparent outliers. After removing these outliers, we create a new dataset called

WUSTL-EHMS-CLEAN-2022 that can be publicly accessed at [32] for the secondary use of the original dataset when conducting the performance evaluation of machine learning algorithms for intrusion detection when attackers infiltrate the IoMT network and (2) Existing security mechanisms in this domain are mainly developed using network traffic data, that would work well for most of the traditional intrusion detection systems. They, however, would need to include the other essential part of the information collected by medical sensors when applying the concept of intrusion detection systems into the domain of the IoMT network. The proposed research is the first attempt to combine multiple misuse detectors into one hybrid model, where we fuse decisions from the MLR detector and XGBoost detector, resulting in the number of attacks detected improving significantly from 111 to 305 over the total 423 attack instances.

The rest of the paper is organized as follows. Section 2 presents the WUSTL-EHMS-CLEAN-2022 dataset. Section 3 is feature selection, where we introduce the process of selecting six significant features from the 26 network traffic features and how the five medical sensors features are chosen from the seven raw medical sensors features. Section 4 then conducts comparative studies for selecting the best-fit model to build misuse detectors based on network traffic and medical sensors using the WUSTL-EHMS-CLEAN-2022 dataset. Section 5 presents the hybrid detection system using a simple voting machine to decide on intrusion detection in the IoMT network. Finally, in Sect. 6, we make some concluding remarks and discuss future work.

2   WUSTL-EHMS-CLEAN-2022 Dataset

The WUSTL-EHMS-CLEAN-2022 dataset is a cleaned version of the original WUSTL-EHMS-2020 dataset [31, 32]. We conducted descriptive statistics on the medical sensor features and found that some of the vital features were outliers or contaminated, likely due to malfunctioning medical sensors during data collection, such as disconnection of the EKG pad or incorrect initial temperature settings. Figure 1 shows descriptive statistics for the eight vital features collected from medical sensors.

	Temp	SpO2	Pulse_Rate	SYS	DIA	Heart_rate	Resp_Rate	ST
count	16318.000000	16318.000000	16318.000000	16318.000000	16318.000000	16318.000000	16318.000000	16318.000000
mean	26.906815	97.808861	76.723741	142.846611	80.094190	75.443927	19.695551	0.258007
std	0.919766	1.496269	7.431914	8.493933	6.125289	6.609102	7.325856	0.103980
min	23.600000	0.000000	0.000000	0.000000	0.000000	0.000000	0.000000	-0.300000
25%	26.600000	98.000000	73.000000	142.000000	76.000000	73.000000	18.000000	0.200000
50%	27.000000	98.000000	73.000000	144.000000	83.000000	73.000000	19.000000	0.300000
75%	27.600000	98.000000	79.000000	148.000000	84.000000	79.000000	24.000000	0.300000
max	29.200000	100.000000	194.000000	149.000000	95.000000	119.000000	73.000000	1.000000

Fig. 1. Descriptive statistics of features from medical sensors

We created a new dataset called WUSTL-EHMS-CLEAN-2022 by cleaning the original WUSTL-EHMS-2020 dataset and removing 929 polluted data instances. The new dataset includes 15,389 data instances with 40 features and is publicly available in [32] for use as a benchmark for comparative studies on different machine learning models. Table 1 shows the observations for each label category in the new dataset.

Table 1. Frequency table of target variable label in new dataset

Label	Count	Proportion of observations
1 (attack behavior)	2,046	13.3%
0 (normal behavior)	13,343	86.7%

3 Feature Selection

There are 40 raw features in the newly constructed WUSTL-EHMS-Clean-2022 dataset, where SrcAddr, DstAddr, Sport, Dport, SrcMac, and DstMac are pseudo features that are used to define the direction of network traffic, and their corresponding source and destination logical and hardware medium addresses. Therefore, those features need to be excluded when we select significant features. Moreover, the feature Label is not included because it is the target variable. As a result, the total number of features we use for feature selection is 33. Figure 2 illustrates the framework of our feature selection methodology. The input includes 33 features split into two groups; one is network traffic based with 26 features, and the other is medical sensors based, including seven features. Details on the description of these raw features can see [13].

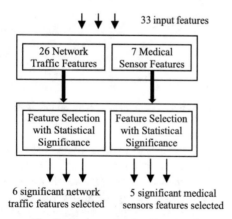

Fig. 2. Feature selection framework

In order to select features for analysis of network traffic, we used a regression model to fit 26 different network traffic features and set the p-value significance threshold to 0.05. The results of this modeling process are shown in Fig. 3. Upon examination of the results, we found that eight features (SrcGap, DstGap, DIntPktAct, sMaxPktSz, dMaxPktSz, dMinPktSz, Trans, and TotBytes) were constant, with no change recorded during data collection. These features were therefore eliminated, and the feature selection process was repeated. After repeating the process, we identified pScrLoss as having the largest p-value (0.9983), which was above the significance threshold of 0.05. As a result, pScrLoss was also eliminated. The process was repeated once more, resulting in the removal of pDstLoss, due to its high p-value of 1.0. We continued this process of identifying and eliminating features with p-values above the significance threshold, ultimately arriving at a final model featuring only the significant features of DstBytes, DstLoad, SIntPkt, SrcJitter, DstJitter, and Rate.

Term		Estimate	Std Error	ChiSquare	Prob>ChiSq
Intercept	Biased	-52547.071	214969.98	0.06	0.8069
SrcBytes	Biased	-0.1555626	0.2200081	0.50	0.4795
DstBytes	Biased	2.88640974	5.7690252	0.25	0.6168
SrcLoad		0.00974739	0.0803395	0.01	0.9034
DstLoad		-0.0655831	0.069986	0.88	0.3487
SrcGap	Zeroed	0	0	.	.
DstGap	Zeroed	0	0	.	.
SIntPkt		0.50911938	0.6862652	0.55	0.4582
DIntPkt		0.0240749	0.1957492	0.02	0.9021
SIntPktAct		9.2014652	8583.4871	0.00	0.9991
DIntPktAct	Zeroed	0	0	.	.
SrcJitter		-0.1829262	0.017944	103.92	<.0001*
DstJitter		0.03552641	0.1960344	0.03	0.8562
sMaxPktSz	Biased	-2.6297575	0	.	.
dMaxPktSz	Zeroed	271.734071	0	.	.
sMinPktSz		589.588253	3582.7067	0.03	0.8693
dMinPktSz	Zeroed	0	0	.	.
Dur	Biased	-134.14857	228.62209	0.34	0.5574
Trans	Zeroed	0	0	.	.
TotPkts	Biased	-58.391596	151.15367	0.15	0.6993
TotBytes	Zeroed	0	0	.	.
Load		-0.0022517	0.0725707	0.00	0.9752
Loss		-7617.2002	10810140	0.00	0.9994
pLoss		1970.32387	3619417.5	0.00	0.9996
pSrcLoss		-726.52525	1634754	0.00	0.9996
pDstLoss		-824.25381	1713024.5	0.00	0.9996
Rate		7.49550858	18.75941	0.16	0.6895

Fig. 3. Parameter estimates with all 26 raw features.

Using the same modeling process as before, we applied it to seven medical sensor features and set the p-value significance threshold to 0.05. The results of this process are shown in Fig. 4. After examining the results, we identified Heart_rate as having a p-value of 0.3607, which was above the significance threshold of 0.05. As a result, we eliminated Heart_rate and repeated the feature selection process. We then removed SYS, as it did not meet the significance threshold, and were left with a final model featuring only the significant features of SpO2, Pulse_rate, DIA, Resp_Rate, and ST.

Term	Estimate	Std Error	ChiSquare	Prob>ChiSq
Intercept	10.2395766	2.9615371	11.95	0.0005*
SpO2	-0.1161181	0.0278507	17.38	<.0001*
Pulse_Rate	0.02473574	0.0038391	41.51	<.0001*
SYS	-0.0067777	0.005852	1.34	0.2468
DIA	-0.0165456	0.0069858	5.61	0.0179*
Heart_rate	0.00361046	0.0039501	0.84	0.3607
Resp_Rate	-0.0430564	0.0048645	78.34	<.0001*
ST	0.85345197	0.2458511	12.05	0.0005*

Fig. 4. Parameter estimates with seven raw features.

4 Selecting Best-Fit Machine Learning Models

Instead of using a single model to fit a combination of two types of input features without performing feature selection, our approach uses statistical significance to determine the best-fit model for each group of features. To do this, we compared the performance of six machine learning algorithms (MLR, DT, RF, SVM, ANN, and XGBoost) using the same benchmark dataset. The results of this comparison showed that MLR was the best fit for the network traffic features, while XGBoost was the best fit for the medical sensor features. This allows us to select the most appropriate model for each group of features, ensuring that our analysis is as accurate as possible.

According to Table 2, the MLR model is the best-performing model in terms of both interpretability and accuracy. This is because, in this case, we prioritize simplicity and ease of interpretation, and the MLR model meets these criteria while still achieving a high accuracy of 92.50% on the testing dataset. Additionally, the MLR model has a low discrepancy value of 0.0011 between training and testing data, indicating that it maintains a high level of accuracy across both datasets. Overall, the MLR model is a strong choice for this analysis due to its combination of interpretability and high accuracy. According to Table 3, the XGBoost model is the best-performing model when analyzing the five significant medical sensor features. This model has the highest accuracy of 89.64% on the testing dataset and a low discrepancy value of 0.0083 between training and testing data. These results indicate that the XGBoost model is able to maintain a high level of accuracy across both datasets, making it a strong choice for this analysis.

5 Hybrid Detection Using a Simple Voting Machine

By comparing six typical machine models, we conclude that the MLR model is the best-fit approach to handle network traffic-based features, and XGBoost is the best-fit model to take medical sensors features. Table 4 illustrates the confusion matrix generated by the MLR model using the six network traffic features with significance, where 92.5% of 3,078 data instances in the testing dataset are classified accurately. Table 5 illustrates the confusion matrix generated by the XGBoost model using the five medical sensors features with significance, where 89.64% of data instances in the testing dataset are classified accurately. Figure 5 illustrates the hybrid detection framework, where the results of the two individual detectors are logically *ORed* into the final output.

Table 2. Comparative studies among six different models with network traffic features

Model	Performance metrics		
	Accuracy with training data	Accuracy with testing data	The discrepancy in training and testing
MLR	92.61%	92.50%	0.11%
DT	92.55%	92.33%	0.22%
SVM	92.49%	92.49%	0.00%
RT	92.41%	92.27%	0.14%
XGBoost	92.35%	92.24%	0.11%
ANN	92.37%	92.00%	0.37%

Table 3. Comparative studies among six different models with medical sensor features

Model	Performance metrics		
	Accuracy with training data	Accuracy with testing data	The discrepancy in training and testing
MLR	86.82%	86.26%	0.56%
DT	87.44%	86.94%	0.50%
SVM	86.12%	84.93%	1.19%
RT	87.35%	87.03%	0.32%
XGBoost	90.47%	89.64%	0.83%
ANN	86.81%	86.25%	0.56%

Table 4. Confusion matrix using MLR with network traffic features

Actual label	Predicted	
	Count	
	0	1
0	2,654	1
1	230	193

Table 6 shows the truth table of this simple voting machine using the logical function OR with a digital circuit design in mind when deploying the system in hardware-based detection sensors. As illustrated in Table 6, if any one of two detectors (i.e., MLR-based detector and XGBoost-based detector) votes 1, where 1 means to attack, then the output will be 1, i.e., attack identified. When both detectors vote 0, where 0 means normal, the

Table 5. Confusion matrix using XGBoost with medical sensor features

Actual label	Predicted	
	Count	
	0	1
0	2,648	7
1	312	111

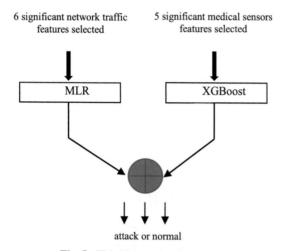

Fig. 5. Hybrid detection framework

output will be 0, i.e., normal behavior identified. Using this voting machine, our hybrid model achieves an accuracy of 93.54% on the same benchmark of the testing dataset.

Table 6. Truth table of the simple voting machine

MLR	XGBoost	Output
0	0	0
0	1	1
1	0	1
1	1	1

Table 7 illustrates the confusion matrix of the proposed hybrid model, where the number of attacks detected is improved from 111 to 232 over the total 423 attack instances, doubling the sensitivity score compared to the two individual detectors. These evaluation results show that our hybrid model provides a solution for getting different detection technologies to interoperate effectively and efficiently when building an operational hybrid intrusion detection system.

Table 7. Confusion matrix using hybrid model

Actual label	Predicted	
	Count	
	0	1
0	2,647	8
1	191	232

To further evaluate our hybrid detection model's effectiveness, we increase the number of estimators in the XGBoost model and then rerun the detection experiment with the same testing dataset. Table 8 illustrates the confusion matrix for this evaluation, where the model achieves an accuracy of 95.4% on the testing data.

Table 8. Confusion matrix using hybrid model with enhanced XGBoost

Actual label	Predicted	
	Count	
	0	1
0	2,631	24
1	118	305

Based on the confusion matrix, we summarize a set of performance metrics in Table 9, including true positives (TP), true negatives (TN), false positives (FP), false negatives (FN), Precision, Recall, and Accuracy [34]. TP is the number of records in which the model correctly predicts that the record is an attack. TN is the number of records in which the model correctly predicts that the record is normal. FP is the number of records in which the model incorrectly predicts that the record is an attack (when it is actually normal). FN is the number of records in which the model incorrectly predicts that the record is normal (when it is actually an attack). Precision measures the model's accuracy when it predicts that a record is an attack. It is calculated using the following formula:

$$Precision = \frac{TP}{TP + FP}$$

And Recall is a measure of the model's ability to identify all of the attack records in the dataset correctly. It is calculated using the following formula:

$$Recall = \frac{TP}{TP + FN}$$

Finally, Accuracy is a measure of the overall correctness of the model. It is calculated using the following formula:

$$Accuracy = \frac{TP + TN}{TP + FP + FN + TN}$$

Table 9. Performance metrics for the hybrid detector

TP	TN	FP	FN	Precision	Recall	Accuracy
305	2,631	24	118	0.9271	0.721	0.9540

6 Conclusions and Future Work

The IoMT refers to a network of interconnected medical devices, software applications, and digital health services that work together to create, collect, analyze, and transport health data and medical images. The increasing use of IoMT in the healthcare industry has brought about significant changes and has also made medical devices more vulnerable to cyberattacks, raising concerns about the security and privacy of sensitive medical information. These issues are particularly important in the context of life-critical information and require careful consideration by healthcare professionals.

In this paper, we present a new hybrid intrusion detection system that combines two misuse detectors: one based on the MLR model using network traffic features, and the other based on the XGBoost model using medical sensor features. Existing security mechanisms in the Internet of Medical Things (IoMT) are primarily prevention-based and are often insufficient due to their vulnerability to widely available exploits and their resource-intensive nature. Our proposed system offers a more effective and efficient alternative to these prevention-based approaches, which can be easily bypassed and may be difficult to implement on resource-constrained IoMT devices. Our hybrid intrusion detection system outperforms existing approaches in terms of both accuracy and the number of features used. The system achieves an accuracy of 95.4% using only 11 features, compared to an existing solution that achieves an accuracy of 92.98% using 40 features. Furthermore, by combining the decisions of the two individual detection systems, we were able to increase the number of detected attacks from 111 to 305 out of a total of 423 attack instances, almost tripling the sensitivity score. These results demonstrate the potential of our hybrid system to effectively and efficiently integrate different detection technologies when building an operational hybrid intrusion detection system for the IoMT.

During our experimental evaluation using the WUSTL-EHMS-Clean-2022 dataset, we found that all six supervised machine learning models incorrectly classified around 100 attack instances out of 423. In future work, we plan to further investigate this issue in a practical intrusion detection system [35, 36] and conduct more comprehensive comparative studies on the detection performance of these machine learning approaches, including the application of intrusion responses [37] and alert correlation to reduce false alarms [38]. This will help to improve the overall effectiveness of our system in detecting and responding to attacks.

Acknowledgments. This research is supported by New Hampshire - INBRE through an Institutional Development Award (IDeA), P20GM103506, from the National Institute of General Medical Sciences of the NIH.

References

1. Anand, A., Rani, S., Anand, D., Aljahdali, H.M., Kerr, D.: An efficient CNN-based deep learning model to detect malware attacks (CNN-DMA) in 5G-IoT healthcare applications. Sensors **21**(19), 6346 (2021)
2. How connected medical devices are transforming health care (2018). https://www2.deloitte.com/content/dam/Deloitte/global/Documents/Life-Sciences-Health-Care/gx-lshc-medtech-iomt-brochure.pdf. Accessed 23 Nov 2022
3. Chen, H.C.: Smart health and wellbeing [trends & controversies]. IEEE Intell. Syst. **26**(5), 78–90 (2011)
4. Connectivity Standards Alliance. https://csa-iot.org/. Accessed 23 Nov 2022
5. Bhagwat, P.: Bluetooth: technology for short-range wireless apps. IEEE Internet Comput. **5**(3), 96–103 (2001). https://doi.org/10.1109/4236.935183
6. Wi-Fi Alliance. https://www.wi-fi.org/. Accessed 23 Nov 2022
7. Filkins, B.: Health Care Cyberthreat Report: Widespread Compromises Detected, Compliance Nightmare on Horizon. SANS Institute (2014)
8. Li, C.X., Raghunathan, A., Jha, N.K.: Hijacking an insulin pump: security attacks and defenses for a diabetes therapy system. In: 2011 IEEE 13th International Conference on e-Health Networking, Applications and Services, pp. 150–156 (2011). https://doi.org/10.1109/HEALTH.2011.6026732
9. Halperin, D., et al.: Pacemakers and implantable cardiac defibrillators: software radio attacks and zero-power defences. In: 2008 IEEE Symposium on Security and Privacy, pp. 129–142 (2008). https://doi.org/10.1109/SP.2008.31
10. Medical devices hit by ransomware for the first time in us hospitals. https://www.forbes.com/sites/thomasbrewster/2017/05/17/wannacry-ransomware-hit-real-medical-devices/?sh=67f42679425c. Accessed 23 Nov 2022
11. Sehatbakhsh, N., Alam, M., Nazari, A., Zajic, A., Prvulovic, M.: Syndrome: spectral analysis for anomaly detection on medical IoT and embedded devices. In: 2018 IEEE International Symposium on Hardware Oriented Security and Trust (HOST), pp. 1–8 (2018). https://doi.org/10.1109/HST.2018.8383884
12. Gope, P., Hwang, T.: BSN-care: a secure IoT-based modern healthcare system using body sensor network. IEEE Sens. J. **16**(5), 1368–1376 (2016). https://doi.org/10.1109/JSEN.2015.2502401
13. Lu, W.: Detecting malicious attacks using principal component analysis in medical cyber-physical systems. In: Traore, I., Woungang, I., Saad, S. (eds.) Artificial Intelligence for Cyber-Physical Systems Hardening, vol. 2, pp. 203–215. Springer, Cham (2023). https://doi.org/10.1007/978-3-031-16237-4_9
14. Ghorbani, A.A., Lu, W., Tavallaee, M.: Network attacks. In: Ghorbani, A.A., Lu, W., Tavallaee, M. (eds.) Network Intrusion Detection and Prevention. ADIS, vol. 47, pp. 1–25. Springer, Boston (2010). https://doi.org/10.1007/978-0-387-88771-5_1. ISBN-10: 0387887709

15. Ghorbani, A.A., Lu, W., Tavallaee, M.: Theoretical foundation of detection. In: Ghorbani, A.A., Lu, W., Tavallaee, M. (eds.) Network Intrusion Detection and Prevention. ADIS, vol. 47, pp. 73–114. Springer, Boston (2010). https://doi.org/10.1007/978-0-387-88771-5_4. ISBN-10: 0387887709

16. Garant, D., Lu, W.: Mining botnet behaviors on the large-scale web application community. In: Proceedings of 27th IEEE International Conference on Advanced Information Networking and Applications, Barcelona, Spain, 25–28 March 2013 (2013)

17. Ghorbani, A.A., Lu, W., Tavallaee, M.: Detection approaches. In: Ghorbani, A.A., Lu, W., Tavallaee, M. (eds.) Network Intrusion Detection and Prevention. ADIS, vol. 47, pp. 27–53. Springer, Boston (2010). https://doi.org/10.1007/978-0-387-88771-5_2. ISBN-10: 0387887709

18. Lu, W., Ghorbani, A.A.: Bots behaviors vs. human behaviors on large-scale communication networks (extended abstract). In: Lippmann, R., Kirda, E., Trachtenberg, A. (eds.) RAID 2008. LNCS, vol. 5230, pp. 415–416. Springer, Heidelberg (2008). https://doi.org/10.1007/978-3-540-87403-4_33

19. Lu, W., Miller, M., Xue, L.: Detecting command and control channel of botnets in cloud. In: Traore, I., Woungang, I., Awad, A. (eds.) ISDDC 2017. LNCS, vol. 10618, pp. 55–62. Springer, Cham (2017). https://doi.org/10.1007/978-3-319-69155-8_4. ISBN 978-3-319-69154-1

20. Lu, W., Ghorbani, A.A.: Botnets detection based on IRC-community. In: 2008 IEEE Global Telecommunications Conference, IEEE GLOBECOM 2008, pp. 1–5 (2008). https://doi.org/10.1109/GLOCOM.2008.ECP.398

21. Lu, W., Mercaldo, N., Tellier, C.: Characterizing command and control channel of mongoose bots over TOR. In: Woungang, I., Dhurandher, S.K. (eds.) WIDECOM 2020. LNDECT, vol. 51, pp. 23–30. Springer, Cham (2020). https://doi.org/10.1007/978-3-030-44372-6_2

22. Tavallaee, M., Lu, W., Ghorbani, A.: Online classification of network flows. In: Proceedings of the 7th Annual Conference on Communication Networks and Services Research (CNSR 2009), Moncton, New Brunswick, Canada, 11–13 May 2009, pp. 78–85 (2009)

23. Lu, W., Xue, L.: A heuristic-based co-clustering algorithm for the internet traffic classification. In: 2014 28th International Conference on Advanced Information Networking and Applications Workshops, pp. 49–54 (2014). https://doi.org/10.1109/WAINA.2014.16

24. Lu, W.: An Unsupervised Anomaly Detection Framework for Multiple-connection-Based Network Intrusions. Ottawa Library and Archives Canada (2007). ISBN 9780494147795

25. Lu, W., Traore, I.: A new unsupervised anomaly detection framework for detecting network attacks in real-time. In: Desmedt, Y.G., Wang, H., Yi, M., Li, Y. (eds.) Cryptology and Network Security, pp. 96–109. Springer, Heidelberg (2005). https://doi.org/10.1007/11599371_9. ISBN 978-3-540-32298-6

26. Lu, W., Traore, I.: An unsupervised approach for detecting DDoS attacks based on traffic based metrics. In: Proceedings of IEEE Pacific Rim Conference on Communications, Computers and Signal Processing (PACRIM 2005), Victoria, B.C., pp. 462–465 (2005)

27. Lu, W., Traore, I.: Determining the optimal number of clusters using a new evolutionary algorithm. In: Proceedings of IEEE International Conference on Tools with Artificial Intelligence (ICTAI 2005), Hongkong, pp. 712–713 (2005)

28. Lu, W., Tong, H.: Detecting network anomalies using CUSUM and EM clustering. In: Cai, Z., Li, Z., Kang, Z., Liu, Y. (eds.) ISICA 2009. LNCS, vol. 5821, pp. 297–308. Springer, Heidelberg (2009). https://doi.org/10.1007/978-3-642-04843-2_32. ISBN 978-3-642-04843-2

29. Lu, W., Traore, I.: Unsupervised anomaly detection using an evolutionary extension of K-means algorithm. Int. J. Inf. Comput. Secur. **2**(2), 107 (2008). https://doi.org/10.1504/IJICS.2008.018513

30. Lu, W., Traore, I.: A new evolutionary algorithm for determining the optimal number of clusters. In: Proceedings of IEEE International Conference on Computational Intelligence for Modeling, Control and Automation (CIMCA 2005), vol. 1, pp. 648–653 (2005)

31. WUSTL EHMS Dataset. https://www.cse.wustl.edu/~jain/ehms/index.html. Accessed 23 Nov 2022

32. WUSTL EHMS Clean. https://unh.box.com/s/qja9cnmvtbyr0ctsw6p6fx8y8vr2z8oo. Accessed 23 Nov 2022

33. Ghorbani, A.A., Lu, W., Tavallaee, M.: Data collection. In: Ghorbani, A.A., Lu, W., Tavallaee, M. (eds.) Network Intrusion Detection and Prevention. ADIS, vol. 47, pp. 55–71. Springer, Boston (2010). https://doi.org/10.1007/978-0-387-88771-5_3. ISBN-10: 0387887709

34. Ghorbani, A.A., Lu, W., Tavallaee, M.: Evaluation criteria. In: Ghorbani, A.A., Lu, W., Tavallaee, M. (eds.) Network Intrusion Detection and Prevention. ADIS, vol. 47, pp. 161–183. Springer, Boston (2010). https://doi.org/10.1007/978-0-387-88771-5_7. ISBN-10: 0387887709

35. Nunley, K., Lu, W.: Detecting network intrusions using a confidence-based reward system. In: 2018 32nd International Conference on Advanced Information Networking and Applications Workshops (WAINA), pp. 175–180 (2018). https://doi.org/10.1109/WAINA.2018.00083

36. Ghorbani, A.A., Lu, W., Tavallaee, M.: Architecture and implementation. In: Ghorbani, A.A., Lu, W., Tavallaee, M. (eds.) Network Intrusion Detection and Prevention. ADIS, vol. 47, pp. 115–127. Springer, Boston (2010). https://doi.org/10.1007/978-0-387-88771-5_5. ISBN-10: 0387887709

37. Ghorbani, A.A., Lu, W., Tavallaee, M.: Intrusion response. In: Ghorbani, A.A., Lu, W., Tavallaee, M. (eds.) Network Intrusion Detection and Prevention. ADIS, vol. 47, pp. 185–198. Springer, Boston (2010). https://doi.org/10.1007/978-0-387-88771-5_8. ISBN-10: 0387887709

38. Ghorbani, A.A., Lu, W., Tavallaee, M.: Alert management and correlation. In: Ghorbani, A.A., Lu, W., Tavallaee, M. (eds.) Network Intrusion Detection and Prevention. ADIS, vol. 47, pp. 129–160. Springer, Boston (2010). https://doi.org/10.1007/978-0-387-88771-5_6. ISBN-10: 0387887709

Early-Stage Ransomware Detection Based on Pre-attack Internal API Calls

Filippo Coglio, Ahmed Lekssays[(✉)], Barbara Carminati, and Elena Ferrari

Università degli Studi dell'Insubria, Varese, Italy
{fcoglio,alekssays,barbara.carminati,elena.ferrari}@uninsubria.it

Abstract. Ransomware attacks have become one of the main cyber threats to companies and individuals. In recent years, different approaches have been proposed to mitigate such attacks by analyzing ransomware behavior during the infection and post-infection phases. However, few works focused on early-stage ransomware detection. The analysis of recent ransomware has shown that they are designed to perform sensing activities to evade detection by known anti-viruses and anti-malware software. This paper proposes an early-stage ransomware detector based on a neural network model for multi-class classification. Our model achieves 80.00% accuracy on our dataset and 93.00% on another state-of-the-art dataset [10]. We show that our model performs better than the state-of-the-art approaches, especially on a challenging, large, and varied dataset we made publicly available.

1 Introduction

Ransomware is a type of malware that encrypts user data or restricts access to infected devices and their resources. A ransomware exploits secure communication channels with C&C (Command and Control) servers to encrypt the victims' systems and force them to pay a ransom [8]. If the attacked entity refuses to pay the ransom, data is deleted or published on the web. Ransomware attacks have become one of the main cyber threats to both companies and individuals. In 2021, the average cost of a ransomware attack for companies was $4.62 million, with an increase of 148% in the number of ransomware attacks from 2020 to 2021[1]. This increase was expected due to ransomware-as-a-service (RaaS) growth, where attackers sell their ransomware in underground markets, and accept payments in cryptocurrencies to preserve their anonymity [9]. This has turned ransomware into a lucrative tool for attackers who look for financial gains [10].

In recent years, different approaches have been proposed to mitigate such attacks using dynamic or static analysis to understand ransomware's code structure and behavior during infection and post-infection phases. Despite all the work, the defense against ransomware is challenging due to the lack of knowledge of newly detected ransomware.

---

[1] https://www.pandasecurity.com/en/mediacenter/security/ransomware-statistics/.

L. Barolli (Ed.): AINA 2023, LNNS 654, pp. 417–429, 2023.
https://doi.org/10.1007/978-3-031-28451-9_36

Therefore, there is a need to investigate effective approaches for detecting ransomware, keeping in mind their constant evolution. In this paper, we focus on early-stage ransomware detection. The analysis of recent ransomware has shown that they are programmed to execute some functions and operations to evade detection by known anti-viruses and anti-malware software. These *paranoia activities* aim to sense the environment to understand whether the ransomware can run the malicious code [10]. Thus, based on pre-attack activities, we aim to detect ransomware before the encryption phase. We dynamically analyzed over 11,000 ransomware samples and 1,200 benign samples from 23 different families to extract key API calls used by ransomware before launching attacks. These API calls help in classifying samples into their corresponding ransomware families or benign ones. We have developed a neural network model for multi-class classification that achieves 80.00% accuracy on our dataset and 93.24% on another state-of-the-art dataset [10]. We show that our model performs better than the state-of-the-art approaches, especially on a challenging, large, and varied dataset. In addition, we show the effectiveness and feasibility of the proposed approach compared to previous work.

The contributions of this work can be summarized as follows: (i) we have compiled a dataset of 5203 benign and ransomware samples from 12 different families; to the best of our knowledge, it is the largest dataset available for ransomware detection; (ii) we have developed a neural network model that achieves an accuracy of 80.00% in a challenging, large, and varied dataset, outperforming the state-of-the-art; (iii) we have made our source code and dataset publicly available[2] to reproduce the results. The remainder of this paper is organized as follows. We discussed state-of-the-art approaches in Sect. 2. Section 3 presents background knowledge on ransomware detection. In Sect. 4, we discuss our methodology and the building blocks of our solution. Section 5 shows the obtained results and the comparison with state-of-the-art approaches. Finally, Sect. 6 concludes the paper.

2   Related Work

In the last years, different techniques have been proposed for ransomware classification. [10] presents several machine-learning models for early-stage ransomware classification based on pre-attack paranoia activities using API calls as features. They have used different techniques for data representation: Occurrence of Words (OoW), representing the presence/absence of a feature, Bag of Words (BoW), expressing the frequency of a feature, and Sequence of Words (SoW), building a chain of API calls to take into consideration the order in which an API is executed.

The work in [2] presents an ML model for ransomware detection by comparing algorithms like Random Forest, Logistic Regression, Stochastic Gradient Descent, etc. After performing a dynamic analysis using the Intel PIN tool's dynamic binary instrumentation (DBI), features are extracted according to the CF-NCF (Class Frequency - Non-Class Frequency) technique. According to the

---

[2] https://github.com/Ph1l99/RansomwareEarlyDetection.

authors, this process provides higher accuracy during classification experiments. [6] proposes a behavioral classification method by analyzing 150 samples and extracting a set of features and attributes based on reports from   [3].

The authors of [8] presented a two-stage detection method based on dynamic analysis. The first stage relies on Markov chains, whereas the second relies on Random Forest.

[16] relies on the Term Frequency-Inverse Document Frequency (TF-IDF) of the N-grams extracted from opcodes. They analyze different N-gram feature dimensions using various machine learning models. Similarly, [15] extracts N-grams features from opcodes; but it only uses a Self-Attention Convolutional Neural Network (SA-CNN) to test the approach, which worked well for some long sequences of opcodes.

Despite the promising results of the above-mentioned papers, they have several limitations. For instance, the usefulness of the obtained results may be distorted by the limited number of analyzed samples, and the low variability of families included in the training phase may not represent the current ransomware landscape. We try to address these limitations by analyzing a more representative number of samples from 12 different families. Another limitation is that some solutions (i.e., [15,16]) use a static analysis approach for extracting N-grams for their models. This needs to deal with obfuscated ransomware samples, making reverse engineering the most complex step. We resolve this obfuscation problem by using a dynamic analysis approach to capture the pre-attack activities performed by ransomware samples.

Furthermore, our proposal focuses exclusively on ransomware detection, unlike the work in [2,15,16]. Second, it focuses on early-stage ransomware detection, whereas all other works, with the exception of [10], focus on later stages of detection. However, our work differs from [10] in the choice of the API calls considered in the detection phase. In addition, we tested our solution on, to the best of our knowledge, the largest ransomware detection dataset including 12 different ransomware families, whereas the work on [10] has only been tested on a dataset of 5 ransomware families.

3   Background

We introduce the ransomware and give a background on neural networks.

3.1   Ransomware

Ransomware is a type of malware that denies access to user files and demands a ransom from users to regain access to the system and stored information [7]. Ransomware are mainly of two types:

Locker: it prevents the victim from reaching their files by denying access to computing resources (e.g., locking the desktop or blocking the logging in) [7].

---

[3] https://www.virustotal.com.

Crypto: it encrypts data on the target machine, holding it hostage until the victim pays the ransom and obtains the decryption key from the attacker. Some variants of crypto-ransomware will progressively delete hostage files or release them to the public if the victim fails to pay the ransom on time.

Ransomware can be organized into families depending on their behavior and the type of operations they perform. In the following, we present the main characteristics of well-known ransomware families:

Cerber: it infects computers using common attack vectors, such as phishing e-mails. It comes bundled with free online software. Cerber mainly utilizes malicious Microsoft Office files with macros to spread and encrypt victim files.

CryptoWall: it writes its registry autorun keys in the Windows registry to maintain its persistence through reboots. It then searches for all system restore points and Volume Shadow Copy files and destroys them to prevent the victim from restoring any file. Then, it begins encrypting files using the RSA-2048 encryption algorithm.

WannaCry: it is a crypto-ransomware that spreads by exploiting a Windows Server Message Block (SMB) vulnerability that provides unrestricted access to any computer running Windows. WannaCry is also able to propagate throughout corporate LANs automatically. It encrypts files on the infected device and tries to affect other devices in the network.

Locky: the most common technique used by Locky to infect systems is through receiving an e-mail with a malicious Microsoft Word attachment. When this attachment is opened, an executable is downloaded from a C&C server, a private key is generated, and the ransomware starts encrypting files by infecting all connected devices.

All the above families target a single operating system, that is, Windows, which has been shown to be the most targeted operating system[4]. There are also ransomware that target different OSs, like macOS, GNU/Linux, and Android, but the percentage of attacks that target Windows-based machines is very high, compared to other operating systems.

3.2   Artificial Neural Networks

We rely on Artificial Neural Networks (ANN) for multi-class classification, as these are lightweight and give a good detection rate. ANNs are mainly composed of many interconnected computational nodes (aka neurons), working in a distributed fashion to collectively learn from the input. ANN nodes are divided into layers: input and output layers, as well as various hidden layers between them. Nodes in the input layer take a multidimensional vector as input and send it to the hidden layer. Here, nodes perform nonlinear transformations of inputs

---

[4] https://www.statista.com/statistics/701020/major-operating-systems-targeted-by-ransomware/.

and send the results to another hidden layer or to the final output layer. In order to output a value, a neuron in the hidden layer takes the weighted sum of all its inputs and passes it through the activation function to obtain the results [13]. The role of the activation function is to decide whether a neuron's input is important or not in the process of prediction. The most commonly used activation functions are: ReLu, Linear, Sigmoid, and SoftMax [13]. Having multiple hidden layers stacked upon each other is commonly called *deep learning* [11].

4 Methodology

We discuss the building blocks of our methodology: data collection, feature extraction, and classification.

4.1 Data Collection

Ransomware Samples. Sample collection has been challenging for different reasons. First, since there is no unique online repository that contains all existing ransomware, we had to merge all repositories to avoid duplication. Second, repositories use security vendors' scores and sandbox results to map ransomware to their respective families. However, these classifications may be incorrect or not accurate, since some ransomware may have similar behavior but a completely different name. Finally, some families, like TeslaCrypt (and its variants, like AgentTesla), may contain ransomware samples together with malware that affect the choice of features used for ransomware detection. Thus, they should not be considered for this research since they will affect its effectiveness. We obtained a total of 11,523 samples detected in the last few years (i.e., 2018–2022) by using different online repositories (i.e.,   [5],   [6], and   [7]). Moreover, to have a balanced dataset, the 11,523 collected samples are evenly split into 23 families namely *Ako, BB, Cerber, Conti, Cryptolocker, Cryptowall, Erica, Expiro, Gandcrab, Hive, Kryptik, Lockbit, Lockfile, Locky, Matrix, Matsnu, Shade, Stop, TeslaCrypt, Trik, Virlock, Wannacry, and Winlock*, where 501 samples represent each family.

Benign Software. To properly identify ransomware, we need to have some benign software for the classification tasks.

   We downloaded 1,111 benign samples from various sources (i.e.,   [8] and   [9]). We focused on benign software that have similar behavior to ransomware and use a large number of API calls, such as file compressors, disk analyzers, anti-viruses, and password managers. Table 1 shows the distribution of the benign samples.

---

[5] https://www.virustotal.com.
[6] https://bazaar.abuse.ch/browse/.
[7] https://virusshare.com/.
[8] https://www.portablefreeware.com/.
[9] https://portableapps.com/.

Table 1. Benign samples distribution

Category	Software	Samples	Category	Software	Samples
Anti-viruses	McAfee	100	Disk analyzers	CrystalDiskMark	36
	Others	3		Others	4
Compressors	7-Zip	28	Browsers	Google Chrome	20
	PeaZip	99		Others	5
	Others	3	Miscellaneous	Audacity	48
Graphics	GIMP	79		FileZilla	73
	Blender	50		VeraCrypt	15
	JPEGView	21		Others	102
	ScribusPortable	19	Messaging clients	TelegramDesktop	100
	Others	4	Media players	VLC	80
Text editors	AkelPad	36	Mail clients	Various	3
	Geany	18	Password managers	KeePassXC	23
	Notepad 2	33	PDF managers	Various	9
	Notepad++	100		**Total**	**1111**

4.2   Features Extraction

We used widely known tools and methods for running ransomware in a controlled environment to perform a dynamic analysis of samples. We select the features to be used for the classification task from the reports returned by dynamic analysis. In this step, we are interested in studying the usage of API calls that software use to communicate with the kernel. In our context, it is worth noting that a feature is a binary vector that shows how the analyzed sample uses the specific APIs. In this step, the main challenge is the selection of representative features that could help distinguish different families. The similarity between samples belonging to different families leads to a set of similar features that affect the effectiveness of the developed ML models.

From the 12,634 analyzed samples, we removed the ones that failed to execute. Moreover, we removed the ones that belong to underrepresented families (i.e., less than 200 samples), which were 11 families out of 23. We removed these families to keep the dataset balanced. The final dataset contains 5203 samples from 12 ransomware families, and one benign family (see Table 2). These samples contain at least one occurrence of the API calls specified in Table 3.

Table 2. Ransomware curated dataset

Family	Samples
Cerber	450
CryptoWall	450
Matsnu	450
Shade	450
Teslacrypt	450
Benign	450
Hive	443
Ako	432
Erica	377
Conti	359
Matrix	331
Gandcrab	295
Expiro	266
Total	**5,203**

We have chosen these API calls, shown in Table 3, based on the most used evasion techniques adopted by ransomware, namely process injection, environment sensing, and unpacking. We present each of the evasion techniques in what follows.

Process Injection. Code injection is the process of copying the code from an injecting entity ϵ_{inject} into a victim entity ϵ_{victim} and executing this code within the scope of ϵ_{victim} [3]. The definition of a code injection does not specify the place of residence of ϵ_{inject} and ϵ_{victim}. We can have two cases: if the attacker and the victim reside on the same system, we refer to Host-Based Code Injection, while if they reside on different systems, the process is called Remote Code Injection.

Environment Sensing. Before executing the malicious payload, usually, an attacker wants to determine if the environment is a virtual one or not [1]. Ransomware use different techniques for evading sandboxes and virtual analysis environments. The first one is fingerprinting, which aims to detect the presence of sandboxes by looking for environmental artifacts that could indicate a virtual/emulated machine. These signs can range from device drivers, overt files on disk, and registry keys, to discrepancies in emulated/virtualized processors. Another technique used in environment sensing is *Reverse Turing Test* which checks for human interaction with the system. This tactic capitalizes on the fact that sandboxes are automated machines with no human or operator directly interacting with them. Thus, if malware does not observe any human interaction, it presumes to be in a sandbox. The malware waits indefinitely for any form of user input to test whether it is running on a real system. In a real system, eventually, a key would be pressed, or the user would move a mouse. If that occurs a specific number of times, the malware executes its malicious payload [1].

Unpacking. Packing is a common way for attackers to hide their code when they create malware. Malware is then transmitted in a "scrambled" form, which is then restored to its original form just before execution using unpacking techniques [5]. Packers use different techniques for obfuscating malicious code. First, they use multi-level compression to obfuscate the payload of an executable, making it hard to perform reverse-engineering tasks on the executable [4]. Moreover, packers can achieve malware polymorphism by producing different binaries, i.e., different hash signatures for the same payload [4, 12]. Encryption is widely used to conceal some parts of the code, which are then decrypted during unpacking by using the encryption keys provided within the packed malware; finally, packers may use techniques like dead code insertion and instruction permutation that aim at making the unpacked malicious executable more challenging to analyze [12].

Table 3. Evasion APIs

Category	Evasion techniques	Evasion API	Description
Data access and storage	Unpacking	MoveFileWithProgressW	Move a file or directory, including its children
	Environment Sensing	NtCreateFile	Creates a new file or directory or opens an existing file
	Process Injection	NtWriteFile	Write data to an open file
		SetFileAttributesW	Sets the attributes for a file or directory
		GetDiskFreeSpaceExW	Retrieve information about the amount of space available on a disk
		GetDiskFreeSpaceW	Retrieves information about the specified disk
		ShellExecuteExW	Perform an operation on a specified file
		DeviceIoControl	Send a control code directly to a specified device driver
Generic OS queries	Environment Sensing	GetComputerNameW	Retrieve the name of the local computer
		NtQuerySystemInformation	Retrieve the specified system information
Memory management	Unpacking	GlobalMemoryStatusEx	Retrieve information about the system memory usage
	Environment Sensing	NtAllocateVirtualMemory	Reserve a region of pages within the user-mode virtual address space
	Process Injection	NtMapViewOfSection	Map specified part of Section Object into process memory
		NtProtectVirtualMemory	Change the protection on a region of committed pages
		NtUnmapViewOfSection	Unmap a view of a section from the virtual address space
		WriteProcessMemory	Writes data to an area of memory in a specified process
		LdrGetDllHandle	Loads a file in memory
Network	Unpacking	GetAdaptersAddresses	Retrieve the addresses associated with the adapters
	Environment Sensing	InternetOpenA	Initialize an application's use of the WinINet functions

(*continued*)

Table 3. (*continued*)

Category	Evasion techniques	Evasion API	Description
Process	Process Injection	CreateProcessInternalW	Create a new process and its primary thread
		NtGetContextThread	Return the user-mode context of the specified thread
		NtResumeThread	Map specified part of Section Object into process memory
		NtSetContextThread	Set the user-mode context of the specified thread
		NtTerminateProcess	Terminate a process and all of its threads
		Process32NextW	Retrieve information about the next process recorded in a snapshot
		NtLoadDriver	Load a driver into the system
Registry	Process Injection	NtSetValueKey	Create or replaces a registry key's value entry
	Environment Sensing	RegOpenKeyExW	Open the specified registry key
		RegQueryValueExW	Retrieve the type and data for the specified value name of a key
		RegSetValueExW	Set the data and type of a specified value under a registry key
		NtCreateKey	Create a new registry key or opens an existing one
Security	Process Injection	CryptGenKey	Generate a random cryptographic session key or a key pair
		CryptExportKey	Export a cryptographic key or a key pair
		LookupPrivilegeValueW	Retrieve the identifier used to represent the specified privilege name
		CryptHashData	Add data to a specified hash object
Services	Environment Sensing	CreateServiceW	Create a service object and adds it to the specified service manager
		EnumServicesStatusW	Enumerate services in the specified service control manager database
UI artifacts	Environment Sensing	SetWindowsHookExW	Install an application-defined hook procedure into a hook chain
		FindWindowW	Retrieve a handle to the top-level window

In Table 3, APIs that end with W have twin API that ends with A with a similar goal. The difference in the names is due to the encoding. The APIs that end with W work with Unicode strings and the ones that end with A work with ANSI strings. For the sake of brevity, we included only the Unicode ones.

4.3 Classification

Since each ransomware family has unique characteristics, we model ransomware detection as a multi-class classification problem where the classifier determines which class (i.e., family) the ransomware belongs to. The state-of-the-art classifiers for this problem are Random Forest, Bernoulli Naive Bayes, k-Nearest Neighbors, and Artificial Neural Networks (ANNs). In this paper, we use an

ANN (see Sect. 5.2), since it is lightweight and gives good accuracy. Our artificial neural network has three layers with ReLu as an activation function. We use dropout on the input and hidden layers to drop nodes and reduce overfitting randomly. We also add a hidden layer with the Softmax activation function to the network's end.

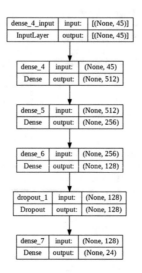

Fig. 1. Classification model architecture

Figure 1 depicts the model architecture. The hyperparameters for this network are: the number of epochs (i.e., 50), indicating how many times the model will iterate over the whole dataset, and the batch size (i.e., 15), denoting the number of samples after which the network will adjust its internal parameters.

5   Experimental Results

As an environment, we used an Ubuntu virtual machine with installed the (version 2.0.7)[10] to execute ransomware samples. It is one of the most widely used tools for analyzing the behaviors of malicious executables. We set up the sandbox with Windows 7 and basic software (e.g., Internet Explorer, Windows Media Player), as well as sample files (e.g., Word documents, PowerPoint slides). The ransomware is run in a controlled virtual machine, keeping track of everything it does, like API calls, files opened, registry keys and files dumped. All the behavioral characteristics are then saved into a JSON report, which also contains the machine name, operating system, internet access, and many other parameters. All the analyzed ransomware samples had access to the internet to contact, if required, their C&C servers for downloading additional malicious payloads.

---

[10] https://cuckoosandbox.org/.

5.1   Datasets

Our dataset (cfr. Table 2) consists of 5,203 samples distributed across 13 families (including a benign family). In addition, we have used the dataset provided by [10], which is composed of 2,994 ransomware samples from 5 families and 438 benign samples resulting in a total of 3,432 samples (cfr. Table 4).

Table 4. Description of [10] dataset

Family	Reveton	TeslaCrypt	Cerber	Locky	Yakes	Benign	*Total*
Samples	600	600	600	600	594	438	*3432*

5.2   Multi-class Classification

We tested several state-of-the-art classifiers, i.e., RF, BNB, KNN, and ANN (cfr. Table 5. With ANN, we reached good accuracy, especially in top-k accuracy ($k = 2$) (cfr. Table 5. ANN scores 80.00% in accuracy and 90.41% in top-2 categorical accuracy. We took the weighted average of all individual scores of the classes (i.e., families) we have. Similarly to [10], we used the default scikit-learn metrics[11].

Table 5. Multi-class classification results

Model	Precision	Recall	F1-score	Accuracy	Top-k Acc. ($k=2$)
Random forest	81.23%	78.38%	78.28%	78.38%	85.82%
Bernoulli Naïve bayes	61.41%	56.38%	55.94%	56.38%	67.33%
K-nearest neighbors	78.39%	75.98%	76.07%	75.98%	82.03%
Artificial neural network	82.00%	80.00%	81.00%	**80.00%**	**90.41%**

We then compared our approach with [10], since it is the only work with a public dataset and source[12] We ran their Random Forest classifier 5 times on their dataset. Then, we used their dataset to train our Artificial Neural Network model. The obtained results are promising since the ANN performs very well even with a completely different dataset. The accuracy is 93.00%, and the top-2 categorical accuracy is 98.62% (see Table 6).

---

[11] https://scikit-learn.org/stable/modules/model_evaluation.html.
[12] Available on Github https://github.com/Rmayalam/Ransomware_Paranoia.

Table 6. Comparison of our work with [10]

Approach	Dataset	Precision	Recall	F1-score	Accuracy	Top-k Acc. (k = 2)
[10]	[10]	92.36%	92.30%	92.19%	92.30%	97.82%
Our approach	[10]	93.00%	93.00%	93.00%	**93.00%**	**98.62%**
[10]	Our approach	79.29%	78.77%	78.78%	78.77%	86.74%
Our approach	Our approach	82.00%	80.00%	81.00%	**80.00%**	**90.41%**

6 Conclusion

In this paper, we proposed an early-stage ransomware detector based on a neural network model that achieves an accuracy of 80.00% in a challenging, large, and varied dataset, outperforming the state-of-the-art. The dataset we have compiled consists of 4753 ransomware samples from 12 different families and 450 benign samples. To the best of our knowledge, it is the largest dataset available for ransomware detection. We have made publicly available our source code and dataset, to reproduce the results. This work can be extended in many directions. First, we aim to make a decentralized version of it that runs over a blockchain. Second, we plan to explore the effect of adding other features, such as the registry and memory dumps, as input to our model. Third, we aim to explore other ML techniques, like transformers [14] that perform well with huge amounts of data.

Acknowledgements. The authors would like to thank the authors of [10] for their responsiveness and support. In addition, we would like to thank VirusTotal, VirusShare, and Bazaar for providing us with the ransomware samples. This work has received funding from the Marie Skłodowska-Curie Innovative Training Network Real-time Analytics for Internet of Sports (RAIS), supported by the European Union's Horizon 2020 research and innovation programme under grant agreement No 813162. Additionally, it has been partially supported by CONCORDIA, the Cybersecurity Competence Network supported by the European Union's Horizon 2020 research and innovation programme under grant agreement No 830927.

References

1. Afianian, A., Niksefat, S., Sadeghiyan, B., Baptiste, D.: Malware dynamic analysis evasion techniques: a survey. ACM Comput. Surv. **52**(6), 1–28 (2019)
2. Bae, S.I., Lee, G.B., Im, E.G.: Ransomware detection using machine learning algorithms. Concurr. Comput. Pract. Exp. **32**(18), e5422 (2020)
3. Barabosch, T., Gerhards-Padilla, E.: Host-based code injection attacks: a popular technique used by malware. In: 2014 9th International Conference on Malicious and Unwanted Software: The Americas (MALWARE), pp. 8–17. IEEE (2014)
4. Chakkaravarthy, S.S., Sangeetha, D., Vaidehi, V.: A survey on malware analysis and mitigation techniques. Comput. Sci. Rev. **32**, 1–23 (2019)
5. Coogan, K., Debray, S., Kaochar, T., Townsend, G.: Automatic static unpacking of malware binaries. In: 2009 16th Working Conference on Reverse Engineering, pp. 167–176. IEEE (2009)

6. Daku, H., Zavarsky, P., Malik, Y.: Behavioral-based classification and identification of ransomware variants using machine learning. In: 2018 17th IEEE International Conference on Trust, Security and Privacy in Computing and Communications/12th IEEE International Conference on Big Data Science and Engineering (TrustCom/BigDataSE), pp. 1560–1564. IEEE (2018)

7. Hassan, N.A.: Ransomware families. In: Ransomware Revealed, pp. 47–68. Apress, Berkeley, CA (2019). https://doi.org/10.1007/978-1-4842-4255-1_3

8. Hwang, J., Kim, J., Lee, S., Kim, K.: Two-stage ransomware detection using dynamic analysis and machine learning techniques. Wirel. Pers. Commun. **112**(4), 2597–2609 (2020). https://doi.org/10.1007/s11277-020-07166-9

9. Kharraz, A., Robertson, W., Balzarotti, D., Bilge, L., Kirda, E.: Cutting the Gordian knot: a look under the hood of ransomware attacks. In: Almgren, M., Gulisano, V., Maggi, F. (eds.) DIMVA 2015. LNCS, vol. 9148, pp. 3–24. Springer, Cham (2015). https://doi.org/10.1007/978-3-319-20550-2_1

10. Molina, R.M.A., Torabi, S., Sarieddine, K., Bou-Harb, E., Bouguila, N., Assi, C.: On ransomware family attribution using pre-attack paranoia activities. IEEE Trans. Netw. Serv. Manag. **19**(1), 19–36 (2021)

11. O'Shea, K., Nash, R.: An introduction to convolutional neural networks. arXiv preprint arXiv:1511.08458 (2015)

12. Rad, B.B., Masrom, M., Ibrahim, S.: Camouflage in malware: from encryption to metamorphism. Int. J. Comput. Sci. Netw. Secur. **12**(8), 74–83 (2012)

13. Sharma, S., Sharma, S., Athaiya, A.: Activation functions in neural networks. Towards Data Sci. **6**(12), 310–316 (2017)

14. Vaswani, A., et al.: Attention is all you need. Adv. Neural Inf. Process. Syst. **30** (2017)

15. Zhang, B., et al.: Ransomware classification using patch-based CNN and self-attention network on embedded N-grams of opcodes. Futur. Gener. Comput. Syst. **110**, 708–720 (2020)

16. Zhang, H., et al.: Classification of ransomware families with machine learning based on N-gram of opcodes. Futur. Gener. Comput. Syst. **90**, 211–221 (2019)

User-Oriented Cybersecurity Systems

Marek R. Ogiela[✉] and Urszula Ogiela

AGH University of Science and Technology, 30 Mickiewicza Ave, 30-059 Kraków, Poland
mogiela@agh.edu.pl, ogiela@agh.edu.pl

Abstract. In this paper will be presented factors connected with creation of secure user-oriented security systems. In particular the ways of application personal features and characteristics in defining modern security protocols will be presented. Using personal information allow to make such systems useful in prevention of information leakage as well as resist against cyberattacks.

1 Introduction

Cybersecurity has recently been one of the leading areas in the development of modern computing. Such issues include guaranteeing the confidentiality of data in distributed systems and services, as well as the security of cryptographic protocols. Both such protocols and computer systems can be subject to cyber-attacks that exploit weaknesses in computer systems or the security features used in them. This paper will present methods of preventing cyber-attacks that will be based on the exploitation of user characteristics and will target individuals [1]. Examples of such protocols may include steganographic techniques for hiding information in digital media that exploit the user's biometric traits, or methods for distributing secrets based on personal characteristics. In particular, protocols will be presented that take advantage of hash functions used to create personalized vectors of user characteristics of a specific length [2]. The hash sequence will then be the key to the operation of the chosen cryptographic protocol, e.g. hiding a secret, or dividing information, etc.

2 Personal Information Record

We can use personal information in the form of biometric data or behavioral traits in security systems and cryptographic protocols targeting a specific system participant or user. The collection of such characteristics allows us to create personal data records, which can then be used in cryptographic protocols or web services oriented to a specific user. Such data can be used, for example, to determine a user's access rights to computer resources, or be used as keys in protocols for encryption, information sharing, authorization or hiding secret information.

In order to create a personal data record, it is necessary to use a range of sensory devices that allow the recording of personal data in the form of biometrics and behavioral traits. Such a record may include DNA patterns and other biometric traits, as well as unique personalized movement sequences (hand movement parameters, or

L. Barolli (Ed.): AINA 2023, LNNS 654, pp. 430–433, 2023.
https://doi.org/10.1007/978-3-031-28451-9_37

gait parameters) of a given user. The creation of such records with personal data allows their subsequent use in any algorithms or protocols for guaranteeing the security of data, users and computer systems [3, 4].

The methodology for using such data is to select one or more personal characteristics, and then create a personal feature vector of a certain length using a selected hash function. For this purpose, functions such as SHA2 or SHA3 can be used [5]. The way of application of personal features for security purposes is presented in Fig. 1.

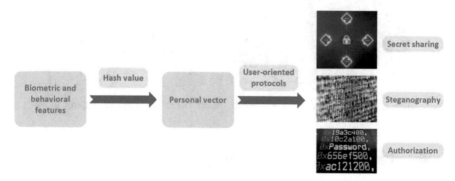

Fig. 1. Application of personal features in different security protocols.

3 Personalized Security Protocols

Application of personal characteristics is possible in practical cybersecurity solutions. Among such applications, protocols for hiding information and granting access to resources or services in distributed systems seem particularly interesting [6, 7]. In the former area, i.e. hiding and transmitting secret data in imaging media, the vector of personal characteristics produced by means of a hash function can be used as a key for distributing secret data in imaging media.

The hash of personal features obtained using the SHA hash function of 256 or 512 bits can be divided into 2-bit sequences that will determine how to embed successive bits of secret data in the pixels of the container. The 2-bit blocks of the key will allow the selection of one of the three color components of the RGB model of the pixels of the information carrier, and depending on these values it will be possible to embed the secret bit of information by the LSB method in the selected color component. This method is illustrated in Fig. 2.

The second application mentioned is the generation and allocation of user access privileges in distributed systems. In this case, the user's feature vector is the key used in protocols for reconstructing shared information previously shared using threshold secret sharing algorithms [8]. In such systems, a user can use his personal data to create and then recreate secret data that has been shared among authorized users located at a certain access level in hierarchical structures. In such structures, different privileges related to access to information can be established through the use of secret sharing methods with privileged shares [4].

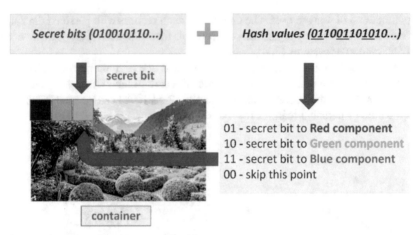

Fig. 2. Hash-based secret hiding procedure in visual container with application of personal features.

4   Security of User-Oriented Protocols

The use of personal traits in cryptographic protocols can be important for enhancing the security of computer systems. Protocols using user traits target specific individuals which means they can only be used by authorized users. In order to use them, it is necessary to have selected personal characteristics (biometrics or behavioral traits), as well as a hash sequence obtained using an appropriate hash function. Possible cyberattacks on such protocols must focus on acquiring personal characteristics, which is extremely difficult when they are used in direct mode. In addition, the use of hash functions means that personal features are not used directly only in the form of a generated hash sequence, which will look different for different features.

5   Conclusions

This paper describes methods of using personal features to create cryptographic protocols used in cybersecurity systems. In particular, the idea of using hash functions to create an encoded vector of personal characteristics, which can then be used in message hiding systems and distributed information processing systems, has been presented. The methods described enhance the security of traditional cryptographic systems and solutions, and develop the area of cryptography known as cognitive and personalized cryptography [8], and can be used in lightweight cryptography [9].

Acknowledgments. Research project supported by program „Excellence initiative – research university" for the AGH University of Science and Technology. This work has been partial supported by the funds of the Polish Ministry of Education and Science assigned to AGH University of Science and Technology.

References

1. Ogiela, L., Ogiela, M.R.: Cognitive security paradigm for cloud computing applications. Concurr. Comput.: Pract. Exp. **32**(8), e5316 (2020). https://doi.org/10.1002/cpe.5316
2. Ogiela, M.R., Ogiela, L., Ogiela, U.: Biometric methods for advanced strategic data sharing protocols. In: 2015 9th International Conference on Innovative Mobile and Internet Services in Ubiquitous Computing IMIS 2015, pp. 179–183 (2015). https://doi.org/10.1109/IMIS.2015.29
3. Ogiela, M.R., Ogiela, U.: Secure information splitting using grammar schemes. In: Nguyen, N.T., Katarzyniak, R.P., Janiak, A. (eds.) New Challenges in Computational Collective Intelligence. SCI, vol. 244, pp. 327–336. Springer, Heidelberg (2009). https://doi.org/10.1007/978-3-642-03958-4_28
4. Ogiela, M.R., Ogiela, U., Ogiela, L.: Secure information sharing using personal biometric characteristics. In: Kim, T.-H., Kang, J.-J., Grosky, W.I., Arslan, T., Pissinou, N. (eds.) FGIT 2012. CCIS, vol. 353, pp. 369–373. Springer, Heidelberg (2012). https://doi.org/10.1007/978-3-642-35521-9_54
5. Nezami, Z.I., Ali, H., Asif, M., Aljuaid, H., Hamid, I., Ali, Z.: An efficient and secure technique for image steganography using a hash function. PeerJ Comput. Sci. **8**, e1157 (2022). https://doi.org/10.7717/peerj-cs.1157
6. Koptyra, K., Ogiela, M.R.: Steganography in IoT: information hiding with APDS-9960 proximity and gestures sensor. Sensors **22**(7), 2612 (2022). 1–11. https://doi.org/10.3390/s22072612
7. Ogiela, M.R., Koptyra, K.: Visual pattern embedding in multi-secret image steganography. In: ICIIBMS 2015 - International Conference on Intelligent Informatics and BioMedical Sciences, 28–30 November 2015, Okinawa, Japan, pp. 434–437. IEEE (2015). ISBN 978-1-4799-8562-3/15
8. Ogiela, L.: Transformative computing in advanced data analysis processes in the cloud. Inf. Process. Manag. **57**(5), 102260 (2020)
9. Pizzolante, R., Carpentieri, B., Castiglione, A., Castiglione, A., Palmieri, F.: Text compression and encryption through smart devices for mobile communication. In: Seventh International Conference on Innovative Mobile and Internet Services in Ubiquitous Computing, pp. 672–677 (2013). https://doi.org/10.1109/IMIS.2013.121

Cybersecurity of Distributed Systems and Dispersed Computing

Urszula Ogiela[1], Makoto Takizawa[2], and Lidia Ogiela[1(✉)]

[1] AGH University of Science and Technology, 30 Mickiewicza Ave, 30-059 Kraków, Poland
{ogiela,logiela}@agh.edu.pl
[2] Research Center for Computing and Multimedia Studies, Hosei University, 3-7-2, Kajino-cho, Koganei-shi, Tokyo 184-8584, Japan
makoto.takizawa@computer.org

Abstract. This paper will describe the most important aspects of cybersecurity systems oriented for distributed communication as well as dispersed computation. Especially the most important ways how to guarantee the highest level of security will be describes, and issues how to prevent cyberattacks. Especially human factors which have influence for general security levels also will be considered.

Keywords: Cybersecurity · Distributed systems · Dispersed computing · Data security · Linguistic methods

1 Introduction

Cybersecurity processes are used to ensure protection against various attacks directed both at protected data, resources, as well as infrastructure, network or modern technologies [1–3]. The variety of currently protected resources and entities makes the created security systems more and more effective while at the same time high precision of detecting possible threats. Cyberspace protection covers all possible types of attacks, both inside and outside the systems.

Cybersecurity is currently perceived as a comprehensive, most effective and complete (comprehensive) way to protect infrastructure and resources, based on:

- use of the most modern threat identification tools,
- cryptographic data protection protocols,
- data processing technology,
- methods of secrecy of protected data,
- data transmission protocols, as well as,
- construction of secure IT systems.

The main task of cybersecurity is undoubtedly the effective protection of systems and data stored in various types of IT infrastructures. One of the most important and most extensive types of systems are distributed systems. They allow for the concurrency of various tasks. A special feature of distributed systems is the ability to perform tasks and processes both at different levels and by different entities.

Distributed systems are effective in multi-level structures [4, 5]. An example of such structures are hierarchical structures:

- with a multi-level, independent hierarchy and
- about multi-level dependent connections between given levels of the structure.

An example of a hierarchical structure with dependent levels and dependencies between them are organizational structures (enterprises, companies, organizations, etc.). A much broader functional solutions are structures characteristic of multi-level distributed systems.

An example of this type of solution are structures of the Edge-Fog-Cloud type [6, 7], functioning effectively both in full dependence of individual levels, and completely independently on each of them. The universality of this type of structures results from the fact that at each level it is possible to perform specific (fully defined) tasks, independent of the implementation of other tasks characteristic of other levels of the structure. At the same time, it is possible to define such processes (e.g. data protection and management) that will be implemented in connection between individual levels of the structure.

2 Linguistic Methods of Data Interpretation in Distributed Systems

Distributed systems allow for the implementation of cybersecurity processes in various aspects of data protection. An important novelty in terms of enriching the functionality of distributed systems is the inclusion of interpretation and reasoning processes in their operation. These processes, as a result of linguistic and semantic methods of interpreting the analyzed data and protected information, allow for proper analysis of threats and the effects of their occurrence in cybersecurity processes.

Linguistic methods widely known for their applications in semantic analysis processes can be adapted to the tasks of reasoning and analysing data of special importance (subject to protection) also in the area of cybersecurity [8–10].

Data protection processes in distributed structures can be implemented at individual levels of the structure:

- basic level – the use of linguistic methods to describe and mean confidential data in the processes of securing them, distributing them at the edge level and dividing the secret and sending its parts to the superior level,
- higher level – the use of methods of meaning interpretation of data for the analysis of the basic level of the part of the shared secret and the use of linguistic methods for the proper description of the secreted data at the higher level and the distribution of the part of the secret to the highest level,
- the highest level – the use of linguistic techniques to analyse the parts of the shared secret received from the lower level and the use of linguistic description methods in the process of data security at the highest level.

Data Protection in Distributed Structures

```
//structure level//

  Edge:=Level1 //k1 characteristics of the disclosed
data//
                //linguistic description//
                //(m1, n1) secret sharing threshold
scheme//

  Fog:=Level2 //k2 characteristics of the disclosed
data//
                //linguistic description//
                //(m2, n2) secret sharing threshold
scheme//

  Cloud:=Level3 //k3 characteristics of the disclosed
data//
                //linguistic description//
                //(m3, n3) secret sharing threshold
scheme//
  ;
```

3   Dispersed Computing in Distributed Systems

Computational processes in distributed structures are classified as dispersed computing, which are performed at many different levels of the structure's operation. In terms of cybersecurity, the most important stages of dispersed computing are:

- data protection and security,
- secure data transmission and processing,
- distribution of all parts of secret information,
- semantic analysis, the results of which are secret,
- processes of semantic reasoning shaping the decision-making processes regarding the implemented solutions,
- selection and structure of an optimal distributed system that implements the processes of securing, distributing, processing and managing data.

Dispersed computing is carried out independently at each stage of the hierarchical multi-level and multi-layer structure. They can also be implemented from the superior level in order to ensure full control of the security and correct operation of the distributed system.

The calculations of this class include:

- proper data analysis (statistical, descriptive),

- linguistic description and interpretation of data,
- semantic (meaningful) analysis of protected data,
- information security and confidentiality protocols,
- cryptographic threshold schemes,
- calculations recording data from various levels of the structure,
- calculation methods related to the creation of appropriate system solutions,
- confidential data transmission and distribution protocols,
- techniques for defining distributed systems.

4 Conclusions

This paper describes the methodology of creating distributed systems based on the meaning interpretation of protected data, functioning on the basis of dispersed computing carried out at various levels of multi-level structures. The proposed solution is an enrichment of cybersecurity techniques in terms of both data protection and infrastructure, allowing for the protection of various elements of the cybernetic world and cyberspace. The proposed solution allows for a fairly flexible choice of the optimal method of data security using both cryptographic techniques and modern computational methods supported by linguistic description and interpretation techniques and artificial intelligence.

Acknowledgments. Research project supported by program "Excellence initiative – research university" for the AGH University of Science and Technology. This work has been supported by the funds of the Polish Ministry of Education and Science assigned to AGH University of Science and Technology.

References

1. Menezes, A., van Oorschot, P., Vanstone, S.: Handbook of Applied Cryptography. CRC Press, Waterloo (2001)
2. Ogiela, L.: Computational intelligence in cognitive healthcare information systems. Bichindaritz, I., Vaidya, S., Jain, A., Jain, L.C. (eds.) Computational Intelligence in Healthcare 4. Studies in Computational Intelligence, vol. 309, pp. 347–369. Springer, Heidelberg (2010). https://doi.org/10.1007/978-3-642-14464-6_16
3. Ogiela, M.R., Ogiela, L., Ogiela, U.: Biometric methods for advanced strategic data sharing protocols. In: 2015 9th International Conference on Innovative Mobile and Internet Services in Ubiquitous Computing IMIS 2015, pp. 179–183 (2015). https://doi.org/10.1109/IMIS.2015.29
4. Nakamura, S., Ogiela, L., Enokido, T., Takizawa, M.: Flexible synchronization protocol to prevent illegal information flow in peer-to-peer publish/subscribe systems. In: Barolli, L., Terzo, O. (eds.) Complex, Intelligent, and Software Intensive Systems. Advances in Intelligent Systems and Computing, vol. 611, pp. 82–93. Springer, Cham (2018). https://doi.org/10.1007/978-3-319-61566-0_8
5. Yan, S.Y.: Computational Number Theory and Modern Cryptography. Wiley, Hoboken (2013)
6. Ogiela, L.: Transformative computing in advanced data analysis processes in the cloud. Inf. Process. Manage. **57**(5), 102260 (2020)

7. Gil, S., et al.: Transformative effects of IoT, Blockchain and artificial intelligence on cloud computing: evolution, vision, trends and open challenges. Internet Things **8**, 100118 (2019)
8. Ogiela, M.R., Ogiela, U.: Secure information splitting using grammar schemes. In: Nguyen, N.T., Katarzyniak, R.P., Janiak, A. (eds.) New Challenges in Computational Collective Intelligence. Studies in Computational Intelligence, vol. 244, pp. 327–336. Springer, Heidelberg (2009). https://doi.org/10.1007/978-3-642-03958-4_28
9. Ogiela, L.: Data management in cognitive financial systems. Int. J. Inf. Manage. **33**(2), 263–270 (2013). https://doi.org/10.1016/j.ijinfomgt.2012.11.008
10. Ogiela, M.R., Ogiela, U., Ogiela, L.: Secure information sharing using personal biometric characteristics. In: Kim, T.-H., Kang, J.-J., Grosky, W.I., Arslan, T., Pissinou, N. (eds.) FGIT 2012. CCIS, vol. 353, pp. 369–373. Springer, Heidelberg (2012). https://doi.org/10.1007/978-3-642-35521-9_54

A Dynamic Machine Learning Scheme for Reliable Network-Based Intrusion Detection

Eduardo K. Viegas[1(✉)], Everton de Matos[1], Paulo R. de Oliveira[2],
and Altair O. Santin[2]

[1] Secure Systems Research Center, Technology Innovation Institute (TII),
Abu Dhabi, United Arab Emirates
{eduardo,everton}@ssrc.tii.ae

[2] Pontifícia Universidade Católica do Paraná, PUC-PR, Curitiba, Brazil
{paulo.oliveira,santin}@ppgia.pucpr.br

Abstract. Several works have proposed highly accurate machine learning (ML) techniques for network-based intrusion detection over the past years. However, despite the promising results, proposed schemes must address the high variability of network traffic and need more reliability when facing new network traffic behavior. This paper proposes a new dynamic and reliable network-based intrusion detection model implemented in two phases. First, the behavior of to-be-classified events is assessed through an outlier detection scheme to reject potentially new network traffic, thus, keeping the system reliable as time passes. Second, classification is performed through a dynamic selection of classifier to address the high variability of network traffic. Experiments performed in a new dataset composed of over 60 GB of network traffic have shown that our proposed scheme can improve detection accuracy by up to 33% when compared with traditional approaches.

1 Introduction

Over the last few years, the number of network-based attacks significantly increased and is still on the rise. As an example, according to a Kaspersky report [11], the occurrence of distributed denial-of-service attacks has rose by 48% in the third quarter of 2022, showing that current security mechanisms have been unable to adequately secure systems. In general, network operators resort to network-based intrusion detection systems (NIDS) to detect this growing number of threats, typically implemented through either two approaches [12]. On the one hand, *misuse-based* techniques detect attacks based on a previously defined set of well-known attack patterns; hence, it can only detect previously known threats, leaving systems insecure against new kinds of attacks. On the other hand, *behavior-based* approaches signals misconducts according deviations of the normal expected behavior, thus, it is able to detect new attacks as long as they behave significantly differ from the normal behaviors.

L. Barolli (Ed.): AINA 2023, LNNS 654, pp. 439–451, 2023.
https://doi.org/10.1007/978-3-031-28451-9_39

As a consequence, due to the ever increasing number of newly discovered attacks, significant research efforts have been conducted on *behavior-based* intrusion detection, wherein authors typically resort to machine learning (ML) techniques implemented as a pattern recognition task. To achieve such a goal, proposed detection mechanisms are implemented through a three-phase process, namely model *training*, *validation*, and *testing* [18]. The model *training* goal is to extract a behavioral ML model from a training dataset, whereas the *validation* phase allows researchers to fine-tune their scheme, such as performing feature selection and hyperparameter optimization. Finally, the model accuracy performance is estimated through a *test* dataset, which is then assumed to be experienced when it is deployed on production environments.

Unfortunately, networked environments presents a plethora of challenges when compared to areas that ML have been successfully applied. The behavior of network traffic is highly variable, while also changes over time, a situation that can be caused either due to the discovery of new attacks or even due to the provision of new services [5]. The non-stationary behavior of network traffic presents a significant challenge to the reliability of designed ML-based intrusion detection schemes. On the one hand, due to the high variability of network traffic behavior, the building of a realistic training dataset becomes an unfeasible task. On the other hand, even if the model is able to generalize the behavior of the network traffic, due to the its non-stationary behavior, hard to-be-conducted ML model updates will need to be periodically executed.

Therefore, ensuring that designed ML-based intrusion detection schemes are able to generalize the behavior of network traffic, while also withstanding long periods of time without frequent model updates is a must. In contrast, current proposed techniques usually pursues the highest accuracy on a given dataset, in general, incurring on significant tradeofs on model generalization capabilities [3]. As a result, ML-based intrusion detection techniques, despite the promising reported results, remains mostly as a research topic, rarely being deployed in real-world applications.

In light of this, this paper proposes a new dynamic and reliable network-based intrusion detection model implemented in two stages. First, the classification task is performed through a dynamic selection of classifier scheme, Thus, only the most suitable subset of classifiers are used for the labeling task. Second, unreliable classifications are properly identified through a novelty detection module, which signals unknown network event behaviors that should be used for model retraining purposes. As a consequence, our proposed model is able to address the highly variable nature of network traffic, through the dynamic selection of classifier, while also adequately identify new network behavior through our novelty detection, maintaining the system accurate and reliable.

The main contributions of this paper are as follows:

- An evaluation of the detection reliability of traditional ML-based techniques in face of variable and new network behavior. The experiments shows that current approaches are unreliable for intrusion detection task in real-world network conditions.

- A new dynamic and reliable network-based intrusion detection model able to withstand and identify unknown network traffic behavior.

The remainder of this paper is organized as follows. Section 2 introduces the fundamentals on ML-based intrusion detection and its challenges, while Sect. 3 provides the related works. Section 4 evaluates the reliability of current ML-based techniques. Section 5 describes our proposed model, while Sect. 6 evaluates it. Section 7 concludes our work.

2   Preliminaries

A typical network-based intrusion detection system (NIDS) is composed by four sequential modules [12]. The first module namely *Data Acquisition* collects network events that will be subsequently evaluated. The behavior of the collected data is extracted by a *Feature Extraction* module. In general, network traffic behavior is depicted through network flow features, which summarizes the communication between network entities in a given time window. For instance, the number of exchanged network packets between two hosts over the last 2 seconds. The built feature vector is used as input by a *Classification* module, which adequately label the network flow as either *normal* or *intrusion*, e.g. by applying a ML model. Finally, *intrusion*-classified events are signaled to the network operator by the *Alert* module.

Several approaches can be used for the classification task in NIDS, wherein the most promising approach resort to machine learning (ML) techniques, as a pattern recognition task [6,17]. In such a case, the network operator must collect huge amounts of network traffic, and adequately label them as either *normal* or *attack* for the model building task. However, event labeling of network traffic is not easily achievable, and typically demands human-assistance, which is not always available [16]. Therefore, building a realistic training dataset for ML-based NIDS demands significant network operator effort and time.

Notwithstanding, the behavior of network traffic is highly variable while also changes significantly as time passes, a situation caused either by the discovery of new attacks, or even the provision of new services. As a result, the built ML model becomes outdated as time passes, increasing its error rates and affecting its reliability to the network operator [10]. To address such a challenge, the ML model must be periodically updated, which often demands several days or even weeks to be conducted, due to the challenges related to the training dataset building task. Therefore, although widely used in several fields, such as image classification, fraud detection, and medical diagnosis, ML is rarely used in production for network-based intrusion detection, remaining mostly as a research topic.

3   Related Works

Over the last years, a plethora of works have proposed highly accurate ML-based techniques for network traffic classification tasks [9,12]. In general, authors focus their efforts on achieving the highest accuracy on a given intrusion dataset, often neglecting how that reflects under real-world conditions. For instance, F. Salo *et al.* [15] make use of an ensemble of classifiers coped with a feature reduction technique to improve the system's accuracy on a variety of intrusion detection datasets. The authors were able to improve accuracy when compared to related approaches, however, no research effort was given on the variability and changes on network traffic. Another ensemble-based approach was proposed by J. Gu *et al.* [7] that makes use of support vector machine (SVM) coped with feature augmentation. Similarly, their scheme is able to improve accuracy while over-looking the applicability of their techniques. N. Moustafa *et al.* [13] makes use of an ensemble of three widely used ML classifiers to improve accuracy in network traffic classification. The authors shows that ensemble-based techniques can improve accuracy, however, they do not address the network traffic variability neither how new network traffic behavior affects their scheme.

In general, to address environments with non-stationary behavior, related works resort to concept drift detection mechanisms. G. Andresini *et al.* [2] proposes a concept-drift-based technique to detect new network traffic behavior in a well-known intrusion dataset. Their approach is able to signal new network traffic that can affect the reliability of deployed intrusion detection techniques. Unfortunately, the authors assume a supervised setting, wherein the network operator can provide the event label as needed. O. Abdel Wahab [21] identifies concept drift making use of a feature reduction technique coped with a variance detection scheme. The author is able to identify new network traffic behavior in a unsupervised setting, however, no action is taken over the newly network traffic behavior. G. Andresini *et al.* [1] proposed a concept drift detection mechanism to identify new network traffic behavior. Their proposed model makes use of a deep neural architecture integrated with a concept drift detection approach to signal new network traffic. Unfortunately, the authors assume a supervised setting for identification of new traffic.

The high variability of network traffic behavior is rarely addressed by related works. Z. Chkirbene *et al.* [4] proposes a dynamic intrusion detection scheme through feature selection. The authors make use of a feature ranking approach to build new ML models over time, while discarding outdated ones as time passes. Although their approach works in unsupervised setting, the authors neglect how network traffic behavior changes affects their technique, as they make use of an outdated dataset for evaluation purposes. R. Heartfield *et al.* [8] proposed a self-configurable intrusion detection scheme based on reinforcement learning. The author's approach makes use of a reinforcement learning technique to continuously adjust the underlying model's parameters to address new network traffic behavior. Unfortunately, their proposal assume a supervised setting to identify new network traffic. F Pinage *et al.* [14] proposed a dynamic classifier selection for drift detection that can be applied for intrusion detection. The author's approach uses a semi-supervised drift detector that makes use of an ensemble of classifiers. However, the authors do not evaluate their technique on a realistic intrusion dataset.

Table 1. Testbed behavior variations, over time, according to each considered scenario. Both normal and attacker behaviors vary as time passes.

Scenario	Time window (minutes)	Attacker behavior	Normal behavior
Attack (Serv. Scan)	*zero* to 5	*OS* and *service fingerprint*	100 benign clients performs periodic queries on HTTP, and SNMP services
Attack (Portscan)	5 to 10	*udpscan, synscan, nullscan, finscan, xmasscan,* and *ackscan*	
Attack (Vuln. Scan)	10 to 15	*vulnerability scan*	
Normal (Content)	15 to 20	*OS* and *service fingerprint*	Requested service content differs from previous scenario
Normal (Service)	20 to 25		100 benign clients performs periodic queries on SMTP, NTP and SSH services

As a result, although widely studied in the literature, ML-based intrusion detection remains mostly as a research topic, despite the promising results reported by the related works. This is because, in general, authors neglect how their proposed schemes can be used in real-world conditions, wherein the network traffic behavior is highly variable while also changes as time passes.

4   Problem Statement

In this section we investigate how widely used ML-based intrusion detection techniques perform when being used under real-world network conditions. More specifically, we first introduce our used dataset that presents real-world network characteristics, then, we evaluate the detection performance when ML-based techniques are used for intrusion detection over it.

The next subsections present our dataset and our evaluation.

4.1   A Fine-Grained Intrusion Detection Dataset

Realistic intrusion detection datasets are a must in order to ensure that proposed schemes are adequately built and evaluated [12]. Unfortunately, making use of outdated datasets with known flaws have become a common practice in the literature [19]. Yet, apart from making use of a realistic dataset, with real and valid network traffic behavior, researchers must ensure that they consider the non-stationary behavior of network traffic. Thus, to adequately evaluate ML-based intrusion detection approaches, our work make use of the *Fine-grained Intrusion Dataset* (FGD), which was built using the methodology presented in [20]. The FGD is made of a variety of network flows that usually is evidenced in real network environments, wherein designed ML models must be able to address. As a result, designed techniques can be adequately assessed considering the network traffic variability of production environments.

To achieve such a goal, the FGD dataset was built making use of a controlled environment with 100 client machines, which are responsible to generate the normal network traffic according to a number of application protocols, as well as several attackers that generates the malicious network traffic using standardized pentesting tools [20]. Each machine executed the service requests and the attacks on a honeypot server used to generate adequate response and also collect the generated network traffic. Table 1 shows the overview of the testbed behavior variation as time passes, concerning the normal and attack network traffic. A detailed description of each network profile can be found in [20]. At total, the used testbed is composed by $\approx$ 60 GB of network traffic, account for a total of 5 different network traffic behavior.

4.2   Chasing a Moving Target

In this section, we evaluate how changes in the network traffic behavior can affect ML-based intrusion detection schemes. To achieve such a goal, we take into account a training setting composed of network-related probing attacks, such as service scanning (Table 1, *Attack (Serv. Scan)*). To build the selected ML classifiers, we use as input the data that occurs on the first dataset minute. Therefore, the classifiers are built considering application-level network probing attacks, such as OS and service scanning. Each network flow is represented by a feature vector composed by 50 features [20]. Each feature summarizes the network data exchanged between a two given hosts in our testbed and their related services considering a 2-second time window.

We select three widely used ML classifiers for network-based intrusion detection task, namely *Bagging*, *Decision Tree*, and *Random Forest*. The Bagging and Random Forest classifiers were implemented making use of 100 decision trees as their base-learners, where each base learner makes use of *gini* as node split quality metric. The Decision Tree classifier was implemented using *gini* as node split quality metric. A random undersampling without replacement is used at the training procedure to balance the occurrence between the classes. The classifiers were implemented through *scikit − learn* API v0.24.

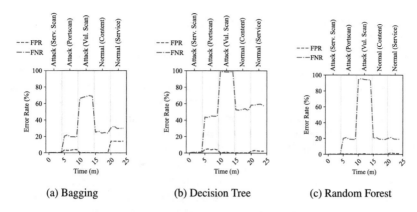

Fig. 1. The accuracy behavior, over time, on the evaluated testbed of widely used ML classifiers. Classifiers are trained using only *Attack (Serv. Scan)* data occurred at time window ranging from *zero* to 1 min.

The classifiers are evaluated concerning their false-positive (FPR) and false-negative (FNR) rates, where the FP denotes the ratio of normal events misclassified as an attack. In contrast, the FN represents the ratio of attack events misclassified as normal ones.

Figure 1 shows the error rate variation as time passes for the selected classifiers on our dataset. It can be noted that the selected techniques were able to provide significantly low error rates when facing behaviors already present during the training phase (Fig. 1, 0th to 5th minute). In practice, the selected approaches presented an average of 0.2% and 0.1% of FPR and FNR rates respectively. Unfortunately, over time as the behavior of the underlying network traffic also evolves, the selected approaches becomes unable to cope the intrusion detection. In practice, the selected techniques were only able to present low error rates (false rates < 5%) when being subject to the same kinds of network traffic behavior they were trained with.

Consequently, evaluated approaches were not able to deal with new kinds of network traffic behavior, demanding model updates to be frequently conducted. Such a situation is caused by changes in the network traffic, either caused by new services or by new attacks beig generated, and as a result, increasing the ratio of false alarms the system produces as time passes.

5   A Dynamic and Reliable Network-Based Intrusion Detection Model

In light of this, to address the non-stationary and highly variable nature of real-world networks, we propose a dynamic and reliable network-based intrusion detection model. The overview of our proposed scheme is shown on Fig. 2 and is implemented in two phases.

First, the highly variable nature of network environments is addressed through a dynamic selection of classifier approach. As a consequence, our proposed scheme is able to proactively select the most suitable subset of classifiers that should be used for classification purposes, based on the to-be-classified event behavior. The main insight of making use of dynamic selection of classifier is to enable our system to adapt according to the network event behavior, which is usually highly variable in production environments. Second, to address new network traffic behavior as time passes, our proposed scheme performs the classification assessment of to-be-classified events. The classification assessment goal is to ensure that our system is only used to classify events that are similar to those used during the training phase, thus, ensuring that only reliable classifications are performed by our model. Consequently, our proposed model is able to address the highly variable nature of network environments, while also keeping its reliability in face of new network traffic behavior.

The next subsections further describe our proposed model including the modules that implements it.

5.1   Classification Assessment

The behavior of real-world network traffic is highly variable while also changes as time passes. As a result, the building of a reliable ML-based technique for intrusion detection becomes a challenging and unfeasible task, demanding that designed techniques are able to cope with unseen network traffic. Unfortunately, ML-based techniques performs a decision for every given input, regardless if it can be reliably made by the underlying ML model, leaving ML-based intrusion detection techniques unreliable in such situations.

To address such a shortcomming, our proposed scheme makes use of a *Assessment* module, as shown in Fig. 2. The module goal is to ensure that our system only performs the classification task on events behaves similarly to those used during our system training phase. Our main assumption is that, in order to keep the system reliability, only events that are known to the underlying ML model should be used for classification task. To achieve such a goal, our scheme builds an outlier detector to signal events that are not similar to the events used for training purposes. The outlier detector is build upon the training dataset, hence, it flags as an anomaly events that are significantly different from the training dataset. Signaled events are not used for classification purposes, and, as a consequence, are rejected by our system in order to keep its reliability.

Therefore, our model is able to identify new network traffic behavior in an unsupervised fashion, as events that are identified as anomalies are rejected by our system. As a main contribution our scheme is able to identify the network events that should be evaluated by the network operator, e.g. to update the ML model.

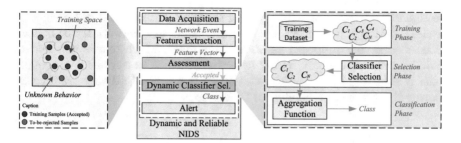

Fig. 2. Proposed dynamic and reliable network-based intrusion detection system model.

5.2 Classification Pipeline

Reliable network-based intrusion detection schemes must ensure their generalization in face of the highly variable network traffic in production environments. However, widely used ML-based techniques are static after the training phase, in practice, they are not able to select which model should be used for classification based on the newly input data. To address such a challenge, our proposed scheme performs the classification task through a dynamic selection of classifier approach. As a result, only the most suited subset of classifiers are used to perform the event labeling task, hence, adjusting our classification procedure according to the network traffic behavior.

Our proposed model classification pipeline is shown in Fig. 2, and it starts with a to-be-classified network event collected by a *Data Acquisition* module. The collected event behavior is extracted by a *Feature Extraction* module which compounds a related feature vector to be used by our system. The built feature vector is first evaluated by our *Assessment* module, which reject outliers based on the used training data (see Sect. 5.1), while inliers are forwarded for the classification task. Accepted events are used as input by our dynamic classifier selection module.

Our classification procedure considers a traditional dynamic selection of classifiers as shown in Fig. 2. Therefore, a pool of classifiers (C) of size N is built over a training dataset. At classification phase (Fig. 2, *Selection Phase*), a subset of classifiers are selected to be used for the classification task according to the input event, e.g. through a competence region map. The output of the selected subset is used by an aggregation function, which outputs a corresponding event class (Fig. 2, *Classification Phase*). Finally, intrusion-detected events are signaled to the operator by an *Alert* module.

5.3 Discussion

Our proposed model aim to address the challenges of real-world networked environments in which the network behavior is highly variable while also changes over time. On the one hand, to address the highly variable nature of network behavior, our proposed scheme performs the classification task through a dynamic selection of classifier approach, selecting the most suited subset of classifiers for the classification of each event. On the other hand, new network traffic behavior is identified by our proposed assessment approach, which rejects new network traffic behavior through an outlier detection scheme. As a result, our proposed model is able to address the non-stationary behavior of network traffic, while ensuring the systems's reliability as time passes.

6   Evaluation

In this section we further investigate the performance of our proposed model when compared to traditional ML-based intrusion detection schemes. More specifically, our evaluation aims at answering the following research questions (RQ):

- **(RQ1)** *What is the classification performance of our proposed dynamic selection of classifier approach?*
- **(RQ2)** *How does our proposed model perform with our assessment technique?*
- **(RQ3)** *How does our proposed model perform when compared to traditional approaches?*

The next subsections further describes our model building procedure and the performed evaluations.

6.1   Model Building

Our proposed model was built and evaluated making use of the same settings evaluated previously (see Sect. 4.2). More specifically, we build our scheme using the FGD (Table 1) dataset, and evalute its performance while changing the scenario behavior. To achieve such a goal, our *Assessment* module (Fig. 2, *Assessment*) was implemented making use of the Isolation Forest outlier detector. The Isolation Forest was implemented with 100 base estimators, and 256 samples per estimator. The outlier detector was trained using the whole training dataset, i.e., first minute of the FGD dataset, and implemented through the $scikit-learn$ API v0.24. The dynamic selection of classifier was implemented with the k-*Nearest Oracle Union* (KNORA-U) method, with a Bagging classifier as base pool, and 7 neighbors for the competence region computation. The KNORA-U was implemented through the *deslib* API $v.0.3$.

6.2   A Reliable and Dynamic NIDS

Our first experiment aims at answering RQ1, and evaluates our proposed model accuracy performance on FGD dataset while not making use of our assessment technique. More specifically, we first evaluate how our dynamic selection of classifier technique performs when being subject to our dataset behavior variations. Figure 3a shows the classification accuracy of our model without assessment.

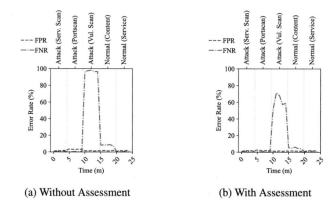

(a) Without Assessment   (b) With Assessment

Fig. 3. The accuracy behavior, over time, on the evaluated testbed of our proposed scheme. Our approach is trained using only *Attack (Serv. Scan)* data occurred at time window ranging from *zero* to 1 min.

It is possible to note that our dynamic selection of classifier technique is able to significantly decrease the intrusion detection error rate when compared to traditional techniques (Fig. 3a *vs.* 1). More specifically, our scheme presented a high error rate only on one scenario (*Attack (Vul. Scan)*), while the traditional approaches significantly decrease their accuracy as soon as a new behavior is evidenced. Therefore, the dynamic selection of classifier can be used to address the network traffic behavior variability.

Our second experiment aims at answering RQ2, and evaluates our proposed model with our assessment technique. In such a case, we only evaluate the events that are deemed as inlier by our assessment model (Fig. 2, *Assessment*). Figure 3b shows the classification accuracy of our model while rejecting outlier events identified by our assessment technique. It can be noted that our assessment approach is able to decrease further the error rate on the *Attack (Vul. Scan* scenario by up to 35%. As a consequence, the assessment approach can be used to improve the system's reliability when a new network traffic behavior is experienced. In addition, the network operator can use the assessment technique to trigger the model update, addressing the increased error rate when the system is deployed in production.

Finally, to answer RQ3, we compare the accuracy performance of our proposed scheme versus the traditional approaches. Table 2 shows the error rate performance of the selected techniques. Our proposed model significantly improve the accuracy when compared to other approaches. For instance, at the most difficult scenario (*Attack (Vul. Scan)*) our scheme was able to decrease the error rate by up to 33.1%, while also presenting a significantly low error rate on the other scenarios. Therefore, our proposed model was able to address the high variability of network traffic, while also being able to identify new network traffic behavior that should be used for model updates.

Table 2. Accuracy of evaluated techniques according to the testbed behavior

Testbed behavior	Detection measure	Detection approach error rate (%)			
		Bagging	Decision tree	Random forest	Proposed approach
Attack (Serv. Scan)	FPR	0.8	0.3	0.4	0.0
	FNR	1.2	1.3	0.3	0.4
Attack (Portscan)	FPR	3.8	4.5	1.2	2.3
	FNR	22.2	45.3	23.1	0.4
Attack (Vuln. Scan)	FPR	0.3	0.4	0.7	0.3
	FNR	68.2	97.3	92.1	64.2
Normal (Content)	FPR	0.3	0.1	0.2	0.4
	FNR	25.2	53.2	22.1	4.5
Normal (Service)	FPR	18.3	3.4	2.1	1.3
	FNR	28.2	58.2	21.2	3.2

7   Conclusion

Network-based intrusion detection through machine learning techniques have been a widely explore topic in the literature over the past years. However, despite the promising reported results proposed scheme are rarely deployed in production settings. This paper proposed a new technique for ML-based intrusion detection to address the high variability of network traffic and new behavior's as time passes. Our proposed model addressed network traffic behavior variability through a dynamic classifier selection scheme, while tackled new network traffic behavior through an assessment technique. As future works, we plan on extending the proposed model to incorporate into the deployed model the identified new network traffic behavior.

References

1. Andresini, G., Appice, A., Loglisci, C., Belvedere, V., Redavid, D., Malerba, D.: A network intrusion detection system for concept drifting network traffic data. In: Soares, C., Torgo, L. (eds.) Discovery Science, pp. 111–121. Springer International Publishing, Cham (2021). https://doi.org/10.1007/978-3-030-88942-5_9
2. Andresini, G., Pendlebury, F., Pierazzi, F., Loglisci, C., Appice, A., Cavallaro, L.: INSOMNIA. In: Proceedings of the 14th ACM Workshop on Artificial Intelligence and Security. ACM (2021). https://doi.org/10.1145/3474369.3486864
3. Arp, D., et al.: Dos and don'ts of machine learning in computer security. In: 31st USENIX Security Symposium (USENIX Security 22), pp. 3971–3988. USENIX Association, Boston, MA (2022). https://www.usenix.org/conference/usenixsecurity22/presentation/arp
4. Chkirbene, Z., Erbad, A., Hamila, R., Mohamed, A., Guizani, M., Hamdi, M.: Tidcs: a dynamic intrusion detection and classification system based feature selection. IEEE Access **8**, 95864–95877 (2020)

5. Gates, C., Taylor, C.: Challenging the anomaly detection paradigm: a provocative discussion. In: Proceedings of the Workshop on New Security Paradigms (NSPW), pp. 21–29 (2006)
6. Geremias, J., Viegas, E.K., Santin, A.O., Britto, A., Horchulhack, P.: Towards multi-view android malware detection through image-based deep learning. In: 2022 International Wireless Communications and Mobile Computing (IWCMC). IEEE (2022). https://doi.org/10.1109/iwcmc55113.2022.9824985
7. Gu, J., Wang, L., Wang, H., Wang, S.: A novel approach to intrusion detection using SVM ensemble with feature augmentation. Comput. Secur. **86**, 53–62 (2019). https://doi.org/10.1016/j.cose.2019.05.022
8. Heartfield, R., Loukas, G., Bezemskij, A., Panaousis, E.: Self-configurable cyber-physical intrusion detection for smart homes using reinforcement learning. IEEE Trans. Inf. Forensics Secur. **16**, 1720–1735 (2021)
9. Horchulhack, P., Viegas, E.K., Santin, A.O.: Detection of service provider hardware over-commitment in container orchestration environments. In: GLOBECOM 2022–2022 IEEE Global Communications Conference. IEEE (2022)
10. Horchulhack, P., Viegas, E.K., Santin, A.O., Geremias, J.: Intrusion detection model updates through GAN data augmentation and transfer learning. In: GLOBECOM 2022–2022 IEEE Global Communications Conference. IEEE (2022). https://doi.org/10.1109/globecom48099.2022.10000666
11. Kaspersky Lab.: Kaspersky press release 2022–q3 (2022). https://www.kaspersky.com/about/press-releases/2022-hacktivists-step-back-giving-way-to-professionals-a-look-at-ddos-in-q3-2022
12. Molina-Coronado, B., Mori, U., Mendiburu, A., Miguel-Alonso, J.: Survey of network intrusion detection methods from the perspective of the knowledge discovery in databases process. IEEE Trans. Netw. Serv. Manag. **17**(4), 2451–2479 (2020)
13. Moustafa, N., Turnbull, B., Choo, K.K.R.: An ensemble intrusion detection technique based on proposed statistical flow features for protecting network traffic of internet of things. IEEE Internet Things J. **6**(3), 4815–4830 (2019)
14. Pinagé, F., dos Santos, E.M., Gama, J.: A drift detection method based on dynamic classifier selection. Data Min. Knowl. Discov. **34**(1), 50–74 (2019). https://doi.org/10.1007/s10618-019-00656-w
15. Salo, F., Nassif, A.B., Essex, A.: Dimensionality reduction with IG-PCA and ensemble classifier for network intrusion detection. Comput. Netw. **148**, 164–175 (2019)
16. dos Santos, R.R., Viegas, E.K., Santin, A.O.: Improving intrusion detection confidence through a moving target defense strategy. In: 2021 IEEE Global Communications Conference (GLOBECOM). IEEE (2021)
17. dos Santos, R.R., Viegas, E.K., Santin, A.O., Cogo, V.V.: Reinforcement learning for intrusion detection: more model longness and fewer updates. IEEE Trans. Netw. Serv. Manag. 1–17 (2022)
18. Sommer, R., Paxson, V.: Outside the closed world: on using machine learning for network intrusion detection. In: 2010 IEEE Symposium on Security and Privacy. IEEE (2010)
19. Tavallaee, M., Stakhanova, N., Ghorbani, A.A.: Toward credible evaluation of anomaly-based intrusion-detection methods. IEEE Trans. Syst. Man Cybern. **40**(5), 516–524 (2010)
20. Viegas, E.K., Santin, A.O., Oliveira, L.S.: Toward a reliable anomaly-based intrusion detection in real-world environments. Comput. Netw. **127**, 200–216 (2017)
21. Wahab, O.A.: Intrusion detection in the IoT under data and concept drifts: online deep learning approach. IEEE Internet Things J. **9**(20), 19706–19716 (2022)

A Roadmap to Blockchain Technology Adoption in Saudi Public Hospitals

Adel Khwaji[1]([✉]), Yaser Alsahafi[2], and Farookh Khadeer Hussain[1]

[1] School of Computer Science, Faculty of Engineering and Information Technology, University of Technology Sydney, Sydney, Australia
Adel.Khwaji@student.uts.edu.au, Farookh.Hussain@uts.edu.au
[2] School of Information Technology, University of Jeddah, Jeddah, Saudi Arabia
yaalsahafi@uj.edu.sa

Abstract. Blockchain has great potential and promise in the healthcare sector, even though it has not yet seen widespread adoption. One of the main challenges with blockchain adoption is IT employees' awareness and knowledge. IT knowledge reduces adoption uncertainty and a basic understanding of blockchain and its features is necessary. This study evaluated the knowledge of IT employees who work in Saudi public hospitals and suggested a roadmap to blockchain adoption in hospitals.

1 Introduction

Blockchain technology can be defined as a time-stamped and immutable group of records of data which is stored on a wide network of participating computers which are not owned by any single central authority [1]. There are various advantages of blockchain, for example immutability, decentralisation, anonymity, traceability, transparency and security [2]. This technology can be used to provide verification without depending on third parties and there is append-only data in blockchain which makes it impossible to modify or delete data [3].

Healthcare is one industry that can gain significant benefits from blockchain technology. There are several areas of the healthcare system where blockchain technology can be used as it offers accessibility and security. A few of its applications have been used in the medical sector, particularly for the storage and sharing of data pertaining to insurance and medical records [4]. Mobile applications as well as remote monitoring systems (RMSs) also make use of blockchain technology [5]. Furthermore, it helps in solving traceability issues in the drug supply chain [6] and facilitates interoperability between healthcare organisations [7].

There continues to be a low adoption rate of health information system (HIS) in Saudi healthcare organisations. A few of the major obstacles for the successful implementation of HIT are insufficient health informatics experts, inadequate experience using computer applications, weak leadership [8], poor information system infrastructure and technical support [9], lack of policies and standards [10], and lack of an implementation strategy [11]. Before a new technology is adopted, it is critical to determine the knowledge and

© The Author(s), under exclusive license to Springer Nature Switzerland AG 2023
L. Barolli (Ed.): AINA 2023, LNNS 654, pp. 452–460, 2023.
https://doi.org/10.1007/978-3-031-28451-9_40

awareness of IT staff. If the IT staff have ample knowledge and the skills required to adopt blockchain technology, then the hospital will exhibit greater confidence in the adoption process. Generally, a lack of technical knowledge is perceived as an obstacle for e-health development [12].

1.1 Objective

The purpose of this study is to evaluate IT employees' knowledge about blockchain technology and to provide a roadmap to blockchain technology adoption in Saudi public hospitals.

2 Methods

2.1 Data Collection

The data collection took place between August 2020 and May 2021 in Saudi Arabia. A cross-sectional design was used to collect the quantitative data. The targeted population includes all the IT employees (e.g., CIOs, CTOs, IT technicians, programmers, etc.) who work in Saudi public or governmental hospitals because they play a crucial role in decision-making at the strategic and operational levels. They are familiar and knowledgeable about the technology and these employees would be the ones to introduce the frontline adoption of new technology.

Snowballing is the most appropriate sampling technique for the current research because the population in this study is difficult to reach due to COVID-19 restrictions [13]. The researcher distributed the questionnaire to his contacts working in Saudi hospitals as an invitational link to the online questionnaire. Those contacted were asked to participate in the survey and invite other relevant people to participate. The researcher used his contacts on social networking websites (such as WhatsApp, Twitter, LinkedIn) to increase the response rate by inviting the targeted participants to complete the questionnaire. The researcher submitted the relevant ethics form to the Human Research Ethics Committee at the University of Technology, Sydney and approval was obtained in August 2020 (NO. ETH20-5126).

2.2 Data Analysis

It is important to thoroughly screen and clean the collected data to ensure its accuracy. Following Hair et al. [14], the data were screened for missing values, checked for errors, unengaged responses. The data collected were analysed using descriptive statistics. Descriptive statistics are commonly generated and used to define the essential characteristics of data in research [15].

3 Results

449 respondents participated in the survey, and after the data screening process was completed, 363 responses were considered valid. Table 1 presents the demographic

Table 1. Demographic summary of respondents

	Participants		IT employees in Saudi Arabia [16]	
	No.	%	No.	%
Gender				
Male	253	69.7%	3111	67.4%
Female	110	30.3%	1506	32.6%
Age				
18–24	82	22.6%		
25–34	170	46.8%		
35–44	69	19%		
45–54	32	8.8%		
55+	10	2.8%		
Computer skills				
Very poor	–	–		
Poor	–	–		
Moderate	3	0.8%		
Good	48	13.2%		
Very good	312	86%		
Blockchain knowledge				
Very poor	91	25.1%		
Poor	163	44.9%		
Moderate	88	24.2%		
Good	16	4.4%		
Very good	5	1.4%		
Blockchain use				
Yes	46	12.7%		
No	317	87.3%		
Blockchain experience				
Less than 1 year	36	78.3%		
1–3 years	10	21.7%		

information on the IT employees and includes information on their gender and age, and also provides information on the level of their computer skills, blockchain knowledge, and experience using blockchain services or applications.

About 44.9% and 25.1% of respondents indicated that their overall knowledge of blockchain technology is poor and very poor, respectively. 24.2% stated that they have moderate blockchain knowledge. A further 4.4% stated that their knowledge is good, and only 1.4% indicated that they had excellent blockchain knowledge. Of the 363 participants, only 46 had experience using blockchain. However, the results indicate that almost 78.3% of those 46 respondents had less than one year of experience, while 21.7% had 1 to 3 years of experience using any blockchain services or applications.

4 Discussion

The questionnaire asked participants to rate their knowledge of blockchain technology. As a result of the descriptive analysis, it was determined that about 70 percent of the participants had little knowledge about blockchain technology. This indicates that there is a lack of knowledge of blockchain among IT employees in Saudi public hospitals. One reason for this may be that blockchain technology is relatively new [17]. This is supported by the participants' comments in the last part of the questionnaire, which revealed that 19% of respondents had never heard of blockchain but had been informed about it from the link provided with the questionnaire explaining the technology. For example, of the comments received, one participant's response was, "I am wondering why the Ministry of Health does not put adequate effort into educating and training the staff about this technology". Another participant stated, "It is imperative that the staff become aware of what blockchain technology is and the benefits it brings".

The lack of knowledge of blockchain is preventing small and medium-sized businesses from adopting the technology in Italy [18]. In a study of German consumers, Knauer and Mann [19] have revealed that a lack of knowledge might be one of the barriers to blockchain adoption. IT employees should have a thorough understanding of how blockchains work and how they are built. They should be familiar with distributed ledgers, hash functions, cryptography, smart contracts and other related concepts. Accordingly, the Ministry of Health in Saudi Arabia should take advantage of the best practices that have previously been implemented in developed countries and apply these to avoid such obstacles by educating staff about blockchain. These practices could include providing staff with paid online or face-to-face courses to gain a better understanding of the technology; encouraging them to read important white papers or scientific publications; providing them with information explaining a topic related to blockchain; providing them with reading materials such as books in the hospital's library; and organising blockchain meetups, events or seminars. In doing so, the likelihood of blockchain technology being adopted meaningfully would increase.

4.1 Roadmap

Providing resources and efforts to enhance the healthcare sector is a priority of the Saudi government represented by Ministry of Health. Healthcare provider decision makers and the IT team need to seriously consider adopting blockchain to move forward more rapidly technologically and economically than traditional technologies are currently able to do. This may be achieved by the following:

- It is essential that the Saudi Ministry of Health conducts awareness-raising activities to educate employees about the benefits of blockchain technology. For ensure these awareness campaigns are effective, the Ministry of Health should utilise communication channels such as television, social media, and e-mail to reach all employees and IT employees in particular. Internal awareness campaigns can be organised within the hospital by distributing printed and digital materials.

- The lack of internal IT expertise caused hospitals to perceive themselves as being unable to adopt blockchain. Providing comprehensive and extensive training programs for IT employees is likely to increase the technical competence to successfully implement blockchain technology in Saudi hospitals. It is necessary for hospitals to have the knowledge and tools required to understand and build their own blockchain. The training materials should include an introduction to blockchain technology which provides a broad overview of the fundamental concepts, such as decentralised consensus, smart contracts, immutable distributed ledgers and mining. It also should include types of blockchain consensus algorithms such as Proof of Work (PoW), Proof of Stake (PoS), Proof of Elapsed Time (PoET) and Practical Byzantine Fault Tolerance (PBFT) [20]. IT employees should learn how to use various tools for blockchain development that can facilitate the development process such as Mist, Geth, Hyperledger Caliper, Remix, Metamask, Solidity Compiler and Blockchain Testnet [21]. There are several platforms for developing blockchain applications which be beneficial to learn about such as Ethereum, Hyperledger Fabric, EOS and Hyperledger Sawtooth [22]. Technical employees and particularly developers should have knowledge of at least one high-level programming language applicable to blockchain development, such as Python, JavaScript, Solidity, C and C++. After training, an ROI model can be created for training evaluation [23].

- It is imperative that decision makers and managers have a comprehensive understanding of blockchain technology to develop strategies and plan effectively for the introduction of blockchain technology and to make use of it. Hospitals should develop a blockchain strategy that integrates the current hospital strategy based on their resources. The strategy may include identifying a priority area for blockchain implementation. Furthermore, blockchain is a technology that needs collaboration at all levels of the healthcare organisation including all stakeholders. It is therefore important to identify and engage all stakeholder involved in the healthcare domain such as patients, IT professionals, medical professionals, the Ministry of Health and health insurance companies. A key element of the strategic plan should also include blockchain governance, which is not simply about who controls the blockchain, but also relates to establishing mechanisms for resolving issues as they arise [24]. Governing blockchain involves how technical decisions are made and how the database features are modified, what data should be stored on- and off-chain and whether permissionless or permissioned blockchain will be developed. In addition, it is recommended that hospitals hire external qualified consultants with experience in blockchain to provide strategic advice based on research and inventive evaluation on implementing it effectively to decrease the possibility of failure and reduce costs.

- Before planning for blockchain implementation, certain technical specifications should be taken into consideration or should be available to ensure hospital readiness. The basic requirements that should be available are high performance computers and a stable internet connection. Other considerations should be taken into account such as the type of processing power, storage capacity for the nodes (computers or servers), memory (RAM/SRAM) as well as the environmental conditions, such as shock and vibration, temperature fluctuations, and humidity (Byers, 2017). The higher the complexity of the system, the more specifications and requirements are needed. The first consideration in planning what type of blockchain (public, private, consortium) to implement is to select the most suitable consensus protocol (e.g. PoW, PoS, PoET, PBFT) and to choose the most suitable platform (Ethereum, Hyperledger Fabric, EOS and Hyperledger Sawtooth). An application programming interface (API) needs to be considered and the user interface needs to be designed using a programming language (e.g. Python, JavaScript, Solidity, C and C++). There is no restriction on the language used for the front-end of the application, but smart contracts must be used in the backend. Examining current healthcare blockchain projects will assist with the development process. For example, Medicalchain is a platform for the fast and secure exchange and utilisation of medical information on blockchain and access to medical records is controlled by permissions [25]. Two blockchains are used by Medicalchain, one built on Hyperledger Fabric and the other based on Ethereum. GitHub provides access to the Medicalchain repository, which contains both smart contracts and APIs. There are also other healthcare applications and platforms built on blockchain such as MedRec for medical data management [4], Akiri and Patientory for data sharing [26, 27], MediLedger for the medical supply chain [28], FarmaTrust for pharmaceutical traceability [29] and Bloqcube for clinical trials [30]. It is important to note that there are many reputable companies that provide blockchain consulting services and solutions such as Amazon, IBM, HPE and Accenture. These companies could also help with the timeline and cost as these two factors are critical for developing a successful blockchain strategy and plan.

- Once the necessary groundwork has been laid, pilot projects should be introduced in non-critical or sensitive applications that will not interfere with hospital operations. Examples of this are staff credential verification and medication tracking. A credential data solution on blockchain is already being used in pilot projects by companies such as ProCredEx [31]. Also, in this testing stage, it should be ensured that the blockchain solution integrates seamlessly with legacy systems as this is one of the key aspects of the blockchain implementation strategy. Finally, it is worth noting that the implementation of blockchain is a long-term transformation process which may take several years to complete. It will be necessary to conduct further research to develop effective implementation strategies that can be used to encourage the use of blockchain in hospitals (Fig. 1).

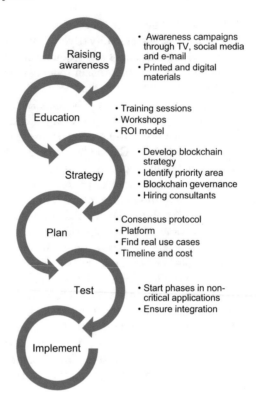

Fig. 1. A roadmap to blockchain adoption in hospitals

5   Conclusions, Limitations and Future Direction

This research evaluated the knowledge of IT staff working in Saudi public hospitals and determined that there is a lack of knowledge of blockchain technology among them. Therefore, it provided a guideline for raising awareness activities and a roadmap for blockchain technology adoption in hospitals.

This study has certain limitations that need to be considered in future research. Snowballing was the most appropriate sampling technique because the population was difficult to reach due to COVID-19 restrictions. So, there could be a sampling bias in this study due to the use of a non-probability sampling method to recruit participants. It is worth considering in future research a probability sampling method such as simple random sampling as it results in an unbiased representation of target population.

The researcher initially intended to conduct a focus group interview with IT employees at Saudi hospitals for the purpose of qualitative data collection, but this method was significantly inhibited by the ongoing COVID-19 situation during the time of data collection. Therefore, future research may consider the applicability of mixed methods research. This research covers public and government hospitals and does not take into account private hospitals. Further research to investigate the private sector will be beneficial.

References

1. Sharples, M., Domingue, J.: The blockchain and kudos: a distributed system for educational record, reputation and reward. In: Verbert, K., Sharples, M., Klobučar, T. (eds.) EC-TEL 2016. LNCS, vol. 9891, pp. 490–496. Springer, Cham (2016). https://doi.org/10.1007/978-3-319-45153-4_48
2. Abeyratne, S.A., Monfared, R.P.: Blockchain ready manufacturing supply chain using distributed ledger (2016)
3. Crosby, M., et al.: Blockchain technology: beyond bitcoin. Appl. Innov. 2(6–10), 71 (2016)
4. Azaria, A., et al.: MedRec: using blockchain for medical data access and permission management. In: 2016 2nd International Conference on Open and Big Data (OBD). IEEE (2016)
5. Chen, H.S., et al.: Blockchain in healthcare: a patient-centered model. Biomed. J. Sci. Tech. Res. 20(3), 15017 (2019)
6. Mettler, M.: Blockchain technology in healthcare: the revolution starts here. In: 2016 IEEE 18th International Conference on e-Health Networking, Applications and Services (Healthcom). IEEE (2016)
7. Linn, L.A., Koo, M.B.: Blockchain for health data and its potential use in health it and health care related research. In: ONC/NIST Use of Blockchain for Healthcare and Research Workshop. Gaithersburg, Maryland, United States: ONC/NIST (2016)
8. Alsulame, K., Khalifa, M., Househ, M.S.: eHealth in Saudi Arabia: current trends, challenges and recommendations. In: ICIMTH, vol. 213, pp. 233–236 (2015)
9. Alkraiji, A., Jackson, T., Murray, I.: Barriers to the widespread adoption of health data standards: an exploratory qualitative study in tertiary healthcare organizations in Saudi Arabia. J. Med. Syst. 37(2), 9895 (2013)
10. Aldosari, B.: Causes of EHR projects stalling or failing: a study of EHR projects in Saudi Arabia. Comput. Biol. Med. 91, 372–381 (2017)
11. Khalifa, M.: Barriers to health information systems and electronic medical records implementation. A field study of Saudi Arabian hospitals. Procedia Comput. Sci. 21, 335–342 (2013). https://doi.org/10.1016/j.procs.2013.09.044
12. Boonstra, A., Broekhuis, M.: Barriers to the acceptance of electronic medical records by physicians from systematic review to taxonomy and interventions. BMC Health Serv. Res. 10(1), 231 (2010)
13. Biernacki, P., Waldorf, D.: Snowball sampling: problems and techniques of chain referral sampling. Sociol. Methods Res. 10(2), 141–163 (1981)
14. Hair, J.F., et al.: Multivariate Data Analysis, vol. 6. Pearson Prentice Hall, Upper Saddle River (2006)
15. Liu, R.Y., Parelius, J.M., Singh, K.: Multivariate analysis by data depth: descriptive statistics, graphics and inference, (with discussion and a rejoinder by Liu and Singh). Ann. Stat. 27(3), 783–858 (1999)
16. MinistryofHealth. Statistical Yearbook (2020). https://www.moh.gov.sa/en/Ministry/Statistics/book/Pages/default.aspx. Accessed 11 Dec 2021
17. Tapscott, D., Tapscott, A.: How blockchain will change organizations. MIT Sloan Manag. Rev. 58(2), 10 (2017)
18. Bracci, E., et al.: Knowledge, diffusion and interest in blockchain technology in SMEs. J. Knowl. Manag. 26(5), 1386–1407 (2021). https://doi.org/10.1108/JKM-02-2021-0099
19. Knauer, F., Mann, A.: What is in it for me? Identifying drivers of blockchain acceptance among German consumers. J. Br. Blockchain Assoc. 3(1), 1–16 (2020). https://doi.org/10.31585/jbba-3-1-(1)2020
20. Baliga, A.: Understanding blockchain consensus models. Persistent 4(1), 14 (2017)

21. Aarti, E.: A review of blockchain technology. In: Smart City Infrastructure: The Blockchain Perspective, pp. 225–246 (2022)
22. Suvitha, M., Subha, R.: A survey on smart contract platforms and features. In: 2021 7th International Conference on Advanced Computing and Communication Systems (ICACCS). IEEE (2021)
23. Phillips, J.J.: Level four and beyond: an ROI model. In: Brown, S.M., Seidner, C.J. (eds.) Evaluating Corporate Training: Models and Issues. EEHS, vol. 46, pp. 113–140. Springer, Dordrecht (1998). https://doi.org/10.1007/978-94-011-4850-4_6
24. Maggiolino, M., Zoboli, L.: Blockchain governance: the missing piece in the competition puzzle. Comput. Law Secur. Rev. **43**, 105609 (2021)
25. Medicalchain. Medicalchain Whitepaper (2018). https://medicalchain.com/Medicalchain-Whitepaper-EN.pdf
26. Akiri. About Akiri and building trusted networks for health data (2018). https://akiri.com/about
27. McFarlane, C., et al.: Patientory: a healthcare peer-to-peer EMR storage network v1. Entrust Inc.: Addison, TX, USA (2017)
28. Chronicled. Trust & Automation Between Companies (2022). https://www.chronicled.com
29. FarmaTrust. Digitising and Innovating the Pharmaceutical and Healthcare Business (2017). https://www.farmatrust.com
30. Bloqcube. Decentralized Software for Clinical Trials (2017). https://bloqcube.com
31. ProCredEx. A Digital Revolution for Professional Credentials Data (2022). https://procredex.com

Zero-Knowledge Multi-transfer Based on Range Proofs and Homomorphic Encryption

Emanuele Scala[1](✉), Changyu Dong[2], Flavio Corradini[1], and Leonardo Mostarda[1]

[1] Computer Science, University of Camerino, Camerino, Italy
{emanuele.scala,flavio.corradini,leonardo.mostarda}@unicam.it
[2] Guangzhou University, Guangzhou, China
changyu.dong@gzhu.edu.cn

Abstract. Zero-knowledge proofs are widely adopted in Confidential Transactions (CTs). In particular, with these proofs, it is possible to prove the validity of transactions without revealing sensitive information. This has become an attractive property in public blockchain where transactions must be publicly verifiable. However, several challenges must be addressed in order not to alter important properties of the blockchain, such as not introducing trusted third parties and/or circuit-dependent trusted setups. Moreover, there are limited proposals working on the standard account model and considering extended payment models where multiple payees are involved in one transaction. With this paper, we first present our concept of *Multi-Transfer* (MT) in CTs settings, i.e., a transfer that involves multiple payees in a single transaction with privacy guarantees for balances and transfer amounts. Inspired by the work of Zether, we design the MT zero-knowledge proof system, named *MTproof*, by combining the aggregate version of Bulletproofs and several Σ-Protocols to prove that an MT transaction is legit. We provide concrete evaluations of the *MTproof* in terms of proof size, prover and verifier execution time.

1 Introduction

With the rise of the first digital currency of Bitcoin [17], blockchain technology has become a growing active field of study in both academia and industry. Ethereum [5] has extended the functionality of the blockchain with smart contracts. The transactions enclosed in the blocks are executed according to the logic specified in smart contracts. The state updates produced in a smart contract are propagated to all nodes of the network and immutably stored on the blockchain. This public announcement brings the transaction information publicly visible. Further, a known problem is that an adversary can link the user's real-world identity and disclose transaction information by analysing the public transaction graph [7, 13, 19]. In light of this, an increasing body of research is focusing on privacy applications such as *Confidential Transactions* (CTs), where guarantees on the anonymity of the entities involved in a financial transfer and the confidentiality of the transfer amounts are considered. *Zero-Knowledge Proofs* (ZKPs) are spreading widely in CTs, as they are capable of revealing nothing except the validity of

This work was supported by HD3FLAB project the Nationally funded by POR MARCHE FESR.

the statements being proved relating to the transactions. Therefore, ZKPs are adopted in many privacy protocols, among the most cited are MimbleWimble [18], Zerocash [22], Lelantus [15] and Monero [1]. Such proposals are designed on the UTXO (unspent transaction outputs) model and move around a simple equation for a valid transaction: the total of the input amounts must equal the total of the output amounts. To keep the amounts hidden, this equation is then proved through the *commitment to zero* or a private computation linked to an *arithmetic circuit* (e.g., zk-SNARK). On another side, there are privacy protocols that move to the Ethereum account model: a model in which a transfer of funds between two addresses has the effect that the payer's account balance is debited and the payee's account balance is credited. In this setting, proposals of Zether [3] and Quisquis [11] design schemes in which the balances kept in an encrypted form are updated homomorphically. Moreover, Zether and Quisquis share the use of proof systems such as Bulletproofs [4] for *range proofs* and Σ-Protocols to prove algebraic statements under the DLOG assumption. Other solutions rely on proof systems for more general computations inspired by Zerocash, examples are Blockmaze [14], which has designed a specific *private computation circuit* for each transaction, and ZETH [20] that has developed the logic of UTXO on top of the account model using zk-SNARK to prove the transaction correctness.

Research Question. Complex smart contracts can have more than two users, opening scenarios for a new model of transaction taking place amongst a set of participants. A similar concept is the "redistribution of wealth" introduced in Quisquis [11], where multiple account balances can be updated at the same time as the effect of one UTXO transaction. However, there are few proposals for CTs in the standard account model, and none consider such an extended payment model. Hence, we want to ask the question: *how can we design a zero-knowledge transfer in the account model, where there are multiple payees in one transaction?* Rather than having the traditional payment model, we introduce our concept of *Multi-Transfer* (MT) in CTs, i.e., a transfer that involves multiple payees with a single confidential transaction with hidden amounts and balances. We first formulate the statements over an MT zero-knowledge relation for which our proof is constructed, and then we design our zero-knowledge proof system, called *MTproof*, satisfying that relation. Inspired by the Σ-Bullets of Zether [3], our system is based on homomorphic primitives and combines two well-known proof systems: the *Aggregated Bulletproofs* [4] to construct one range proof of m aggregated values and several Σ-protocols [6] to prove statements on the encryption of multiple transfer amounts together with the balance of the sender. Since the range proof is the most expensive component of the proof system even for a single value, it turns out that a single range proof for m aggregated values is more efficient.

Our Contribution. In this paper, we initiate the formalization of the *Multi-Transfer* (MT) in the account model. We design the *MTproof* proof system for the corresponding MT relation, which is provided by the following properties: (i) Multi-Transfer - multiple payees receive funds from a single transaction; (ii) Confidential Transfers - hide the amounts in the transaction and sender/recipients balances values; (iii) Zero-Knowledge - verifier learns nothing about the secrets and witnesses of the transaction; (iv) Non-Interactive - can be transformed into a non-interactive version, i.e., no interactions from the verifier to the prover; (v) Trustless - no added trust derived from the proof systems or trusted executors; (vi) Aggregation - single aggregate proof for many statements related to the MT relation. As part of our contribution, we implement the *MTproof* proof system using `arkworks` [2] Rust ecosystem. We evaluate the *MTproof* concretely in terms of proof size, prover and verifier execution time.

2 Preliminaries

In what follows, we denote with pp the public parameters and with λ the security parameter of a scheme. We write $a \xleftarrow{\$} S$ when a random variable a is uniformly sampled from a set S.

Groups. Let G be a *group-generation algorithm* that on input 1^λ outputs the tuple $(\mathbb{G}, p, g)$, where $\mathbb{G}$ is a description of a *cyclic group*, p is a prime number and is the order of the group and $g \in \mathbb{G}$ is a generator of the group. We consider groups in which the *discrete logarithm problem* is computationally intractable. We refer to the DLOG assumption and its variant of the Decisional Diffie-Hellman (DDH) assumption for groups in which the discrete logarithm problem is hard.

Pedersen Commitments. A Pedersen commitments is defined over a finite cyclic group $\mathbb{G}$ of prime order. The message space is from the set of integers modulo p, $\mathbb{Z}_p$ where p is a safe prime. Let $g, h \in \mathbb{G}$ be two random public generators, $m \in \mathbb{Z}_p$ the message and $r \xleftarrow{\$} \mathbb{Z}_p^*$ the blinding factor, the commitment is defined as: $com = \text{Commit}(m;r) = (g^m h^r) \in \mathbb{G}$. We consider a *non-interactive* commitment scheme in which a valid commitment is proved by the knowledge of the secret relating to the commitment. A variant of this scheme is *Pedersen vector commitments* used to commit multiple messages at once. The input messages are gathered in a vector and a vector of generators is defined as well.

ElGamal Homomorphic Encryption. Considering a cyclic group $\mathbb{G}$ of prime order p and generator g of that group, the encryption scheme has a private key in the random integer $x \xleftarrow{\$} \{1,\ldots,p-1\}$ and a public key of the form $y = g^x$, then the public parameters of the scheme are (G, p, g, y). Let r be a randomness uniformly sampled in $r \xleftarrow{\$} \{1,\ldots,p-1\}$, the ciphertext for a message m is: $Enc(m) = (C_L, C_R) = (m \cdot y^r, g^r)$. A more specific use case of the scheme can be considered as in Zether [3]. Here, the message m to encrypt is an integer value $a \in \mathbb{Z}_p$ which can be turned into a group element using the simple mapping $m = g^a$. Hence, the encryption function can be rewritten as $Enc(m) = (C_L, C_R) = (m \cdot y^r, g^r) = (g^a \cdot y^r, g^r)$. An *additive homomorphism* can be derived from the application of the group operator $\cdot$ between two ElGamal ciphertexts: $(C_L \cdot C_L' = g^{a+a'} \cdot y^{r+r'}, C_R \cdot C_R' = g^{r+r'})$.

Zero-Knowledge Proofs. Let $\mathcal{R}$ be a binary relation for an instance x and a witness w and L the corresponding language such that $L = \{x \mid \exists w : (x, w) \in \mathcal{R}\}$. An interactive proof is a protocol between a prover $\mathcal{P}$ and a verifier $\mathcal{V}$ in which $\mathcal{P}$ tries to convince $\mathcal{V}$ that an instance x is in the language L for the given relation $\mathcal{R}$. This can be done through an interactive exchange of messages between $\mathcal{P}$ and $\mathcal{V}$ representing the *transcript*, from which the verifier can accept or reject the conversation (namely the proof). The proof is said to be *zero-knowledge* if it essentially reveals nothing beyond the validity of the proof. An interactive proof is *honest-verifier perfect zero-knowledge* (HVZK) if it has *perfect completeness, special soundness* and *honest-verifier perfect zero-knowledge* properties. The HVZK protocol is called *public coin* if all the verifier's challenges sent to the prover are chosen uniformly at random and are independent of the prover's messages.

Σ-Protocols. HVZK public-coin interactive proofs consisting of three messages (a, c, z) where: a is the *announcement* computed by $\mathcal{P}$ and sent to $\mathcal{V}$, c is the *challenge* randomly sampled by $\mathcal{V}$ and sent to $\mathcal{P}$ and z is the *response* computed by $\mathcal{P}$ based on the challenge and sent back to $\mathcal{V}$. From that structure, one can prove that ciphertexts are well-formed, the knowledge of the opening of a commitment, the knowledge of a secret behind encryption, or other statements relating to the discrete logarithm. Moreover, any interactive zero-knowledge proof can be transformed in a *non-interactive* zero-knowledge proof (NIZK) by means of the Fiat-Shamir heuristic [12]. In such case, an honest prover tries to follow the protocol composed of the messages (a, c, z) where the challenge c is replaced by a *random oracle*.

Zero-Knowledge Relation. The relation is a valid collection of instances-witnesses together with the statements for which the zero-knowledge proof is constructed. We use the notation $Rel : \{(x_1, ..., x_n ; w_1, ..., w_m) : f(x_1, ..., x_n, w_1, ..., w_m)\}$ to specify that prover and verifier know the public instances $x_1, ..., x_n$, and only the prover knows the witnesses $w_1, ..., w_m$ such that $f(x_1, ..., x_n, w_1, ..., w_m)$ is true. f defines the statements for the given instances-witnesses and can be expressed in algebraic form.

2.1 Zero-Knowledge Multi-transfer Relation

The concept of *Multi-Transfer* in the confidential setting has been previously presented in our work *ZeroMT* [8]. The aim is to enable a single payer to transfer currency to multiple payees within a single confidential transaction that has the effect of multiple concurrent transfers. The *ZeroMT* transaction layer builds upon a user program, defined as *MTU*, and a smart contract hosted on the account-model blockchain, the *MTSC* smart contract. Users can utilise the MTU to initiate the multi-transfer transaction, the *MTX* transaction, towards n recipients. Each involved party i has an encrypted balance of the form $b[y_i] = (C_L = g^{b_i} \cdot y_i^r, C_R = g^r)$ obtained by means of an ElGamal public key $y_i = g^{sk_i} \in \mathbb{G}$, derived from a randomly sampled secret key $sk_i \xleftarrow{\$} \mathbb{Z}_p$. Given a list $\mathbf{a} = (a_1, ..., a_n)$ of plaintext amounts to be transferred, what the MTU does is encrypting those values with the public key y of the sender and the public keys $\bar{\mathbf{y}} = (\bar{y}_1, ..., \bar{y}_n)$ of the recipients. The result are respectively the lists $\mathbf{C} = (C_1, ..., C_n)$ and $\bar{\mathbf{C}} = (\bar{C}_1, ..., \bar{C}_n)$, where $\forall i \in [1, n]$, $\mathrm{Enc}_y(a_i) = C_i = g^{a_i} y^r$ and $\mathrm{Enc}_{\bar{y}_i}(a_i) = \bar{C}_i = g^{a_i} \bar{y}_i^r$. The two lists of ciphertexts are associated with the same randomness value $D = g^r$, where $r \xleftarrow{\$} \mathbb{Z}_p^*$ is a value uniformly sampled from the set of inverses in $\mathbb{Z}_p$. In order to initiate a transaction, the MTU forwards the $\mathbf{C}$ and $\bar{\mathbf{C}}$ lists to the MTSC along with a zero-knowledge proof π satisfying a *zero-knowledge multi-transfer relation*, which (informally) states that: (i) the sender knows the secret key sk for which the respective public key y encrypts the values in $\mathbf{C}$; (ii) the sender knows the randomness r used in the encryption process;

(iii) the sender balance cannot be overdraft; (iv) the i-th ciphertexts in both $\mathbf{C}$ and $\bar{\mathbf{C}}$ are well-formed and encrypt the same amounts; (v) each of the transfer amounts in $\mathbf{a}$ is non-negative; (vi) the sender remaining balance b' is non-negative. For these statements, the MTU acts as a prover and the MTSC as a verifier. Both share the public instances $(y, \bar{y}, C_L, C_R, \mathbf{C}, \bar{\mathbf{C}}, D, g)$, while the witnesses $(sk, \mathbf{a}, b', r)$ are known only by the prover. Further, we express the statement for the range proofs in which we prove that $\forall a_i \in (a_1, \ldots, a_n)$ each transfer amount a_i and the remaining sender's balance b' fall within the range of admissible values $[0, MAX]$, where MAX is the upper limit equal to $2^n - 1$ with n the bit length of the values. After the proof π is successfully verified by the MTSC, all the requested transfers are executed. The updates on the balances of the sender and each of the recipients are made through the additive homomorphism of the ElGamal scheme on the underlying group $\mathbb{G}$: $b[y] = b[y] \circ (C_{tot}^{-1}, D^{-1})$ where $b[y]$ is the reference of the sender's balance, and $\forall i \in [1, n], b[\bar{y}_i] = b[\bar{y}_i] \circ (\bar{C}_i, D)$ is the i-th receiver's balance increased by the corresponding amount in the $\bar{\mathbf{C}}$ list.

3 MTproof: Multi-transfer Zero-Knowledge Proof System

In our context of the previous section, a zero-knowledge proof must guarantee that the MTX transaction is well-formed and legit, i.e., the balance transfer goes to the right payee and the payer has enough money to spend in his/her wallet. To this end, we design the interactive zero-knowledge proof system, called *MTproof*, satisfying the *multi-transfer relation*. The main idea is to combine the *Aggregated Inner Product Range Proof* and several Σ−Protocols as follows. We generate one aggregated range proof for $m = n + 1$ values, such that the $m - 1$ values correspond to each transfer amount a_i and one value b' to the remaining balance; we use Σ−Protocol to prove the statements for the two lists of n encrypted transfer amounts $(C_i, D)_{i=1}^n$ and $(\bar{C}_i, D)_{i=1}^n$ and for the sender's remaining balance. Using the above combination of proof systems, we generalise the Σ-Bullets protocol [3] of Zether to the case of n transfers per epoch instead of one. Our proof system benefits from short and logarithmic-sized proofs as well as trustless property. We organize this section as follows: we present the key concepts of the Bulletproofs theory; we show how we modify Bulletproofs in our MTproof proof system; we present the Σ-protocols that compose the remaining part of MTproof.

Bulletproofs Review. First, we summarise some notations of Bulletproofs [4], used in our proof system. $\mathbb{Z}_p$ is the ring of integers modulo p prime ($\mathbb{Z}_p^*$ is $\mathbb{Z}_p / \{0\}$). g and h are generators of a cyclic group $\mathbb{G}$ of prime order p. Capitalised are *commitments* and Greek letters are *blinding factors*, e.g., $A = g^a \cdot h^\alpha$ is a *Pedersen commitment* to the value a with blinding factor α. In bold are vectors, e.g., $\mathbf{a} \in \mathbb{Z}_p^n$ is a vector with elements in $\mathbb{Z}_p$ of dimension n. The *inner-product* of two vectors having size n is $\langle \mathbf{a}, \mathbf{b} \rangle = \sum_{i=1}^n a_i \cdot b_i$. The *Hadamard product* of two vectors having size n is $\mathbf{a} \circ \mathbf{b} = (a_1 \cdot b_1, \ldots, a_n \cdot b_n)$. The following is useful for *Pedersen vector commitments*: let $\mathbf{g} = (g_1, \ldots, g_n) \in \mathbb{G}^n$ be a vector of generators then $A = \mathbf{g}^{\mathbf{a}} = \prod_{i=1}^n g_i^{a_i}$ is *binding* (but not *hiding*) commitment to the vector $\mathbf{a} \in \mathbb{Z}_p^n$. Slices of vectors are denoted with $\mathbf{a}_{[:k]} = (a_1, \ldots, a_k) \in \mathbb{F}^k$ and $\mathbf{a}_{[k:]} = (a_{k+1}, \ldots, a_n) \in \mathbb{F}^{n-k}$. A vector polynomial $p(X) \in \mathbb{Z}_p^n[X]$ is defined as $p(X) = \sum_{i=0}^d \mathbf{p_i} \cdot X^i$, where $\mathbf{p_i} \in \mathbb{Z}_p^n$ and d is the degree of the polynomial. For a complete description, refer to Sect. 2.3 of [4]. With Bulletproofs, a prover can convince a verifier that a value v is in range, in particular, $0 \leq v < 2^n$. Given as public parameters the generators g and h and the Pedersen commitment $V = h^\gamma g^v$ of the value v using randomness γ, the system ends up proving that the inner-product of two committed vectors $\mathbf{l}, \mathbf{r}$ is a certain value $\hat{t}$, then the equality $\hat{t} = \langle \mathbf{l}, \mathbf{r} \rangle$ is valid if and only if $v \in [0, 2^n - 1]$. This is the final step of the range proof in which the prover and verifier engage in the *Inner-Product Argument* (IPA) protocol. We now present a high-level overview of the previous steps. The prover first creates two commitments, A commitment to the vectors $\mathbf{a}_L$ and $\mathbf{a}_R$ where $\langle \mathbf{a}_L, 2^n \rangle = v$ and $\mathbf{a}_R = \mathbf{a}_L - \mathbf{1}^n$, S commitment to blinding terms $\mathbf{s}_L$ and $\mathbf{s}_R$. The prover receives from the verifier two challenge points y and z and creates T_1 and T_2 commitments to the coefficients t_1 and t_2 of the polynomial $t(X)$. The prover honestly constructs the polynomial $t(X)$ from the inner product of two vector polynomials $l(X)$ and $r(X)$, which in turn have a special form derived from a linear combination of the vectors $\mathbf{a}_L$ and $\mathbf{a}_R$ respectively and the verifier's challenges. The prover does not commit to the zero-coefficient t_0 of $t(x)$ and sends to the verifier only $T_{1,2}$. Instead, the prover proves the opening of the commitments, sending the evaluation $\hat{t} = t(x)$ at a random point x from the verifier. The verifier can calculate the zero-coefficient himself from the commitment V of the value v and the challenges. After receiving the challenge x, together with $\hat{t}$ the prover also sends to the verifier a blinding value τ_x for $\hat{t}$, the two blinded vectors $\mathbf{l} = l(x)$ and $\mathbf{r} = r(x)$ and a blinding factor μ for the commitments A and S. Now the verifier can verify the Pedersen commitment V of the value v, that A and S are valid and that $\hat{t} = \langle \mathbf{l}, \mathbf{r} \rangle$ is correct. Transmitting $\mathbf{l}$ and $\mathbf{r}$ has a linear cost in n (bits of ranges). Using the IPA protocol with $\mathbf{l}$ and $\mathbf{r}$ becoming witnesses and some adjustments explained in Sect. 4.2 of [4], we obtain a logarithmic proof size in n. Moreover, an *aggregated range proof* can be used to perform one proof for m values. This implies some modifications, in particular the prover computes $\mathbf{a}_L \in \mathbb{Z}_p^{m \cdot n}$ such that $\langle 2^n, \mathbf{a}_{L[(j-1) \cdot n : j \cdot n - 1]} \rangle = v_j$ for all $j \in [1, m]$. The prover modifies the constructions of $l(X)$ and $r(X)$ such that they stay in $\mathbb{Z}_p^{m \cdot n}[X]$ and τ_x to include the randomness of each V_j commitment of the v_j value. Finally, verifier modifies the verification of $\hat{t}$ to include all the V_j commitments and the verification of A, S commitments respect to the new $\mathbf{r} \in \mathbb{Z}_p^{m \cdot n}$.

Using Bulletproofs in MTproof. We now outline the modifications of the Bulletproofs protocol required in *MPproof*, in order to prove that each transfer amount a_i of the vector $\mathbf{a} = (a_1, \ldots, a_n)$ is non-negative and the sender remaining balance b' after the transfer is also non-negative. More specifically, we generate an aggregated range proof valid for a vector of m values, where the first value is b' and the $m - 1$ remaining values are those in $\mathbf{a}$. The range domain within which these values must be proved valid is $[0, 2^n - 1]$. The protocol is initiated by the prover creating the commitments $A = h^\alpha \cdot \mathbf{g}^{\mathbf{a}_L} \cdot \mathbf{h}^{\mathbf{a}_R} \in \mathbb{G}$ and $S = h^\rho \cdot \mathbf{g}^{\mathbf{s}_L} \cdot \mathbf{h}^{\mathbf{s}_R} \in \mathbb{G}$, where $\mathbf{g}, \mathbf{h}$ are vectors of generators, $h \in \mathbb{G}$ is the blinding generator and α, ρ the blinding random values. The commitment A to the vectors $\mathbf{a}_L$ and $\mathbf{a}_R$ is generated from the binary representation of b' and $\mathbf{a}$, respectively: $\langle \mathbf{a}_{L[:n]}, 2^n \rangle = b'$ and $\langle \mathbf{a}_{L[(j-1) \cdot n : j \cdot n]}, 2^n \rangle = a_{j-1} \ \forall j \in [2, m]$. Then, by subtracting from $\mathbf{a}_L$ the vector of powers of 1 we obtain: $\mathbf{a}_R = \mathbf{a}_L - \mathbf{1}^{m \cdot n} \in \{0, -1\}^{m \cdot n}$. The commitment S is a commitment to the randomly sampled blinding terms in $\mathbf{s}_L, \mathbf{s}_R \in \mathbb{Z}_p^{m \cdot n}$. The prover then sends these commitments to the verifier, which responds with the random challenges y, z. The prover then constructs the commitments: $T_1 = g^{t_1} \cdot h^{\tau_1} \in \mathbb{G}$ and $T_2 = g^{t_2} \cdot h^{\tau_2} \in \mathbb{G}$. Both commitments consist of two uniformly random values $\tau_1, \tau_2 \in \mathbb{Z}_p$ and the coefficients t_1, t_2 of the polynomial $t(X) = t_0 + t_1 \cdot X + t_2 \cdot X^2 \in \mathbb{Z}_p[X]$. Such $t(X)$ polynomial is derived from the inner product of two polynomials $l(X), r(X) \in \mathbb{Z}_p^{m \cdot n}[X]$, both built from a linear combination of the challenges of the verifier and the vectors $\mathbf{a}_L, \mathbf{a}_R, \mathbf{s}_L, \mathbf{s}_R$. After receiving the commitments T_1 and T_2 from the prover, the verifier responds with the randomly sampled challenge x. The prover calculates the polynomial evaluation $\hat{t} = t(x)$, and two blinding factors, one for $\hat{t}$ and one for the commitments A and S, respectively: $\tau_x = \tau_1 \cdot x + \tau_2 \cdot x^2 \in \mathbb{Z}_p$ and $\mu = \alpha + \rho \cdot x \in \mathbb{Z}_p$. Simply replacing Pedersen commitment with ElGamal encryption is not sufficient at this point. In Bulletproofs, the equality to be verified, which includes commitments V, T_1 and T_2, exploits the additive homomorphic property of the Pedersen commitment, and this cannot be done with ElGamal encryptions under different keys. We follow the same intuition of Zether to combine Bulletproofs with a Σ-Protocol: instead of giving to the verifier the opening of polynomial commitments to $t(X)$, knowledge of the opening is proved. This corresponds to proving knowledge of the blinding value τ_x for the commitments $T_{1,2}$. To include this statement, our proof system now generates a proof for the following *zero-knowledge multi-transfer relation* (formally):

$$Rel_{ConfMultiTransfer} : \{ (y, \bar{y}, C_L, C_R, \mathbf{C}, \bar{\mathbf{C}}, D, g, z, \hat{t}, \delta(y,z); sk, \mathbf{a}, b', r, \tau_x) :$$
$$(C_i = g^{a_i} y^r \wedge \bar{C}_i = g^{a_i} \bar{y}_i^r \wedge D = g^r)_{i=1}^n \wedge$$
$$C_L / \prod_{i=1}^n C_i = g^{b'} (C_R / \prod_{i=1}^n D)^{sk} \wedge y = g^{sk} \wedge \tag{1}$$
$$g^{\hat{t} - \delta(y,z) - b' \cdot z^2 - \sum_{i=1}^n a_i \cdot z^{2+i}} h^{\tau_x} = T_{1,2} \}$$

Hence, the prover also calculates a blinding commitment $A_t = g^{-k_{ab}} \cdot h^{k_\tau}$, where k_{ab} and k_τ are two uniformly random scalars. At this point, the prover sends the values $\hat{t}, \mu$ and A_t to the verifier, and after receiving the random challenge c from the verifier, the prover generates and sends the scalars s_{ab} and s_τ such that: $s_{ab} = k_{ab} + c \cdot (b' \cdot z^2 + \sum_{i=1}^{m-1} a_i \cdot z^{2+i}) \in \mathbb{Z}_p$ and $s_\tau = k_\tau + c \cdot \tau_x \in \mathbb{Z}_p$. The verifier uses them in the final proof of knowledge check: $g^{c(\hat{t} - \delta(y,z)) - s_{ab}} \cdot h^{s_\tau} \overset{?}{=} A_t T_1^{c \cdot x} T_2^{c \cdot x^2}$ where $\delta(y,z) = (z - z^2) \cdot \langle \mathbf{1}^{m \cdot n}, \mathbf{y}^{m \cdot n} \rangle - \sum_{j=1}^m z^{2+j} \cdot \langle \mathbf{1}^{m \cdot n}, \mathbf{2}^{m \cdot n} \rangle$. From this point, *MTproof* follows with the IPA protocol to prove that the inner product $\hat{t} = \langle \mathbf{l}, \mathbf{r} \rangle$ of the committed vectors $\mathbf{l}$ and $\mathbf{r}$ is valid. We use the logarithmic-sized proof optimization and the verifier *multi-exponentiation* technique from the IPA protocol of Bulletproofs. Hence, the prover and verifier engage in the IPA protocol on the inputs $(\mathbf{g}, \mathbf{h}', Ph^{-\mu}, \hat{t}; \mathbf{l}, \mathbf{r})$, where $\mathbf{h}' = (h_1, h_2^{y^{-1}}, h_3^{y^{-2}}, ..., h_{m \cdot n}^{y^{-m \cdot n+1}})$ and $P = AS^x \cdot \mathbf{g}^{-z} \cdot \mathbf{h}^{z \cdot \mathbf{y}^{m \cdot n}} \cdot \prod_{j=1}^m h_{[(j-1) \cdot n: j \cdot n]}^{z^{1+j} \cdot \mathbf{2}^n}$.

Using Σ-Protocol in MTproof. We present the remaining part of *MTproof* relating to the Σ-protocols. In particular, we design four Σ-proofs, each for one of the equalities in conjunction in relation (1), except for the last one (the Bulletproofs proof of knowledge of the opening of the polynomial commitment). In the following, we parse the statements and we show the corresponding Σ-proof.

Σ-proof-sk for statement: the sender knows a secret key sk for which the respective public key y encrypts the values in $\mathbf{C}$ and the public key is well-formed

---

1: **input**: $(g, y \in \mathbb{G};\ sk \in \mathbb{Z}_p) : y = g^{sk}$
2:     $\mathcal{P}$'s input : (g, sk)
3:     $\mathcal{V}$'s input : (g, y)
4: **output**: $\{\mathcal{V}$ accepts or $\mathcal{V}$ rejects$\}$
5: $\mathcal{P} \rightarrow \mathcal{V} : A_y$ where $A_y = g^{k_{sk}} \in \mathbb{G}$ and $k_{sk} \xleftarrow{\$} \mathbb{Z}_p$
6: $\mathcal{V} \rightarrow \mathcal{P} : c \xleftarrow{\$} \mathbb{Z}_p$
7: $\mathcal{P} \rightarrow \mathcal{V} : s_{sk}$ where $s_{sk} = k_{sk} + c \cdot sk \in \mathbb{Z}_p$
8: $\mathcal{V} : g^{s_{sk}} \overset{?}{=} A_y y^c \in \mathbb{G}$ if yes, $\mathcal{V}$ accepts; otherwise rejects

---

Σ-proof-r for statement: the sender knows a randomness r used in the encryption

---

1: **input**: $(g, D \in \mathbb{G};\ r \in \mathbb{Z}_p) : D = g^r$
2:     $\mathcal{P}$'s input : (g, r)
3:     $\mathcal{V}$'s input : (g, D)
4: **output**: $\{\mathcal{V}$ accepts or $\mathcal{V}$ rejects$\}$
5: $\mathcal{P} \rightarrow \mathcal{V} : A_D$ where $A_D = g^{k_r} \in \mathbb{G}$ and $k_r \xleftarrow{\$} \mathbb{Z}_p$
6: $\mathcal{V} \rightarrow \mathcal{P} : c \xleftarrow{\$} \mathbb{Z}_p$
7: $\mathcal{P} \rightarrow \mathcal{V} : s_r$ where $s_r = k_r + c \cdot r \in \mathbb{Z}_p$
8: $\mathcal{V} : g^{s_r} \overset{?}{=} A_D D^c \in \mathbb{G}$ if yes, $\mathcal{V}$ accepts; otherwise rejects

---

Σ-proof-ab for statement: the sender balance cannot be overdraft, i.e., the sender remaining balance is equal to the subtraction between the sender balance and all of the $(m-1)$ transfer amounts in $\mathbf{C}$

1: **input**: $(g, D, C_L, C_R \in \mathbb{G},\ \mathbf{C} \in \mathbb{G}^{m-1};\ sk, b' \in \mathbb{Z}_p, \mathbf{a} \in \mathbb{Z}_p^{m-1})$:
    $C_L / \prod_{i=1}^{m-1} C_i = g^{b'} (C_R / \prod_{i=1}^{m-1} D)^{sk}$
2:     $\mathcal{P}$'s input : $(g, D, C_R, sk, \mathbf{a}, b')$
3:     $\mathcal{V}$'s input : $(g, D, C_L, C_R, \mathbf{C})$
4: **output**: $\{\mathcal{V}$ accepts or $\mathcal{V}$ rejects$\}$
5: $\mathcal{V} \rightarrow \mathcal{P} : z \xleftarrow{\$} \mathbb{Z}_p^*$
6: $\mathcal{P}$ computes:
7:     $k_{sk}, k_{ab} \xleftarrow{\$} \mathbb{Z}_p$
8:     $A_{ab} = ((C_R / \prod_{i=1}^{m-1} D)^{z^2} \cdot \prod_{i=1}^{m-1} D^{z^{i+2}})^{k_{sk}} \cdot g^{k_{ab}} \in \mathbb{G}$
9: **end** $\mathcal{P}$
10: $\mathcal{P} \rightarrow \mathcal{V} : A_{ab}$
11: $\mathcal{V} \rightarrow \mathcal{P} : c \xleftarrow{\$} \mathbb{Z}_p$
12: $\mathcal{P}$ computes:
13:     $s_{sk} = k_{sk} + c \cdot sk \in \mathbb{Z}_p$
14:     $s_{ab} = k_{ab} + c \cdot (b' z^2 + \sum_{i=1}^{m-1} (a_i z^{i+2})) \in \mathbb{Z}_p$

15: **end** $\mathcal{P}$

16: $\mathcal{P} \to \mathcal{V} : s_{sk}, s_{ab}$

17: $\mathcal{V} : g^{s_{ab}}((\frac{C_R}{\prod_{i=1}^{m-1} D})^{z^2} \cdot \prod_{i=1}^{m-1} D^{z^{2+i}})^{s_{sk}} \stackrel{?}{=} A_{ab}((\frac{C_L}{\prod_{i=1}^{m-1} C_i})^{z^2} \cdot \prod_{i=1}^{m-1} C_i^{z^{2+i}})^c \in \mathbb{G}$

18:   if yes, $\mathcal{V}$ accepts; otherwise rejects

Σ-proof-y for statement: the i-th values in both $\mathbf{C}$ and $\bar{\mathbf{C}}$ are well-formed and correspond to the encryption of the i-th amount to be transferred

1: **input**: $(y \in \mathbb{G}, \bar{\mathbf{y}}, \mathbf{C}, \bar{\mathbf{C}} \in \mathbb{G}^{m-1}; \ r \in \mathbb{Z}_p)$:
   $(C_i = g^{a_i} y^r \wedge \bar{C}_i = g^{a_i} \bar{y}_i^r \wedge D = g^r)_{i=1}^{m-1}$

2:      $\mathcal{P}$'s input : $(y, \bar{\mathbf{y}}, r)$

3:      $\mathcal{V}$'s input : $(y, \bar{\mathbf{y}}, \mathbf{C}, \bar{\mathbf{C}})$

4: **output**: $\{\mathcal{V}$ accepts or $\mathcal{V}$ rejects$\}$

5: $\mathcal{P} \to \mathcal{V} : A_{\bar{y}}$ where $A_{\bar{y}} = \prod_{i=1}^{m-1}(y \cdot \bar{y}_i^{-1})^{k_r} \in \mathbb{G}$ and $k_r \stackrel{\$}{\leftarrow} \mathbb{Z}_p$

6: $\mathcal{V} \to \mathcal{P} : c \stackrel{\$}{\leftarrow} \mathbb{Z}_p$

7: $\mathcal{P} \to \mathcal{V} : s_r$ where $s_r = k_r + c \cdot r \in \mathbb{Z}_p$

8: $\mathcal{V} : \prod_{i=1}^{m-1}(y \cdot \bar{y}_i^{-1})^{s_r} \stackrel{?}{=} A_{\bar{y}} \cdot (\prod_{i=1}^{m-1} C_i / \bar{C}_i)^c$

9:   if yes, $\mathcal{V}$ accepts; otherwise rejects

---

4   MTproof Implementation and Evaluation

In this section, we evaluate *MTproof* using our code implementation in Rust and the arkworks [2] library suites. The source code of *MTproof* can be found on GitHub [10]. The elliptic curve, underlying all the operations on elements of $\mathbb{G}$, is the Barreto-Naehrig curve **BN-254**. The modular structure of *MTproof* allows us to conduct accurate and modular benchmarks to estimate each execution time for the prover and verifier functions and the associated proofs' sizes. The results of the evaluations are shown in Table 1. The benchmarks are executed multiple times with different values for n and m, equal to powers of two. All the measurements are carried on a machine with an *Intel Core i7-10750H* (12 threads and 6 cores at 2.60GHz, with turbo frequencies at 5.00GHz) CPU and 16 GB of RAM, running the Rust compiler.

Comparison with Concurrent Work and Results. We consider one concurrent work, Anonymous Zether in [9]. Anonymous Zether enhances the Basic Zether scheme by introducing the *many-out-of-many* primitive to build an anonymity set for the unlinkability of addresses. Proof size, proving time, and verifying time are provided with respect to the growth of the anonymity set. These evaluations are carried out by setting a fixed number of two aggregate 32-bit range values. Compared with *MTproof*, it is possible to notice the same logarithmic growth with the increase of m (aggregate values) and the increase of the anonymity set size, as shown in Fig. 1 for the proof size. Similar observation also applies to the proving and verifying times. Hence, for the considerations that follow, we consider our proof system evaluation as the lower limit $\Omega(\log M)$,

Table 1. *MTproof* evaluation results

n	m	Proving time (ms)	Verifying time (ms)	Proof size (bytes)
16	2	534	202	1,584
16	4	1,000	379	1,712
16	8	2,030	718	1,840
16	16	3,941	1,389	1,968
16	32	7,815	2,680	2,096
16	64	15,192	5,327	2,224
32	2	967	348	1,712
32	4	1,891	686	1,840
32	8	3,753	1,329	1,968
32	16	7,431	2,626	2,096
32	32	14,844	5,259	2,224
32	64	30,052	10,460	2,352
64	2	1,899	679	1,840
64	4	3,762	1,320	1,968
64	8	7,496	2,619	2,096
64	16	14,980	5,218	2,224
64	32	29,794	10,478	2,352
64	64	61,430	21,533	2,480

Proof size (log)

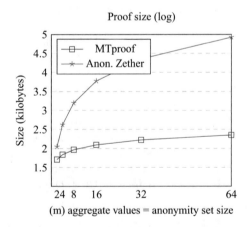

Fig. 1. Proof size comparison *MTproof* and *Anon. Zether*.

where $M = m * k$ with m the number of aggregate values and $n = k$ the bit range constant factor. The benchmark results in Table 1 highlight that as the number of aggregated values grows, hence m, the execution times and proof sizes become more and more convenient than those resulting from multiple and separate executions of the proving or verifying functions. Therefore, a consideration arises of how convenient it is to provide an aggregate proof for multiple values instead of providing proof for each of those values. For instance, considering a 64 bit range, the time required to generate a proof for multiple aggregated values (from a number of 2 up to 64 values) is between $33,97\%$ and $48,65\%$ smaller than the time required to generate one proof for each of such values. In the same way, the time required to verify a proof for multiple aggregated values is between $35,20\%$ and $49,66\%$ smaller than the time required to verify the proofs individually. Furthermore, considering the same bit range, the size of an aggregated proof is between $64,35\%$ and $97,86\%$ smaller than all single proofs combined together.

5   Conclusion and Future Work

MTproof is the zero-knowledge proof system aimed at realizing the Multi-Transfer, proposing aggregation, correctness and confidentiality in transactions. The system is based on Bulletproofs and Σ-protocol proof systems and has been implemented according to the tools of the research community. The evaluations highlight benefits from the aggregation, in terms of generation and verification times, and size of the proof. As future work, an optimization can be integrated into MTproof in its IPA component, which may lead to significant savings in execution time. Moreover, we will evaluate MTproof in real scenarios involving streams of sensor data [16,21,23].

Acknowledgment. We acknowledge the master student Francesco Pio Stelluti for his contribution to the codebase of MTproof.

References

1. Alonso, K.M., et al.: Zero to monero (2020)
2. arkworks rs. arkworks
3. Bünz, B., Agrawal, S., Zamani, M., Boneh, D.: Zether: towards privacy in a smart contract world. In: Bonneau, J., Heninger, N. (eds.) FC 2020. LNCS, vol. 12059, pp. 423–443. Springer, Cham (2020). https://doi.org/10.1007/978-3-030-51280-4_23
4. Bünz, B., Bootle, J., Boneh, D., Poelstra, A., Wuille, P., Maxwell, G.: Bulletproofs: short proofs for confidential transactions and more. In: 2018 IEEE Symposium on Security and Privacy (SP), pp. 315–334. IEEE (2018)
5. Buterin, V., et al.: A next-generation smart contract and decentralized application platform
6. Butler, D., Aspinall, D., Gascón, A.: On the formalisation of σ-protocols and commitment schemes. In: POST, pp. 175–196 (2019)
7. Chan, W., Olmsted, A.: Ethereum transaction graph analysis. In: 2017 12th International Conference for Internet Technology and Secured Transactions (ICITST), pp. 498–500. IEEE (2017)
8. Corradini, F., Mostarda, L., Scala, E.: ZeroMT: multi-transfer protocol for enabling privacy in off-chain payments. In: Barolli, L., Hussain, F., Enokido, T. (eds.) AINA 2022. LNNS, vol. 450, pp. 611–623. Springer, Cham (2022). https://doi.org/10.1007/978-3-030-99587-4_52
9. Diamond, B.E.: Many-out-of-many proofs and applications to anonymous zether. In: 2021 IEEE Symposium on Security and Privacy (SP), pp. 1800–1817. IEEE (2021)
10. EmanueleSc. Zeromt

11. Fauzi, P., Meiklejohn, S., Mercer, R., Orlandi, C.: Quisquis: a new design for anonymous cryptocurrencies. In: Galbraith, S.D., Moriai, S. (eds.) ASIACRYPT 2019. LNCS, vol. 11921, pp. 649–678. Springer, Cham (2019). https://doi.org/10.1007/978-3-030-34578-5_23
12. Fiat, A., Shamir, A.: How to prove yourself: practical solutions to identification and signature problems. In: Odlyzko, A.M. (ed.) CRYPTO 1986. LNCS, vol. 263, pp. 186–194. Springer, Heidelberg (1987). https://doi.org/10.1007/3-540-47721-7_12
13. Fleder, M., Kester, M.S., Pillai, S.: Bitcoin transaction graph analysis. arXiv preprint arXiv:1502.01657 (2015)
14. Guan, Z., Wan, Z., Yang, Y., Zhou, Y., Huang, B.: Blockmaze: an efficient privacy-preserving account-model blockchain based on zk-snarks. IEEE Trans. Dependable Secure Comput. (2020)
15. Jivanyan, A.: Lelantus: towards confidentiality and anonymity of blockchain transactions from standard assumptions. IACR Cryptol. ePrint Arch. **2019**, 373 (2019)
16. Mehmood, N.Q., Culmone, R., Mostarda, L.: Modeling temporal aspects of sensor data for MongoDB NoSQL database. J. Big Data **4**(1) (2017)
17. Nakamoto, S.: Bitcoin: a peer-to-peer electronic cash system. Technical report, Manubot (2019)
18. Poelstra, A.: Mimblewimble (2016)
19. Ron, D., Shamir, A.: Quantitative analysis of the full bitcoin transaction graph. In: Sadeghi, A.-R. (ed.) FC 2013. LNCS, vol. 7859, pp. 6–24. Springer, Heidelberg (2013). https://doi.org/10.1007/978-3-642-39884-1_2
20. Rondelet, A., Zajac, M.: Zeth: on integrating zerocash on ethereum. arXiv preprint arXiv:1904.00905 (2019)
21. Russello, G., Mostarda, L., Dulay, N.: A policy-based publish/subscribe middleware for sense-and-react applications. J. Syst. Softw. **84**(4), 638–654 (2011)
22. Sasson, E.B., et al.: Zerocash: decentralized anonymous payments from bitcoin. In: 2014 IEEE Symposium on Security and Privacy, pp. 459–474. IEEE (2014)
23. Vannucchi, C., et al.: Symbolic verification of event–condition–action rules in intelligent environments. J. Reliable Intell. Environ. **3**(2), 117–130 (2017)

MTA Extension for User-Friendly Enforcement of Mandatory TLS Encryption

Max Körber[1], Leo V. Dessani[2], and Ronald Petrlic[1(✉)]

[1] Nuremberg Institute of Technology, Nürnberg, Germany
{koerberma64258,ronald.petrlic}@th-nuernberg.de
[2] Saarland University, Saarland Informatics Campus, Saarbrücken, Germany
leo.dessani@uni-saarland.de

Abstract. Since TLS is not yet available on all Mail Transfer Agents (MTA), mandatory TLS encryption for SMTP connections could result in single emails not being delivered. Therefore, TLS encryption for SMTP connections is mainly negotiated opportunistically using STARTTLS. As a missing or weak TLS configuration of the target MTA cannot be influenced by the sender in practice, some emails may be sent unencrypted or poorly encrypted without additional protective measures such as end-to-end encryption. This is problematic when transmitting sensitive content by email, such as personal data, and the reason why laws, e.g. privacy laws as the General Data Protection Regulation (GDPR), require encrypted transmission of such content. Furthermore, STARTTLS does not provide the sender with the ability to specify the technical security requirements for email delivery. We therefore propose a concept to extend an MTA that allows senders to select from predefined confidentiality levels for emails with sensitive content in a user-friendly way. Each confidentiality level is linked to technical security requirements that must be met when establishing an SMTP connection for sending an email to the target MTA. In this way, the legal requirements are met. To this end, we present an implementation of our concept for the Postfix MTA which does not require any modifications to an email client.

1 Introduction

According to Google's statistics [6], only 87% of all outgoing emails were transmitted TLS-encrypted via Gmail servers in 2022. Some MTAs are still configured without TLS. Enforcing TLS encryption for SMTP connections is therefore currently not practical, as some emails might not be delivered.

TLS encryption in SMTP connections is typically negotiated opportunistically, i.e. using STARTTLS [2]. If a target MTA is not configured to use TLS or uses deprecated TLS versions, like TLS 1.0 and 1.1 [11], an email is usually delivered unencrypted or poorly TLS-encrypted with weak algorithms. When transmitting sensitive content via email, this is problematic.

© The Author(s), under exclusive license to Springer Nature Switzerland AG 2023
L. Barolli (Ed.): AINA 2023, LNNS 654, pp. 473–486, 2023.
https://doi.org/10.1007/978-3-031-28451-9_42

One of the key challenges in email security is that ensuring confidentiality cannot be influenced by the sender or the source MTA only. Both the source and the target MTA need to implement secure TLS versions, like TLS 1.2 and 1.3, to be able to negotiate a strongly encrypted connection. Since senders cannot influence the configuration of the target MTA, to ensure confidentiality, they must prevent emails with sensitive content from being sent to target MTAs that do not support (strong) TLS encryption. End-to-end encryption (E2EE) with S/MIME or PGP is not always practical because it is rarely used and requires that the receiver has it set up [13,17]. Therefore, another approach is needed. Encrypting communication is not only in the sender's own interest to ensure confidentiality during email transmission, but is also required by law, e.g. by the European GDPR [4]. According to Art. 32 (1) GDPR, technical security requirements to secure communications must be taken as soon as personal data is processed. This applies to all controllers within the meaning of Art. 4 (7) GDPR that process personal data.

2   Related Work

Both SMTP via DNS-based Authentication of Named Entities (DANE) [1] and MTA Strict Transport Security (MTA-STS) [9] ensure confidentiality in SMTP connections by preventing downgrade attacks and session hijacking in TLS encryption. However, the sender cannot influence the configuration of the target MTA. If a target MTA does not implement DANE or MTA-STS, there is no guarantee that an email will be transmitted TLS-encrypted.

Some email providers, like Posteo [14,15], offer TLS guarantees for receiving and sending emails. The providers' mail servers accept incoming and outgoing SMTP connections only if they are TLS-encrypted. However, these guarantees do not allow the sender to distinguish between emails with sensitive and non-sensitive content. They can only be configured globally for all incoming and outgoing emails. If a sender wants to send a non-sensitive email to a target MTA without TLS configured, this will no longer be possible.

Require TLS [5] is very similar to our approach and allows senders to define which emails must be transmitted using TLS encryption. If the email contains sensitive content, the sender flags it in the email client before sending it to the source MTA. The source MTA then negotiates a TLS encrypted SMTP connection to the target MTA and delivers the email. Require TLS requires the MX record of the receiving domain to be validated via a DNSSEC signature or by a MTA-STS policy. The X.509 certificate presented by the target MTA must be successfully verified by a trust chain or using DANE. In contrast to our approach, not only the source MTA must be extended, but also the target MTA and the mail client where the email can be flagged as sensitive. However, only a few implementations exist for little-known MTAs, such as for *MDaemon* [10], which makes Require TLS currently impractical. Also, there is no popular mail client for which Require TLS is implemented.

3   Our Contribution

Similar to Require TLS, our approach allows senders to define a (fine-grained) confidentiality level for an email, which the source MTA guarantees by requiring the proper mechanisms to support the chosen level. However, in contrast to Require TLS, our approach neither requires the target MTA to be adapted, nor the email client. Our approach best suits the requirement for email security in practice. Alice, who wants to send an email with sensitive content to Bob, needs to have assurance beforehand that her email will be transmitted to Bob encrypted. However, she wants to keep the whole process as simple as possible, i.e. she does not want to deal with S/MIME or PGP, or any other means that require further actions on her or on Bob's side. If she is lucky, Bob's MTA is properly configured and she can send the email without any further means. It is fully *transparent* beforehand to her whether this is the case or not. Only if it is not, i.e. Bob's MTA is not properly configured, she needs to perform further steps in order for the email to be transmitted in a secure way. Our approach is an alternative to E2EE for sensitive content that meets the GDPR's requirements.

4   Concept

Our concept allows the sender to select from predefined confidentiality levels for emails in a user-friendly way. Each confidentiality level is linked to technical security requirements that must be met when establishing an SMTP connection from the source MTA to the target MTA. Our approach focuses on sending emails with sensitive content and assumes that the email has one sender, one receiver, and the receiver does not forward the email. The sending process is performed by a source MTA trusted by the sender without using any email gateway or proxy. No assumption can be made about the target MTA.

4.1   Confidentiality Levels

We define two confidentiality levels and their technical requirements:

Secure: Requires the use of state-of-the-art TLS encryption in SMTP connections, currently TLS 1.2 and 1.3 [11], with a valid X.509 certificate, and SMTP via DANE. We do not include MTA-STS because it does not require DNSSEC signatures for resource records and is therefore vulnerable to DNS spoofing attacks [9]. This confidentiality level allows to renounce E2EE even for sending sensitive personal data via email.[1]

Normal: Default confidentiality level. Requires the same requirements as secure, but may also downgrade to TLS 1.0 or 1.1 and STARTTLS if the target MTA does not support stronger encryption. If TLS encryption is not possible at all, emails marked as normal may also be sent unencrypted. This confidentiality level represents the default case for email delivery today.

---

[1] This is approved by the German Data Protection Conference [12].

4.2 Confidentiality Level Selection

The choice of the confidentiality level can be made at different points. To keep the implementation efforts and the dependencies between email client and source MTA as low as possible, we describe different approaches for specifying the confidentiality level and argue why a keyword in the subject line is best:

SMTP Service Extension: The widespread establishment of an SMTP service extension is unlikely. In addition, both the email client and the source MTA must be extended. Since the Require TLS standard has not found widespread establishment yet, we do not develop a new SMTP service extension.

Email Headers: Implementing the confidentiality level with email headers requires changes to existing email clients. Thus, we do not favor this approach.

File Attachments: In case that content encryption with S/MIME or PGP is used, the confidentiality level selection cannot be made via a file attachment, since it will be unreadable for the source MTA. Also, it is cumbersome to add an attachment to every email.

Keyword in the Subject: Adding a keyword to the subject is a quick and easy approach. No changes need to be made to the email client. Therefore, we decided to use a keyword in curly brackets in the subject, as shown in Listing 1.1.

```
From: alice@example.com
To: bob@example.org
Subject: {secure} Confidential document

Please find the super confidential document attached
    ↪ to this email.
```

Listing 1.1. The confidentiality level is selected using the keyword *secure* in curly brackets at the beginning of the subject.

4.3 Checking the Target MTA

The check of the target MTA can be done at different points in the sending process. Due to the possible different configurations of the email client and the source MTA and because the source MTA is responsible for email delivery, we decided to perform the check in the source MTA instead of the email client. For example, if there are multiple MX records for a target domain, the email client and source MTA may differ in that different MX records are checked. In this case, the result of a check in the email client is useless because the source MTA checked another target MTA and will deliver the email at the end. Thus, implementing the check at the source MTA is the safer option.

4.4   Procedure of the Concept

Figure 1 shows the concept's structure. The single steps are described below.

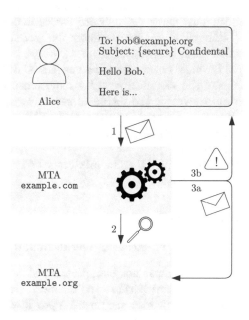

Fig. 1. Our concept allows the sender to specify a confidentiality level for an email.

Step 1: First, the sender specifies the confidentiality level in the subject line in the email client. The sender can enter the keyword *secure* before the subject or enter the subject directly to choose the default confidentiality level (normal).

Step 2: After the email has been sent to the source MTA, the latter evaluates the confidentiality level and starts checking the target MTA.

Step 3a: If the technical security requirements of the selected confidentiality level are met, the email will be sent to the target MTA.

Step 3b: If the technical security requirements of the selected confidentiality level are not met, the email will not be sent to the target MTA and the sender will be notified. For this purpose, a notification email is sent to the sender, like the SMTP service extension DSN does. The notification should contain a meaningful reason and refer to the original email. If the DSN extension is used for this purpose, code 550 must be used. This results from the description of the codes in RFC 5321 [7], since in this case the transmission will fail permanently.

4.5 Functional Requirements

The following functional requirements result from the previous sections.

- **(F1)** The sender must be able to select a confidentiality level for an email. Therefore, the sender must enter the keyword *secure* in curly brackets before the real subject (Listing 1.1) or directly enter the real subject to select the default confidentiality level (*normal*).
- **(F2)** The source MTA must ensure that the negotiated SMTP connection to the target MTA meets the technical security requirements according to the selected confidentiality level. If the technical security requirements are not met, the email will not be delivered to the target MTA.
- **(F3)** An extension of the confidentiality levels must be supported, e.g. to cover further developments.
- **(F4)** The sender must be notified if an email cannot be sent because technical security requirements are not met.

4.6 Non-functional Requirements

The following non-functional requirements also result from the previous sections.

- **(N1)** The sending of emails must not be noticeably slowed down.
- **(N2)** If an email cannot be sent because the technical security requirements are not met, a meaningful error message must be provided in the notification to the sender that refers to the original email.
- **(N3)** The implementation must be developed for a widely used and open source MTA.

5 Implementation

In this section we describe the implementation of the concept. We decided to implement our concept as an extension instead of a source code change, because an extension can easily be added to existing MTA installations.

5.1 Postfix Characteristics

We implemented our extension for the Postfix MTA since it is a popular open-source MTA [3] and offers several extension options. Postfix consists of small programs and components, which work independently. Figure 2 shows the rough structure of Postfix [16]. In general, an email client delivers the email via the *smtpd* component. The *cleanup* component then routes the email to the *incoming queue*. The *smtp* component sends the email from the queue to the target MTA. The three components *smtpd*, *sendmail*, and *smtp* represent interfaces to the outside, whereas the others are internal components of Postfix.

The architecture of Postfix provides two possible filter methods for our extension. One filter is called *Mail filter (Milter)* [16], which is executed while an email

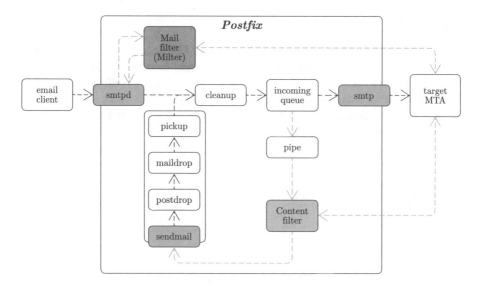

Fig. 2. Architecture of postfix: various filtering methods for checking. The Mail filter (Milter) filter is executed during the reception of an email (orange). The Content filter is executed as a standalone component (green).

is received. Therefore, the entire email cannot be used as input for the filter because parts of the email are still being transmitted. The filter has a limited execution time of 30 s [16]. To check the target MTA, our extension needs to query several resource records and needs to establish an SMTP connection to the target MTA to verify if it meets the technical security requirements. Normally, this will be completed within the timeout of 30 s, but there may be some cases, especially when extending the concept to multiple receivers, where the limit may be exceeded. Therefore, the filter is not suitable for our concept.

Another filter for Postfix is called *Content filter* [16] which is used for content-based checks. Emails are passed to the filter via *pipe* before being sent by *smtp*. After the check of the target MTA, emails are returned to the queue via *sendmail* and other internal components like *postdrop*. The disadvantage for our concept is that the email client first successfully transfers the email to the source MTA and only then the check of the target MTA is performed. In case of a Milter, the transmission to the source MTA is aborted and the email client can directly display an error message.

We decided to perform the check of the target MTA using a Content filter. This offers the advantage of a more flexible solution, since no time limits have to be met. The notice to the sender can still be sent via an email.

5.2   Filter Script

Our Postfix extension runs a script on a source MTA that checks a target MTA for technical security requirements. If the requirements are met, the email will be delivered, otherwise the delivery will be stopped. By using a filter, the script is executed for each email. Our filter script is written in Bash and can be found on Github: https://github.com/email-security/scripts. The procedure of the script is shown in Fig. 3 and is explained below.

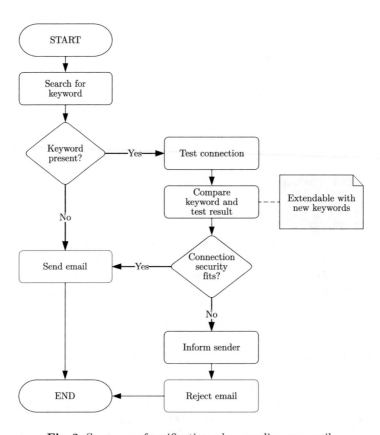

Fig. 3. Sequence of verification when sending an email.

When a new email is sent from the email client to the source MTA, the subject is searched for a confidentiality level keyword. The keyword is then removed and not transmitted to the target MTA. If the source MTA uses Domain Keys Identified Mail (DKIM), the DKIM signature is generated by the source MTA after the filter script was executed (after the keyword was removed), so the email can contain a valid DKIM signature. In case that no keyword is specified, the source MTA assumes the default confidentiality level (*normal*).

Next, the source MTA establishes an SMTP connection to the target MTA using the program *posttls-finger*. This program is part of Postfix and provides information about the specific encryption details of the target MTA. These details contain information about supported TLS versions, X.509 certificate validity, and if the target MTA supports SMTP via DANE. *posttls-finger* is used because it is available on every system with Postfix.

If the technical security requirements are met, the email is returned to Postfix via *sendmail*. Otherwise, an error message is sent to the sender and email delivery is stopped by indicating an error code. The error message contains a detailed description with reasons why the email could not be sent.

The script can be flexibly extended with new keywords. Therefore, it can cover different regulatory or area-specific requirements. For example, the TLS parameters can be checked more in detail or TLS encryption can be accepted without SMTP via DANE. In our implementation, the experimental keyword *encrypted* was added to demonstrate the extension possibilities. If this keyword is specified, emails can be sent without SMTP via DANE. However, TLS 1.0 or newer and a trusted X.509 certificate must be used for email delivery.

5.3   X.509 Certificate Check

The content of the certificate and its chain of trust is validated by the filter script. The trust chain is validated with a list of valid certificate authorities (CAs) passed with the *-P* parameter. In our implementation, the list originates from the *ca-certificates* debian package[2], which contains the CAs trusted by Mozilla. In case a confidentiality level requires DANE, the authenticity of the X.509 certificate is ensured by comparing the received certificate information with the data of the TLSA record. As already described in Sect. 4.1, a valid X.509 certificate must be used by the target MTA for every confidentiality level except for *normal*. Otherwise, the email delivery will be canceled.

6   Evaluation

6.1   Practical Tests

We have tested our implementation using different scenarios.

Secure Delivery with Successful Test. Alice wants to send an email with sensitive content and therefore wants to ensure confidentiality during transmission. To do this, she inserts the keyword {*secure*} into the subject and sends the email to the source MTA. The source MTA receives Alice's email. The *posttls-finger* program on the source MTA then establishes an SMTP connection to the target MTA and checks whether it meets the technical security requirements

---

[2] see https://packages.debian.org/en/sid/ca-certificates.

according to Sect. 4.1. As shown in Listing 1.2, the SMTP connection to the target MTA of the domain *example.org* is classified as a verified TLS connection. This means that the connection is secured with SMTP via DANE. In addition, TLS 1.3 is used. Therefore, Alice's email is sent because the technical security requirements for the selected confidentiality level are met.

```
posttls-finger -l dane-only -P /etc/ssl/certs/ -L summary -c --
    ↪ example.org
# Output:
# posttls-finger: Verified TLS connection established to mail.
    ↪ example.org[198.51.100.11]:25: TLSv1.3 with cipher
    ↪ TLS_AES_256_GCM_SHA384 (256/256 bits) key-exchange ECDHE
    ↪ (P-256) server-signature RSA-PSS (2048 bits) server-
    ↪ digest SHA256
```

Listing 1.2. Evaluation of the email server of `example.org` with DANE.

Secure Delivery with Negative Test. Alice again wants to send an email with sensitive content and therefore wants to ensure confidentiality during transmission. To do this, she inserts the keyword {*secure*} into the subject and sends the email to the source MTA. The source MTA receives Alice's email. The *posttls-finger* program on the source MTA then establishes an SMTP connection to the target MTA and checks whether it meets the technical security requirements according to Sect. 4.1. As shown in Listing 1.3, an encrypted TLS connection to the target MTA *example.net* could be established again. But, the target MTA does not support SMTP via DANE, which is why *posttls-finger* does not find a TLSA entry. Since Alice requested the email to be delivered with SMTP via DANE only, the source MTA does not deliver it to the target MTA.

Alice receives a notification about the failed delivery. Listing 1.4 shows the content of this notification. In addition to a general description, the individual reason for the failed attempt is listed for each receiver.

```
posttls-finger -l dane-only -P /etc/ssl/certs/ -L summary -c --
    ↪ example.net
# Output:
# posttls-finger: no TLSA records found, resorting to "secure"
# posttls-finger: Verified TLS connection established to mail.
    ↪ example.net[203.0.113.5]:25: TLSv1.3 with cipher
    ↪ TLS_AES_256_GCM_SHA384 (256/256 bits) key-exchange X25519
    ↪ server-signature RSA-PSS (2048 bits) server-digest
    ↪ SHA256
```

Listing 1.3. Evaluation of the email server of `example.net` without DANE.

```
This is a message from the mail server of example.com.

Unfortunately your email could not be sent to one or more
↪ receivers. Your email is attached below.

Please also note the following message(s).

For further assistance please send an email to
↪ postmaster@example.com. Please attach this email. You can
↪ remove your text from the undeliverable email.

                    Your mail server

<Bob@example.net>: service unavailable. Command output:
  It is not possible to establish a secure connection to the
    ↪ destination server of the domain example.net. Therefore
    ↪ , the email was not sent. Details: Domain example.net;
    ↪ Rating normal high; Keyword secure; Connection Verified
    ↪ ; TLS Version TLSv1.3
```

Listing 1.4. Feedback in case of a negative evaluation.

6.2 Fulfillment of Functional Requirements

The functional requirements from Sect. 4.5 have been fullfilled:

- **(F1)** Our implementation offers the sender the option of selecting a confidentiality level for the email by using the keyword *secure* in curly brackets at the beginning of the subject or by directly entering the real subject without additional keyword to address the default confidentiality level (*normal*).
- **(F2)** The filter script ensures that emails marked as *secure* are transmitted using state-of-the-art encryption, currently TLS 1.2 or 1.3, and SMTP via DANE. Emails marked as *normal* are transmitted using the same measures as *secure* if possible, but may also use TLS 1.0 or 1.1 and STARTTLS. If TLS encryption is not possible, they are sent unencrypted.
- **(F3)** Other confidentiality levels can be added to the filter script so that requirements by authorities from different countries and for different areas (e.g. special requirements for financial or health institutions) can be met.
- **(F4)** The filter script notifies the sender by email if an email cannot be sent due to non-fulfillment of confidentiality requirements.

6.3 Fulfillment of Non-functional Requirements

The non-functional requirements from Sect. 4.6 have been fullfilled, too:

- **(N1)** The filter script does not noticeably slow down the sending of emails. The average duration for sending 500 emails of different confidentiality levels from the email client to the target MTA was determined. It was ensured that the receivers could comply with the required technical security requirements. Otherwise, the durations would not be comparable. The timestamps logged in the emails by the email client and the target MTA were used for the calculation. The averaged time was 1.15 s. Therefore, we have no slowdown of the filter script. However, these tests did not take into account different email sizes, etc.
- **(N2)** The filter script provides the sender with a meaningful error message if an email cannot be sent due to non-fulfillment of confidentiality requirements and the error message refers to the original email.
- **(N3)** The filter script is compatible with the widely used and open source Postfix MTA.

7   Discussion

7.1   Separation of Requirements Checking and Email Delivery

The current filter script implementation has the disadvantage that the check of the technical security requirements is done before the email delivery. This leads to some problems:

If the target email domain has multiple MX records or uses different target MTAs with load balancing, it would be possible for an email to be sent to a different target MTA than the checked one. The other MTA could have a weak configuration and would not be suitable for transmitting a sensitive email. However, with load balancing, the target MTAs are usually configured identically.

Changes to the target MTA configurations exactly in the time between the check of the technical security requirements and the email delivery cannot be taken into account. On the one hand, emails could therefore be sent to a target MTA with a weaker configuration than tested. On the other hand, it could happen that a better target MTA configuration is not detected and the email is not delivered. However, it is very unlikely that a configuration change will occur in the short time between checking and sending.

To address these problems, it would be better to implement the check of the technical security requirements directly in the delivery process. This saves resources for verification of the target MTA before each email delivery. However, this optimization is not possible with a filter script, but only with a source code modification. In our eyes, this prevents widespread adoption in practice.

Additionally, it can happen that an X.509 certificate of a target MTA is valid when the technical security requirements are checked and expires or is revoked shortly thereafter. However, this is not particularly problematic because the certificate is normally checked with each SMTP connection, especially when using SMTP via DANE, and the email is normally not delivered in case of an expired or revoked certificate.

7.2   Extension with Additional Features

Our concept can be extended with additional features:

- An extension of the email client can be used for error-free use of the keywords for the confidentiality level. It can also be used to explain the meaning of the keywords to the sender.
- The MTA extension can check emails for content encryption and use this as a classifier for the technical security requirements of confidentiality levels. If the content is encrypted with S/MIME or PGP, the other encryption functions can be dispensed with, as confidentiality is already better ensured.
- The logging of the check of the target MTA can be archived in an audit-proof manner. In short, this means that the log is stored and protected from changes, so integrity is ensured. This is helpful if it has to be clarified at a later point in time whether an email was sent TLS encrypted, for example, in the case of accountability according to Art. 5 (2) GDPR.
- X.509 certificates of a specific target MTA can be manually marked as trustworthy so that this target MTA is trusted even without DANE. For this purpose, the certificates should be verified by other means.
- The selection of the confidentiality level could be automated by searching for sensitive content in emails.

8   Conclusion

Until mandatory TLS encryption (with SMTP via DANE) becomes standard in SMTP connections [8], the sender must ensure that sensitive content in emails is transmitted securely. For this reason, we have introduced a concept that allows the sender to select from predefined confidentiality levels, each linked to specific technical security requirements, such as TLS encryption with TLS 1.2 or 1.3 and SMTP via DANE. We found that a standardized approach with an SMTP service extension is not necessarily the best way to achieve an improvement in practice as we have seen that the Require TLS extension is not widely used. A filter extension for the Postfix MTA as well as the selection of the confidentiality level with a keyword in the subject line proves to be best suitable. The advantage of this approach is that it requires little implementation effort for the postfix MTA.

References

1. Dukhovni, V., Hardaker, W.: SMTP security via opportunistic DNS-based authentication of named entities (DANE) transport layer security (TLS). RFC 7672 (2015). https://doi.org/10.17487/RFC7672, https://www.rfc-editor.org/info/rfc7672
2. Durumeric, Z., et al.: Neither snow nor rain nor MITM... an empirical analysis of email delivery security. In: Proceedings of the 2015 Internet Measurement Conference, pp. 27–39 (2015)

3. E-Soft Inc.: Mail (MX) server survey (2022). http://www.securityspace.com/s_survey/data/man.202205/mxsurvey.html. Accessed 24 June 2022

4. European Commission: Regulation (EU) 2016/679 of the European Parliament and of the Council of 27 April 2016 on the protection of natural persons with regard to the processing of personal data and on the free movement of such data, and repealing Directive 95/46/EC (General Data Protection Regulation) (Text with EEA relevance) (2016). https://eur-lex.europa.eu/eli/reg/2016/679/oj

5. Fenton, J.: SMTP require TLS option. RFC 8689 (2019). https://doi.org/10.17487/RFC8689, https://www.rfc-editor.org/info/rfc8689

6. Google LLC: Email encryption in transit. https://transparencyreport.google.com/safer-email/overview. Accessed 22 June 2022

7. Klensin, D.J.C.: Simple mail transfer protocol. RFC 5321 (2008). https://doi.org/10.17487/RFC5321, https://www.rfc-editor.org/info/rfc5321

8. Lange, C., Chang, T., Fiedler, M., Petrlic, R.: An email a day could give your health data away. In: Garcia-Alfaro, J., Navarro-Arribas, G., Dragoni, N. (eds.) DPM CBT 2022 2022. LNCS, vol. 13619, pp. 53–68. Springer, Cham (2023). https://doi.org/10.1007/978-3-031-25734-6_4

9. Margolis, D., Risher, M., Ramakrishnan, B., Brotman, A., Jones, J.: SMTP MTA strict transport security (MTA-STS). RFC 8461 (2018). https://doi.org/10.17487/RFC8461, https://www.rfc-editor.org/info/rfc8461

10. MDaemon Technologies Ltd.: SMTP extensions. http://help.altn.com/mdaemon/en/ssl_starttls-required-list_2.html. Accessed 22 June 2022

11. Moriarty, K., Farrell, S.: Deprecating TLS 1.0 and TLS 1.1. RFC 8996 (2021). https://doi.org/10.17487/RFC8996, https://www.rfc-editor.org/info/rfc8996

12. Orientierungshilfe der Konferenz der unabhängigen Datenschutzaufsichtsbehörden des Bundes und der Länder vom 27. Mai 2021: Maßnahmen zum Schutz personenbezogener Daten bei der Übermittlung per E-Mail. https://www.datenschutzkonferenz-online.de/media/oh/20210616_orientierungshilfe_e_mail_verschluesselung.pdf

13. Petrlic, R.: The general data protection regulation: from a data protection authority's (technical) perspective. IEEE Secur. Priv. 17(6), 31–36 (2019). https://doi.org/10.1109/MSEC.2019.2935701

14. Posteo e.K.: What is the TLS-receiving guarantee and how do i activate it?. https://posteo.de/en/help/activate-tls-receiving-guarantee. Accessed 22 June 2022

15. Posteo e.K.: What is the TLS-sending guarantee and how do i activate it?. https://posteo.de/en/help/activating-tls-sending-guarantee. Accessed 22 June 2022

16. Postfix: Postfix documentation. https://www.postfix.org/documentation.html. Accessed 20 June 2022

17. Stransky, C., Wiese, O., Roth, V., Acar, Y., Fahl, S.: 27 years and 81 million opportunities later: investigating the use of email encryption for an entire university. In: 43rd IEEE Symposium on Security and Privacy, IEEE S&P 2022, 22–26 May 2022. IEEE Computer Society (2022)

Secure and Privacy Preserving Proxy Biometric Identities

Harkeerat Kaur[1]([⊠]), Rishabh Shukla[1], Isao Echizen[2], and Pritee Khanna[3]

[1] Indian Institute of Technology Jammu, Jagti, India
{harkeerat.kaur,2021rcs2012}@iitjammu.ac.in
[2] National Institute of Informatics, Tokyo, Japan
iechizen@nii.ac.jp
[3] Indian Institute of Information Technology Design
and Manufacturing Jabalpur, Jabalpur, India
pkhanna@iiitdmj.ac.in

Abstract. With large-scale adaption to biometric based applications, security and privacy of biometrics is utmost important especially when operating in unsupervised online mode. This work proposes a novel approach for generating new artificial fingerprints also called '**proxy fingerprints**' that are natural looking, non-invertible, revocable and privacy preserving. These proxy fingerprints can be generated from the original ones only with the help of a user-specific key. Instead of using the original fingerprint, these proxy templates can be used anywhere with same convenience. The manuscripts walks through an interesting way in which proxy fingerprints of different types can be generated and how they can be combined with use-specific keys to provide revocability and cancelability in case of compromise. Using the proposed approach a proxy dataset is generated from samples belonging to Anguli fingerprint database. Matching experiments were performed on the new set which is 5 times larger than the original, and it was found that their performance is at par with 0% FAR and 0% FRR in the stolen key/safe key scenarios. Other parameters on revocability and diversity are also analyzed for protection performance.

1 Introduction

Recent advances have witnessed huge growth in biometric enabled online identity verification and authentication. We can now experience all forms of life activities like teaching, learning, shopping, interacting, playing, working in the virtual world, popularly also called as the Metaverse which was once just an imaginary concept. We have already begun doing biometric based online monetary transactions, access control, possession of virtual real estate, marriage registry and digital avatars. As one application of the metaverse, why not envision being able to posses a unique proxy biometric (face, eyes, fingerprint, etc.) in the cyberspace?

A proxy biometric is a virtual agent of an actual user which can act on its behalf and represent him/her in the metaverse with a unique and privacy

L. Barolli (Ed.): AINA 2023, LNNS 654, pp. 487–500, 2023.
https://doi.org/10.1007/978-3-031-28451-9_43

preserving identity. An identity that can link us from the physical to the digital world, ensure our presence, and at the same time safeguard the revealing of our personal information. In the present day scenario we are directly converting our original biometric to digital ones. Although, coupled with various protection mechanisms, once your 'digital biometric' template leaves your device, the owner has no information over where and how it is stored, shared or sold for profit, surveillance, or has been victimized to data hacks. As the expected number of metaverse applications will increase, it becomes imperative to question how can we exist online with digital versions of our biometrics and possibility of biometric avatars?

This work proposes the concept of '**Proxy Biometrics**' for creating new digital fingerprint biometric templates that will operate on behalf of user's actual biometric. These artificial fingerprints will be generated from original ones coupled with a second factor key/PIN using deep neural networks (DNNs). Most importantly they will look like real templates so that online applications can extract discriminative information from them just like original templates. Third parties can process and encrypt proxy biometric in the way they want without user having to be worried about its loss of personal information. In case of a compromise, only the proxy fingerprint will be lost which cannot be inverted to original one. Unlike only ten fingerprints, a user can have multiple proxy fingerprints by coupling different keys.

The organization of this work is as follows. Section 2 discuss the timeline of biometric template protection techniques, followed by the model design of the proposed approach in Sect. 3. The proxy fingerprints are analyzed for performance in Sect. 4 followed by conclusions in Sect. 5.

2   Biometric Template Protection Timeline

A biometric authentication system is susceptible to various types of attacks where the identity can be compromised at various points between collection, transmission, storage and matching [17]. Biometric data stored in remote

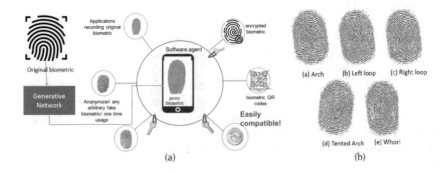

Fig. 1. Overview of original to proxy fingerprints for metaverse applications.

databases is highly vulnerable to hacking, personal data mining, reconstruction attacks, and sharing for covert surveillance and tracking. Templates stored in database once leaked renders the user to a continuous risk of identity thefts and exposure of sensitive information. In order to address this problem various template protection techniques have been proposed which can be majorly categorized as - biometric cryptosystems, cancelable biometrics, and secret sharing techniques [10]. Soutar et. al (1999) proposed biometric encryption technique to provide security to stored templates [20]. Fuzzy commitment, fuzzy vaults, homomorphic encryptions and secure sketches were other design techniques proposed under the category of biometric cryptosystems [9]. Over the period of time various correlation and reconstruction attacks have been proved on these approaches and the schemes underwent various improvisations to combat them [18]. However, one major drawback is that these schemes heavily rely on security of encryption keys and do not provide the ability to revoke.

Ratha et al. (2001) proposed the concept of cancelable biometrics [17]. It subjects biometric template to various systematic and non-linear distortions combined with a user-specific key to generate protected template. The transformed templates are non-invertible and cannot be traced back to the original even if the key and the transformation function is known. The combination of user-specific key provides multi-factor security and easy revocation. One of the most popular transformation technique is BioHashing (2004), which belongs to the category of random projection (RP) based transformations. The data is projected on user-specific orthonormal vectors followed by non-linear quantization to generate a protect code known as BioCode [7]. Multiple random projections, multispace random projections (2007), dynamic random projections (2010), feature adaptive random projections (2021), and a number of RP based transformations have been proposed along this line to improved security and matching performance [22,24,25]. Random Convolution based transforms is another category under which biometric feature is convolved with user specific random kernels. Bio-Convolving (2010) and curtailed circular convolutions (2014) are two important approaches under this transform [14,23]. Apart from that a number of techniques have been proposed that combine biometric data with synthetic noise or patters to generate transformed versions. Geometric transformations driven by random noise signals (2006), BioPhasor (2007), Gray-Salting/Bin-Salting (2008), locality sensitive hashing like Indexing-First-One (2017) and Index-of-Max (2018) have been proposed [8,16,21,26].

With advances in deep learning from 2018–19, a number of DNN techniques have been proposed that transform the features simultaneously or separately using one of the transformation techniques described above. Liu et al. (2018) proposed FVR-DLRP for generating secure fingervein templates which used random projection methods to distort the extracted features later trained over Deep Belief Networks (DBN) for accurate matching [13]. In 2018, another work by Singh et al. proposes cancelable knuckle prints which are generated by computing local binary pattern over CNN architecture [19]. Jo and Chang (2018) proposed CNN-based features extracted from face templates into binary code by using a Deep Table-based Hashing (DTH) framework [6]. Lee et al. (2021) proposed SoftmaxOut Transformation-Permutation Network (SOTPN) which is a neural version of popular cancelable transform known as Random Permutation Maxout (RPM) transform [11].

However one of the major drawback of all the above mentioned approached is that the transformed template has a noise like appearance. These unstructured templates makes matching process application specific and effects performance. Some of the service providers would prefer cryptosystem approach, some would like cancelable biometrics, while there may be some who would deploy none due to technical or real time challenges. Literature also proposed various approaches to generate fake biometric using GANS (generative adversarial networks) [2,15]. However they are useful only for large scale database generation for training purposes only.

The major motivation of this work is to provide a protection layer at user-level which allows generation of artificial fingerprints from original fingerprints. These artificial fingerprints can then act as cyberproxies of original and interact with third parties applications for storage and transmission purposes. Natural looks makes them independent of the third party security choice and guarantees user-level protection without affecting performance. Figure 1 shows an illustration which depicts proxy template generated at smartphone interacting with various third parties. Overall, the major contributions of this work are:

1. Generation of proxy fingerprint biometrics, that are natural looking, revocable, posses good descriptive characteristics and can only be derived from original fingerprints.
2. Generation of new fingerprint samples of different class from same original by using random projection technique and key.
3. A framework which allows enrollment, authentication, and easy revocation of proxy fingerprints by changing keys in the metaverse.

3   Proposed Approach

Overview: Based upon their dominant patterns the fingerprints belong to five different type of classes namely - whorl, arch, tented, left loop, right loop [1], Fig. 1. The proposed work is divided into four important steps as shown in Fig. 2. The encoder takes an original fingerprint image as input from the user and extract a latent vector. It is then mixed with some user-specific random key to generated a transformed feature as called as salted latent vector. The salted latent vector is projected on an orthonormal basis which is finally passed through a decoder to generate a new fingerprint image that belongs to one of the five classes mentioned above.

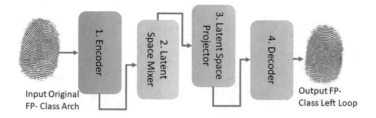

Fig. 2. Overview of the proposed approach.

3.1 Design Autoencoder to Extract Latent Spaces

At the first step we train a CNN based auto-encoder and decoder. The encoder $f_{enc}(x, \theta)$ takes the original fingerprint sample x and maps it to latent space representation z. The decoder $f_{dec}(z, \phi)$ reconstructs the fingerprint image $\hat{x}$ from latent representation z. Given that the input image sample $x \sim \mathbb{X}^{H \times W \times 3}$ and latent vector $z \sim \mathbb{Z}$, the image reconstruction loss L_{rec} for the network is $L_{rec}(f_{enc}, f_{dec}) = \mathbb{E}_{x \sim X}[\| x - f_{dec}(f_{enc}(x)) \|]$.

The encoder of a trained autoencoder outputs discriminative latent vector that is often used as input to various other tasks like image denoising, classification, anomaly detection etc. Figure 3 shows the encoder and decoder architecture. The input image $x \in \mathbb{X}^{200 \times 136 \times 3}$ is re-scaled in the range $[0,1]$ before passing. The output taken from the last layer of encoder is flattend to give latent vector $z \in \mathbb{Z}^{13600}$. A database consisting of 30,000 synthetic fingerprints spanning across five classes is generated using Anguli software out of which 27000 images were used for training [1].

3.2 Visualizing and Projecting Latent Spaces

Once trained for perfect reconstruction, the latent vector z outputted from 'encoder' and their corresponding class and subject labels were collected. Although, the encoded representations are dense and contain discriminative features but they are also highly entangled and not well separated. Figure 4(a) shows the the result of conducting principal component analysis over the original latent space. It can be easily observed that overlapping distributions are inseparable in terms of their class patterns.

Our model requires to have a highly discriminative feature space which can be used to generate new samples of different class. The proposed approach tries to maximize the separations between these overlapping distributions using a technique called Random Projection. By projecting data on orthonormal basis, the pairwise distances of the points before and after projection are not changed thus retaining their statistical properties [3]. The proposed approach utilizes this property to make latent vectors belonging to a specific class follow a certain distribution pattern. The transformation process for the same is outlined as follows.

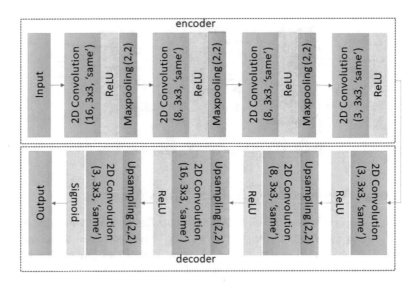

Fig. 3. Encoder and Decoder Architecture.

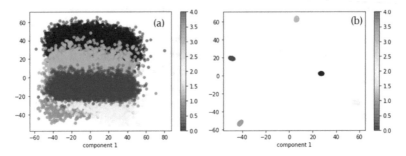

Fig. 4. Visualizing PCA over latent space (a) Original Space (b) After Projection.

Step 1: Step 1. For each class $'i'$, generate an orthonormal random matrix $\mathbb{M}^i$, of dimension 136×136 using the Gram-Schmidt process, where $i \in [1, 5]$ [12].

Step 2: The latent vector zp is reshaped into a 2D such that $zp \in \mathbb{Z}^{100 \times 136}$.

Step 3: Project zp belonging onto matrix $\mathbb{M}_i$ as $Pj_i = z.M_i$.

Step 4: Reshape Pj_i again in one dimension $Pj_i \in \mathbb{Z}_1^{13600}$.

PCA is again performed over the projected latent spaces and results are visualized Fig. 4(b). It can be seen that the projected spaces of latent vectors are well separated.

3.3   Preparing Projected Vectors for Decoder

This section aims to propose an approach which trains a decoder model on features belonging to different classes but projected on random matrix $\mathbb{M}^i$ to generate new samples belonging of class i.

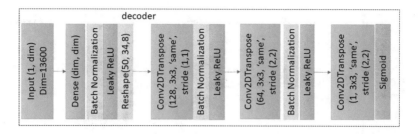

Fig. 5. Final decoder architecture.

As mentioned, the original database D comprises of 30000 fingerprint images belonging to 7500 subjects equally distributed over five classes. With 6000 images per class and with 4 images per subject, the images with subjects numbered 1 to 1500 belong to class 1, 1501 to 3000 to class 2, and so on till 6001 to 7500 to class 5. We take ith sample of each class and assign it to new subset DM_j where $j = mod(i, 6)$.

Thus we have five new subsets of mixed samples named as Dm_1, $Dm_2,...,Dm_5$. The data in subset Dm_i is projected on random matrix M_i to form a projected subset Pj_i. The decoder takes input the latent vector $Pj_i \sim \mathbb{Z}_i^{13600}$, where input Pj_i consisting of salted latent vectors belonging to any class but projected over matrix and targeted output $x_{dec} \sim \mathbb{X}^{200 \times 136 \times 1}$ is a fingerprint image belonging to set D_i. The decoder model architecture is depicted in Fig. 5. Since the images are in range [0,1] the final activation is sigmoid. The model is optimized using Adam optimizer to minimize MSSIM (mean structure similarity index) loss between targeted and predicted output images, $\mathcal{L}_{dec} = minimize(1- MSSIM(true, predicted))$.

3.4   Generating New Fingerprint Images

Once the encoder and decoder models are trained, it can be used to generate new fingerprint samples.

This section explains how an existing individual logs on to the proposed architecture to generate a proxy biometric and enroll/register it. Also, how it can generate the probe version of the proxy biometric to gain access into the system.

Enrollment: The process is illustrated in Fig. 6 and explained step wise below.

Step 1: At the time of enrollment a new user inputs its original biometric B to the encoder which extracts its latent vector z.

Step 2: A random key generation unit generates key k with distribution $\mathcal{N}(\mu = 0, \sigma = 0.5)$.

Step 3: The key is added to latent vector z and the resultant salted vector zp is generated as $zp = z + k$

Step 4: The target class is selected, let say class C_i. The salted vector zp is projected on the orthonormal random matrix M_i for the class Ci as $Pj = zp.M_j$

Step 5: The Projected vector Pj is decoded to output a new proxy biometric PB as $PB = decoded(Pj)$

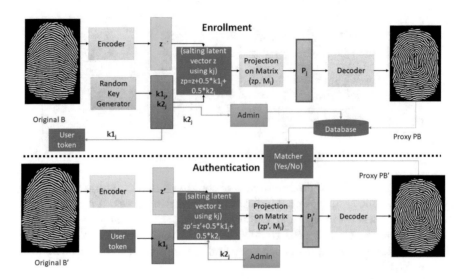

Fig. 6. Enrollment and authentication process using proxy biometrics.

At the end of this process this new proxy biometric PB is recorded as reference template and stored for further matching. The user-specific random key k is divided into two shares $k1$ and $k2$, such that $K = 0.5 * k1 + 0.5 * k2$. The first key share $k1$ and class C_i is provided to the user in tokenized manner. The second key share $k2$ is maintained at the generation site by the system administrator with respect to an identification number provided to the user.

Authentication: The steps for authentication are:

Step 1: The user presents its original probe biometric sample B' and tokenized key $k1$ and target class C_i.

Step 2: The latent vector z' is extracted, second key $k2$ is fetched and salted vector zp' is generated as $zp' = z' + 0.5 * k1 + 0.5 * k2$

Step 3: The salted vector is projected on targeted class C_i as $Pj' = zp'.M_i$

Step 4: Finally, the projected vector is decoded to form the new proxy probe biometric PB'.

Instead of matching the original ones, the proxy samples PB and PB' are matched to grant and deny access. This offers great security from revealing your actual fingerprint yet at the same time ensuring your presence. Due to non-linearity imparted by various model, even if the proxy biometric and key are revealed to the attacker, he/she cannot use it to recover the original ones.

4   Experimental Results and Analysis

Proposed approach is evaluated on various performance and protection parameters to verify the effectiveness of the generated artificial/proxy fingerprint templates.

4.1 Matching Performance

As mentioned this work uses synthetically generated fingerprint form Anguli software since most of the publicly available fingerprint dataset as quite small in size for training DNNs. Given a synthetic database of total 7500 subjects, 25 subjects are selected randomly from each class, thus giving a total database of 125 subjects with 4 samples per subject. Let this data subset be called as "original_db". The new biometrics are generated by proposed combination of original and some random keys after passing through various layers of convolutions neural networks. The matching performance is evaluated keeping in account of the generating parameters under two scenarios, namely -*worst case or stolen token scenario* and *best case or safe token scenario*. It is expected that these proxy biometric must exhibit same uniqueness as well as intra and inter-user discriminating characteristics. To match the fingerprint, one of the most common brute force based keypoint matching algorithm is used [4]. The brute force algorithm matches keypoints of one fingeprint with another in one to all mode. Apart from that since fingerprint are highly structured in terms of constituting of ridges and bifurcations, Structure Similarity Index (SSIM) has also been proven to be an effective metric to record matching similarity between two images [5]. SSIM is computed locally over small image patches (here of size 10×10) and overall mean is computed as Mean Structural Similarity Index (MSSIM), $MSSIM(X,Y) = \frac{1}{M}\sum_{j=1}^{M} SSIM(x_j, y_j)$. Usually the index values are normalized to range between [0 to 1], where 1 indicates perfect similarity and value 0 otherwise.

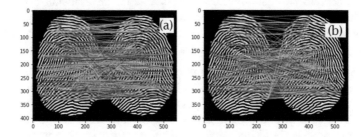

Fig. 7. Matching using keypoint brute force (a) Intra-user (b) Inter user.

Table 1. Matching performance in the best and worst case scenario.

Scenario	Database	Brute force	MSSIM + Brute force
Original	*original_db*	18.65% EER	0% EER
Best Case	*proxy_db*	21.35% EER	0% EER
Worst Case	*proxy_db*	25.35% EER	0% EER

a) Best Case/Safe Key Scenario: In this scenario, each user j in *original_db* is assigned a distinct user-specific and system key $k_j = (k1_j, k2_j)$. Five new samples are generated by projecting the salted vector zp_j on matrix $M_1, M_2,...,M_5$,

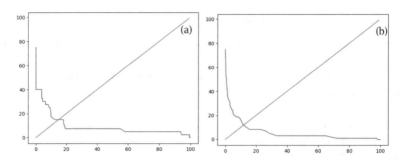

Fig. 8. ROC curves for brute force technique (a) Original database (b) Proxy batabase in best case.

where each sample belongs to particular class. This results in a new dataset called '*proxy_db*' having 625 subjects, with 4 samples per subject. This is exactly a five fold increase in population. The proxy samples of each subject PB_m^x and PB_n^y are matched to output a score, where $1 \leq m, n \leq 625$ (no. of subjects) and $1 \leq x, y \leq 4$ (samples per subject). If $m = n$, PB_m^x and PB_n^y are expected to match and result in genuine score, otherwise imposter score. Matching performance on'*proxy_db*' and similar size subset of *original_db* is computed using brute force, and a combination of brute force and MSSIM values. The results in terms of Equal Error Rates are reported in Table 1, Fig. 8. Figure 7 illustrates brute force keypoint matching between two proxy fingerprints belonging to same user and different users. The matching scores for MSSIM+brute force were recorded and the genuine and impostor score distribution over *proxy_db* is depicted in Fig. 9(a). A clear separation is observed. Thus indicating that the proxy templates preserve good discriminating characteristics and perform desirably with 0% False Accept Rate and 0% False Reject Rate.

b) **Worst Case/Stolen Key Scenario:** The scenario consider the possibility that the user key is stolen and is used by the impostor to gain access to the system. Under this scenario the impostor generates proxy biometric using its own/others biometric and stolen key. It is expected that this attacker generated proxy biometric should not match with the legitimate key holder generated proxy biometric or with templates of other enrolled legitimate entities. To simulate this scenario, all the users are assigned same user specific key. For each subject j, the extracted latent vectors are salted, projected, and decoded to form a new dataset called *proxy*_db*. Similarly to above, this dataset has 625 subjects. Samples PB_m^{*x} and PB_n^{*y}, where $1 \leq m, n \leq 625$, $1 \leq x, y \leq 4$ are matched and genuine and imposter scores are recorded (refer Table 1). The score distribution for MSSIM+brute force is depicted in Fig. 9(b) and once again a distinct separation can be observed indicating 0% EER even in the worst case scenario.

4.2   Revocability and Diversity

This section overviews the revocability property of the proposed approach by noting the effect of changing user specific key and projection matrices on new sample generation.

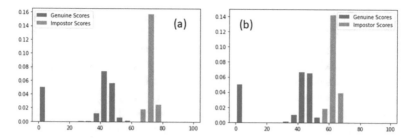

Fig. 9. Genuine imposter score distribution (a) Best case scenario and (b) Worst case scenario.

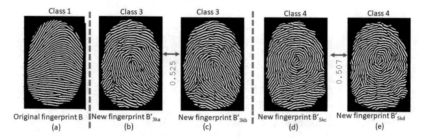

Fig. 10. Visualizing the effect of changing both keys.

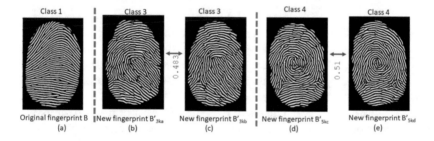

Fig. 11. Visualizing the effect of changing only one of the key.

Changing One or Both Keys: The latent feature z is extracted from original fingerprint B. Then it is mixed with a user specific key K consisting of two parts $k1$ and $k2$ to form basis for new sample generation. This section analyzes the effect of changing one or both key parts on new sample generation while keeping the projection matrix same. For *original* having 125 subjects two key sets are generated $ka=(k1a, k2a)$ and $kb=(k1b, k2b)$ and each is projected on class matrix M_i two create two sample sets *proxy_db_key1* and *proxy_db_key2* having 625 subjects and 4 samples each. Then corresponding samples from *proxy_db_key1* and *proxy_db_key2* are matched by computing MSSIM between them. Figure 10 illustrates an original fingerprint sample B and its proxies belonging to class 3 and 4 obtained using key ka and kb and the MSSIM between them. Similarly, the MSSIM between 2500 samples of two proxy set is computed and its mean value is found to be 0.49383. Thus indicating same biometric samples when salted with different keys and projected over same matrix show significant diversity.

On the same line figure Fig. 11 shows the effect by only changing the user key part $(k1a)$ and $(k1b)$ keeping the second key part same for classes 3 and 5. The generated samples and the MSSIM values between them show that new samples can also be generated by only changing one part of the key. Thus in case of any comprise the one/both keys can be changed and combined with the latent feature to generate new proxy biometric samples as required. The average MSSIM between proxy dataset generated this way is reported to be 0.493738. In both cases average MSSIM value is around 0.5 thus indicating significant in change in overall structure with change in keys ka and kb.

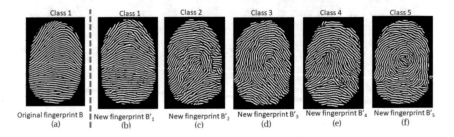

Fig. 12. Visualizing the effect of changing projection matrix.

Changing the Projection Matrix: This section analyzes the effect of changing projection matrix M_i keeping the salted vector same. Given an original fingerprint image B of any class 1, its latent vector z is extracted and salted with a key to get zp. New samples B'_i each class C_i is generated by decoding projection of zp on the respective class matrix M_i as:

$B'_i = decoded((zp).M_i)$ The effect of using different projection on matrix M_i on same zp is depicted in Fig. 12. Figure 12 (a) shows an original biometric sample belonging to class 1 and 12 (b–f) shows proxy samples belonging to class 1 to 5. For each subject in original database five new samples are generated by keeping the"subject" key fixed and changing projection matrix M_i. MSSIM is computed between the generated samples of different classes having same underlying vector and the average value is reported to be 0.49263. Thus, depicting the trained model is capable of generating new samples having patterns belonging to the projected class c_i, when same latent zp is projected on different matrices M_i, i =1 to 5.

4.3 Non-invertibility

Non-invertibility implies that an attacker should not be able to recover the original biometric from its transformed version even if the key and transformation functions are know. Following that the proposed architecture uses DNNS for the transformation, non-invertibility is implicit here even if the user-specific key and projection matrices are known to the attacker.

5   Conclusions and Future Directions

The proposed work provides proof of concept for producing proxy fingerprint templates from the original ones combined with user-specific key. Samples across various classes can be easily generated by projection along specific orthonormal random matrices for a particular class. With a five fold increase in the population size, the effect of performance on proxy samples under stolen token and safe token scenario were analysed and reported to be satisfactory.

The future work aims at using more realistic dataset for proxy generation as well as testing for other different fingerprint matching schemes. Further increase in population size and improvement in generation model is expect to open new avenues towards secure and privacy preserving biometric authentication in the upcoming meta-world. Overall this work gives a good working proof of how AI will be able to actually manifest natures' ability to generate new biometric samples using DNNs.

Acknowledgements. This work was partially supported by JSPS Bilateral Program (JSPS-DST) JSPSBP120217719, JSPS KAKENHI Grants JP16H06302, JP18H04120, JP20K23355, JP21H04907, and JP21K18023, and by JST CREST Grants JPMJCR18A6 and JPMJCR20D3, Japan.

References

1. Ansari, A.H.: Generation and storage of large synthetic fingerprint database. ME Thesis (2011)
2. Bamoriya, P., Siddhad, G., Kaur, H., Khanna, P., Ojha, A.: DSB-GAN: generation of deep learning based synthetic biometric data. Displays **74**(102), 267 (2022)
3. Dasgupta, S., Gupta, A.: An elementary proof of a theorem of Johnson and Lindenstrauss. Random Struct. Algorithms **22**(1), 60–65 (2003)
4. Howse, J., Hua, Q., Puttemans, S., Sinha, U.: Fingerprint detection using OpenCV3 (2015). https://hub.packtpub.com/fingerprint-detection-using-opencv/
5. Jain, A.K., Cao, K.: Fingerprint image analysis: role of orientation patch and ridge structure dictionaries. Geom. Driven Stat. **121**(288), 124 (2015)
6. Jang, Y.K., Cho, N.I.: Deep face image retrieval for cancelable biometric authentication. In: 2019 16th IEEE International Conference on Advanced Video and Signal Based Surveillance (AVSS), pp 1–8. IEEE (2019)
7. Jin, A.T.B., Ling, D.N.C., Goh, A.: Biohashing: two factor authentication featuring fingerprint data and tokenised random number. Pattern Recogn. **37**(11), 2245–2255 (2004)
8. Jin, Z., Hwang, J.Y., Lai, Y.L., Kim, S., Teoh, A.B.J.: Ranking-based locality sensitive hashing-enabled cancelable biometrics: index-of-max hashing. IEEE Trans. Inf. Forensics Secur. **13**(2), 393–407 (2017)
9. Juels, A., Wattenberg, M.: A fuzzy commitment scheme. In: Proceedings of the 6th ACM Conference on Computer and Communications Security, pp. 28–36 (1999)
10. Kaur, H., Khanna, P.: Biometric template protection using cancelable biometrics and visual cryptography techniques. Multimed. Tools Appl. **75**(23), 16333–16361 (2015). https://doi.org/10.1007/s11042-015-2933-6

11. Lee, H., Low, C.Y., Teoh, A.B.J.: SoftmaxOut transformation-permutation network for facial template protection. In: 2020 25th International Conference on Pattern Recognition (ICPR), pp. 7558–7565. IEEE (2021)

12. Leon, S.J., Björck, Å., Gander, W.: Gram-Schmidt orthogonalization: 100 years and more. Numer. Linear Algebra Appl. **20**(3), 492–532 (2013)

13. Liu, Y., Ling, J., Liu, Z., Shen, J., Gao, C.: Finger vein secure biometric template generation based on deep learning. Soft. Comput. **22**(7), 2257–2265 (2018)

14. Maiorana, E., Campisi, P., Neri, A.: Bioconvolving: cancelable templates for a multi-biometrics signature recognition system. In: 2011 IEEE International Systems Conference, pp. 495–500. IEEE (2011)

15. Minaee, S., Abdolrashidi, A.: Finger-GAN: Generating realistic fingerprint images using connectivity imposed GAN. arXiv preprint arXiv:1812.10482 (2018)

16. Ratha, N., Connell, J., Bolle, R.M., Chikkerur, S.: Cancelable biometrics: a case study in fingerprints. In: 18th International Conference on Pattern Recognition (ICPR 2006), vol. 4, pp. 370–373. IEEE (2006)

17. Ratha, N.K., Connell, J.H., Bolle, R.M.: Enhancing security and privacy in biometrics-based authentication systems. IBM Syst. J. **40**(3), 614–634 (2001)

18. Scheirer, W.J., Boult, T.E.: Cracking fuzzy vaults and biometric encryption. In: 2007 Biometrics Symposium, pp. 1–6. IEEE (2007)

19. Singh, A., Hasmukh Patel, S., Nigam, A.: Cancelable knuckle template generation based on LBP-CNN. In: Proceedings of the European Conference on Computer Vision (ECCV) Workshops (2018)

20. Soutar C, Roberge D, Stoianov A, Gilroy R, Kumar BV (1999) Biometric encryption. In: ICSA guide to Cryptography, vol 22, McGraw-Hill New York, p 649

21. Teoh, A.B., Ngo, D.C.: Biophasor: token supplemented cancellable biometrics. In: 2006 9th International Conference on Control, Automation, Robotics and Vision, pp 1–5. IEEE (2006)

22. Teoh, A.B.J., Yuang, C.T.: Cancelable biometrics realization with multispace random projections. IEEE Trans. Syst. Man Cybern. Part B (Cybern.) **37**(5), 1096–1106 (2007)

23. Wang, S., Hu, J.: Design of alignment-free cancelable fingerprint templates via curtailed circular convolution. Pattern Recogn. **47**(3), 1321–1329 (2014)

24. Yang, B., Hartung, D., Simoens, K., Busch, C.: Dynamic random projection for biometric template protection. In: 2010 Fourth IEEE International Conference on Biometrics: Theory, Applications and Systems (BTAS), pp. 1–7. IEEE (2010)

25. Yang, W., Wang, S., Shahzad, M., Zhou, W.: A cancelable biometric authentication system based on feature-adaptive random projection. J. Inf. Secur. Appl. **58**(102), 704 (2021)

26. Zuo, J., Ratha, N.K., Connell, J.H.: Cancelable iris biometric. In: 2008 19th International Conference on Pattern Recognition, pp. 1–4. IEEE (2008)

Expectation-Maximization Estimation for Key-Value Data Randomized with Local Differential Privacy

Hikaru Horigome[1], Hiroaki Kikuchi[1(✉)], and Chia-Mu Yu[2]

[1] Meiji University, 4-21-1 Nakano, Nakano, Tokyo 164-8525, Japan
{cs212030,kikn}@meiji.ac.jp
[2] National Yang Ming Chiao Tung University,
1001 University Rd., Hsinchu 300, Taiwan

Abstract. This paper studies the local differential privacy (LDP) algorithm for key-value data that are pervasive in big data analysis. One of the state-of-the-arts algorithms, PrivKV, randomizes key-value pairs with a sequence of LDP algorithms. However, most likelihood estimation fails to estimate the statistics accurately when the frequency of the data for particular rare keys is limited. To address the problem, we propose the expectation-maximization-based algorithm designed for PrivKV. Instead of estimating continuous values $[-1, 1]$ in key-value pairs, we focus on estimating the intermediate variable that contains the encoded binary bit $\in \{1, -1\}$. This makes the problem tractable to estimate because we have a small set of possible input values and a set of observed outputs. We conduct some experiments using some synthetic data with some known distributions, e.g., Gaussian and power-law and well-known open datasets, MoveLens and Clothing. Our experiment using synthetic data and open datasets shows the robustness of estimation with regards to the size of data and the privacy budgets. The improvement is significant and the MSE of the proposed algorithm is 602.83×10^{-4} (41% of PrivKVM).

1 Introduction

A key-value is a primitive data structure used for many applications and is pervasive in big data applications such as mobile app activity analysis. If we can collect daily usage data of smartphone apps, the data can be applied for optimizing battery management, personalized services, digital contents delivery and prediction diseases for healthcare. However, daily active usage is confidential data, and many people deny access to their personal data.

Local differential privacy (LDP) [1,5,9] is a state-of-the-art private data anonymization mechanism. Erlingsson et al. at Google proposed a LDP algorithm [6]. It has been deployed by major platformers including Apple [7], Google [6], and Microsoft [8]. Variation of LDP schemes have been studied in order to expand the domain of LDP applications. Ye et al. [4] proposed the key-value data collection mechanisms that can satisfy LDP and estimate the key frequencies and the mean values from the sophisticatedly randomized key-value pairs.

L. Barolli (Ed.): AINA 2023, LNNS 654, pp. 501–512, 2023.
https://doi.org/10.1007/978-3-031-28451-9_44

Their algorithm combines two LDP mechanisms known as primitives, the Randomized Response (RR) for keys [2] and the Value Perturbation Primitive (VPP) [3] for values. PCKV [14] is also proposed to collect the key-value pairs in an LDP manner.

The estimation accuracy is a one of current issues in LDP schemes. The size of users is also known as one of the factors to determine the estimation accuracy. The more perturbed key-value pairs, the more accurate the estimate. Most LDP algorithms, e.g., RAPPOR and PrivKV, estimate the statistics by solving the expected relationship between the observed randomized value and the true statistics. It is estimated *by means of a single point of expected value of randomized output*, which is a kind of Most Likelihood Estimation (MLE). Therefore, it suffers lower estimation accuracy when many values are randomized far from the theoretical expected value.

In order to address the lower estimation accuracy when the frequency is limited, we propose an iterative approach to improve the estimation accuracy of perturbed data in the LDP algorithm. Our idea is based on Bayes' theorem and the Expectation-Maximization (EM) algorithm [10]. The iterative process updates the posterior probabilities so that the all elements are consistent with the given observed data. Hence, it is more stable and more robust than the data that contain values for rear keys. However, it is not trivial to apply the EM algorithm to PrivKV because of its sequential combination of randomized key and continuous value $v \in [-1, 1]$. Instead of naïvely estimating v, we attempt to estimate the probability of intermediate values $v^+ \in \{1, -1\}$ in randomizing process of PrivKV. It makes the problem simple and tractable.

We conducted some experiments using synthetic data with some known distributions, e.g., Gaussian and Power-law. Then, we compared our proposed algorithm with PrivKV and PrivKVM (three iterations with responders) [4] to explore the accuracy improvement in terms of privacy budget ϵ and the number of responders n. We also evaluate the estimation accuracy using some open datasets, MovieLens and Clothing. It demonstrates that the proposed scheme performs well in general cases.

Our contribution are as follows.

- We propose a new algorithm to estimate the key frequencies and the mean values in key-value data that randomized in local differential private algorithm PrivKV (Sect. 3).
- The experimental results using synthetic data with three major probability distributions Gaussian, power-law and linear, and well-known open datasets, MovieLens and Clothing, demonstrate that our proposed algorithm overperforms state-of-the-art LDP schemes in estimation of frequency and mean value. Our experiment using synthetic data shows the robustness of estimation with regard to the size of data and the privacy budgets (Sect. 4).

Our paper is organized as follows. In Sect. 2, we provide some necessary fundamental definitions of LDP and the baseline estimation algorithms. In Sect. 3, we propose our algorithm and prove useful property in estimation of frequency and mean values. We report our experiments using synthetic and open data in

Sect. 4. Our experiments show that the performance and the efficiency of the proposed algorithm in comparison to the existing LDP schemes. Section 5 shows some related works in this study. We conclude our study in Sect. 6.

2 Local Differential Privacy

2.1 Fundamental Definition

Suppose that users periodically submit their location data to a service provider. Differential privacy guarantees that the randomized data do not reveal any privacy disclosure from them. By contrast, LDP needs no trusted party. LDP is defined as follows.

Definition 1. A randomized algorithm Q satisfies ϵ-local differential privacy if for all pairs of values v and v' of domain V and for all subset S of range Z $(S \subset Z)$, and for $\epsilon \geq 0$, $Pr[Q(v) \in S] \leq e^{\epsilon} Pr[Q(v') \in S]$.

2.2 PrivKV

Multidimensional data are one of the big challenges for perturbations. Several randomization mechanisms with LDP have been proposed so far.

Ye et al. [4] addressed the issue using two variables that are perturbed accordingly in their proposed LDP algorithm, PrivKV. PrivKV takes inputs in the form key-value data, two-dimensional data structure of discrete (key) and continuous (value) variables, and estimates the key frequencies and the mean values. Their idea combines two LDP protocols, RR for randomizing keys and VPP for perturbing values. The dimension is restricted to two, but the key-value is known as a primitive data structure commonly used for several applications. For example, a movie evaluation dataset consists of ratings for movies, which are stored in a key-value database in which keys are movie titles and the values are ratings for titles. In a smartphone survey, users indicate their favorite apps such as $\langle \texttt{YouTube}, 0.5 \rangle$, $\langle \texttt{Twitter}, 0.1 \rangle$, $\langle \texttt{Instagram}, 0.2 \rangle$, by stating their total time using those apps.

Let S_i be a set of key-value tuple $\langle k, v \rangle$ owned by i-th user. In PrivKV, the set of tuples is encoded as a d-dimensional vector, where d is the cardinality of the domain of keys K and missing key is represented as $\langle k, v \rangle = \langle 0, 0 \rangle$. For instance, a set of key-value $S_i = \{ \langle k_1, v_1 \rangle, \langle k_4, v_4 \rangle, \langle k_5, v_5 \rangle \}$ is encoded as $d = 5$ dimensional vector $\boldsymbol{S}_i = (\langle 1, v_1 \rangle, \langle 0, 0 \rangle, \langle 0, 0 \rangle, \langle 1, v_4 \rangle, \langle 1, v_5 \rangle)$ where keys k_1, k_4 and k_5 are specified implicitly with 1 at the corresponding location.

Perturbation in PrivKV is performed by random sampling one element $\langle k_a, v_a \rangle$ from $\boldsymbol{S}_i$. It has two proceeding steps, perturbing values and keys. It uses VPP used in Harmony [3] for the chosen tuple. A value of the tuple $\langle 1, v_a \rangle$ is replaced by $v^+{}_a = VPP(v_a, \epsilon_2)$, where ϵ_2 is a privacy budget for values. A value of the "missing" tuple $\langle 0, 0 \rangle$ is replaced by $v^+{}_a = VPP(v'_a, \epsilon_2)$, where v'_a is chosen uniformly from $[-1, 1]$.

It uses RR with privacy budget ϵ_1. A tuple $\langle 1, v_a \rangle$ is randomized as

$$\langle k_a^*, v_a^+ \rangle = \begin{cases} \langle 1, v_a^+ \rangle \ w/p \ p_1 = \frac{e^{\epsilon_1}}{1+e^{\epsilon_1}}, \\ \langle 0, 0 \rangle \ \ w/p \ q_1 = \frac{1}{1+e^{\epsilon_1}}, \end{cases}$$

where v_a^+ is perturbed as mentioned. A "missing" tuple $\langle 0, 0 \rangle$ is randomized as

$$\langle k_a^*, v_a^+ \rangle = \begin{cases} \langle 0, 0 \rangle \ \ w/p \ p_1 = \frac{e^{\epsilon_1}}{1+e^{\epsilon_1}}, \\ \langle 1, v_a^+ \rangle \ w/p \ q_1 = \frac{1}{1+e^{\epsilon_1}}. \end{cases}$$

Responder in PrivKV submits the perturbed tuple $\langle k_a^*, v_a^+ \rangle$ with the index a of the tuple.

3   Proposed Algorithm

3.1   Idea

The drawback of PrivKV and PrivKVM is their low estimation accuracy. Because PrivKV uses the MLE of frequencies and means, the estimate accuracy reduces when the values are not uniformly distributed or sparse data are given. An iterative approach like PrivKVM consumes a privacy budget for every iteration, and the optimal assignment is not trivial.

MLE used in PrivKV works well for some cases but has low estimate accuracy for a biased distribution. Instead, we attempt to address this limitation by using an iterative estimate approach known as the EM (EM) algorithm. Because EM estimates posterior probabilities iteratively so that the estimated probabilities are more consistent with all observed values, it can improve accuracy when the number of users n increases and many observed data are given. However, it is challenging to estimate exact continuous value $v \in [-1, 1]$. Instead of estimating v directly, we focus to estimate the encoded v in binary, $v^+ \in \{1, -1\}$. It is tractable to estimate because we have a small set of possible value of input as $X = \{\langle 1, 1 \rangle, \langle 1, -1 \rangle, \langle 0, 1 \rangle, \langle 0, -1 \rangle\}$ and a set of observed output $Z = \{\langle 1, 1 \rangle, \langle 1, -1 \rangle, \langle 0, 0 \rangle\}$. Estimated marginal probability of X allows the mean values to be estimated accurately.

3.2   EM Algorithm for PrivKV

EM algorithm performs an iterative process for which posterior probabilities are updated through Bayes' theorem [10]. Each iteration estimates the best probabilities θ^j for all possible values in a domain. First, we show the EM algorithm generally and then modify it for PrivKV.

Let $X = \{x_1, \ldots, x_d\}$ be a set of input values and $Z = \{z_1, z_2, \ldots, z_{d'}\}$ a set of output values. A responder owning private value $x_i \in X$ uses a randomized algorithm to output $z_i \in Z$. Given n observed values $z_1, \ldots, z_n$, we iterate estimating posterior probabilities for $x_1, \ldots, x_d$ as $\Theta^{(t)} = (\theta_1^{(t)}, \theta_2^{(t)}, \ldots, \theta_d^{(t)})$ until

Table 1. Conditional probabilities of observed tuple Z given tuple X

$X = \langle k' , v^+\rangle$		$Z = \langle k^* , v^*\rangle$		$Pr[z|x]$	$X = \langle k' , v^+\rangle$		$Z = \langle k^* , v^*\rangle$		$Pr[z|x]$
1	1	1	1	$p_1 p_2$	0	1	1	1	$q_1 p_2$
1	1	1	-1	$p_1 q_2$	0	1	1	-1	$q_1 q_2$
1	1	0	0	$q_1(p_2 + q_2)$	0	1	0	0	$p_1(p_2 + q_2)$
1	-1	1	1	$p_1 q_2$	0	-1	1	1	$q_1 q_2$
1	-1	1	-1	$p_1 p_2$	0	-1	1	-1	$q_1 p_2$
1	-1	0	0	$q_1(p_2 + q_2)$	0	-1	0	0	$p_1(p_2 + q_2)$

converged. We start iteration with the initialized values assigned to probabilities uniformly as $\Theta^{(0)} = (\frac{1}{d}, \frac{1}{d}, \ldots, \frac{1}{d})$.

The conditional probability of input x_i given output z_j is given as $Pr[z_j|x_i] = \frac{Pr[z_j, x_i]}{Pr[x_i]}$. Bayes' theorem gives the posterior probability of $X = x_i$ given z_j as $Pr[x_i|z_j] = \frac{Pr[z_j|x_i]Pr[x_i]}{\sum_{s=1}^{|X|} Pr[z_j|x_s]Pr[x_s]}$. By letting $\theta_i^{(t-1)} = Pr[x_i]$ be the $(t-1)$-th estimate of marginal probability of $x_i \in X$, we have the t-th estimate of conditional probability for the u-th responder who responds $z_u \in Z$ as

$$\hat{\theta}_{u,i}^{(t)} = Pr[x_i|z_u] = \frac{Pr[z_u|x_i]\theta_i^{(t-1)}}{\sum_{s=1}^{|X|} Pr[z_u|x_s]\theta_s^{(t-1)}}, \tag{1}$$

which follows the t-th estimate of marginal probability by aggregating all n estimates of responders as, $\theta^{(t)} = \frac{1}{n}\sum_{u=1}^{n} \hat{\theta}_u^{(t-1)}$. This process iterates until updating converges as $|\theta_i^{(t)} - \theta_i^{(t-1)}| \leq \eta$, where η is predetermined precision.

In PrivKV, a randomization of tuple $\langle k, v\rangle$ is performed in sequential algorithms. A value $v \in [-1, 1]$ is encoded into $v^* \in \{-1, 1\}$ in a probability depending on v. Then, it is randomized as v^+ in $RR(v^*, \epsilon_2)$ (a part of VPP) using probabilities $p_2 = (e^{\epsilon_2})/(1+\epsilon_2)$, and $q_2 = 1/(1+\epsilon_2) = 1-p_2$. Finally, it is randomized in $RR(v^*, \epsilon_1)$ using probabilities $p_1 = (e^{\epsilon_1})/(1+\epsilon_1)$, and $q_1 = 1/(1+\epsilon_1) = 1-p_1$, as a part of key randomization. Hence, if we perturb a given tuple $\langle k, v^+\rangle = \langle 1, 1\rangle$ in PrivKV, the output $\langle k^*, v^*\rangle = \langle 1, 1\rangle$ is observed with probability $p_1 p_2$ as the consequence of VPP and RR. Similarly, another tuple happens as

$$\langle k^*, v^*\rangle = \begin{cases} \langle 1, 1\rangle & \text{w/p } p_1 p_2, \\ \langle 1, -1\rangle & \text{w/p } p_1 q_2, \\ \langle 0, 0\rangle & \text{w/p } q_1(p_2 + q_2). \end{cases}$$

Thus, we have the conditional probability $Pr[z_1 = \langle k^*, v^*\rangle = \langle 1, 1\rangle | x_1 = \langle k', v^+\rangle = \langle 1, 1\rangle]$ is $p_1 q_1$. Other conditional probabilities are given in Table 1.

Using these probabilities with Bayes' theorem, we have the posterior probability of input variable x_1 being $\langle 1, 1\rangle$ given observed z_1 as follows:

$$Pr[x_1|z_1] = \frac{Pr[z_1|x_1]Pr[x_1]}{\sum_{s=1}^{4} Pr[z_1|x_s]Pr[x_s]} = \frac{Pr[z_1|x_1]\theta_1^{(0)}}{\sum_{s=1}^{4} Pr[z_1|x_s]\theta_s^{(0)}}$$

$$= \frac{\frac{1}{4}p_1 p_2}{\frac{1}{4}p_1 p_2 + \frac{1}{4}p_1 q_2 + \frac{1}{4}q_1 p_2 + \frac{1}{4}q_1 q_2} = p_1 p_2 = \frac{e^{\epsilon_1} e^{\epsilon_2}}{(1 + e^{\epsilon_1})(1 + e^{\epsilon_2})}.$$

Algorithm 1 EM algorithm for PrivKV

$S_1, \ldots, S_n \leftarrow$ key-value data for n responders.
for all $u \in [n]$ **do** sample a tuple $\langle k_a', v_a' \rangle$ from a vector $\boldsymbol{S}_i$
    $v_a^+ \leftarrow VPP(v_a', \epsilon_2)$ and $k_a^* \leftarrow RR(k_a', \epsilon_1)$
end for
$\Theta^{(0)} \leftarrow$ a uniform probability for $X = \{\langle 1, 1 \rangle, \langle 1, -1 \rangle, \langle 0, 1 \rangle, \langle 0, -1 \rangle\}$.
repeat(E-step)
    $t \leftarrow 1$
    Estimate posterior probability $Pr[V_i = 1 | Z_i]$ in Eq. (1).
    (M-step) Update marginal probability $\theta_i^{(t+1)}$.
until $|\theta_i^{(t+1)} - \theta_i^{(t)}| \leq \eta$
for all $a \in K$ **do** estimate
    Estimate $\hat{f}_a$ and $\hat{m}_a$ in Eq. 2.
end for return $\hat{f}_1, \hat{m}_1, \ldots, \hat{f}_d, \hat{m}_d$

With privacy budgets $\epsilon_1 = \epsilon_2 = 1/2$ and $\epsilon = \epsilon_1 + \epsilon_2 = 1$, we estimate $\hat{\theta}_{1,u}^{(1)} \approx$ 0.387455. Posterior probabilities for input $x_2 = \langle 1, -1 \rangle, x_3 = \langle 0, 1 \rangle, x_4 = \langle 0, -1 \rangle$ can be computed similarly.

3.3 Frequency and Mean Estimation

After the EM algorithm estimates the marginal probabilities for binary vector v_a^+, we need to identify the key frequency and mean values in the original key-value data. To estimate these quantities, we show the following property.

Theorem 1 (frequency and mean). *Let* $\Theta^{(t)} = (\theta_{\langle 1,1 \rangle}^{(t)}, \theta_{\langle 1,-1 \rangle}^{(t)}, \theta_{\langle 0,1 \rangle}^{(t)},$ $\theta_{\langle 0,-1 \rangle}^{(t)})$ *be marginal probabilities of the binary-encoded tuples* $\langle k_a^*, v_a^+ \rangle$ *in PrivKV. Then, the expected values for the frequency for key* k_a *and the mean values are*

$$\hat{f}_a = n \left(\theta_{\langle 1,1 \rangle}^{(t)} + \theta_{\langle 1,-1 \rangle}^{(t)} \right), \hat{m}_a = \frac{\theta_{\langle 1,1 \rangle}^{(t)} - \theta_{\langle 1,-1 \rangle}^{(t)}}{\theta_{\langle 1,1 \rangle}^{(t)} + \theta_{\langle 1,-1 \rangle}^{(t)}}. \tag{2}$$

If we have an accurate estimation of marginal probabilities via the EM algorithm, the theorem means that we can estimate the frequency and the mean as well.

Algorithm 1 shows the overall processes in the proposed EM algorithm for estimating frequency and mean for key-value data.

4 Experiment

4.1 Objective

The objective of the experiment is to explore the accuracy improvement in terms of privacy budget ϵ and the number of responders n. Using synthetic data with some common distributions, we compare our proposed algorithm with some conventional ones.

Table 2. Statistics of the synthetic data (# responders $n = 10^5$, and # keys d)

model	$E(f_k/n)$	$Var(f_k/n)$	$E(m_k)$	$Var(m_k)$
Gaussian	0.49506	0.10926	−0.00987	0.43702
Power-law	0.20660	0.062901	−0.58681	0.25160
Linear	0.51	0.08330	0	0.34694

Table 3. Open datasets

item	MovieLens [15]	Clothing [16]
# ratings	10,000,054	192,544
# users	69,877	9,657
# items	10,677	3,183
value range	0.5–5	1–10

Table 4. $MSE(f)[\times 10^{-4}]$ with regard to ϵ

ϵ	Gauss			Power-Law			Linear		
	EM	PrivKV	PrivKVM	EM	PrivKV	PrivKVM	EM	PrivKV	PrivKVM
0.1	756.682	1921.743	1472.772	671.251	2170.253	1851.214	602.837	1885.284	1462.740
0.5	63.996	84.629	75.394	55.478	84.833	62.403	70.346	92.988	82.795
1	18.076	22.588	26.213	18.579	19.274	23.183	16.023	20.174	18.440
3	2.018	2.324	2.508	1.591	2.587	2.420	2.523	2.790	2.597
5	1.147	1.320	1.173	0.973	1.019	0.992	1.283	1.429	1.280

4.2 Data

We use some synthetic data and open datasets for our analysis. For synthetic data, we generate keys and values according to three known probability distributions, Gaussian ($\mu = 0, \sigma = 10$) power-law ($F(x) = (1 + 0.1x)^{-\frac{11}{10}}$), and linear ($F(x) = x$). Table 2 shows the mean and the variance of frequency f_k for key and mean f_k for values, where the number of users is $n = 10^5$.

Table 3 shows the specifications of open datasets used for our analysis. Both datasets have a large number of items, e.g., movie titles and clothing brands. Hence, the use-item matrices are sparse. The values of ratings are distribute normally and the frequency of items follows power-law distributions. Therefore, the synthetic data are models of the real open data.

4.3 Method

We perform the proposed and the conventional algorithms PrivKV and PrivKVM to estimate the key frequency $\hat{f}_k$ and the mean values $\hat{m}_k$ of given n-responder synthetic data. The mean errors for key and value are evaluated by Mean Square Error (MSE) defined as $MSE(f) = \frac{1}{|K|}\sum_{i=1}^{|K|}\left(\frac{\hat{f}_i}{n} - \frac{f_i}{n}\right)^2$, $MSE(m) = \frac{1}{|K|}\sum_{i=1}^{|K|}(\hat{m}_i - m_i)^2$, where f_i and m_i are true statistics. We repeat the measurements for 10 times and take the mean.

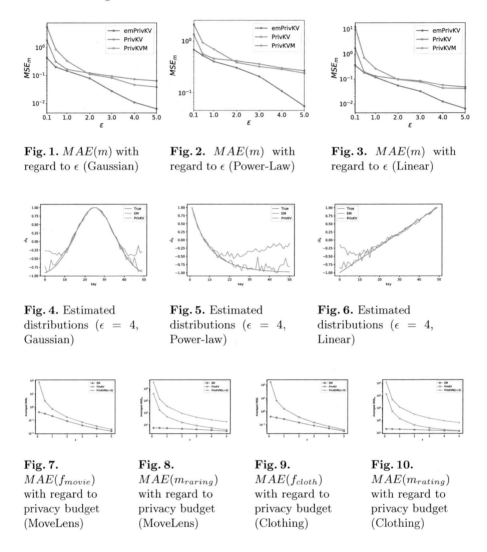

Fig. 1. $MAE(m)$ with regard to ϵ (Gaussian)

Fig. 2. $MAE(m)$ with regard to ϵ (Power-Law)

Fig. 3. $MAE(m)$ with regard to ϵ (Linear)

Fig. 4. Estimated distributions ($\epsilon = 4$, Gaussian)

Fig. 5. Estimated distributions ($\epsilon = 4$, Power-law)

Fig. 6. Estimated distributions ($\epsilon = 4$, Linear)

Fig. 7. $MAE(f_{movie})$ with regard to privacy budget (MoveLens)

Fig. 8. $MAE(m_{raring})$ with regard to privacy budget (MoveLens)

Fig. 9. $MAE(f_{cloth})$ with regard to privacy budget (Clothing)

Fig. 10. $MAE(m_{rating})$ with regard to privacy budget (Clothing)

4.4   Results

4.4.1   Privacy Budget ϵ

Table 4 shows the MSE of frequency estimation of synthetic data generated in Gaussian, Power-law and Linear distributions. We use the MSE with regard to the privacy budget ϵ ranging from 0.1 to 5, the number of responders $n = 10^5$, and the number of keys $d = 50$.

The estimation accuracy of the proposed EM algorithm overwhelms the conventional PrivKV and PrivKVM for all distributions and all privacy budgets. The improvement is significant for $\epsilon = 0.1$, and the MSE of EM algorithm is 602.83×10^{-4} (41% of PrivKVM).

Table 5. $MSE(f)$ $[(\times 10^{-4})]$ with regard to n

$n[10^4]$	Gauss			Power-Law			Linear		
	EM	PrivKV	PrivKVM	EM	PrivKV	PrivKVM	EM	PrivKV	PrivKVM
1	476.259	527.912	612.903	346.709	543.281	424.728	404.467	538.943	733.468
5	107.263	110.045	137.269	72.823	99.442	95.245	89.922	118.344	113.040
10	36.087	51.430	62.515	39.235	51.116	62.976	54.166	69.999	56.959
100	4.636	4.766	6.249	4.876	5.498	5.254	5.619	7.404	6.158
1000	1.410	1.597	1.556	0.760	1.181	1.094	0.974	1.493	1.041

Figures 1, 2, and 3 show the MSE of mean estimations of synthetic data generated in Gaussian, Power-law, and Linear distributions, respectively. The proposed EM algorithm (blue) has the smallest MSE for all algorithms and all cases. The accuracy is improved well as the privacy budget ϵ increases.

We show the estimated mean distributions for keys synthesized in Gaussian, Power-law and Linear distribution, in Figs. 4, 5, and 6, respectively. We find that PrivKV suffers large estimation errors for both edges in Fig. 4, where the frequencies of values are less than that of the center. MLE is not robust when not enough samples are given, whereas EM performs well for even small samples by iterative processes.

Figures 7 and 8 shows the distribution of MAE of frequency of items and that of mean values in the MovieLens dataset [15], respectively. Similarly, the MAEs of frequency and mean values in the Clothing datasets [16] are in Figs. 9 and 10, respectively.

All results show that the proposed ME algorithm estimate the frequencies and the means the most accurately. The improvements of accuracy by the EM estimation are consistent with the results of the synthesized data.

4.4.2   Number of Responders

We evaluate the estimation error with regard to the number of responders (users). We estimate with the fixed privacy budget $\epsilon = 2$ and the number of keys $d = 50$ for the number of users from 10^4 to $30 \cdot 10^4$.

Table 5 shows the MSE for frequency estimation of synthetic data generated in Gaussian, Power-law, and Linear distributions. The EM algorithm has the smallest MSE for all algorithms and all distributions regardless of the number of users. The larger, the less error in general. The estimate improvement is significant when smaller data such as $n = 10^4$ are used. With the result, the EM should be used for the use case where a confidential and rare data are sampled, such as epidemiological study of rare diseases.

Figures 11, 12, and 13 show the distribution of MAE for means with regard to the number of responders n, of the synthetic data generated in Gaussian, Power-law, and Linear distributions, respectively. Similar to the frequency estimation, the EM algorithm always outperforms than any other algorithm for all estimations and distributions. For example, the accuracy improves 31.6% in the Gaussian distribution where there are many less-frequent keys.

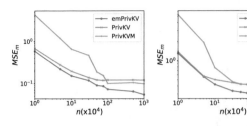
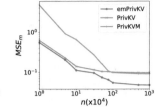

Fig. 11. $MAE(m)$ with regard to number of responders (Gaussian)

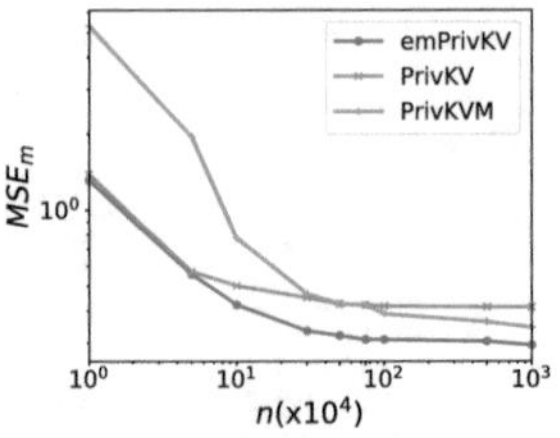

Fig. 12. $MAE(m)$ with regard to number of responders (Power-law)

Fig. 13. $MAE(m)$ with regard to number of responders (Linear)

5 Related Works

The idea to preserve the privacy of input with randomization has been studied so far. Agrawal and Srikant [13] proposed a privacy-preserving collaboration filtering and an estimation algorithm based on Bayes' theorem, called reconstruction.

Chen et al. [12] proposed the notion of LDP to provide a privacy guarantee for the user. Compared with the conventional differential privacy studies, LDP has been used for many real-world applications. For example, Erlingsson et al. introduced RAPPOR [6] to use a Bloom filter to encode input as a bit of a vector.

Ren et al. proposed a multidimensional joint distribution estimation algorithm that satisfies LDP [11]. Their proposed method is also based on the EM and Lasso regression. They reported the experimental results on real-world datasets and showed that the proposed algorithm outperforms the existing estimation schemes such as support vector machine and random forest classifications.

Gu et al. [14] proposed a locally differentially private key-value data collection that utilizes correlated perturbation. Their protocol uses an advanced Padding-and-Sampling with two primitives, PCKV-UE (Unary Encoding) and PCKV-GRR (Generalized Randomized Response) to improve the accuracy of mean estimation and does not require further interaction between responders and collector.

6 Conclusion

We study the LDP algorithms for key-value data that estimate the key frequencies and the mean values. We propose an algorithm based on the EM algorithms to improve the estimation data accuracy perturbed in the LDP algorithms PrivKV and PrivKVM. Our proposed algorithm estimates the marginal probability of variable X that is a tuple of binary-encoded keys and values and hence can reduce the conditional probabilities needed for iterative processes. With some synthetic data generated in Gaussian, Power-law, and Linear distributions, we conduct experiments that show the proposed estimation has higher accuracy than the PrivKV algorithm. The estimate is robust for privacy budgets. The improvement was 69.5% on average when the number of responders was $n = 10^4$ with $\epsilon = 0.1$.

Major open datasets, MovieLens (10,000,000 records) and Clothing (192,000 records), are used to ensure the performance of the propose algorithm as estimated in the synthetic data. The experimental results confirm that the proposed EM algorithm outperforms any of state-of-the-art LDP schemes, PrivKV and PrivKVM, for all privacy budgets and for both of key frequencies and mean values. The improvement in estimation is especially significant when small privacy budget is used for randomization (strong privacy level). The EM algorithm estimates the mean well even when the frequency of the key-value record is limited. Hence, we conclude that the proposed EM algorithm is appropriate for private data analysis in epidemiological purposes that requires dealing with rare decease.

For future works, we plan to compare the utility improvement with PCKV [14], the latest version of PrivKV family which improves accuracy without expensive iterations incurred by PrivKVM.

Acknowledgment. Part of this work was supported by JSPS KAKENHI Grant Number JP18H04099 and JST, CREST Grant Number JPMJCR21M1, Japan.

References

1. Kairouz, P., Oh, S., Viswanat, P.: Extremal mechanisms for local differential privacy. In: NIPS, pp. 2879–2887 (2014)
2. Warner, S.L.: Randomized response: a survey technique for eliminating evasive answer bias. J. Am. Stat. Assoc. **60**, 63–69 (1965)
3. Nguyên, T.T., Xiao, X., Yang, Y., Hui, S.C., Shin, H., Shin, J.: Collecting and analyzing data from smart device users with local differential privacy (2016). arXiv:1606.05053
4. Ye, Q., Hu, H., Meng, X., Zheng, H.: PrivKV : key-value data collection with local differential privacy. In: IEEE S&P, pp. 294–308 (2019)
5. McSherry, F.: Privacy integrated queries: an extensible platform for privacy-preserving data analysis. In: SIGMOD, pp. 19–30 (2009)
6. Erlingsson, Ú., Pihur, V., Korolova, A.: RAPPOR: randomized aggregatable privacy-preserving ordinal response. In: ACM Conference on Computer and Communications Security, pp. 1054–1067 (2014)
7. Differential Privacy Team: Learning with privacy at scale. Apple Mach. Learn. J. **1**(8) (2017)
8. Learning with Privacy at Scale. https://machinelearning.apple.com/. Accessed 2019
9. Dwork, C., Roth, A.: The algorithmic foundations of differential privacy. Found. Trends Theor. Comput. Sci. **9**(3–4), 211–407 (2014)
10. Miyagawa, M.: EM algorithm and marginal applications. Adv. Stat. **16**(1), 1–19 (1987). (in Japanese)
11. Ren, X., et al.: LoPub?: high-dimensional crowdsourced data publication with local differential privacy. IEEE Trans. Inf. Forensics Secur. **13**(9), 2151–2166 (2018). https://doi.org/10.1109/TIFS.2018.2812146

12. Chen, R., Li, H., Qin, A.K., Kasiviswanathan, S.P., Jin, H.: Private spatial data aggregation in the local setting. In: Proceedings IEEE ICDE, pp. 289–300 (2016)
13. Agrawal, R., Srikant, R.: Privacy-preserving data mining. ACM SIGMOD **2000**, 439–450 (2000)
14. Gu, X., Li, M., Cheng, Y., Xiong, L., Cao, Y.: PCKV: locally differentially private correlated key-value data collection with optimized utility. In: 29th USENIX Security Symposium (USENIX Security 20), pp. 967–984 (2020)
15. MovieLense 10M Dataset. https://grouplens.org/datasets/movielens/. Accessed 2022
16. Clothing Fit Dataset for Size Recommendation

MPolKA-INT: Stateless Multipath Source Routing for In-Band Network Telemetry

Isis de O. Pereira[1]([✉]), Cristina K. Dominicini[1], Rafael S. Guimarães[2],
Rodolfo S. Villaça[3], Lucas R. Almeida[3], and Gilmar Vassoler[1]

[1] Federal Institute of Education, Science and Technology of Espírito Santo,
Serra, Brazil
isisolip@gmail.com
[2] Federal Institute of Education, Science and Technology of Espírito Santo,
Cachoeiro de Itapemirim, Brazil
[3] Federal University of Espírito Santo, Vitoria, Brazil

Abstract. Real-time monitoring and measurement are the basis for most management operations, such as traffic engineering, quality of service assurance, and anomaly detection. Nevertheless, collecting measurements from all network devices with sampling and polling-based methods is not scalable. To tackle this issue, the *P4 Language Consortium* proposed the *In-band Network Telemetry (INT)* framework that provides real-time and fine-grain measurements in the data plane using telemetry packets. However, the task of specifying the route taken by a telemetry probe still relays on the traditional routing protocols, which require state changes in routing tables and fail to achieve the required accuracy, coverage and latency. In this context, this article investigates how to combine INT with a source routing method based on the stateless Multipath Polynomial Key-based Architecture (M-PolKA). We implemented this multipath telemetry solution as MPolKA-INT using the P4 language, and the experimental results showed low overhead in the data and control planes with agile and flexible path (re)configuration, since it does not require state changes in the routing tables of the devices in the network core.

1 Introduction

Monitoring the state of the network provides a view of its behavior, which is the basis for other management operations, such as traffic engineering, quality of service assurance and anomaly detection [12]. These operations require dynamic services, because the network state frequently changes due to modifications in the traffic patterns or in the underlay network. As a result, real-time network monitoring has become essential to identify events related to failures, performance, and security [7].

Traditional network management protocols, such as SNMP (Simple Network Management Protocol), fail to achieve the novel requirements of accuracy, coverage and latency [4]. Conversely, software-defined networks (SDN) can provide a more efficient monitoring solution, since a logically centralized controller

© The Author(s), under exclusive license to Springer Nature Switzerland AG 2023
L. Barolli (Ed.): AINA 2023, LNNS 654, pp. 513–524, 2023.
https://doi.org/10.1007/978-3-031-28451-9_45

can request various measurements from the network nodes. However, collecting measurements from each node is not scalable as it increases control traffic overhead [11].

To tackle this problem, the P4 Language Consortium proposed the In-band Network Telemetry (INT) framework for collecting real-time data plane information from programmable switches [8]. Specific INT packets, called probes, are forwarded along a predetermined path, and each node in that path extends the packet by attaching its own measurements (e.g., queuing latency, queue depth, and packet metering). Thus, it may incur significant bandwidth overhead, and the packet length grows linearly with the path size. The last node in the telemetry path receives collective information from multiple nodes in a single INT packet, which is sent to the controller to provide real-time data about the network devices and the traffic state of the path.

However, the INT framework does not specify the route of the telemetry probes, which depends on traditional routing mechanisms. For example, if the INT header is embedded in an IP packet, the probe forwarding will be passively decided by its destination IP address according to the routing tables of the network devices [2]. Due to this uncontrollable probing path, it is not easy to cover all network nodes with minimal overlap, requiring redundant flows. Additionally, the control plane has to process the telemetry workload and constantly update the routing tables.

In this context, an interesting strategy is to use source routing (SR), which allows the sender of a packet to specify its path through the network, by adding a routing label into the packet header [2]. Related works incorporated SR into INT probes [9], and proposed path planning algorithms to generate less overlapping probe paths [10]. Nevertheless, as these proposals represent the route as a list of nodes, they are not able to represent multipath trees and cause redundancy of telemetry data. On the other hand, other works explore multipath telemetry [13] to eliminate the redundancies caused by replicating packets in unicast probes, but, as they do not use SR, they increase the number of network states required to configure flow tables.

This work proposes a multipath telemetry solution for programmable networks, ensuring the collection of information from all nodes of interest, with low overhead in the control and data planes. This is achieved by reducing the redundancy of telemetry data with multicast trees, as well as eliminating the routing tables with a stateless core network. To this end, we investigate how to combine INT with a multipath SR solution based on the Multipath Polynomial Key-based Architecture (M-PolKA) [3]. We implemented a prototype of our proposal, named MPolKA-INT, in the P4 language and showed its feasibility through functional and scalability experiments with real-world topologies. Experimental results demonstrate that MPolKA-INT guarantees low overhead without using routing tables and allows flexible and agile path (re)configuration for the INT probes.

Table 1. Comparison of MPolKA-INT with related works

Related works	Source routing	Multipath telemetry	Fix. length SR header	No tables
INT-Path [9]	✓			✓
INT-Probe [10]	✓			✓
SR-INT [1]	✓		✓	
MPINT [13]		✓	✓	
MPolKA-INT	✓	✓	✓	✓

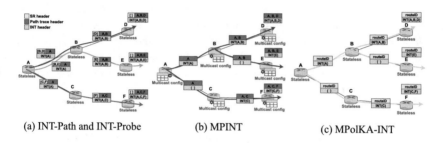

(a) INT-Path and INT-Probe   (b) MPINT   (c) MPolKA-INT

Fig. 1. Comparison of INT solutions

The paper is structured as follows. In Sect. 2, we review the related works. In Sect. 3, we specify our proposal. We provide our results and discussions in Sect. 4, with conclusions and future works in Sect. 5.

2   Related Works

INT-Path [9] was a pioneering study that incorporated SR into INT to allow the specification of probing paths as a list of nodes included in the packet header. However, its performance depends on the number of odd-degree vertices in the network, limiting applicability and scalability, and the probe connection points are topology sensitive. INT-Probe [10] solves the problem in a similar way, but proposes an INT path planning algorithm for full network coverage, under the restriction of stationary probes. It aims to generate less overlapping probing, but it uses a variable-length SR header, which can lead to problems related to the MTU limits of the network.

On the other hand, SR-INT [1] uses a fixed length header and mitigates the INT and SR overheads by exploring the Segment Routing protocol, but it still requires the configuration of routing tables in the network. MPINT [13] innovates in applying INT to multicast traffic. In this way, it reduces the redundancy of telemetry data during the replication of packets, decreasing the overhead for both data and control planes. However, it does not use SR, and also depends on routing tables, which increases the number of network states. As shown in the Table 1, these related works either implement SR or multipath telemetry. Thus, MPolKA-INT differs from them as it combines INT with a stateless (no tables) multipath SR based on M-PolKA to perform efficient network monitoring. To the best of our knowledge, this dilemma had not yet been addressed in the literature.

Figure 1 details the differences between the related works. INT-Path and INT-Probe (Fig. 1a) generate overhead in the data control planes, due to the amount of telemetry and path data generated by replicating the INT probes for each fork (i.e., A-B-D, A-B-E and A-C-F). MPINT can mitigate such overhead, but rely on routing tables (Fig. 1b). In addition, it increases the number of network states, because it needs to configure the multicast groups in each switch.

In contrast, MPolKA-INT (Fig. 1c) does not suffer from the limitations of table-based per-hop methods, such as the scalability of the number of states and the latency for path modification, neither the limitation of list-based SR solutions, which cannot represent multicast trees in a topology-agnostic manner [3]. This is achieved by embedding a route label within the packets, which represents the multicast probing path and can be decoded in an arithmetic form by the programmable switches (more details in Sect. 3). Likewise, there is a reduction in the INT overhead, due to the elimination of INT data replication at each fork. Also, as the route label already represents the path, there is no need to record path trace information.

MPolKA-INT also solves the problem of agile configuration of probing paths (e.g., for failover or quality of service assurance). In table-based approaches, there is a huge latency for path (re)configuration and convergence, since the controller needs to apply the changes to all affected nodes along the path. With MPolKA-INT, the path can be quickly reconfigured just by setting a new route label at the edge.

3   Proposal

This section details the extension of M-PolKA, a multipath source routing paradigm, to enable support for the INT framework.

3.1   M-PolKA: Multipath SR

M-PolKA is a multipath SR solution based on the Residue Number System (RNS) encoding and the polynomial arithmetic using Galois field (GF) of order 2 [3]. It was selected for this work, because is the only SR method that can encode a route label representing a multipath cycle or tree in a topology agnostic manner with a full stateless network core and a fixed header [3]. In this scheme, at core nodes, the transmission states of the output ports are given by the remainder of the binary polynomial division (i.e., a *mod* operation) of the route identifier of the packet by the node identifier of the network device. Its implementation in programmable switches rely on a packet cloning mechanism and the reuse of the cyclic redundancy check (CRC) hardware support, which enables the *mod* operation.

As shown in Fig. 2a, the M-PolKA architecture is composed by: (i) edge nodes (yellow), (ii) core nodes (blue) and (iii) an SDN controller (green), responsible for the node configuration. The SR has three polynomial identifiers: (i) nodeID: a fixed identifier assigned to core nodes by the controller in a network configuration phase; (ii) portID: an identifier assigned to the transmission state of the output ports in each core node; and (iii) routeID: a multipath route identifier,

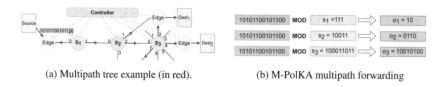

(a) Multipath tree example (in red).   (b) M-PolKA multipath forwarding

Fig. 2. M-PolKA routing scheme.

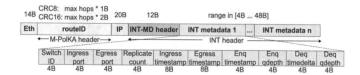

Fig. 3. Packet format for MPolKA-INT.

calculated by the controller using the chinese remainder theorem and embedded in the packet by the edge nodes. In the example of Fig. 2a, after calculating the routeID, the controller installs flow entries at the edge to embed the routeID 10101100101100 in the packets of that flow. Therefore, each node calculates its portID by dividing this routeID by its nodeID, as shown in Fig. 2b. For example, the remainder of $R(t) = 10101100101100$ when divided by $s_2(t) = 10011$ is $o_2(t) = 0110$. Thus, in s_2, ports 1 and 2 forward this flow to the next hop, while ports 0 and 3 do not forward it.

3.2 MPolKA-INT: Multipath SR for In-Band Network Telemetry

As M-PolKA does not originally support telemetry, MPolKA-INT proposed how to combine its multipath SR with the INT framework to provide a stateless multipath telemetry solution that can reduce the redundancy of the telemetry information. The architecture is composed of three elements: (i) firstly, the INT generator (root node) creates the telemetry probe and inserts the routeID, which was previously obtained from the controller, into the packet header of the probe; (ii) then, the INT forwarders add the telemetry data into the INT header stack, and perform the routing of INT packets according to the routeID; and (iii) finally, in the leaf nodes, the INT collectors forward the INT packets to the controller.

Figure 3 illustrates the packet format for MPolKA-INT, which follows the INT specification [8] for the INT header. The INT-MD field contains INT instructions and the metadata (MD) that will be inserted into packets. The INT metadata stack collects internal device information at each node in the telemetry path. The M-PolKA header contains the routeID, calculated as described in Sect. 3.1, and the implementation of the *mod* operation rely on the embedded CRC operation, which usually supports CRC8, CRC16 and CRC32 algorithms. The length of the routeID depends on the size of the topology and is given by multiplying the maximum number of hops by 1B (CRC8, topology size up to 30 nodes), or 2B (CRC16, topology size up to 4080 nodes) [3]. The MPolKA-

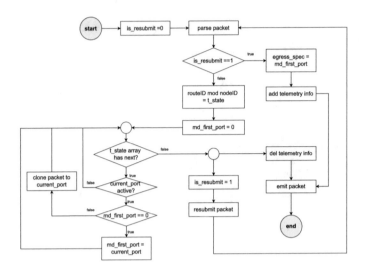

Fig. 4. MPolKA-INT's P4 pipeline

INT's P4 pipeline is detailed in Fig. 4, where packets are cloned to all active ports of the transmission state vector. To eliminate the redundancy, we use the resubmit feature of P4 programmable switches to keep the INT metadata stack only in the packet that is steered in the first active output port, as shown in Fig. 1c (e.g., forks at A-B-D, A-B-E and A-C-F). To enable the addition of the telemetry data in the packet header with minimal redundancies, this pipeline is significantly different from the original pipeline proposed by MPolKA [3], which is only concerned with cloning the packets according to the route label.

When the packet arrives in the ingress control, the switch checks if the packet has already been resubmitted. If false, the switch performs a *mod* operation between the packet's routeID and its own nodeID to get the transmission state (*t_state*) of the ports. Then, the switch iterates the (*t_state*) using a chain of *if*(*<condition>*) statements. When identifying the first port, its value is stored in the *md_first_port* metadata, and the *is_resubmit* metadata is changed to 1. Next, after resubmitting the packet at the ingress, the switch changes *egress_port* metadata to the value of *md_first_port* and forwards to egress control. At the same time, in the first run, if there are other active ports in the state vector, other pipelines are created to replicate the packets via the clone function and directed to egress control. In the egress, the metadata *is_resubmit* is verified. If true, the telemetry metadata is pushed to the stack, and the packet is emitted. If false, a pop operation is performed on the entire INT metadata stack and the packet is emitted without telemetry information.

Analyzing the operation of MPolKA-INT (Fig. 1c) and MPINT (Fig. 1b), we can see that MPINT has the path trace header, which records the nodes traversed in the path. Otherwise, in MPolKA-INT, all information related to the path is obtained via the routeID, which remains unchanged during the entire path. Regarding the multipath forwarding, in MPINT, each switch executes a table lookup to perform the multicast operation, while M-PolKA only requires an arithmetic operation on the routeID to calculate the transmission state. MPINT depends on configuring multicast groups in P4, which results in the following state entries in each switch of the network [13]: one state for creating a multicast group, and two states for each transmitting port (one for creating the port node, and other for associating the port to the multicast group). In MPolKA-INT, the packet is cloned to all ports with an active transmission state, and changing a multicast tree is only a matter of changing the routeID in the ingress (the network edge) of the probe packet.

4   Evaluation

Two experiments were executed to evaluate its main functionalities and overhead in comparison to MPINT: (i) an emulated prototype to validate the P4 implementation, and (ii) a scalability analysis with real-world topologies.

4.1   Emulated Prototype

We built a emulated prototype in Mininet [6] using the $P4_{16}$ language, the BMv2 software switch, and the *simple_switch* (v1.15.0) with the v1model architecture as target. The hardware configuration consists of a Dell PowerEdge T640 server, with an Intel Xeon Silver 4114 2.2 GHz processor and 96 GB of RAM. Although our tests could use CRC8 for topologies up to 30 nodes, since the *simple_switch* only supports the definition of 16-bit and 32-bit CRC polynomials, the P4 implementation adopted CRC16. However, when comparing data and control plane overheads, the calculations considered both CRC8 and CRC16 polynomials. The Original INT approach in the tests refers to the INT specification [8]. For all the approaches, we used the INT headers according to this specification (12 Bytes: INT-MD header, 48 Bytes per hop: INT metadata stack, 1 Byte: path header, 4 Bytes: path metadata stack), in addition to the Ethernet (14 Bytes) and IP (20 Bytes) headers.

Data Plane and Control Plane Overhead: Figure 5 shows a 7-node topology used as an example to compare the Original INT, MPINT and MPolKA-INT approaches. As the topology size is 7, the size of the M-PolKA header was set to 7 Bytes for CRC8, and 14 Bytes for CRC16. The total data plane overhead is the sum of the Bytes sent in the links that connect the nodes (in red). The control plane overhead is the sum of the Bytes sent to the controller (in gray).

Figure 5a analyzes the Original INT approach. SW1 receives three probing packets (to cover all nodes of interest, including the branch in SW6) with Ethernet, IP, INT-MD and path headers (47B per packet). For each packet, it updates the path header and adds the telemetry and path information stacks from SW1 (99B in total). Then, it sends one packet to SW5, and two packets to SW2. In a similar way, at each next hop, 52B of telemetry information is added to the probing packet. Finally, each collector switch (SW3, SW4 and SW7) extract the INT and path metadata stacks and send them to the control plane for further analysis.

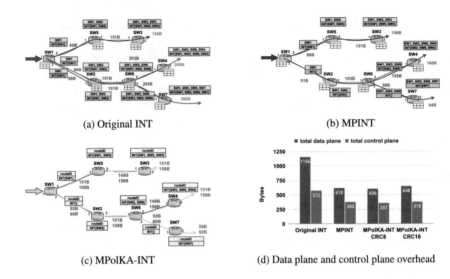

(a) Original INT

(b) MPINT

(c) MPolKA-INT

(d) Data plane and control plane overhead

Fig. 5. Comparison of the overhead in data and control planes.

Figure 5b analyzes the MPINT approach. SW1 receives a single probing packet, adds the SW1 switch ID to the path metadata stack, adds telemetry data, and replicates the packet to SW5 and SW2. Before sending the packet to SW2, it pops the telemetry metadata stack, because the INT metadata have already been sent to SW5. As a result, 51B is sent to SW2 and 99B is sent to SW5. Then, at each next hop, 52B of information is incremented, referring to the telemetry and path information stacks. However, when there is a replication, as in the case of SW6 for output port 2, the INT metadata stacks are removed, increasing only 4B per hop of the path stack with the current switch ID information. As in the Original INT approach, collector switches (SW3, SW4 and SW7) send telemetry information to the control plane.

Figure 5c shows the approach of MPolKA-INT. In addition to the Ethernet, IP and INT headers, the M-PolKA header is included, with the routeID information. SW1 receives a single probing packet, adds telemetry data, and clones the packet to SW5 and SW2. Since it is not necessary to add path tracing headers, it only adds the INT metadata stack, which is incremented by 53B (CRC8) or 60B (CRC16), resulting in 101B (CRC8) or 108B (CRC16). Similar to MPINT, the telemetry metadata stack is popped before sending the packet to SW2 in order to eliminate telemetry redundancy. So, 53B (CRC8) or 60B (CRC16) are sent to SW2. At each next hop, 48B of telemetry information is added to the probing packet. On nodes with replication, as in the case of SW6, the telemetry stacks are incremented by 48B for the first port, while the INT metadata stack is removed for the other ports.

In the results shown in Fig. 5d, it is possible to observe that the number of probes and the telemetry redundancy directly affect the overhead generated in data and control planes, as the overheads in Original INT is much greater

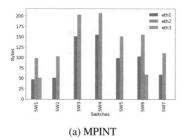

(a) MPINT

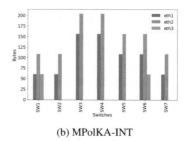

(b) MPolKA-INT

Fig. 6. Data transferred on input and output switch ports

than the other approaches. Also, there is no significant difference in the overhead generated in the data and control planes when comparing MPINT and MPolKA-INT for this specific topology. However, MPolKA-INT gains in guaranteeing a full stateless and flexible core, where telemetry probing can be easily configured at the source.

Functional Test: The goal of this experiment is to measure the bytes transmitted on each input and output port of the switches of the topology shown in Fig. 5, for the MPINT and MPolKA-INT solutions, in order to validate if the P4 implementations are correctly representing the pipelines of both approaches. For each test run, the *bwm-ng* tool captured the traffic on the switch ports. One probing packet was sent via a *scapy* Python script, and the result is plotted in Fig. 6. This proves the correct operation of the P4 implementations, validating the numerical value results for all the network links, as calculated in Fig. 5d.

4.2 Scalability Analysis in Real-World Topologies

This section aims to evaluate the scalability of the MPolKA-INT solution with respect to two different variables: the number of nodes in the network and the number of probes required to cover all nodes in the network. Both variables depend on the network topology, and we evaluate the packet size overhead in both data and control planes, measured in bytes, and the number of state entries in each network device. The lower the overhead and number of states, the better the solution.

As different topologies are required to perform these evaluations, we used a set of topologies of different network operators publicly available at the Topology Zoo [5]. The software tools developed in this paper for the scalability analysis were made available on Github[1] for reproducibility reasons. To improve justice in the comparisons between MPolKA-INT, MPINT, and the Original INT, we limited the evaluations to topologies with at most 30 nodes, resulting in 88 topologies.

Overhead Due to INT Probes: Each INT probe adds an overhead to the packets responsible for collecting data from the nodes under measurement, and,

---

[1] https://github.com/nerds-ufes/mpolka-int.

consequently, to the network. As the M-PolKA header is different from the MPINT and Original INT headers (see Fig. 5), and they use different path encodings, it is expected to observe different overheads between them in the control and data planes.

The overhead in the data plane counts (in bytes) the INT header and the space required for the path encoding scheme for M-PolKA, MPINT, and the Original INT. This overhead increases at each node in the path from ingress to egress in the network. On the other hand, the overhead in the control plane counts (in bytes) only the messages to collect these probes at the egress, from the data plane to the control plane. For MPolKA-INT, two different evaluations were considered for both control and data planes: one using CRC8 for path encoding and the other using CRC16.

Figure 7a shows the results of this evaluation, considering 88 networks of different sizes from the Topology Zoo dataset. In this figure, we modeled the topologies as a graph and calculate the Minimum Spanning Tree (MST) using Prim's algorithm. For each MST given, we choose the optimal node sender for the Original INT approach as our baseline. As expected, the overhead highly depends on the network size and on its topology, presenting an almost linear increase for both data and control plane. It is also important to notice that both MPolKA-INT and MPINT have lower overhead than the Original INT, while the overhead of our MPolKA-INT proposal, especially with the CRC8 encoding, is very close to the overhead of the MPINT solution, which is the lowest for the selected topologies.

Number of State Entries in each Network Device: The most important characteristic of the MPolKA-INT solution is the complete absence of state configurations in the core, while the MPINT solution requires routing table (re)configurations for each new probe launched in the network. Each new (re)configuration task in each network device (each node in the topology) is counted as 1 state entry. In this section, we aim to evaluate how the number of state entries in each node varies according to the number of nodes and the probe replication average in each node. In this evaluation, we used the same dataset with 88 topologies.

In Fig. 7b, we present the results of this evaluation as a heatmap. The x-axis represents the replication average of each node, which is highly dependent on the average node degree of the network, while the y-axis represents the number of nodes in the network. Some selected topologies in the dataset are plotted over the heatmap to improve the comprehension of the results in real network topologies. It is also important to highlight that **the number of state entries for the MPolKA-INT solution is equal to zero**, so, it is not depicted in the figure. As expected, this state overhead in the MPINT solution increases linearly according to the number of nodes and the replication average of each node in the network. This is a very important result for MPolKA-INT, because it enables path (re)configuration for each probe without the need to modify any state in the core of the network under monitoring.

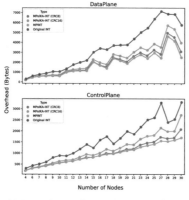

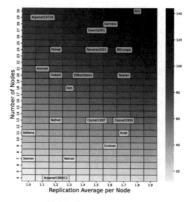

(a) Data plane and control plane overhead

(b) MPINT: Nr. of state entries

Fig. 7. Scalability analysis in real topologies

5   Conclusion

In this article, we presented MPolKA-INT, an M-PolKA extension for the INT framework that explores the stateless multipath SR capabilities of M-PolKA [3] to mitigate the network state maintenance overhead for agile path (re)configuration of INT probes. MPolKA-INT was implemented using the P4 network programming language and evaluated in Mininet with the BMv2 software switch. The scalability of MPolKA-INT was evaluated using graph theory insights and a set of heterogeneous network topologies. Our experimental results showed that MPolKA-INT adds no significant overhead in the data and control plane when compared to related work in the literature. However, the MPolKA-INT approach offers a complete stateless and agile multipath network solution with SR capabilities, enabling network managers to express the complete path to be followed by the telemetry probe. Additionally, the results show that, when compared to the traditional approaches that use unicast routing for the INT probes, MPolKA-INT has the advantage of eliminating redundancy in both the number of INT probes in the network, and also in the number of monitoring data collected from the nodes that appear in multiple paths. In future works, we intend to improve the P4 implementation of MPolKA-INT in order to reduce the overhead generated in the network and, also to develop control plane applications to optimize the monitoring task with the best path and node selection for the INT probes according to the network topology.

Acknowledgment. This work was a recipient of the 2021 Google Research Scholar Award, and the 2022 Intel Fast Forward Initiative. It was also supported by: FAPES (941/2022, 1026/2022), CAPES (Programa de Desenvolvimento da Pós-Graduação - PDPG - Parcerias Estratégicas nos Estados) and FAPESP/MCTI/CGI.br (2020/05182-3).

References

1. Bosshart, P., et al.: P4: programming protocol-independent packet processors. ACM SIGCOMM Comput. Commun. Rev. **44**(3), 87–95 (2014)
2. Dominicini, C., et al. Deploying polka source routing in p4 switches : (invited paper). In: 2021 International Conference on Optical Network Design and Modeling (ONDM), pp. 1–3 (2021)
3. Guimarães, R.S., et al.: M-PolKA: multipath polynomial key-based source routing for reliable communications. IEEE Trans. Netw. Serv. Manage. **19**(3), 2639–2651 (2022)
4. Hare, C.: Simple network management protocol (SNMP) (2011)
5. Knight, S., Nguyen, H.X., Falkner, N., Bowden, R., Roughan, M.: The internet topology zoo. IEEE J. Sel. Areas Commun. **29**(9), 1765–1775 (2011)
6. Lantz, B., Heller, B., McKeown, N.: A network in a laptop: rapid prototyping for software-defined networks. In: Hotnets-IX, New York, NY, USA. Association for Computing Machinery (2010). ISBN 9781450304092
7. Lin, W.-H., et al.: Network telemetry by observing and recording on programmable data plane. In: 2021 IFIP Networking Conference (IFIP Networking), pp. 1–6 (2021)
8. P4.org. In-band network telemetry (INT) dataplane specification, November 2020. https://p4.org/p4-spec/docs/INT_v2_1.pdf
9. Pan, T., et al.: Int-path: towards optimal path planning for in-band network-wide telemetry. In: IEEE INFOCOM 2019-IEEE Conference on Computer Communications, pp. 487–495. IEEE (2019)
10. Pan, T., et al.: Int-probe: lightweight in-band network-wide telemetry with stationary probes. In: 2021 IEEE 41st International Conference on Distributed Computing Systems (ICDCS), pp. 898–909 (2021)
11. Simsek, G., Ergenç, D., Onur, E.: Efficient network monitoring via in-band telemetry. In: 2021 17th International Conference on the Design of Reliable Communication Networks (DRCN), pp. 1–6 (2021)
12. Tsai, P.-W., Tsai, C.-W., Hsu, C.-W., Yang, C.-S.: Network monitoring in software-defined networking: a review. IEEE Syst. J. **12**(4), 3958–3969 (2018)
13. Zheng, Y., Pan, T., Zhang, Y., Song, E., Huang, T., Liu, Y.: Multipath in-band network telemetry. In: IEEE INFOCOM 2021 - IEEE Conference on Computer Communications Workshops (INFOCOM WKSHPS), pp. 1–2 (2021)

A Vulnerability Detection Method for SDN with Optimized Fuzzing

Xiaofeng Chi[1], Bingquan Wang[1], Jingling Zhao[1(✉)], and Baojiang Cui[2]

[1] School of Computing, Beijing University of Posts and Telecommunications, Beijing, China
{superfeiyou,bingquan.wang,zhaojingling}@bupt.edu.cn
[2] School of Cyber Security, Beijing University of Posts and Telecommunications, Beijing, China
cuibj@bupt.edu.cn

Abstract. In response to the rapid development of the big data era, the concept of Software Defined Networking (SDN) has been introduced and is gradually being applied to business. However, while this new network architecture is flexible enough to handle network data requirements. It is also inherently exposed to a wide range of security threats. In this paper, we propose an intelligent variation strategy to improve the detection efficiency and scope of SDN vulnerability scenarios based on an optimized fuzzing method. It performs vulnerability detection on SDN and prevents possible hazards more effectively. Based on the experimental results, this paper adds two new types of threat scenarios to the existing SDN vulnerability knowledge base and improves the detection efficiency by about 29% .

1 Introduction

Large-scale network service often requires the collaborative efforts of different network data vendors. The management and operations of equipment by each vendor brings a significant drain on human and material resources. This led to the introduction of the Software Define Network (SDN). SDN is a network architecture proposed by the ONF (Open Networking Foundation) and is logically structured into three layers. The key technologies in the SDN include southbound network control and northbound service support. Southbound networking technology is designed for network topology management, forwarding and processing of flow tables, data link exploration and network scheduling policy development, monitoring, and counting network traffic on physical devices at the data layer according to southbound protocols (mainly OpenFlow).

Northbound service support aims to provide an abstract view of network resources for services through relevant interfaces, however, as its design is often related to specific business requirements, it is diversed and no uniform standard exists in the industry yet. Therefore, this paper focuses on the detection and discussion of vulnerabilities for the southbound interfaces of SDN.

Fuzzing is an important method for detecting software threats and uncovering software vulnerabilities by providing unintended inputs to the target and monitoring the target's state to find anomalies. In this paper, we generate test cases by splitting input structure and grouping variants parts of the structure [20]. At the same time, we design a test case matrix generation algorithm based on different variation factors of SDN to evaluate the effectiveness of each test case on the target. Test cases with significant impact on the target are assigned with more effective variation weights in the test case matrix, and such variation factors will normally guide the next variation process more effectively. Test cases with more effective variation weights will receive a higher variation priority in the set. This allows the algorithmic framework to intelligently update the variation policy through self-adaptation and self-evaluation. In addition, the matrix-based high-dimensional data operations in different feedback algorithms tend to consume more time, which limits the feedback efficiency of the testing framework. We propose similarity aggregation calculations on these test case matrices according to different variation factors of SDN, so as to realize the reduction of the test case set.

2 Related Work

Unlike traditional network architectures, the decoupled nature of SDN makes it difficult to find a counterpart to network security in traditional network security technologies. As a result, traditional network vulnerability detection tools also suffer in SDN vulnerability detection.

Traditional network security tools, such as Metasploit, Nessus and Nmap, are equipped with a rich library of vulnerabilities and combinable attack modules. However, as these tools are dedicated to traditional and wide-area networks, they are not suitable for SDN. The detection of SDN vulnerabilities requires layer-by-layer analysis and testing, especially for critical attacks that directly affect network availability and confidentiality. This suggests that there is a level of sophistication in SDN-specific security threats that cannot be detected by existing network security testing tools, as they are unable to sense SDN.

The detection of SDN vulnerabilities has benefited from the work of many national and international scholars on SDN security and vulnerability analysis techniques. Benton and Kreutz have each examined the avenues of attack present in SDN [1, 22]. Benton points out that threats due to the vendor's failure to adopt TLS for the OpenFlow may make man-in-the-middle (MITM) attacks and Kreutz argues that new features and capabilities in SDN such as centralized controllers and programmability of the network introduce new threats. In addition, several researchers such as Dhawan, Hong, Porras have proposed other problems in SDN such as inter-application conflicts, access control, topology manipulation and shared relationships [11, 15, 23]. Röpke and Holz et al. demonstrate that SDN applications can launch stealth attacks and discuss how they can be exploited through third-party SDN app shops to easily distribute these applications [8]. Even if no malicious SDN applications are provided, Dover et al. also show that DoS and spoofing attacks can be launched by exploiting an execution vulnerability presented in the switch management module of the Floodlight controller [11]. Seungsoo Lee et al. propose and design an SDN evaluation framework (AudiSDN), which summarizes the security flaws in different environments and explores the operating principles of attack scenarios by deploying components to replicate publicly available attack scenarios [19]. In addition, unknown security threats in SDN are discovered and evaluated using fuzzing by deploying virtual components and executing different test cases. The evaluation framework tests current publicly available threat scenarios in a more systematic manner.

Existing research efforts have each focused on different SDN components, and vulnerability detection allows for attack detection on different components, including threats such as malicious application protection at the application layer, control flow tampering at the control layer, and communication hijacking at the data layer. There are still limitations in assessing the trustworthiness of applications and the underlying network in SDN. Although several studies have been conducted on SDN security threats, current controllers are still vulnerable to many of these attacks. Several works to assess SDN security have also been proposed by Jero, Ujcich et al. [17, 18]. However, these works have mainly focused on manipulating OpenFlow messages by sniffing the control channel between the controller and the switch.

3 Our Method

We propose a vulnerability detection method for SDN with optimized fuzzing, which consists of two main components. First, based on SDN, we parse the mutation nodes in test cases. In addition, according to different mutation node types, we design mutation dictionaries. Second, we designed matrix generation and clustering algorithms for test cases and prioritized them by weight.

3.1 Variant Node Resolution and Dictionaries Designed for Mutation

The decoupled nature of SDN allows for relative independence between components, making it more difficult for traditional byte-based variants to generate valid input. And it provides developers with a southbound interface and a northbound interface. The southbound interface provides state control and action commands to underlying physical devices such as SDN switches. The OpenFlow protocol has evolved to versions 1.0 to 1.5, with versions 1.0 and 1.3 being the most widely used variant policies for SDN nodes are shown in Fig. 1.

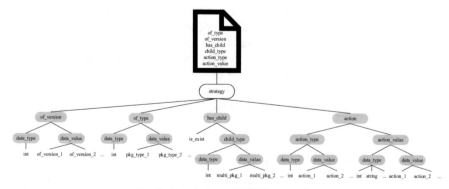

Fig. 1. Resolution of SDN input nodes.

To improve the effectiveness of the input, we decompose its variant nodes into network protocol version nodes (of_version), network message type nodes (of_type), combined message subtype nodes (child_type), message execution type nodes (action_type), and message execution parameter nodes (action_ value). This paper also proposes to make a dictionary for mutation according to the network input characteristics. Some key values and operations are shown in Fig. 2.

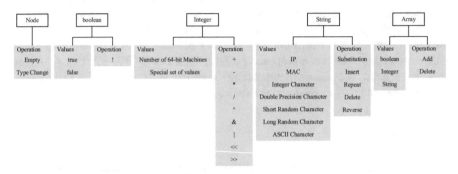

Fig. 2. Key nodes variation values and operations.

According to the variant node parsing above, we classify the input parameters as boolean, Integer, String and Array. The mutation policy selects the node and performs mutation according to the parameter type of the node, for example, the boolean parameter will be randomly mutated in {true, false}; the Integer parameter will be mutated in the range of 64-bit machine numbers or in the special set of values {0, −1,16,32,1024, …}; the String parameter will be mutated in the range of IP, MAC, integer characters, double precision characters, short random characters, long random characters, ASCII characters, and exceptionally loaded characters; the Array parameters are special in that they combine Integer, String, and boolean parameters in the form of containers, so the selection of parameter variants is based on a combination of the above variant processes.

Depending on the specific types of parameters, the variants are different. For boolean parameters, the inverse operation variant is generally used. For Integer parameters, there are basic operation (arithmetic addition, subtraction, multiplication and division), heterogeneous operation, bitwise and operation, bitwise or operation, inverse operation, arithmetic left shift operation and arithmetic right shift operation, etc. For String parameters, there are special character substitution, substring insertion, substring duplication, substring deletion and substring size end bit inversion. For Array parameters, in addition to the basic variants of combining the above elements, there are data variants of adding, deleting and changing elements, in addition, there are variants of duplicating elements, changing types and nulling, etc. For nodes in the variation strategy, there are nulling and type changing.

3.2 Clustering and Prioritization of Test Case Matrix

For the test cases generated by the mutation strategy, we set up a set and evaluate its validity in the form of matrix. The case that completes the test generates a log file that records the running status of processes. We analyze log files, obtaine key data, and compare the state diagram for assessment and matrix generation. The structure of the generated matrix is shown in Fig. 3.

Variant Node 1	Variant Node Type 1	Variant Node Weights 1	Data 1
Variant Node 2	Variant Node Type 2	Variant Node Weights 2	Data 2
Variant Node 3	Variant Node Type 3	Variant Node Weights 3	Data 3
	...		
Variant Node n	Variant Node Type n	Variant Node Weights n	Data n

Fig. 3. The resulting matrix structure.

Each test case matrix contains multiple variant nodes, which correspond to the results of the node resolution. The basic data structure of each variant node contains three parts, which are Variant Node Type, Variant Node Weights, and node-related operation data (Data, such as message execution type, execution parameters, etc.).

Next, we need to perform clustering and prioritization on the large number of matrices generated. The important basis for matrix clustering is the variant node with the highest weight. We cluster the data matrices with the same highest weight variant node into one category and store the matrices in the pool by category. In each cluster, we use the variant node with the highest weight as the representative node of the cluster and calculate the average weight of the clustered nodes. Among all clusters, we rank the clustering nodes according to their average weights, so as to ensure that the clustering nodes with higher weight values have higher variation priority. The next time to select variation nodes, we will sequentially variate the nodes with higher weight values to generate a sequence of variation times to guide the process of generating test cases, such as changing the variation node data type and increasing the variation data length, etc. The specific data structure generation is shown in Fig. 4.

Each matrix cluster contains multiple matrices, and the matrix in the same cluster has the same variation node with the highest weight, that is, the cluster representative node. The value of Times in the sequence of mutation times will be used to guide the selection of mutation nodes. For example, the average weight of the current clustering nodes is {80, 60, 50, ..., 10} in order, and the sequence of mutation times {80, 60, 50, ..., 10} is generated, then the first round of the mutation strategy will be for node 1 mutate 80 times, node 2 mutates 60 times, node 3 50 times... Node n 10 times. After one cycle, the test case matrix will be re-analyzed, the clustering and node weight updates will be completed, and the sequence of mutation times will be regenerated to start the second cycle.

During the calculation of the matrix, the most critical point is the average weight of the clustering nodes. It shows the priority of the mutation strategy to the mutated node, and the way we evaluate this is to parse the control flow of SDN, not to choose a randomized value in some way. According to the operating characteristics of typical

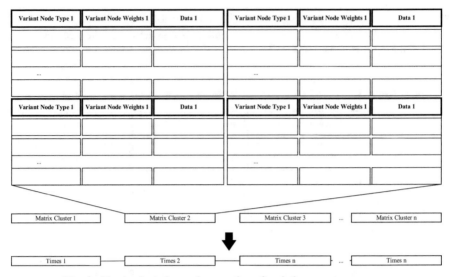

Fig. 4. Matrix clustering and generation of variation count sequences.

controllers based on the OpenFlow, we analyze the operation state of SDN. Based on the symmetric and asymmetric flow analysis, the state diagram can be used with vertices and a directed graph abstract representation of edges. We make the following definition: an operation state graph is a directed graph (G) consisting of a set of vertices (V) representing operation states and a set of edges (E) representing transitions. The important operating states of a typical SDN controller are shown in Fig. 5.

In this operational state diagram, node R represents the initial or ready state of the controller, which is generally a listening state that can send or receive messages at any time. The label of each edge in the directed graph specifies the type of control message and the specific controller behavior that triggers the state transition. For example, when the SDN controller sends an OFPT_HELLO message to a newly connected switch in the initial or ready state, the controller's state moves to S1. In S1, the controller receives an OFPT_HELLO message from the switch, causing the state to transit to S2. If the SDN controller's handshake process with the switch is successful, the state reaches S7, and the controller then updates the topology information. In addition to symmetric control flow, we analyze other states related to control flow, such as asymmetric control flow, internal control flow, inter-controller control flow, and management control flow. This operational state diagram provides a clear description of the points at which the controller receives inputs and how each input induces a state transition. Therefore, based on the results of such operational analysis, we can effectively evaluate the node weights of the test case matrix. For each test case result log, we rate the abnormal test cases based on the operational state diagram. We classify them into the following categories based on the impact of the vulnerability knowledge base (Fig. 6).

We sort them according to the degree of destructiveness to SDN from low to high. Higher destructiveness also means higher weight distribution of mutation nodes. For the impact of basic damage performance, we set up seven basic values as the damage

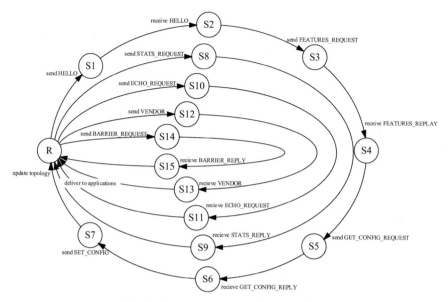

Fig. 5. Important operational state in SDN.

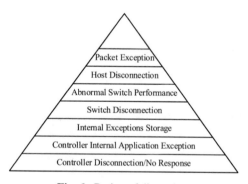

Fig. 6. Rating of disruption.

value (p) of this state. According to the log file analysis, we judge the state of the current architecture and perform inspections such as network connectivity tests, judge the operation status of each component, and calculate the comprehensive destructiveness (c) of the current test case.

$$c = \sum_{i \in [1,7]} k_i \times p_i \tag{1}$$

where, k is the scale coefficient in different destructive scenarios, which affects the proportion of the current foundation damage value. Take the important operation state diagram of a typical SDN controller as an example. When the mutated test case is input into the framework, the SDN switch or host in the virtual component sends the mutated message to the controller in the R state according to the message type and parameter type. The controller processes the mutated message and generates logs. For the status of not receiving relevant response packets or the SDN controller working abnormally, we give rating analysis. The table below shows the final debugging data we used (Table 1).

Table 1. Coefficient distribution used.

K	p
0.38	13
0.75	30
0.26	52
0.88	68
0.65	78
0.54	85
0.70	97

4 Experiment

This section performs tests on the proposed optimized variant policy to verify the consistency, implementation, and provides a theoretical analysis of the new SDN vulnerability scenarios found by the variant policy research.

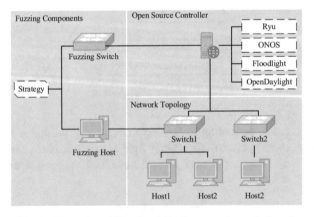

Fig. 7. The vulnerability detection framework we designed.

As shown in Fig. 7, the SDN vulnerability detection framework consists of three parts: Controller, Network Topology and Fuzzing Components. In the controller part, five typical SDN controllers, Ryu, ONOS, Floodlight and OpenDaylight, are deployed for vulnerability detection. Network topology deploys multiple virtual switches and hosts based on Mininet to simulate the actual network environment. The fuzzing components includes intelligent mutation strategy, fuzzing switch and fuzzing host for cases input.

The detection efficiency of the vulnerability knowledge base is one of the evaluation criteria of the optimized fuzzing. In a typical vulnerability scenario test for SDN, the weight nodes guide the variation process and generate a sequence of variation counts with a higher detection efficiency compared to the prototype detection framework due to the inclusion of feedback optimization. We conducted several experiments on the pre-optimized detection framework and the optimized detection framework in a simulation environment to test the running time. The experimental results are shown in Fig. 8.

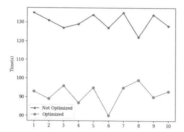

Fig. 8. Comparison of detection times in vulnerability scenarios.

Through several experiments, the average detection time of the prototype detection framework is 130.2 s, while the average detection time of the optimized fuzzing framework is 91.7 s, which improves the detection efficiency by about 29%.

In the SDN vulnerability test, our proposed optimized fuzzing, compared with the unoptimized method, is shown in Fig. 9 for the experiments based on the vulnerability scenario categories of different data streams.

We have tested all the vulnerability scenarios with different data streams, and after the improvement of the optimized fuzzing, the detection time of most scenarios is significantly reduced.

In addition, under the simulation environment testing, our variation policy covers the existing vulnerability detection scenarios, and two new types of SDN vulnerability scenarios are discovered based on the original vulnerability knowledge.

Controller Web Configuration Vulnerability. In the composition of ONOS, a web-based GUI control management terminal is provided to provide a visual user interface for the controller or controller cluster. There are also multiple system application control data layer networks inside ONOS. In the process of constructing test cases, when the message type node and the message execution parameter node are mutated, when cyclic strings such as "%lt" and "0@" are generated, it is found that the control system is abnormal (Fig. 10).

According to the analysis, during the access control lists detection process applied inside ONOS, there is a paragraph such as "/<V? ([a-zA-Z0-9]+)*(. *?)V?>/igm" which is a regular expression match, and the regular expression is at risk of backtracking. For cyclic strings such as "%lt" and "0@", the detection of the input parameters by the control manager is threatened by the inability of its analysis module to correctly parse

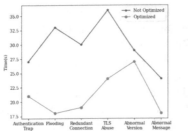
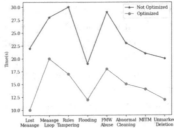

a) Symmetric flow vulnerability scenario detection time and b) asymmetric flow vulnerability scenario detection time.

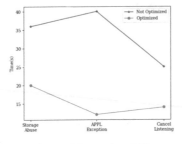
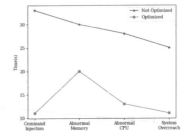

c) Controller internal flow vulnerability scenario detection time and d) System operation flow vulnerability scenario detection time.

Fig. 9. Detection time for different data flow vulnerability scenarios.

Fig. 10. Abnormal load.

the string and generate a loop wait. This loop wait will exhaust the system wait time and system resources, which will lead to service exceptions.

Controller Driver Configuration Vulnerability. The SDN controller's configuration items are often integrated inside the controller, and for network configuration updates there are messages such as OFPT_SET_CONFIG exchanged. In contrast, ONOS has a parameter called "numPorts" in the module of the driver configuration item, which can be configured externally. The variation strategy is used in the construction of test cases, the value of the parameter update part of the variation to a large value, the test found that the overall CPU resource consumption of the controller abnormal, occupancy rate from 3.02% abnormally increased to 332%, and the overall network performance degraded, the controller provides resource management services are not available.

Based on several experiments and analyses, the port number parameter of the device discovery module is somewhat vulnerable. In the driver configuration item, the parameter is parsed without restriction, and the application opens a port list construction loop that establishes the associated port connections and port data structures after reading integer values into the parameter. However, due to the unrestricted value, the parameter generates a large number of data structures and consumes memory resources when parsing large values, resulting in abnormal performance and denial of user service.

5 Conclusion

In this paper, an optimized fuzzing method is proposed by analyzing the structural characteristics of SDN. In the method, the parsing of test case variant nodes and the design of node variant dictionary are carried out first. The variant nodes are decomposed and divided into five basic variant nodes, and the basic variant operations are proposed and designed according to the data types of the nodes. Corresponding fuzzing component input procedures are set. Next, the test case matrix clustering and priority ranking are proposed and designed. By calculating the node weights, the data variation process of fuzzing is optimized and the validity of variation data is improved. We effectively improve the detection efficiency and detection scope compare to existing fuzzing tools for SDN. It increases detection efficiency for 29%, as well as the discovered two new types of threat scenarios.

References

1. Kenton, K., Camp, L.J. and Small, C.: OpenFlow vulnerability assessment (2013)
2. Berde, P., et al.: ONOS: towards an open, distributed SDN OS. In: Proceedings of the third Workshop on Hot topics in Software Defined Networking (HotSDN 2014). ACM (2014)
3. Big Switch Networks, Floodlight. http://www.projectfloodlight.org/floodlight/. BLACK-HAT-USA-2016, Delta: Sdn security evaluation framework. https://www.blackhat.com/us-16/briefings/schedule/index.html
4. BLACK-HAT-USA-2017: Attacking sdn infrastructure: are we ready for the next-gen networking? https://www.blackhat.com/us-17/arsenal/schedule/index.html
5. BLACK-HAT-USA-2018: The finest penetration testing framework for software defined networks. https://www.blackhat.com/us-18/briefings/schedule/index.html
6. Brocade (2016). Brocade SDN Controller. http://www.brocade.com/en/products-services/software-networking/sdn-controllers-applications/sdn-controller.html/
7. Cha, S.K., Woo, M., Brumley, D.: Program-adaptive mutational fuzzing. In: Proceedings of the IEEE Symposium on Security and Privacy (2015)
8. Röpke, C., Holz, T.: SDN rootkits: subverting network operating systems of software-defined networks. In: Bos, H., Monrose, F., Blanc, G. (eds.) Research in Attacks, Intrusions, and Defenses. RAID 2015. LNCS, vol. 9404, pp. 339–356. Springer, Cham (2015). https://doi.org/10.1007/978-3-319-26362-5_16

9. Curtis, A.R., Mogul, J.C., Tourrilhes, J., Yalagandula, P., Sharma, P., Banerjee, S.: Devoflow: scaling flow management for high-performance networks. ACM SIGCOMM Comput. Commun. Rev. **41**, 254–265. ACM (2011)

10. Dhawan, M., Poddar, R., Mahajan, K., Mann, V.: Sphinx: detecting security attacks in software-defined networks. In: NDSS (2015)

11. Dover, J.M.: A denial of service attack against the Open Floodlight SDN controller. Dover Networks, Technical report (2013)

12. Dover, J.M.: A switch table vulnerability in the Open Floodlight SDN controller (2017). http://dovernetworks.com/wpcontent/uploads/2014/03/OpenFloodlight-03052014.pdf

13. Fyodor, P., Levin, M.Y., Molnar, D.A.: Nmap security scanner. Automated whitebox fuzz testing. In: NDSS, vol. 8, pp. 151–166 (2008). http://www.nmap.org.Godefroid

14. Hizver, J.: Taxonomic modeling of security threats in software Defined. Blackhat (2015). https://www.blackhat.com/us-15/briefings/schedule/index.html. Hocevar, S., zzuf. https://git hub.com/samhocevar/zzuf

15. Hong, C.-Y., et al.: Achieving high utilization with software-driven wan. In: ACM SIGCOMM Computer Communication Review, vol. 43, pp. 15–26. ACM (2013)

16. Hong, K., Xu, L., Wang, H., Gu, G.: Poisoning network visibility in software-defined networks: new attacks and countermeasures. In: Proceedings of the 22nd Annual Network and Distributed System Security Symposium (NDSS 2015). HP, HP SDN App Store (2015). https://marketplace.saas.hpe.com/sdn

17. Jero, S., Bu, X., Nita-Rotaru, C., Okhravi, H., Skowyra, R., Fahmy, S.: BEADS: automated attack discovery in OpenFlow-Based SDN systems. In: Dacier, M., Bailey, M., Polychronakis, M., Antonakakis, M. (eds.) Research in Attacks, Intrusions, and Defenses. RAID 2017. LNCS, vol. 10453, pp. 311–333. Springer, Cham (2017). https://doi.org/10.1007/978-3-319-66332-6_14

18. Ujcich, B.E., Thakore, U., Sanders, W.H.: Attain: an attack injection framework for software-defined networking. In: 2017 47th Annual IEEE/IFIP International Conference on Dependable Systems and Networks (DSN), pp. 567–578. IEEE (2017)

19. Lee, S., Kim, J., Woo, S., et al.: A comprehensive security assessment framework for software-defined networks. Comput. Secur. **91**, 101720 (2020)

20. Deng, J., Zhu, X., Xiao, X., et al.: Fuzzing with optimized grammar-aware mutation strategies. IEEE Access **9**, 95061–95071 (2021)

21. Cui, W., Peinado, M., Wang, H.J., Locasto, M.E.: Shieldgen: utomatic data patch generation for unknown vulnerabilities with informed probing. In: Security and Privacy, 2007. SP 2007. IEEE Symposium on IEEE, pp. 252–266 (2007)

22. Kreutz, D., Ramos, F.M.V., Verissimo, P.E., et al.: Software-defined networking: a comprehensive survey. Proc. IEEE **103**(1), 14–76 (2014)

23. Porras, P., Shin, S., Yegneswaran, V., et al.: A security enforcement kernel for OpenFlow networks. In: Proceedings of the first Workshop on Hot topics in Software Defined Networks, pp. 121–126 (2012)

Energy Efficient Virtual Network Function Placement in NFV Enabled Networks

Sudha Dubba$^{(\boxtimes)}$ and Balaprakasa Rao Killi

National Institute of Technology Warangal, Warangal, India
ds720078@student.nitw.ac.in

Abstract. SFCs are an ordered collection of virtual network functions (VNFs) that can be used to provide a specific network service. Service providers started using high performance VMs for deploying their VNFs. Deploying these VNFs in an improper and inefficient way can result in huge resource wastage, high power consumption, prolonged delays in services offered by SPs. Since resources are frequently under-utilized and/or over provisioned, limiting the number of active services improves resource utilization while reducing power consumption. In this paper, we formulated the energy efficient VNF placement problem as an Integer Linear Program (ILP) while considering resource utilization. We proposed an algorithm that selects VNF randomly and deploys it on a power efficient physical server with minimum resource wastage. We compared our proposed algorithm with existing baseline approaches. The proposed approach performs better than existing approaches in terms of number of active servers, CPU utilization, memory utilization, power consumption, resource wastage and delay in the network.

Keywords: Utilization · Power consumption · Resource wastage · Deployment

1 Introduction

Network Virtualization is the process of connecting nodes in a network through virtual links and the functions that are responsible for specific treatment of received packet traffic for example firewalls, proxy etc. Network Function Virtualisation (NFV) is a technology that has made the deployment of devices on the physical servers flexible and efficient as it decouples the software from hardware devices i.e. virtualize the network functions. By disconnecting operations like a firewall or encryption from dedicated hardware and shifting them to virtual servers, network operators can save money and speed up service development.

Virtual Network Functions are software implementations of the network functions that are deployed on a NFV infrastructure. Instead of installing expensive proprietary hardware, service providers can purchase inexpensive switches, storage and servers to run virtual machines that perform network functions.

L. Barolli (Ed.): AINA 2023, LNNS 654, pp. 537–548, 2023.
https://doi.org/10.1007/978-3-031-28451-9_47

This collapses multiple functions into a single physical server, reducing costs and minimizing truck rolls. Following the great success of cloud computing, several service providers (SPs) have begun to deploy their VNFs on high-performance VMs hosted in big data centers owned by public cloud infrastructure providers. However, to fully benefit from NFV, collaboration between different network solution suppliers and network operators is required. These service functions need to be applied on the incoming packet traffic in a particular order. That is, the packets need to be passed through a sequence of VNFs. The sequence of VNFs connected in a predefined order is called a service function chain (SFC). Whenever a SFC request is created there is a need for the deployment of the virtual network functions demanded by the SFC request on the limited physical servers available. Hence, it is necessary to determine optimal placement of these requested VNFs on physical servers in an efficient way.

Cloud computing offers on-demand services like servers, storage, databases, networking, software, and intelligence. But, the availability of physical servers is limited. Service providers have started using high performance VMs for deploying their VNFs. Different users approach service providers with different SFC requests. Therefore, there is a need for placing these VNFs over the servers in the best possible,i.e., optimal and efficient way. Deploying these VNFs in an improper and inefficient way can result in huge resource wastage, high power consumption, prolonged delays in services offered by SPs. In this paper, we formulate the energy efficient VNF placement problem as an Integer Linear Program (ILP) while considering resource utilization. We proposed an algorithm that selects VNF randomly and deploys it on a power efficient physical server with minimum resource wastage. We compared our proposed algorithm with existing baseline approaches. The proposed approach performs better than exisitng approaches in terms of number of active servers, CPU utilization, memory utilization, power consumption, resource wastage and delay in the network.

2   Related Work

Existing literature on the virtual network functions placement in NFV Enabled Networks investigated various performance metrics such as queuing delay, throughput, and energy consumption etc. The authors in [1] presented a 0–1 quadratic fractional program for optimization of end to end delay. The authors in [2] investigated VNF placement for optimization energy consumption and resource utilization. In [3], the authors formulated the VNF placement as an integer linear programming (ILP) model to optimize throughput while considering server resource and bandwidth consumption, and SFC delay requirements. The authors in [4] formulated VNF placement problem as cost-minimizing optimisation problem. The authors in [5] formulated VNF placement problem as cost-minimizing optimisation problem. In [6], the authors formulated a novel reliability aware VNF instances provisioning problem for provisioning reliability aware VNF instances. The authors in [7] proposed the CE-VPS scheme to solve this by considering various dependency factors aiming at reducing the cost paid by SPs. This was formulated as a MILP considering certain constraints. The authors in [8] used deep reinforcement learning for efficiently optimizing the network performance and analyzing the collected data. The authors in [9] proposed the graph cut method for solving the problem of VNF placement while considering performance and the energy consumption in mobile edge-systems. In [10], the authors investigated challenges in routing and placement of the SFC and proposed a solution that considers NFV processing latency and conventional shortest path algorithm to overcome those challenges. The authors of [11] investigated the delay of a flow in partially ordered and totally ordered service function chains. The authors of [12] investigated the load balancing problem for VNF deployment with placement constraints. In [13], the authors investigated the idea of reducing the longest and average completion times. In the paper [14], authors proposed a QoE scheme to support on-the-fly VNFs deployment for OTT video streaming transcoding. The VNF placement problem for mapping user's SFC requests in cloud networks is investigated in [15]. In [16], the authors studied the problem of VNF placement with replications.

3   Problem Formulation

We represent the network of servers as a graph $G(N, E)$ where $N = \{1, 2, \ldots, m\}$ be the set of m physical machines with processing capacities $C_1, C_2, \ldots, C_m$ respectively. Each edge in E represent the communication link between servers. Let $V = \{1, 2, \ldots, n\}$ be the set of n VNFs with processing requirements $R_1, R_2, \ldots, R_n$ respectively. We define a binary decision variables X_{ij} for each pair of physical machine and VNF. We also define a binay decision variable y_i for each server. The decision variable X_{ij} is set to one if VNF V_j is deployed on physical machine N_i, otherwise X_{ij} is set to zero. The binary decision variable Y_i is set to one if the physical machine N_i is switched on, otherwise it is set to zero.

$$X_{ij} = \begin{cases} 1 & \text{if } j^{th} \text{ VNF is deployed on physical machine i} \\ 0 & \text{Otherwise} \end{cases} \tag{1}$$

$$Y_i = \begin{cases} 1 & \text{if the physical machine i is switched on} \\ 0 & \text{Otherwise} \end{cases} \tag{2}$$

Let P_i^{idle} be the ideal power consumed by the server i when the server is active or on. Let P_i^{peak} be the power consumed by the server i when it operates at full capacity (100% utilized). An active server not operating at full capacity, i.e., utilization is strictly less than 100%, then the power consumed by the server is between P_i^{idle} and P_f^{peak}. Moreover, the power consumption of an active server is directly proportional to its resource utilization. Let U_i be the utilization of server i. For an active server, the value of U_i is between zero and one, i.e., $0 < U_i \leq 1$. The following equation computes the utilization of ith server.

$$U_i = \frac{\sum_{j=1}^{n} X_{ij} R_j}{C_i^k} \tag{3}$$

The power consumption of an active server i with utilization U_i is as follows.

$$P_i = P_i^{idle} + (P_i^{peak} - P_i^{idle}) * U_i \tag{4}$$

Hence, the power consumption of all the servers together is as follows:

$$P = \sum_{i=1}^{m} \left(P_i^{idle} + (P_i^{peak} - P_i^{idle}) U_i \right) = \sum_{i=1}^{m} P_i \tag{5}$$

Let Re_i and Rw_i be the residual capacity and resource wastage of ith physical machine. They are defined as below:

$$Re_i = 1 - U_i \tag{6}$$

$$Rw_i = \frac{|Re_i - \min(Re_i|)}{U_i} \tag{7}$$

Hence, the total resource wastage of all servers in the system is as follows:

$$Rw_{total} = \sum_{i=1}^{m} \frac{|Re_i - \min(Re_i|)}{U_i} = \sum_{i=1}^{m} Rw_i \tag{8}$$

Our goal is to deploy VNFs on the servers that jointly minimize the energy consumption and resource utilization. We formulate the problem in the form of a linear programming model.

$$min(P_{total} + Rw_{total}) \tag{9}$$

subject to the following constraints:

$$\sum_{i=1}^{m} X_{ij} = 1 \ \forall j \in V \tag{10}$$

$$\sum_{j=1}^{n} X_{ij} R_j \leq C_i \ \forall i \in N \tag{11}$$

$$X_{ij} \leq Y_i \ \forall i \in N, \ \forall j \in V \tag{12}$$

$$X_{ij}, Y_i \in \{0, 1\} \ \forall i \in N, \ \forall j \in V \tag{13}$$

Equation 10 guarantees that each VNF is deployed on exactly one server. Equation 11 ensures that the sum of processing requirements of VNFs deployed on a server should not exceed its processing capacity. Equation 12 ensures that a VNF should be mapped to an active server. Equation 13 guarantees that the decision variables only takes binary values zero or one.

4   Proposed Algorithm

Algorithm 1 describes the VNF deployment procedure using randomization. The algorithm takes as input network of servers with capacity of resources such as memory, processing capacity, number of cores, number of disks, and requirements of VNFs in given SFC request. The algorithm produces placement solution for each VNF in given SFC request. The algorithm initially calculates power efficiency of every server and sort them in decreasing order of power efficiency. Then, we pick 'd' VNFs uniformly at random from set of VNFs that are not yet deployed. For each physical server, if it accommodate VNFs from set of d VNFs and then we deploy the VNF that results in minimum resource wastage on that server. If none of the d VNFs can be placed on that physical server then choose next physical server from sorted order.

5   Results and Analysis

In this section, we first present the description of dataset used for evaluation, baseline approaches used for comparison, and performance comparision of the proposed algorithm with baseline approaches.

5.1   Simulation Setting

A network of 25 nodes, 57 links are used for the network of physical servers. Each node in networks has memory ranging from 64 to 512 GB, processing capacity from 300 to 600 GB, cores of 16 to 128 and disks of 64 to 128 GB. Latitude and Longitude of each node and bandwidth of links are given. We can calculate weights of links i.e., distances between the links. The delay of each link can be calculated using bandwidth and weight i.e., weight/bandwidth. Focusing then on the power of nodes, we assume the power of a IDLE node (200—300 W) and the power of an ACTIVE node (400—500 W).

Each SFC randomly selects any pair of nodes as its source and destination node, and the amount of VNFs it consisted is uniformly distributed in the range of [3, 7]. Without loss of generality, we disregard the type of VNFs and assume that the memory needed by each VNF is varied between 16 GB to 64 GB, processing capacity needed varied between 100 GB and 300 GB, cores between 16 GB to 64 GB and disks varies between 4 GB to 16 GB.

Algorithm 1: SFCReq

1: **for** each node i∈ V **do**
2:     $P_i^{eff} = C_i^{pc}/P_i^{max}$
3: **end for**
        Sort nodes in the descending order of P_i^{eff}
4: **while** all VNFs are not placed **do**
5:     d = **min**{3,no.of VNFs not placed}
6:     Pick 'd' VNFs uniformly at random from set VNFs that are not placed.
7:     **while** each node i∈ P **do**
8:         **for** j∈ d **do**
9:             **if** j is not placed **then**
10:                 **if** i has enough capacity to accommodate VNF j **then**
11:                     $rw_{i,j}$=Compute Resource wastage when j is placed on i
12:                 **end if**
13:             **end if**
14:         **end for**
15:         $rw_{i,k}$ = **min**{$rw_{i,j}$ **if** j is not placed and **Vnfdef** of i,j is **TRUE**}
16:         place VNF k on node i
17:         update capacities of each resource in node i when VNF k is placed
18:     **end while**
19: **end while**
20: **Return** Placement solution for each VNF.

5.2   Baseline Approaches

We compared the proposed algorithm with following existing algorithms: i) Random first fit algorithm ii) First fit decreasing algorithm iii) Best fit decreasing algorithm iv) Modified best fit decreasing algorithm v) Holu heuristic algorithm. These algorithms are briefly explained below:

5.2.1   Random First Fit Algorithm
The Random first fit algorithm solves VNF placement problem sequentially by selecting a VNF uniformly at random from a set of VNF that are yet to be deployed and place the selected VNF on the first available server. The above process continues until all VNFs of SFC request are placed.

5.2.2   First Fit Decreasing Algorithm
The First fit decreasing algorithm solves VNF placement problem sequentially based on CPU demand of VNF. This algorithm select a VNF with highest CPU demand from a set of VNF that are to be placed and deploy the selected VNF on the first available server. The process continues until all VNFs of SFC request are placed.

5.2.3   Best Fit Decreasing Algorithm

The Best fit decreasing algorithm addresses VNF placement problem sequentially with the objective of minimizing resource wastage. It selects a VNF with highest CPU demand from a set of VNF that are to be placed and deploy the selected VNF on the server which results in minimum resource wastage if the VNF is placed. The process continues until all VNFs of SFC request are placed.

5.2.4   Modified Best Fit Decreasing Algorithm

The Modified best fit decreasing algorithm addresses VNF placement problem with the objective of minimizing power consumption. Modified best fit decreasing acts in a sequential by selectin a VNF with highest CPU demand from a set of VNF that are to be placed and place the selected VNF on the server which results in minimum increase in power consumption if the VNF is placed. The process continues until all VNFs of SFC request are placed.

5.2.5   Holu Heuristic

The Holu heuristic addresses VNF placement with the objective of minimizing power consumption while meeting end-to-end delay. It first calculates the centrality metric for all the nodes in the graph. Then, for each VNF in SFC request, it calculates PM Exploration Matrix by ranking PMs based on centrality value of node, power consumption impact. Since the PMs in each row are sorted, the first (and the best) choice of PM candidates is the first PM from each row in the matrix. If the VNFs cannot be placed on the PM, replace one of the VNF on the next PM from the PM exploration matrix. The VNF that should be replaced is decided by the Largest sub-path delay approach. The process continues until all VNFs of SFC request can be placed.

5.3   Simulation Results

This subsection summarizes the performance evaluation of the proposed algorithm by comparing parameters such as power consumption, number of active servers, resource wastage, memory and cpu utilization, and delay with the existing algorithms presented in Sect. 5.2.

5.3.1   No. of Active Servers

Figure 1 describes performance of the proposed approach along with the baseline approaches in terms of number of active physical servers while varying number of SFC requests. It can be observed that for every algorithm as no. of sfc requests increases we need to use more physical servers. Further, it can be observed that our proposed algorithm is resulting in minimum no. of active servers compared to existing baseline approaches.

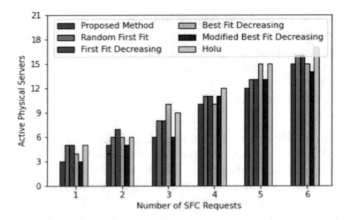

Fig. 1. Impact of number of SFC request on number of active servers

5.3.2   Power Consumption

Figure 2 illustrates performance of the proposed approach along with the baseline approaches in terms of cpu utilization while varying number of SFC requests. This study is aimed primarily at reducing power consumption with minimum resource wastage. Among all the algorithms, it can be observed that our algorithm has the lowest total power consumption except for the modified best fit algorithm. Because modified best algorithm is an algorithm that is based on minimum power consumption only i.e., it is the best algorithm in terms of power consumption whereas our proposed algorithm is aimed primarily at reducing power consumption with minimum resource wastage.

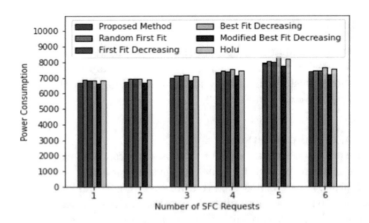

Fig. 2. Impact of number of SFC request on power consumption

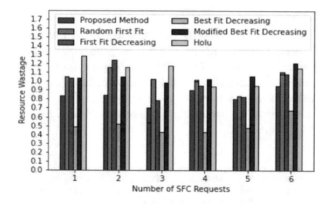

Fig. 3. Impact of number of SFC request on resource wastage

5.3.3 Resource Wastage

Figure 3 describes performance of the proposed approach along with the baseline approaches in terms of resource wastage while varying number of SFC requests. This study is aimed primarily at reducing power consumption with minimum resource wastage. It can be observed that our algorithm has obtained minimum resource wastage except for the best fit algorithm. Because best fit algorithm is an algorithm that is based on minimum resource wastage only i.e.; it is the best algorithm in terms of resource wastage whereas our proposed algorithm is aimed primarily at reducing power consumption with minimum resource wastage. From Figs. 2 and 3, it is observed that the modified best fit is consuming minimum power with high resource wastage and best fit has minimum resource wastage consuming high power whereas our algorithm is able to obtain minimum power consumption with minimum resource wastage simultaneously.

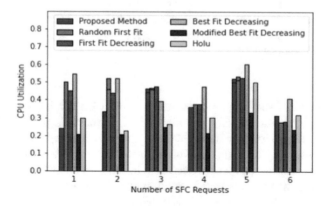

Fig. 4. Impact of number of SFC request on CPU utilization

5.3.4   CPU Utilization

Figure 4 illustrates performance of the proposed approach along with the baseline approaches in terms of cpu utilization while varying number of SFC requests. This study is not aimed to improve CPU utilization, so it is not able to get the best percentage of utilization compared with other algorithms, but able to get a considerable satisfying percentage of cpu utilization. However, the CPU utilization of the proposed algorithm is close to other algorithms for higher number of SFC requests.

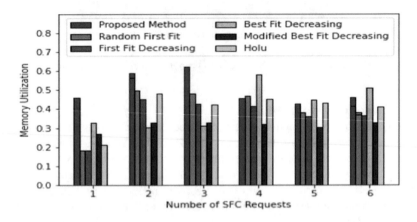

Fig. 5. Impact of number of SFC request on memory utilization

5.3.5   Memory Utilization

Figure 5 illustrates performance of the proposed approach along with the baseline approaches in terms of memory utilization while varying number of SFC requests. This study is not aimed to improve memory utilization, but it is able to get a considerable satisfying percentage of memory utilization. However, the memory utilization of the proposed algorithm is close to other algorithms for higher number of SFC requests.

5.3.6   Delay

The performance of the proposed approach along with the baseline approaches in terms of delay while varying number of SFC requests is presented in Fig. 5. The proposed algorithm results in higher delay when compared with baseline algorithms. However, for higher no. of requests, the gap between the proposed algorithm and baseline approaches decreases in terms of delays for higher number of requests. Holu heuristic is based on minimum power consumption with delay consideration, therefore, results in lesser delay.

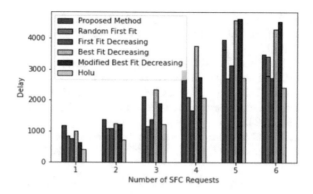

Fig. 6. Delay in the network per user request

6 Conclusion

In this paper, we formulated the energy efficient VNF placement problem as an Integer Linear Program (ILP) while considering resource utilization. We proposed an algorithm that selects VNF randomly and deploys it on a power efficient physical server with minimum resource wastage. We compared our proposed algorithm with existing baseline approaches. The proposed approach performs better than existing approaches in terms of number of active servers, CPU utilization, memory utilization, power consumption, resource wastage and delay in the network.

References

1. Sun, J., Liu, F., Ahmed, M., Li, Y.: Efficient virtual network function placement for poisson arrived traffic. In: ICC 2019-2019 IEEE International Conference on Communications (ICC), pp. 1–7. IEEE (2019)
2. Chen, M., Sun, Y., Hu, H., Tang, L., Fan, B.: Energy-saving and resource-efficient algorithm for virtual network function placement with network scaling. IEEE Trans. Green Commun. Netw. **5**(1), 29–40 (2020)
3. Yue, Y., Cheng, B., Wang, M., Li, B., Liu, X., Chen, J.: Throughput optimization and delay guarantee VNF placement for mapping SFC requests in NFV-enabled networks. IEEE Trans. Netw. Serv. Manag. **18**(4), 4247–4262 (2021)
4. Nemeth, B., Molner, N., Martinperez, J., Bernardos, C.J., De la Oliva, A., Sonkoly, B.: Delay and reliability-constrained VNF placement on mobile and volatile 5G infrastructure. IEEE Trans. Mob. Comput. **21**(9), 3150–3162 (2021)
5. Liu, L., Guo, S., Liu, G., Yang, Y.: Joint dynamical VNF placement and SFC routing in NFV-enabled SDNs. IEEE Trans. Netw. Serv. Manag. **18**(4), 4263–4276 (2021)
6. Huang, M., Liang, W., Shen, X., Ma, Y., Kan, H.: Reliability-aware virtualized network function services provisioning in mobile edge computing. IEEE Trans. Mob. Comput. **19**(11), 2699–2713 (2019)
7. Gao, T., et al.: Cost-efficient VNF placement and scheduling in public cloud networks. IEEE Trans. Commun. **68**(8), 4946–4959 (2020)

8. Pei, J., Hong, P., Pan, M., Liu, J., Zhou, J.: Optimal VNF placement via deep reinforcement learning in SDN/NFV-enabled networks. IEEE J. Sel. Areas Commun. **38**(2), 263–278 (2019)

9. Tao, X., Ota, K., Dong, M., Qi, H., Li, K.: Cost as performance: VNF placement at the edge. IEEE Netw. Lett. **3**(2), 70–74 (2021)

10. Wang, Y., Huang, C.-K., Shen, S.-H., Chiu, G.-M.: Adaptive placement and routing for service function chains with service deadlines. IEEE Trans. Netw. Serv. Manag. **18**(3), 3021–3036 (2021)

11. Yang, S., Li, F., Yahyapour, R., Fu, X.: Delay-sensitive and availability-aware virtual network function scheduling for NFV. IEEE Trans. Serv. Comput. **15**(1), 188–201 (2019)

12. You, C., et al.: Efficient load balancing for the VNF deployment with placement constraints. In: ICC 2019-2019 IEEE International Conference on Communications (ICC), pp. 1–6 (2019)

13. Chen, Y., Wu, J.: Flow scheduling of service chain processing in a NFV-based network. IEEE Trans. Netw. Sci. Eng. **8**(1), 389–399 (2020)

14. Bulkan, U., Iqbal, M., Dagiuklas, T.: Load-balancing for edge QOE-based VNF placement for OTT video streaming. In: 2018 IEEE Globecom Workshops (GC Wkshps), pp. 1–6 (2018)

15. Yue, Y., Cheng, B., Liu, X., Wang, M., Li, B., Chen, J.: Resource optimization and delay guarantee virtual network function placement for mapping SFC requests in cloud networks. IEEE Trans. Netw. Serv. Manag. **18**(2), 1508–1523 (2021)

16. Carpio, F., Dhahri, S., Jukan, A.: VNF placement with replication for LOAC balancing in NFV networks. In: 2017 IEEE International Conference on Communications (ICC), pp. 1–6 (2017)

Rational Identification of Suitable Classification Models for Detecting DDoS Attacks in Software-Defined Networks

Abhirath Anupam Joshi[✉] and K. Haribabu

BITS Pilani, Pilani, India
{f20191136,khari}@pilani.bits-pilani.ac.in

Abstract. Software-Defined Network (SDN) is an approach where the network architecture is divided into 3 planes, namely the control plane, the data plane, and the application plane. It represents a major step forward from traditional, hardware-based networking to software-based networking where a programmable central controller, at the control plane, facilitates controlling the routing of data and allows for easier network management and scalability. On the other hand, the architecture makes the controller a target for many malicious attacks, most common of them being Distributed Denial of Service (DDoS) attacks. Thus, to address cybersecurity issues in SDN architecture, we investigated recent studies and trends that used Machine Learning algorithms to detect DDoS attacks in the control plane. We compared popular ML algorithms - k-Nearest Neighbors (k-NN), Support Vector Machine (SVM), Decision Trees (DT), Artificial Neural Network (ANN) - with different feature selection methods: Neighbourhood Component Analysis (NCA), and minimum Redundancy - Maximum Relevance (mRMR). Considering real-time DDoS attack detection, we have proposed an ensemble learning model that outperforms previously proposed models for detecting DDoS attacks. The proposed model utilizes feature selection and is generalized with a 10-Fold Cross Validation Recall of a 100%, F1-Score of 99.9988%, and Accuracy of 99.9990%.

1 Introduction

A malicious attempt to stop a server, service, or network's normal traffic by flooding the target or its surrounding infrastructure with traffic is known as a distributed denial-of-service (DDoS) attack. DDoS attacks have been a challenge to cyberspace since the first documented DDoS attack in 1996 when a popular internet service provider, Panix, was knocked offline for several days by a SYN flood, a technique that has become a classic DDoS attack. Subsequently, advancement and improvement in cybersecurity became more pertinent as most of the resources migrated from an offline environment to an online environment. Cybersecurity is a major factor in SDNs as they are very susceptible to DDoS attacks because of the sensitive position of the controller. DDoS attacks aimed

L. Barolli (Ed.): AINA 2023, LNNS 654, pp. 549–561, 2023.
https://doi.org/10.1007/978-3-031-28451-9_48

at preventing users from accessing network services are at the top of the attacks on the controller.

To protect SDNs from DDoS attacks, it is important to detect DDoS attacks in real-time and mitigate the attack. In this paper we investigated popular machine learning algorithms to detect DDoS attacks in an attempt to create a novel ML-based solution which acts as a generalized approach towards detecting DDoS attacks using machine learning. The investigation compares different feature selection methods and 4 popular ML algorithms, namely k-Nearest Neighbors, Decision Trees, Support Vector Machine, and Artificial Neural Networks, to detect DDoS attacks.

The paper informs of the related works in the field that form the base of investigation which is followed by a detailed explanation of experiments conducted in the pursuit of identifying the most suitable classification model for detecting DDoS attacks in real-time. The culmination of the experiments is the proposed ensemble model that incorporates feature selection by mRMR algorithm and bagging multiple decision tree classifiers (Random Forest) to create a generalized solution that can be implemented in a SDN architecture and ensure real-time detection of DDoS attacks with 100% recall and F1-Score of 99.9988%.

2   Related Works

In recent years (2017–2021), many studies have been conducted to propose machine learning techniques to detect DDoS attacks. This section discusses several studies of detecting DDoS attacks using machine learning which act as the motivation for this study and the basis of the ensemble model.

Reference [1] proposed a method of detecting DDoS attacks with low-rate using EPS (Expectation of Packet Size). The process works on analyzing the distribution difference of the packet size between the low-rate DDoS attacks and the legitimate traffic. The EPS-based approach shows merit in detecting low-rate DDoS attacks and the method is independent of network topology, arrival patterns, and pulse patterns of attack packets.

Reference [2] evaluated J48, Random Forest (RF), Support Vector Machine (SVM), and K-Nearest Neighbors (K-NN) to detect and block DDoS attack in an SDN network. The study tested the algorithms against a python-script generated dataset on which the J48 performed better than the other evaluated algorithms.

Reference [4] favoured feature selection techniques to shorten training times, facilitate model interpretation, and simplify the models. The method involved building two datasets-one with feature selection and the other without-and using them to train and evaluate various classification models, including Support Vector Machine (SVM), Naive Bayes (NB), Artificial Neural Network (ANN), and K-Nearest Neighbors (k-NN). The test results revealed that the wrapper feature selection method combined with a k-NN classifier had the highest accuracy rate (98.3%) for detecting DDoS attacks. The findings imply that feature selection and machine learning algorithms can improve DDoS attack detection in SDN while promisingly reducing processing load and time.

Reference [3] classified the SDN traffic as normal or attack traffic using machine learning algorithms equipped with Neighbourhood Component Analysis [5]. The study used the NCA algorithm to reveal the most relevant features by feature selection and perform an effective classification. After preprocessing and feature selection stages, the obtained dataset was classified by k-Nearest Neighbor (k-NN), Decision Tree (DT), Artificial Neural Network (ANN), and Support Vector Machine (SVM) algorithms. The experimental results show that DT has a better accuracy rate than the other algorithms with 99.82% classification achievement in a 10 fold cross-validation on the "DDoS Attack SDN Dataset".

Reference [6] proposed novel strategies for diagnosing network metrics and a DDoS algorithm written in P4. They created two major diagnostic metrics: P4LogLog to estimate the flow cardinality, and P4NEntropy to estimate the normalized network traffic entropy. The paper further utilized the novel metric strategies to detect DDoS attacks by means of an entropy-based system. The paper was successful in detecting DDoS attacks but failed to reproduce the results for low-rate DDoS attacks (i.e., with attack traffic proportion $\leq 5\%$) with high accuracy.

Reference [7] describes and implements a two-fold mechanism to detect and mitigate DDoS attacks in the data plane and the control plane. At the data plane, the TLS protocol is integrated with a modified quantum protocol to increase the cipher complexity and enabling secure data transmission using EQTLS handshake at the southbound communication of SDN. At the control plane, an ensemble classifier using fuzzy RF-SVM algorithm is proposed to detect DDoS attacks.

The central theme of the our work is to improve upon the previous works in this field and propose a new benchmark model that outperforms the previous standards. Table 1 compares our work with previous studies.

Table 1. Comparison with related works

Work	Detection algorithms	Feature selection	Ensemble model
Reference [1]	Statistical	None	No
Reference [2]	ML-enabled	None	No
Reference [3]	ML-enabled	NCA	No
This paper	ML-enabled	mRMR	Yes

3 Discussion

From Sect.2, we concluded that:

- Machine learning algorithms with feature selection methods achieve better results with promising reductions in training and testing times.
- Ensemble classifiers provide more flexibility and have more ability to generalize.
- k-NN and DT algorithms outperform most algorithms in detecting DDoS attacks.

3.1   Experiment Process

Our goal is to evaluate different machine learning algorithms with different feature selection methods to understand which permutation has the best performance. For comparison, we considered the permutations with the best F1-scores as it is metric which indicates the model's classification prowess in classifying False Negatives (FN) and False Positives (FP). FN will affect the system as the attack goes undetected and FP will waste limited resources of the system to mitigate a false attack. Figure 1 shows the steps of model comparison and Fig. 2 shows the model training and testing process.

3.1.1   K-Nearest Neighbors (k-NN)

k-Nearest Neighbor Classifier is a lazy learning model for classification based on local approximation. For k-NN classification, a voting policy (majority-based or distance-based) is applied over the set of k nearest datapoints to predict the class. k-NN is a high variance-low bias algorithm.

3.1.2   Decision Trees (DT)

A classification decision tree model is inspired by a tree structure whose branches are created considering objective rules relied on the features of the dataset. To create the tree structure a criterion is used to identify the attribute that will be used to split the node. DT is a low bias-high variance algorithm.

3.1.3   Support Vector Machine (SVM)

Support vector machine is a ML model which works on separating the problem space using hyperplanes. The hyperplane will be an $N-1$ dimensional subspace if there are N features present. The boundary nodes in the feature space are called support vectors. SVM is a low bias-high variance algorithm.

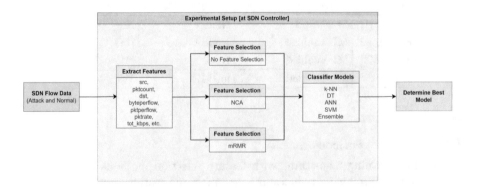

Fig. 1. Steps of model comparison

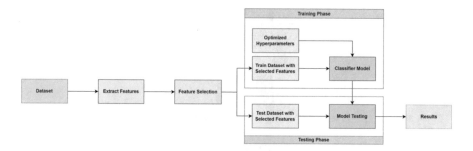

Fig. 2. Model training and testing

3.1.4 Artificial Neural Networks (ANN)

Artificial Neural Network is a computational model for making predictions on non-linear and complicated systems. The computational model comprises of an input layer, one or more hidden layer(s), and an output layer. For the study, Levenberg-Marquardt training algorithm is used because of its stable and fast convergence. The Levenberg-Marquardt algorithm solves the problems existing in both gradient descent method and the Gauss-Newton method for neural-networks training, by the combination of those two algorithms. It is regarded as one of the most efficient training algorithms Fig. 3 shows the Levenberg-Marquardt training algorithm.

3.1.5 Neighbourhood Component Analysis (NCA)

Neighborhood Component Analysis (NCA) is a machine learning algorithm for metric learning. It learns a linear transformation in a supervised fashion to improve the classification accuracy of a stochastic nearest neighbors rule in the transformed space. The NCA weights computed for the dataset are listed in Table 2 obtained from NCA using Stochastic Gradient Descent and automatically determined regularization parameter lambda λ [3]. The top 10 features have been taken into consideration since the remaining attributes' weight is either ≤ 4 or extremely close to 0, if not 0.

3.1.6 Minimum Redundancy – Maximum Relevance (mRMR)

Minimum Redundancy–Maximum Relevance is a minimum-optimal method which seeks to identify a small set of features that, when combined, have the maximum predictive power. It works iteratively, i.e., at each iteration it identifies the best feature and adds it to the list of best k features keeping in mind the relevance of the feature with respect to the target variable and minimum redundancy with respect to the features already added to the best k features list. At each iteration i, a score is computed for each feature f to be evaluated.

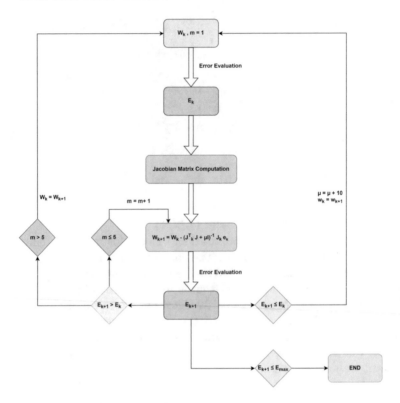

Fig. 3. Levenberg-Marquardt training algorithm

The relevance of a feature f is the F-statistic between the feature and the target variable and the redundancy is the average Pearson Correlation between the feature and all the previously selected features. The 10 top features selected by the mRMR algorithm are listed in Table 2.

$$score_i(f) = \frac{F(f, target)}{\sum_{s \in features\ selected\ until\ (i-1)} \frac{|corr(f,s)|}{(i-1)}} \tag{1}$$

Table 2. The 10 most relevant features selected by mRMR vs NCA

Features	mRMR score	Features	NCA score
pktcount	20099.877787	src	17.87
bytecount	8704.145723	pktcount	15.16
Protocol*	7462.735187	dst*	13.64
flows	3581.738163	byteperflow*	12.97
src	1287.406795	pktperflow*	11.35
dt*	1280.987007	pktrate	11.35
tot_dur*	1022.083847	tot_kbps*	9.68
dur*	1021.848426	rx_kbps*	9.66
pktrate	814.572509	flows	8.95
rx_bytes*	346.009033	bytecount	4.92

* Unique features, i.e., only present in one feature selection method

---

Algorithm 1. Bagging (Bootstrap Aggregating)

1: n labelled training examples with k features

$$(x_1, y_1), (x_2, y_2), ..., (x_n, y_n)$$

   where x_i = Set of selected features $f_1, f_2, ..., f_k$
   y_i = Predicted Classification
2: Let the number of classifiers be j
3: **for all** Classifiers **do**
4:   Select n samples from the dataset with replacement to get the training dataset S_i
5:   Train classifier h_i on S_i
6: **end for**
7: Obtain classifiers $h_1, h_2, ..., h_j$
8: On test Sample t, output Majority($h_1(t), h_2(t), ..., h_j(t)$)

3.2   Ensemble Model

The crux of an ensemble model is to combine the decisions from multiple base classifiers to improve the overall performance. It helps overcome noise, bias, and variance. Ensemble models enables simple models to achieve higher performance and have more ability to generalize. We have created an ensemble model using bootstrap aggregation of multiple decision trees. This approach is famously known as Random Forest and is based on a bootstrapping sampling technique. Bootstrapping creates multiple sets of the original training data with replacement. Replacement enables the duplication of sample instances in a set. Each subset has the same equal size and can be used to train models in parallel. This algorithm is depicted in Algorithm 1. Bagging reduces the variance of the classifier but doesn't help much with the biasness. Thus, decision trees which are inherently high variance models are benefited by this method.

4   Experiments

Our investigation consisted of multiple experiments to compare feature selection methods and machine learning algorithms on different criterias: training time, testing time, classification result, 10–Fold Cross Validation Result.

4.1   Dataset

Building on the learning and the discussion, we will follow the process mentioned in Fig. 1 with an attempt to identify the key strengths of the algorithms. The training and testing of models will be carried out in a Python-based architecture on the publicly accessible dataset: "DDoS attack SDN Dataset" [8] having 23 features. The dataset consists of Transmission Control Protocol (TCP), User Datagram Protocol (UDP), and Internet Control Message Protocol (ICMP) normal and attack traffics. The dataset, including more than 100 thousand recordings, has statistical features such as byte_count, duration_sec, packet rate, and packet per flow, except for features that define source and target machines. Benign traffic has label 0 and malicious traffic has label 1. The dt field shpws the date and time which has been converted into a number and therefore, removed in the preprocessing stage. There are three categorical fields: src, dst, and Protocol. These are converted in labels using label encoding. In our experiments, we have split the dataset into two: Training Dataset (80%) and Test Dataset (20%). The fields rx_kbps and tot_kbps contain NaN (not a number) values and have been replaced with their column's mean value with respect to their corresponding label. All values are standardized (2) during the Data Preprocessing Stage, before passed as input to a model.

$$x = \frac{x - \mu_x}{\sigma_x} \tag{2}$$

4.2   Hyperparameter Tuning

All models' hyperparameters were tuned using grid search to find the best hyperparameter configuration which yielded the best F1–Score. Table 3 shows the grid search results, tested parameters, and the selected parameters.

Table 3. Hyperparamter tuning of ML models

Algorithm	Parameter	Tested values	Selected value
k-NN	No. of neighbors K	{3, 5, 7, 11}	3
	Weight metric	{uniform, manhattan, euclidean}	Manhattan
DT	Splitting criterion	{Gini, Entropy}	Gini
SVM	Kernel	{polynomial, sigmoid, rbf }	rbf
	Regularization Parameter C	$\{1, 10, 10^2, 10^3, 10^4\}$	10^3
	Kernel coefficient γ	$\{10^{-4}, 10^{-3}, 10^{-2}, 10^{-1}, 1\}$	1
ANN*	Initial dampening factor μ_0	0.001	0.001
	Decrease dampening factor μ_{dec}	0.1	0.1
	Increase dampening factor μ_{inc}	10	10
	Maximum dampening factor $\bar{\mu}$	10^{10}	10^{10}
	No. of hidden layers	{1, 2, 3}	3
	No. of epochs	{10, 20, 30, 40, 50}	40

* implemented using Levenberg-Marquardt Training Algorithm

4.3   Performance Metrics and Model Evaluation

A confusion matrix was used in the study to test the performance of the models and evaluate them on the basis of: Accuracy, Sensitivity, Specificity, Precision, and F1-Score. Table 4 shows the confusion matrix and Table 5 lists the performance metrics. F1–Score was given priority over other metrics owing to the fact that a network has limited resources and wastage of resources to mitigate misclassified traffic and undetecetd attacks are more damaging to the system.

Table 4. Confusion matrix

		True class	
		Attack	Normal
Predicted class	Attack	TP	FP
	Normal	FN	TN

Table 5. Performance metrics

Metric	Definition	Formulation	
Accuracy	The overall accuracy of the model	$\frac{TP+TN}{TP+FP+FN+TN}$	(3)
Sensitivity	The performance of the model on detecting attack traffic (also known as Recall)	$\frac{TP}{TP+FN}$	(4)
Specificity	The performance of the model on detecting normal traffic	$\frac{TN}{TN+FP}$	(5)
Precision	The ratio of correctly predicted attack traffic to the total predicted attack traffic	$\frac{TP}{TP+FP}$	(6)
F1-Score	Harmonic mean of precision and sensitivity	$\frac{2\times TP}{2\times TP+FP+FN}$	(7)

4.4   Feature Selection Comparison

mRMR algorithm was the best feature selection method as observed from Table 6. The F1–Scores for all algorithms, when using mRMR, outperformed NCA.

Table 6. Feature selection comparison using F1-scores

	F1-scores				
	k-NN	DT	SVM	ANN	Ensemble
None	98.2338	1	96.9271	99.9326	100
NCA	98.8985	99.9265	99.4113	98.3765	99.9265
mRMR	98.9147	99.9632	99.7055	96.7006	99.9755

4.5   Training vs Testing Time

Table 7 shows the difference in training and testing time of all the algorithms with and without feature selection. Feature selection makes a huge impact in the execution time of the algorithms. The highlight here being that the significant reduction in execution time paves way for certain algorithms to be employed in real-time. Moreover, feature extraction from network traffic is faster since fewer features are needed to be extracted.

Table 7. Training vs Testing time comparison

Model	mRMR		No feature selection	
	Train time (s)	Test time (s)	Train time (s)	Test time (s)
k-NN	0.0673	0.3803	0.6114	2.0933
DT	0.1287	0.0014	0.2696	0.0021
SVM	84.4740	2.8103	482.3296	44.2505
ANN	23	0.0196	34	0.0232
Ensemble	0.34	0.01	0.4052	0.0101

4.6   Classification Results

Tables 8, 9, 10 compare different machine learning models. Table 8 and Table 10 highlight the importance of feature selection in model training as the models with feature selection outperform models trained without feature selection.

Table 8. 10-fold cross validation results of ML models without feature selection

ML	Acc (%)	Se (%)	Sp (%)	Pr (%)	Fsc (%)
kNN	98.6305	98.2689	98.8625	98.2283	98.2485
DT	100	100	100	100	100
SVM	97.8677	96.0843	99.0120	98.4232	97.2394
ANN	99.2127	99.2473	99.1905	98.7455	98.9956
Ensemble	100	100	100	100	100

Table 9. Classification results of ML models with mRMR

ML	Acc (%)	Se (%)	Sp (%)	Pr (%)	Fsc (%)
kNN	99.1519	98.8844	99.3235	98.9450	98.9147
DT	99.9712	99.9632	99.9764	99.9632	99.9632
SVM	99.7699	99.6200	99.8663	99.7912	99.7055
ANN	97.3933	97.7320	97.1759	95.6908	96.7006
Ensemble	99.9808	99.9755	99.9843	99.9755	99.9755

Table 10. 10 fold-cross validation results with mRMR

ML	Acc (%)	Se (%)	Sp (%)	Pr (%)	Fsc (%)
kNN	99.1768	98.8157	99.4084	99.0759	98.9455
DT	99.9952	99.9902	99.9984	99.9975	99.9939
SVM	99.7767	99.6715	99.8442	99.7571	99.7142
ANN	98.2098	98.1096	98.2741	97.3323	97.7190
Ensemble	99.9990	100	100	99.9975	99.9988

5   Results

The investigation shows the importance of feature selection in training the machine learning models. Not only feature selection helps improve model performance but also decreases the number of features that need to be extracted to perform the classification. The ensemble model outperforms precious standards in the field as seen in Table 11. The 10 Fold–Cross Validation Result of the ensemble model on the dataset has 100% sensitivity, i.e., all the attack traffic were classified correctly. This is a major win for the model as it would be executed in a resource–restrained environment. The training and testing time of the ensemble model is feasible to be executed in the control plane of SDN.

Table 11. The comparison of the related studies

Related studies and datasets	Feature selection	ML algorithms	Acc (%)
CIC DoS dataset [9]	None	Random Tree, J48, REP Tree, SVM, RF, MLP	95.00
UNB-ISCX [10]	None	Semisupervised machine-learning algorithm	96.28
NSL-KDD [11]	KPCA	SVM	98.91
InSDN [12]	None	kNN, NB, Adaboost, DT, RF, rbf-SVM, lin-SVM, MLP	99
DDOS attack SDN Dataset [13]	None	CNN, LSTM, CNN-LSTM, SVC-SOM-SAE-MLP	99.75
DDOS attack SDN Dataset [3]	NCA	kNN, ANN, DT, SVM	100^a
DDOS attack SDN Dataset	mRMR	k-NN, DT, SVM, ANN, Ensemble	100

^{a} Uses more features and prone to overfitting and not generalized

6   Conclusion

The proposed ensemble model outperforms precious standards and highlights the importance of feature selection while training the models. The ensemble model shows promise of classifying 100% of the attack traffic in reasonable amount of time.

The study can be extended to investigation and proposing ML techniques employed in the Data Plane of the SDN rather than the Control Plane in an attempt to reduce inference time.

References

1. Zhou, L., Liao, M., Yuan, C., Zhang, H.: Low-rate DDoS attack detection using expectation of packet size. Secur. Commun. Netw. **2017** (2017)
2. Rahman, O., Quraishi, M.A.G., Lung, C.-H.: DDoS attacks detection and mitigation in SDN using machine learning. In: 2019 IEEE World Congress on Services (SERVICES), pp. 184–189 (2019). https://doi.org/10.1109/SERVICES.2019.00051
3. Tonkal, Ö., Polat, H., Başaran, E., Cömert, Z., Kocaoğlu, R.: Machine learning approach equipped with neighbourhood component analysis for DDoS attack detection in software-defined networking. Electronics **10**, 1227 (2021). https://doi.org/10.3390/electronics10111227
4. Polat, H., Polat, O., Cetin, A.: Detecting DDoS attacks in software-defined networks through feature selection methods and machine learning models. Sustainability **12**(3), 1035 (2020). https://doi.org/10.3390/su12031035
5. Goldberger, J., Roweis, S., Hinton, G., Salakhutdinov, R.: Neighbourhood components analysis. In: Proceedings of the 17th International Conference on Neural Information Processing Systems (NIPS 2004), pp. 513–520. MIT Press, Cambridge (2004)

6. Ding, D., Savi, M., Siracusa, D.: Tracking normalized network traffic entropy to detect DDoS attacks in P4. IEEE Trans. Dependable Secure Comput. 1 (2021). https://doi.org/10.1109/tdsc.2021.3116345

7. Saritha, A., Reddy, B.R., Babu, A.S.: QEMDD: quantum inspired ensemble model to detect and mitigate DDoS attacks at various layers of SDN architecture. Wireless Personal Communications **127**, 1–26 (2021). https://doi.org/10.1007/s11277-021-08805-5

8. Ahuja, N., Singal, G., Mukhopadhyay, D.: DDOS attack SDN Dataset. Mendeley Data **V1** (2020). https://doi.org/10.17632/jxpfjc64kr.1

9. Perez-Diaz, J.A., Valdovinos, I.A., Choo, K.K.R., Zhu, D.: A flexible SDN-based architecture for identifying and mitigating low-rate DDoS attacks using machine learning. IEEE Access **8**, 155859–155872 (2020)

10. Ravi, N., Shalinie, S.M.: Learning-driven detection and mitigation of DDoS attack in IoT via SDN-cloud architecture. IEEE Internet Things J. **7**(4), 3559–3570 (2020). https://doi.org/10.1109/JIOT.2020.2973176

11. Sahoo, K.S., et al.: An evolutionary SVM model for DDOS attack detection in software defined networks. IEEE Access **8**, 132502–132513 (2020)

12. Elsayed, M.S., Le-Khac, N.-A., Jurcut, A.D.: InSDN: a novel SDN intrusion dataset. IEEE Access **8**, 165263–165284 (2020). https://doi.org/10.1109/ACCESS.2020.3022633

13. Ahuja, N., Singal, G., Mukhopadhyay, D.: DLSDN: Deep learning for DDOS attack detection in software defined networking. In: Proceedings of the 2021 11th International Conference on Cloud Computing, Data Science & Engineering Confluence, Noida, India, 28–29 January 2021, pp. 683–688 (2021)

Make Before Degrade: A Context-Aware Software-Defined WiFi Handover

Víctor M. G. Martínez[1]([✉]), Rafael S. Guimarães[2], Ricardo C. Mello[1],
Alexandre P. do Carmo[3], Raquel F. Vassallo[1], Rodolfo Villaça[1],
Moisés R. N. Ribeiro[1], and Magnos Martinello[1]

[1] Universidade Federal do Espírito Santo, Vitória, Brazil
victor.martinez@edu.ufes.br
[2] Instituto Federal do Espírito Santo, Cachoeiro de Itapemirim, Brazil
[3] Instituto Federal do Espírito Santo, Guarapari, Brazil

Abstract. Ultra-reliable and low-latency communications are essential
for new services and applications in Industry 4.0. URLLC is also required
to provide traffic density to meet the throughput demands of its clients
across the coverage area via cell densification. However, cell densification implies that handover between serving stations will become more
and more frequent. In this paper, we introduce the *Make Before Degrade*
solution for zero mobility interruption with a single radio connection.
Mobility management is based on contextual information that allows
prudently to shift traffic flows before their propagation paths are compromised. The paper brings a proof-of-principle software-defined WiFi
testbed able to meet 3GPP's traffic density specifications in which contextual information is provided by computer vision. An Industry 4.0 use
case based on cloud robotics demonstrates the practical applicability of
our proposal.

1 Introduction

Industry 4.0 is boosted by ultra-reliable low-latency communications (URLLC),
enabling complex in-factory automation, and envisioning factories as intelligent
environments densely permeated by connected devices, autonomous processes,
and industrial robots. In industrial environments, a significant amount of traffic
will be generated indoors by stationary or (relatively slow) mobile users. In this
context, WiFi may be exploited to fulfill the requirements of many emerging
applications while offloading 5G networks and reducing costs [1]. However, these
cost-effective solutions will require different wireless and wired solutions to be
integrated to access applications in the cloud. Thus, the use of software-defined
networking (SDN) and network function virtualization (NFV) are of great help
in achieving the integration and management of these technologies, with the final
aim of meeting the end-to-end requirements of these services and applications.

The richly connected environment envisioned for Industry 4.0 will also
demand high-traffic density solutions from wireless networks. Small wireless cell
coverage is the way to build ultra-dense networks, where only a few elements

© The Author(s), under exclusive license to Springer Nature Switzerland AG 2023
L. Barolli (Ed.): AINA 2023, LNNS 654, pp. 562–572, 2023.
https://doi.org/10.1007/978-3-031-28451-9_49

share the available bandwidth [7]. However, these networks pose challenges to communication reliability, demanding more frequent migrations of user devices between access stations.

This paper argues that large amounts of connected sensing devices and upcoming intelligent applications can be exploited by context-aware networks. This new degree of freedom shall leverage information beyond communication parameters to enable proactive mobility management. We introduce a context-aware handover solution named *Make Before Degrade* (MBD). The proposal enhances mobility by leveraging contextual information to foresee signal degradation, allowing to (i) proactively shift traffic flows before they are minimally compromised; and (ii) keep true traffic density targets. The work experimentally demonstrates the MBD handover in an SD-Wi-Fi network to achieve high traffic density in industrial environments. The MBD handover is demonstrated by enabling a challenging cloud-enabled mobile robotic application with strict real-time remote control requirements.

2 Background

Traditionally, context is usually restricted to communication-related information, such as spectrum efficiency or specific user-related communications needs [7]. Already presenting high complexity degrees but capable of meeting the demands of the new generations of wireless networks, the handover algorithms can use environment context information to perform user migrations [2]. Within this context approach, specific location-aware solutions have proven to be efficient in mobility management solutions, allowing proactive actions on communication [3]. Such solutions have been proposed to support handover tasks in various scenarios, including indoor industrial scenarios [4]. Even when they report more relevant results than conventional handover schemes [2,4], the use of multi-connectivity through various wireless network interfaces to improve latency and reliability can be prohibitive for some battery-restricted devices. Supported by

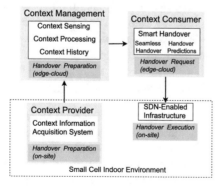

Fig. 1. Logical architecture for the context-aware handover.

an SDN-based approach to obtain the best performance during the handover process, our proposal offers seamless and fast migrations even with a single radio connection.

Solutions that implement trajectory predictions have also been proposed [5] for the execution of handover, mainly focused exclusively on the context regarding the mobile element. For such solutions to be truly efficient in decision-making, highly accurate location solutions are needed [2]. Our solution is not alien to the physical elements that share the space with mobile clients and that can eventually affect communication performance. By leveraging recent technologies such as computer vision, context becomes a much more global concept as it moves towards environmental distribution information. For instance, external sensors can provide localization information with better accuracy than the less-than-one-meter expected in future 5G networks [6]. Thus, not only the information of the context of the clients is used to manage mobility better, but a spatial map with other elements can be built to guarantee communication performance.

Network density in industrial environments reduces the distance between access stations and mobile users [7] in line-of-sight (LoS) transmission, guaranteeing high transfer rates per user. However, more frequent migrations between access stations are necessary, so soft handover schemes are presented as the most efficient solution to face such challenges. For indoor industrial environments using WiFi solutions, performance requirements for mission-critical applications are not always met, especially in communication scenarios dominated by LoS conditions, where communications can be interrupted or degraded by several objects in the space [8]. LoS obstruction leads to SNR degradation, affecting communication bit error rate (BER).

The reduction of the BER causes the WiFi system to switch to lower modulation and coding schemes (MCS) to maintain the system's best performance. Two cases can be analyzed: (i) the WiFi system maintains the same modulation scheme but decreases the coding index since lower coding indexes have a lower corrective capacity, the need for data re-transmissions increases, consequently increasing the communication latency; or (ii) the WiFi system decreases the modulation scheme (with the same or lower coding index), in this case even when the communication reliability can be maintained, the throughput suffers from a decrease, affecting the requirements of traffic density by area and user experience data rate [11]. A design choice is to maintain the WiFi users served by access stations with no degradation of SNR levels, so MCS changes are unnecessary.

3   The Make Before Degrade Handover

This paper introduces the Make Before Degrade (MBD) handover solution, a proactive network-centric context-aware handover mechanism that implements spatial distribution prediction to guarantee 5G-like KPIs. Such a solution outperforms current make-before-brake techniques, which must react to threshold-based triggers such as loss of communication quality or device-only localization information. The MBD can enhance reliability by leveraging context information from several sources. When keeping track of such objects, it is possible to implement algorithms to infer possible obstructions between user devices and access points, detect interference in nearby future, and proactively avoid signal degradation to support traffic density requirements.

3.1   General Architecture

Considering indoor small cell scenarios, a general architecture composed of three fundamental blocks for a context-aware handover solution is shown in Fig. 1. The *context provider* is responsible for obtaining the environmental metrics that will be used as the selection criteria to trigger the handover process. In indoor environments, instantaneous decisions for mobile user migration can be made considering a spatial knowledge of the physical environment. This global view also allows inferring potential LoS obstructions and reacting based on this knowledge. Specific applications can extract information from the scenario to predict these situations and use such context knowledge to trigger a handover to non-occluded access points proactively.

The context information is sent through well-defined interfaces to the *context management* block. This functional block can be located in a centralized or distributed approach at the edge-cloud. The *context management*, made up of several service modules, is responsible for preparing, processing, and storing the context information used by the context consumer block.

Finally, the *context consumer* is in charge of requesting the handover procedure if needed, which is the first stage of the handover process. The context information (already treated by the context management) is used as input in the handover decision algorithms. The handover process is executed through programmable interfaces in an SDN-enabled infrastructure, leveraging the physical and logical network elements responsible for migrations of user's traffic.

Figure 2 illustrates the SDN context-aware handover process. The data traffic from the user device to a cloud application goes through a given access point in a small cell indoor environment. At the same time, the context information is sent

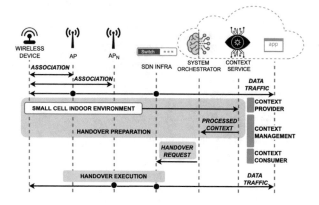

Fig. 2. Context-aware handover process.

via the SDN-enabled infrastructure to the context management service, which is
responsible for processing the information and feeding the system orchestrator.
The orchestrator bears the role of the *context consumer* block hinted in Fig. 1.
The orchestrator determines that a handover is necessary when i) the user device
in the small cell scenario moves to another cell (served by a different access point)
and when ii) potential degradation of the throughput is detected to guarantee
the traffic density per small cell. The orchestrator triggers a handover request to
the SDN infrastructure, informing how to proceed (change OpenFlow rules). The
infrastructure is then responsible for executing the handover process, switching
the traffic in either case through a different access point.

3.2   SDN-Enabled Infrastructure

The infrastructure comprises two fundamental blocks: i) a radio access network
based on WiFi 4 using a novel software-defined WiFi solution and ii) a backhaul
network composed of SDN switches for accessing applications at the edge-cloud.
The infrastructure is based on the wireless solution from previous work [9,10], a
brief description is shown in Fig. 3.

The solution allows several fixed STAs ($fSTA$) to make associations with a
mobile AP (mAP), providing a multi-association scheme that offers communication diversity for the user devices [9]. The multi-connectivity allows the establishment of multiple paths to access the cloud, just requiring decision conditions
to trigger the handover process. First, small coverage zones can be determined

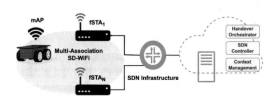

Fig. 3. SDN-based communication architecture.

Algorithm 1: Make Before Degrade

Input: A set $P_O = \{O_k, O_{k-1}, \ldots, O_{k-n}\}$ of object positions, a set
$P_R = \{R_k, R_{k-1}, \ldots, R_{k-n}\}$ of device positions, where k is the current instant, n is
the buffer size

$instructions$;
$t \leftarrow PREDICTION\_WINDOW$;
$objPos, objZone \leftarrow makePrediction(P_O, t)$;
$devicePos, deviceZone \leftarrow makePrediction(P_R, t)$;
if $objZone = deviceZone$ **then**
  if $isInLineOfSight(objPos, devicePos)$ **then**
    $L \leftarrow deviceZone$;
    $performFailure(L, devicePos)$;
  end
else
  $continue$;
end

and associated with each currently working $fSTA$. Then, the system orchestrator decides which $fSTA$ is more suitable to serve the mAP based on the context-aware information. In our case, the context information refers to the physical location of the mAP (user device) in the space. Then, as in the previous works, the handovers (proactive or reactive) are accomplished by modified OpenFlow rules in the SDN-based backhaul to redirect traffic through the chosen $fSTA$.

Assuming a LoS path between the mAP and all the $fSTAs$ associated with it, it is expected that the shortest distance is the connection that guarantees optimal communication. Based on this premise, in a controlled environment only with the mAP, each time the mAP enters a new zone in the space, the system orchestrator obtains this information from a context-aware application (localization service) and triggers the handover process.

The system orchestrator also reacts upon receiving failure notifications from agents installed in the $fSTAs$. Events can be triggered by failures in the $fSTA$ currently proving communication: (i) connection loss (or association loss) between the mAP and the $fSTA$ (i.e., radio link failure); and (ii) downtime in the $fSTA$ wireless interface. The system orchestrator reacts by updating the OpenFlow rules on the backhaul in both cases.

Specific objects, which may be particularly large or prone to cause interference on the wireless signal, can be tracked to generate predictions of their future localization. The objects' positions are compared to estimations of the user device's future position to decide on the chance of the tracked object interfering with the wireless communication (obstructing the LoS path). Algorithm 1 show the MBD handover by leveraging localization-aware information of the objects in the space.

The implementation of such mechanisms results in a centralized network controller able to orchestrate data flow to enable real-time mobile applications demanding zero interruption time and resilience to failure. Moreover, the implementation of the MBD handover pushes the network to a proactive state, rather than reactive, by allowing the handover to be triggered before undesired situations of spatial interference happen, guaranteeing the traffic density values specified by the 3GPP for industrial real-time control applications.

4   Evaluation

Our proposal is deployed into a testbed, which integrates cloud-enabled robots and networked sensors in an intelligent environment connected to an edge cloud via mixed fiber-wireless communication networks [6]. The testbed is set to reproduce a scenario analogous to an everyday use case of Industry 4.0 (i.e., remote control of robots). The physical area in the testbed is divided into several zones of just over $1\,m^2$ (small cells), each of which is served by an fSTA to offer connectivity to wireless robots.

Figure 4a illustrated a general view of the testbed. The testbed is equipped with a network of sensors (cameras), able to gather information about the surroundings, and a network of actuators controlled by different edge-cloud computing services to address precise trajectory control of cloud-enabled robots. Applying computer vision is computed the robot localization, information used by the motion controller service to guide the robot through space. The context-aware handover service is also an edge-cloud service that consumes the robot's location information to trigger the handover process and migrate the robot's controller traffic through the physical WiFi topology. More details about the implementation of the testbed and the mechanisms for detecting objects in space through computer vision can be consulted in [6].

The handover orchestrator was developed in Python and managed three resources: SDN, WiFi, and Computer Vision. To interact between SDN elements in the backhaul and fronthaul, the handover orchestrator uses the REST API as a northbound API. SDN Northbound APIs, like REST API and Message Queuing Protocol (AMQP), communicate between the SDN Controller and the services and applications. The communication between the Handover Orchestrator and WiFi/Computer Vision resources was implemented using RabbitMQ broker, which supports AMQP messages.

4.1   Experimental Protocol

Two experiments were designed on the testbed to evaluate the performance of industrial real-time control applications for cloud-enabled robotics in URLLC scenarios. In both experiments, the robot performs an eight-way course within the space, completing a circuit of two laps in 60 s. The robot follows the trajectory with constant velocities during the straight parts and accelerates/decelerates when steering. The cameras send the captured images to be processed at the edge data center. Then the location information will be computed by the localization service so that the robot controller directs the robot through its trajectory.

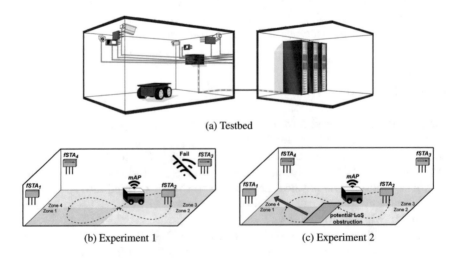

(a) Testbed

(b) Experiment 1   (c) Experiment 2

Fig. 4. Evaluation scenario.

From the backhaul perspective, the traffic is collected from the $fSTAs$, analyzing the migrations between the OpenFlow switch and each connected element on the wireless network. However, we do not have any traffic engineering treatment using queue control to prioritize or deal with latency and jitter. From the application's perspective, the measurements are analyzed directly in the mAP applications to get the throughput and RTT values between mAP and the edge-cloud. Finally, from the user perspective, the quality of experience is analyzed based on the conformity between desired and performed trajectory of the robot.

4.2   Experiment 1: Soft-Handover Scheme

The experiment aims to analyze the real-time control application performance during the handover processes, observing possible interruption time, packet loss, and latency in the communication between the robot and the application at the edge-cloud and evaluate the capacity of failure recovery of the whole system. The experiment is conducted in the 5 GHz band to evaluate the soft-handover approach's communication throughput underpinned by the context-aware localization brought out by the testbed.

The robot moves through different zones (small cells) during its trail, each served by a $fSTA$ as shown in Fig 4b. During the transit of the robot from one zone to another, the orchestrator of the architecture uses the position service to trigger the handover process every time the robot enters a new zone. Additionally, a fault is generated in the $fSTA_3$ in the second lap. To stress out the system, in parallel to the robot control traffic, we have generated a 10 Mbps UDP background traffic using the iperf3 tool. Moreover, a ping is performed between the robot and the edge-cloud to measure the communication latency.

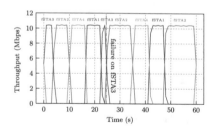 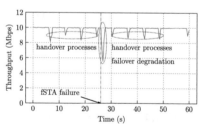

(a) $fSTAs$ throughput: handover and failover   (b) overall throughput: zero mobility interruption time

Fig. 5. Experiment 1 results.

The throughput results obtained in experiment 1 are shown in Fig. 5. Figure 5a shows the throughput at the backhaul interface in each $fSTAs$ (plot colors are in correspondence with the colors of the zones from Fig. 4b). The handover processes start around 6 s once the robot crosses to a new zone (from

$fSTA_3$ to $fSTA_2$). Further, the user downstream traffic usage is restricted to its specific position in a zone, avoiding traffic transmission overlap in the other squares. In the second lap, when the $fSTA_3$ failure causes the robot to lose its association with the WiFi network, the handover orchestrator implements a failover recovery mechanism to keep connectivity with the robot (mAP). Thus, the system is reconfigured, and the traffic to and from the robot is migrated to the fSTA ($fSTA_2$) that attends the communication. This condition is maintained until the system orchestrator receives the recovery information from the failed $fSTA$.

The overall throughput measured in the robot (mAP) is shown in Fig. 5b. With accumulative measurement windows of 1 min, the 1 Mbps/m^2 and 10 Mbps/user requirements suggested by 3GPP are met with minor degradation during the handover processes, and resilience to $fSTAs$ fails. Thus, our context-aware handover solution can guarantee zero interruption, handling the failure event to provide seamless communication to the mAP.

To validate the result, Fig. 6 illustrates the robot's navigational task results. The desired trajectory is observed in dashed lines and the real performed trajectory during the experiment with the red line. The results of the real-time control and guidance of the robot suggest and confirm that our context-aware handover proposal can effectively meet this type of application that demands a stricter WiFi network performance. Given the inherent robot dynamics, slight deviations in the trajectory are expected.

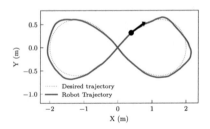

Fig. 6. Robot's navigational task: desired and performed trajectories.

4.3   Experiment 2: Make Before Degrade Handover

The objective of the second experiment is to guarantee a traffic density of 1 Mbps/m^2 (1Tbps/km^2) to provide a client data rate higher than 10 Mbps to one user per 10 m^2 as recommended in [11]. To achieve this goal is necessary to avoid the degradation of WiFi communications supported by the context-aware MBD handover enabled by computer vision for LoS obstruction prediction. The experiment simulates situations that may occur, such as the transfer of iron or steel plates in industrial environments.

The robot continues to perform the same trajectory while migrations occur. However, differently than in the previous scenario, an object with metallic characteristics moves in the space obstructing the LoS path between the robot and the $fSTA_1$ as shown in Fig 4c. The system recognizes the object through the computer vision service. An algorithm infers potential obstructions situations and proactively avoids signal degradation by making a smart handover before signal degradation.

Figure 7a shows the throughput when the robot moves through different zones on its path. An object moving in the testbed space is detected by the testbed's computer vision service (instant 31.2 s) as a potential LoS obstruction between the robot and the $fSTA_1$. This information is passed on to the handover system orchestrator, which triggers a preventive handover to avoid the transmission through $fSTA_1$. The traffic is then migrated to the closest $fSTA$, avoiding a possible transmission degradation in $fSTA_1$, and, therefore, a decrease in throughput for the robot (the user). In the interval from 31.2 s to 41.0 s, the orchestrator, rather than performing the handover for the $fSTA_1$ triggers the handover to communicate through $fSTA_4$.

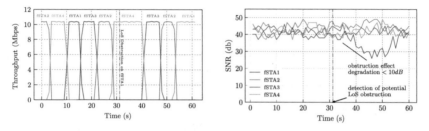

(a) $fSTA$ throughput measurements: Make Before Degrade

(b) SNR perceived by each $fSTA$ during the experiment

Fig. 7. Experiment 2 results.

To better understand the proactivity of the MBD proposal, the SNR levels between the robot and each $fSTA$ throughout the experiment are shown in Fig. 7b. After the potential LoS obstruction is detected by the computer vision service, the system orchestrator act to keep the traffic flowing through a $fSTA$ in which no obstruction was detected. Shortly after, the SNR level of $fSTA_1$ was reduced by more than 10 dB due to the obstruction of the forced LoS in the experiment with the introduction of the metal plate. The result indicates how context-aware handover enabled by SD-WiFi and computer vision techniques make it possible to avoid degradation in communication. Furthermore, the result demonstrates how the MBD solution can maintain throughput without degradation, attending to the 3GPP requirements related to traffic density for industrial environments. The throughput is ensured at 10 Mbps by keeping the robot assigned to $fSTA_4$ (as presented in Fig. 7a).

5   Conclusion

This paper argued for context-aware SDN enabling the wireless handover based on spatial-context information to support new mobility management mechanisms. We developed a proof-of-principle SD-WiFi solution to meet reliable and low-latency communications. We demonstrated the uninterrupted operation of a real-time control robotic application. Furthermore, the proposal can proactively steer the traffic to a different cell avoiding signal degradation and preserving traffic density requirements per cell. Future work involves efforts to support the orchestrator prediction through machine learning to obtain more accurate information about signal degradation.

Acknowledgements. The authors acknowledge the funding received from the Brazilian agencies CNPq, FAPES (515/2021, 284/2021, and 026/2022), CAPES (Finance Code 001), and FAPESP (20/05182-3 and 18/23097-3).

References

1. Ayyash, M., et al.: Coexistence of WiFi and LiFi toward 5G: concepts, opportunities, and challenges. IEEE Commun. Mag. **54**, 64–71 (2016)
2. Sun, L., Hou, J., Shu, T.: Spatial and temporal contextual multi-armed bandit handovers in ultra-dense mmWave cellular networks. IEEE Trans. Mob. Comput. **20**, 3423–3438 (2021)
3. Sand, S., Dammann, A., Mensing, C.: Applications. Positioning in Wireless Communications Systems (2013)
4. Lu, Y., et al.: Feasibility of location-aware handover for autonomous vehicles in industrial multi-radio environments. Sensors **20**, 6290 (2020)
5. Ma, Y., Chen, X., Zhang, L.: Base station handover based on user trajectory prediction in 5G networks. In: 2021 IEEE International Conference on Parallel and Distributed Processing With Applications, Big Data and Cloud Computing, Sustainable Computing and Communications, Social Computing & Networking (2021)
6. Carmo, A., et al.: Programmable intelligent spaces for industry 4.0: indoor visual localization driving attocell networks. Trans. Emerg. Telecommun. Technol. **30**, e3610 (2019)
7. Gotsis, A., Stefanatos, S., Alexiou, A.: UltraDense networks: the new wireless frontier for enabling 5G access. IEEE Veh. Technol. Mag. **11**, 71–78 (2016)
8. Saponara, S., Giannetti, F., Neri, B., Anastasi, G.: Exploiting mm-wave communications to boost the performance of industrial wireless networks. IEEE Trans. Ind. Inform. **13**, 1460–1470 (2017)
9. Martínez, V., et al.: Ultra reliable communication for robot mobility enabled by SDN splitting of WiFi functions. In: 2018 IEEE Symposium On Computers And Communications (2018)
10. Guimaraes, R., Martínez, V., Mello, R., Mafioletti, D., Martinello, M., Ribeiro, M. An SDN-NFV orchestration for reliable and low latency mobility in off-the-shelf WiFi. In: ICC 2020 - 2020 IEEE International Conference On Communications (2020)
11. 3GPP TS 22.261: Service requirements for next-generation new services and markets (2018)

Correction to: Stability and Availability Optimization of Distributed ERP Systems During Cloud Migration

Gerard Christopher Aloysius, Ashutosh Bhatia, and Kamlesh Tiwari

Correction to:
Chapter "Stability and Availability Optimization
of Distributed ERP Systems During Cloud Migration" in:
L. Barolli (Ed.): *Advanced Information Networking*
and Applications, **LNNS 654,**
https://doi.org/10.1007/978-3-031-28451-9_30

In the original version of the book, the following belated correction was updated: an acknowledgement has been added below figures 1 and 2 at pages 348 and 349 respectively in Chapter 30. The book has been updated with the change.

The updated original version of this chapter can be found at
https://doi.org/10.1007/978-3-031-28451-9_30

Author Index

L. Barolli (Ed.): AINA 2023, LNNS 654, pp. 573–575, 2023.
https://doi.org/10.1007/978-3-031-28451-9

Printed in the United States
by Baker & Taylor Publisher Services